ANIONIC SURFACTANTS

SURFACTANT SCIENCE SERIES

CONSULTING EDITORS

1. Nonionic Surfactants, *edited by Martin J. Schick* (see also Volumes 19 and 23)
2. Solvent Properties of Surfactant Solutions, *edited by Kozo Shinoda* (out of print)
3. Surfactant Biodegradation, *by R. D. Swisher* (see Volume 18)
4. Cationic Surfactants, *edited by Eric Jungermann* (see also Volumes 34 and 37)
5. Detergency: Theory and Test Methods (in three parts), *edited by W. G. Cutler and R. C. Davis* (see also Volume 20)
6. Emulsions and Emulsion Technology (in three parts), *edited by Kenneth J. Lissant*
7. Anionic Surfactants (in two parts), *edited by Warner M. Linfield* (out of print)
8. Anionic Surfactants: Chemical Analysis, *edited by John Cross* (out of print)
9. Stabilization of Colloidal Dispersions by Polymer Adsorption, *by Tatsuo Sato and Richard Ruch*
10. Anionic Surfactants: Biochemistry, Toxicology, Dermatology, *edited by Christian Gloxhuber* (out of print; see Volume 43)
11. Anionic Surfactants: Physical Chemistry of Surfactant Action, *edited by E. H. Lucassen-Reynders* (out of print)
12. Amphoteric Surfactants, *edited by B. R. Bluestein and Clifford L. Hilton* (out of print)
13. Demulsification: Industrial Applications, *by Kenneth J. Lissant*
14. Surfactants in Textile Processing, *by Arved Datyner*
15. Electrical Phenomena at Interfaces: Fundamentals, Measurements, and Applications, *edited by Ayao Kitahara and Akira Watanabe*
16. Surfactants in Cosmetics, *edited by Martin M. Rieger*
17. Interfacial Phenomena: Equilibrium and Dynamic Effects, *by Clarence A. Miller and P. Neogi*
18. Surfactant Biodegradation, Second Edition, Revised and Expanded, *by R. D. Swisher*
19. Nonionic Surfactants: Chemical Analysis, *edited by John Cross*
20. Detergency: Theory and Technology, *edited by W. Gale Cutler and Erik Kissa*

21. Interfacial Phenomena in Apolar Media, *edited by Hans-Friedrich Eicke and Geoffrey D. Parfitt*
22. Surfactant Solutions: New Methods of Investigation, *edited by Raoul Zana*
23. Nonionic Surfactants: Physical Chemistry, *edited by Martin J. Schick*
24. Microemulsion Systems, *edited by Henri L. Rosano and Marc Clausse*
25. Biosurfactants and Biotechnology, *edited by Naim Kosaric, W. L. Cairns, and Neil C. C. Gray*
26. Surfactants in Emerging Technologies, *edited by Milton J. Rosen*
27. Reagents in Mineral Technology, *edited by P. Somasundaran and Brij M. Moudgil*
28. Surfactants in Chemical/Process Engineering, *edited by Darsh T. Wasan, Martin E. Ginn, and Dinesh O. Shah*
29. Thin Liquid Films, *edited by I. B. Ivanov*
30. Microemulsions and Related Systems: Formulation, Solvency, and Physical Properties, *edited by Maurice Bourrel and Robert S. Schecter*
31. Crystallization and Polymorphism of Fats and Fatty Acids, *edited by Nissim Garti and Kiyotaka Sato*
32. Interfacial Phenomena in Coal Technology, *edited by Gregory D. Botsaris and Yuli M. Glazman*
33. Surfactant-Based Separation Processes, *edited by John F. Scamehorn and Jeffrey H. Harwell*
34. Cationic Surfactants: Organic Chemistry, *edited by James M. Richmond*
35. Alkylene Oxides and Their Polymers, *by F. E. Bailey, Jr., and Joseph V. Koleske*
36. Interfacial Phenomena in Petroleum Recovery, *edited by Norman R. Morrow*
37. Cationic Surfactants: Physical Chemistry, *edited by Donn N. Rubingh and Paul M. Holland*
38. Kinetics and Catalysis in Microheterogeneous Systems, *edited by M. Grätzel and K. Kalyanasundaram*
39. Interfacial Phenomena in Biological Systems, *edited by Max Bender*
40. Analysis of Surfactants, *by Thomas M. Schmitt*
41. Light Scattering by Liquid Surfaces and Complementary Techniques, *edited by Dominique Langevin*
42. Polymeric Surfactants, *by Irja Piirma*
43. Anionic Surfactants: Biochemistry, Toxicology, Dermatology. Second Edition, Revised and Expanded, *edited by Christian Gloxhuber and Klaus Künstler*

Other Volumes in Preparation

Organized Solutions, *edited by Stig E. Friberg and Bjorn Lindman*

ANIONIC SURFACTANTS

BIOCHEMISTRY, TOXICOLOGY, DERMATOLOGY

Second Edition, Revised and Expanded

edited by

Christian Gloxhuber

Bernau/Chiemsee, Germany

Klaus Künstler

Düsseldorf, Germany

Marcel Dekker, Inc. **New York • Basel • Hong Kong**

Library of Congress Cataloging-in-Publication Data

Anionic surfactants : biochemistry, toxicology, dermatology / edited
by Christian Gloxhuber and Klaus Künstler. - - 2nd ed., rev. and expanded.
p. cm. -- (Surfactant science series ; v. 43)
Includes bibliographic references and index.
ISBN 0-8247-8621-1 (alk. paper)
1. Surface active agents--Toxicology. 2. Surface active agents--Physiological effect. 3. Dermatopharmacology. I. Gloxhuber, Christian. II. Künstler, Klaus. III. Series.
[DNLM: 1. Anions. 2. Skin--drug effects. 3. Surface-Active Agents--toxicity. W1 SU721 v. 43]
RA 1270.S87A54 1992
615.9'51 -- dc20
DNLM/DLC
for Library of Congress 92-4556
CIP

This book is printed on acid-free paper.

MARCEL DEKKER, INC.
270 Madison Avenue, New York, New York 10016

Current printing (last digit):
10 9 8 7 6 5 4 3 2 1

PRINTED IN THE UNITED STATES OF AMERICA

Preface to the Second Edition

Eleven years have passed since the appearance of the first edition of this monograph. During this time much new data on anionic surfactants has been published. Therefore, the publishers decided a second edition was necessary. For this new edition some chapters have been completely rewritten while others have been updated to reflect the present state of knowledge.

In Chapter 1, the interaction of anionic surfactants with the now commonly used detergent enzymes is reviewed. Much new toxicity data on existing or new types of anionic surfactants has been added to Chapters 3, 5, and 7. New methods for the evaluation of effects on skin and mucous membrane are covered in Chapter 5. In Chapter 9, further pharmacological properties of anionic surfactants have been added. To facilitate the quick identification of specific data, the index has been considerably extended.

Most of the contributors to the first edition participated in the second edition. Additionally, some new experts have been added. For the late Dr. Krienke, his co-worker Dr. Oberdisse together with Dr. Hahn and Dr. Ritter have rewritten his chapter.

The editors are particularly grateful to Henkel KGaA for generously supporting this edition and to all experts for their contributions.

We would like to offer our thanks to Marcel Dekker, Inc. for patience and assistance in our work.

Christian Gloxhuber
Klaus Künstler

Preface to the First Edition

This monograph is part of a complete account of the properties of anion-active surfactants. Corresponding literature has been prepared for cation-active and non-ionic surfactants.

Among the chemicals being emitted into the environment, anion-active surfactants are of major importance. It is therefore of great relevance to know their biological properties. From the whole complex of biological properties, this monograph selects for attention those properties that affect warm-blooded animals and man. However, properties relevant from the ecological point of view, such as fish toxicity, effect on plants, etc., are not considered here. The monograph starts with a chapter on biochemistry since the effects observed on the animal as a whole and on man are based on biochemical processes.

The research findings compiled for this monograph on toxicological and dermatological properties and on the amounts absorbed by man are of great practical importance for the calculation of health risks in a great variety of applications, for instance in the fields of detergents and cosmetics. In evaluating the products for these fields, the great benefit these products have, above all, for hygiene must be considered.

In recent years, the number of studies carried out with anionic surfactants has enormously increased. In addition to most modern studies relating to such processes as absorption, metabolism, excretion, teratogenesis, carcinogenesis, tests of dermal and mucous irritation have, of course, been given much attention. Accordingly, these chapters had to be more extensive than others.

As a result of their professional activities the authors of the various chapters are as familiar as is possible with the subjects discussed. I should like to thank them for having been ready to write these chapters despite the multitudinous demands of their everyday work.

I am particularly grateful to Henkel KGaA for their generous help which made it possible for me to undertake this monograph, and the other experts for specialized contributions.

As a large part of the monograph has been contributed by German-speaking authors, there was the problem of translation to be coped with. In this connection, I wish to express my gratitude above all to Mrs. Dagmar Schenck, Düsseldorf, Mrs. Pauline Rose, Düsseldorf, and Dr. Martin Schick, New York, who with outstanding skill made themselves familiar with the terminology used and accomplished excellent translations.

I should also like to offer my thanks to Marcel Dekker, Inc. for having been so patient as to give us sufficient time to write and translate the contributions.

Christian Gloxhuber

Contents

Contributors

Friedhelm G. Bartnik Henkel KGaA, Düsseldorf, Germany

J. Gordon Black Unilever Research, Colworth Laboratory, Bedford, England

Axel Hahn Berlin Poison Control Center, Berlin, Germany

Douglas Howes Unilever Research, Colworth Laboratory, Bedford, England

Werner Kästner Henkel KGaA, Düsseldorf, Germany

W. Matthies Henkel KGaA, Düsseldorf, Germany

Kenkichi Ōba Lion Corporation, Tokyo, Japan

Ursula Oberdisse Berlin Poison Control Center, Berlin, Germany

Matjaž S. Potokar Henkel KGaA, Düsseldorf, Germany

Sigrid Ritter Berlin Poison Control Center, Berlin, Germany

W. Sterzel Henkel KGaA, Düsseldorf, Germany

Reiko Takei Lion Corporation, Tokyo, Japan

ANIONIC SURFACTANTS

1

Interaction of Anionic Surfactants with Proteins, Enzymes, and Membranes

FRIEDHELM G. BARTNIK Henkel KGaA, Düsseldorf, Germany

I. INTERACTION OF ANIONIC SURFACTANTS WITH PROTEINS

A. Introduction

Contact between surfactants and proteins is extremely frequent in our daily lives. Such contact may occur either with the market products themselves or in the secondary region, in which contact with compounds present in sewage or with aquatic organisms and higher mammals can occur. The effects of surfactants on humans are of special interest, particularly in pharmaceutical or cosmetic applications.

Studies concerning the interactions between polymers and surfactants have shown that very intense interactions may occur with polymers, which vary very widely in terms of chemical nature, particularly when the surfactant is anionic. Thus, the physicochemical properties of both partners change even when synthetic polymers of comparatively simple structures are concerned, such as polyvinylpyrrolidone (1-7) and ethylene oxide polymers (8-10).

Despite the comparatively simple structure of such synthetic polymers, many questions relating to the full elucidation of the interaction mechanism remain open. The questions become all the more difficult when the partners are proteins of complex structure. Because of the large number of different protein structures, numerous kinds of interactions are to be expected. Therefore, the present review does not intend to show and interpret the differences among the various types of interactions between anionic surfactants and protein structures of various kinds; rather, it attempts to discuss those principles, which appear to be quite general according to our experience. These principles are mostly discussed without consideration of individual proteins.

B. Formation of Adsorption Complexes

Nonionic surfactants and proteins show either no interaction at all or interactions that are extremely weak and normally close to the limits of sensitivity of the analytic methods used. There are only weak hydrophobic interactions between the hydrophobic sections of proteins, and protein conformation remains essentially unchanged. Therefore, nonionic surfactants will not dissolve sparingly soluble proteins, denature proteins, or contribute to a swelling of the epidermis.

Figure 1 shows the solubility of zein—an almost insoluble protein—as a function of surfactant concentration. Obviously, nonionic surfactants do not change the solubility, whereas there is a significant increase in solubility in the case of anionic surfactants. The solubility of zein in anionic surfactant solutions has been studied in great detail because it correlates well with the skin irritancy of these surfactants (11-14) in the sense that skin irritancy increases

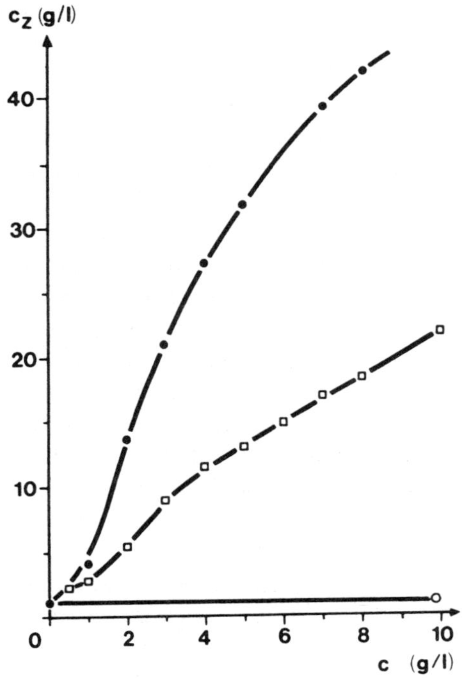

FIG. 1 Zein solubility c_z of saturated solutions as a function of surfactant concentration. Zein concentration, 50 g liter^{-1}; mixing period, 2 hr; temperature, 40°C. ● = sodium dodecyl sulfate; □ = alkyl ether sulfate (2EO); ○ = nonyl phenol (9EO).

with increasing solubility of zein. These increases may be interpreted as a consequence of the formation of protein/surfactant complexes with certain specific properties.

In this respect, it is of interest that a relationship between irritation induced by anionic surfactants and the swelling of isolated skin-epidermal membranes or collagen has been reported in several papers (18-21). The highest levels of swelling of isolated human stratum corneum were induced by the anionic surfactants while very little swelling was induced by the cationic and nonionic surfactants (21). The authors concluded that the swelling of stratum corneum resulted from a reversible conformational change due to cooperative binding of the surfactant. For homologous series of various anionic surfactants (alkyl benzene sulfonates, alpha olefin sulfonates, alkyl sulfates, paraffin sulfonates), maximal swelling occurred at either C_{12} or C_{14} carbon atoms (20). A combination of ionic binding of the hydrophilic group as well as hydrophobic interactions of the alkyl chains with the substrate may be an explanation for the swelling mechanism because nonionic surfactants did not induce swelling. In contrast to most of the anionic surfactants the nonionic surfactant Tween(R) 20 produced little swelling of collagen. No ionic surfactant-substrate interactions are possible in this case (20).

In order to elucidate the irritation phenomena from frequent exposure of keratinous tissue such as epidermis to surfactants, numerous investigations have been performed to study the interaction of surfactants with the epidermis or horny layer (22,23). With human heel stratum corneum a maximum binding of sodium dodecyl sulfate (SDS) at low pH was found. As in the case with zein, the sorption of sodium dodecyl sulfate was accompanied by solubilization of protein and amino acids out of the human heel callus (24).

The results available (15-17) show that both polar and hydrophobic interactions are involved in the formation of adsorption complexes between proteins and surfactants. Polar interactions between the negatively charged hydrophilic group of the surfactant and the positively charged centers along the protein chain are the basic requirement for the development of hydrophobic bonds between the hydrophobic surfactant residue and the protein involved. This may be visualized as a process in which polar interactions lead to a closer approach between surfactant and protein molecules and favor the formation of hydrophobic bonds. There will be no such possibility in the case of nonionic surfactants. The anionic surfactant operates as a bifunctional agent, which explains its different behavior.

The binding between surfactants and protein is a function of the surfactant/protein ratio. At constant protein concentration the tendency of protein/surfactant complex formation increases with increasing concentration of the surfactant present in equilibrium with the protein. The binding power may be shown graphically in a diagram giving the average number of surfactant molecules ($\bar{\gamma}$) per protein molecule as a function of the logarithm of the free ligand concentration ([c]), as shown in Fig. 2.

The binding is subject to the thermodynamic law of equilibrium. A theory of binding in the system surfactants/globular proteins, based on a multiple equilibria model with stoichiometric binding, has been published (47). The number of surfactant molecules bound per protein molecule may be determined directly by equilibrium dialysis (15,28,48-50,52-55,61), gel chromatography (51,56), proteins immobilized on agarose (30), analytical ultracentrifugation techniques (57), surfactant-selective electrode (58), capillary isotachophoresis (31) or microcalorimetry (32). Summaries of experimental procedures can be found in the paper of Steinhard and Reynolds and of Reynolds (15,33).

As shown in Fig. 2, the binding between surfactants and proteins may be subdivided into three regions (47). Only few surfactant molecules (< 10) will be bound to binding sites of high affinity in region a (15,59). This binding may be characterized by the following equilibrium (59):

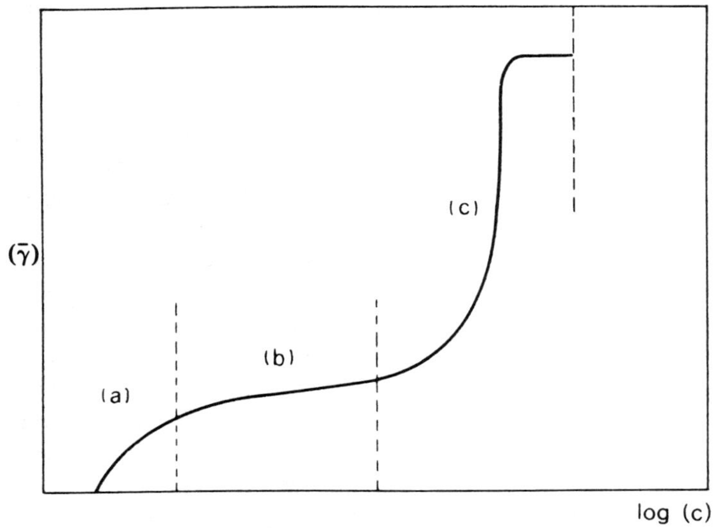

FIG. 2 Schematic plot of the number of bound ligands per protein molecule ($\bar{\gamma}$) as a function of the logarithm of the free ligand concentration ([c]). Region a, specific binding; region b, noncooperative binding; region c, cooperative binding and saturation. (From Ref. 47.)

$$\text{Site (S)} + \text{ligand (L)} \rightleftharpoons \text{SL}$$

Only few proteins are capable of this type of binding; native serum albumin, for example, has binding sites for a considerable number of amphiphilic anions. It is interesting that serum albumin is the transport vehicle of free fatty acids in the circulatory system. The binding of fatty acids to serum albumin is discussed in great detail in a review article (60). However, some of the binding sites which show high affinity to alkyl sulfates and alkyl sulfonates do not show similar affinities to alkyl carboxylates (62). While hydrophobic interactions are responsible for most of the free binding energy (77, 78), this observation indicates that specific interactions of the polar groups are also involved (79,80).

No conformational changes of the proteins are caused by binding of the ligands at high affinity sites; therefore, no significant viscosity changes are observed in this region.

In addition to serum albumin, serum high-density lipoprotein (48), β-lactoglobulin (55,81), and pyruvate oxidase (96) are further examples of water-soluble proteins with specific high-affinity sites for surfactants. There will be further addition of surfactant molecules if the surfactant concentration is higher than required for saturation of the high affinity sites. This further addition will lead to conformational changes of the macromolecule (36). Secondary structure changes of bovine serum albumin were found in solutions of sodium decyl sulfate, SDS and different cationic surfactants (decyl-, dodecyl-, tetradecyl- and hexadecyltrimethylammonium bromide). All of these surfactant caused changes in the helical proportion from 66 to approximately 50%. However, the shorter the hydrocarbon chain was, the higher the concentration necessary to cause the same decrease in the helical proportion (37). Presumably, the conformational changes allow surfactant/protein interaction at previously unexposed hydrophobic side groups of the protein (29). In the cases of dodecyl sulfate, tetradecyl sulfate, and the corresponding sulfonates, this massive addition of surfactant molecules leads to denaturation (56), but sodium oligooxyethylene dodecyl ether sulfate with oxyethylene units of more than 4 has been found not to denature bovine serum albumin in spite of the fact that obviously the surfactant was bound by albumin. Interestingly, this surfactant did not interact with most water-soluble proteins as long as the oxyethylene chain lengths were long enough. In addition to bovine serum albumin β-lactoglobulin was found to have an affinity for these surfactants with 6-8 oxyethylene units. Both proteins are known to have an affinity for nonionic surfactants (38). Ohbu et al. (81b) investigated the denaturation power of different kinds of surfactants. They used circular dichroism spectra as the method to estimate a change of the secondary structure of proteins caused by the surfactants. They obtained the following order of denaturation effects:

Linear alkyl benzene sulfonate > SDS > sodium alpha-olefinsulfonate (hydrocarbon chain length = 14) > sodium polyoxyethylene (3 mol in average) alkyl sulfate.

According to the authors, this order agreed with the degree of desquamation on human skin induced by the various surfactants. In general, proteins that are built up from subunits will dissociate after massive binding of the a.m. surfactants (44,45). Tetradecyl sulfate has a greater affinity for unfolded proteins than dodecyl sulfate and dissociates protein aggregates, which consist of identical peptide chains more easily than dodecyl sulfate does (25).

While few proteins possess high-affinity sites, almost all soluble proteins are capable of this type of massive binding of surfactants. It is not clear as to whether all proteins go through the noncooperative binding shown in region b of the binding isotherm (Fig. 2) or whether there is immediate cooperative binding (region c) between the surfactant molecules and the macromolecule. Unlike sodium dodecyl sulfate, monomeric deoxycholate binds to bovine serum albumin at discrete sites in a noncooperative fashion. This binding behavior suggests that deoxycholate interacts with accessible hydrophobic binding sites present on certain proteins (39). Cooperative binding of surfactant molecules will not always lead to conformational changes of the macromolecules, as shown by experiments with serum albumin and short-chain alkyl sulfates and alkyl sulfonates (50,121). The binding of anionic surfactants to proteins is determined by the thermodynamic activity of the surfactant anion, while surfactant micelles do not play a role in the binding process (50,54). The interaction of *n*-dodecyl sulfate with lysozyme has been studied in some detail. There was a high degree of positive cooperativity on binding SDS to the cationic sites on the lysozyme molecule. The binding of surfactant molecules increased the overall hydrophobicity of the protein surface, facilitating further binding. Ionic binding of the surfactant was weakened and hydrophobic binding strengthened by increasing ionic strength (28).

There is cooperative binding between sodium dodecyl sulfate and almost all globular proteins. Proteins without disulfide bonds bind larger amounts of SDS than proteins with disulfide bonds (53). If the equilibrium concentration of SDS anions exceeds 0.85 mM, all water-soluble proteins have been reported to bind identical amounts of 1.4 g of SDS/g protein (53,54), while below 0.85 mM SDS only 0.4 g SDS/g protein are bound (54) (see also Fig. 3, binding isotherm 7).

However, all proteins do not show cooperative binding with SDS. SDS is hardly bound at all by pepsin, papain, and glucose oxidase, for instance (57). An extensive discussion about this phenomenon can be found in the review of Makino (81a).

In contrast to the results discussed above, Takagi et al. (63) report that the binding isotherms of different proteins with SDS are not identical, but differ, although their shapes are quite similar. Here, two phases were observed (see Fig. 3): an initial slow increase of binding up to 0.3-0.6 g SDS/g protein and a subsequent steep increase to values from 1.2-1.5 g SDS/g protein (63). The isotherms observed for different proteins differ during the first and second phases.

Other authors (52) have also observed different binding isotherms for different native proteins.

C. SDS-Polyacrylamide Gel Electrophoresis

Protein molecular weights in saturated SDS/protein complexes may be determined by measurements relating to transport properties of the complexes. Thus, the logarithm of the intrinsic viscosity (42), the logarithm of the Stokes radii (41), the slow birefringence relaxation time (43), and the electrophoretic mobility in polyacrylamide gels (44-46) are proportional to the logarithm of the polypeptide chain molecular weight for many water-soluble proteins. Recently, the estimation of molecular weight of proteins by low-angle laser light-scattering techniques was published (26).

SDS-polyacrylamide gel electrophoresis is a particularly fast and simple experimental method which requires only small amounts of protein; this

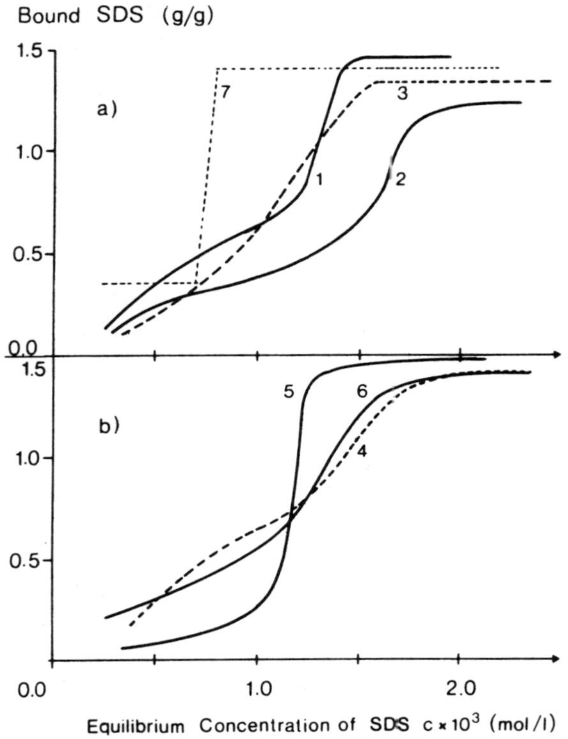

FIG. 3 Binding isotherms of SDS to several proteinpolypeptides at 25°C in 50 mM sodium phosphate buffer pH 7.0; 1, reduced and carboxamidomethylated (RCAM) bovine serum albumin; 2, reduced and carboxymethylated (RCM) bovine serum albumin; 3, RCM Taka-amylase A; 4, RCAM bovine β-lactoglobulin; 5, RCAM bovine pancreatic ribonuclease; 6, RCM bovine pancreatic ribonuclease; 7, the binding isotherm from (54). (From Ref. 63.)

method has therefore found increasing use in the determination of protein molecular weights.

Normally, in the electrophoresis of protein mixtures in polyacrylamide gel, the proteins will be separated according to their intrinsic charges caused by the voltage applied, and according to their sizes, because of the sieving effect of the polyacrylamide gel; the higher the intrinsic charge and the smaller the size of the macromolecule, the larger will be the mobility (64).

However, if saturated SDS/protein complexes are formed by incubation of the proteins in buffer solutions containing 1% SDS and 1% mercaptoethanol (to reduce disulfide bonds in the protein molecule) and subjected to electrophoresis in polyacrylamide gel containing SDS, the relative mobility of the polypeptide chain depends primarily on its molecular weight (44,45). If the relative mobility of various SDS/protein complexes is plotted against the molecular weight of the proteins (in a semilogarithmic plot), a linear relationship will be obtained (Fig. 4) (45).

The relative mobility of the SDS/protein complexes in polyacrylamide gel electrophoresis is given by the following relationship (65,66,70):

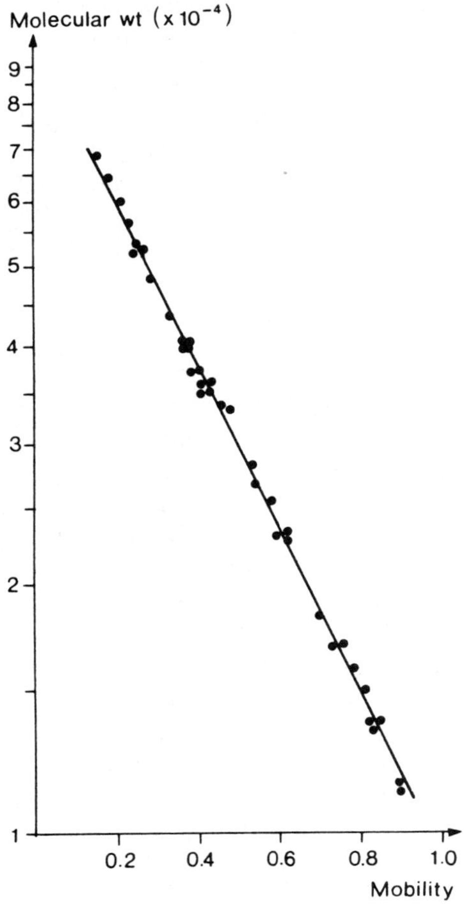

FIG. 4 Comparison of the molecular weights of different proteins with their electrophoretic mobilities on standard 10% gels in the SDS-phosphate system. This plot covers the molecular weight range from 11,000 to 70,000. (From Ref. 45.)

$$\log RM = -K_R T + \log Y_0$$

where

RM = relative mobility of the SDS/protein complex in polyacrylamide gel at gel concentration T

K_R = retardation coefficient, determined by the shape of the protein and the sieving effect of the gel

T = total monomer concentration of the gel; influences the pore width of the gel

Y_0 = apparent free relative mobility in the absence of a molecular sieve effect by the gel, i.e., when the total monomer concentration is 0

RM will depend only on K_R if T and Y_0 are constant, i.e., RM is a function of the size of the protein/SDS complex. Y_0 will be constant and thus independent of RM and K_R only if all SDS/protein complexes have the same relative mobility at $T = 0$. The Y_0-values are obtained by a determination of the relative mobility at various gel concentrations with subsequent extrapolation to $T = 0$. The results show that the Y_0-values are indeed identical for many SDS/protein complexes (67-69) (see Fig. 5).

The determination of the absolute value of electrophoretic mobility of SDS/protein complexes by free-boundary electrophoresis has also shown that the electrophoretic mobilities of the complexes are independent of the polypeptide molecular weight in the molecular weight region between 10,000 and 70,000 (71). These results may be interpreted as indicating that the massive binding of SDS leads to a charge equalization of the proteins.

Y_0 being independent of protein molecular weight and K_R being a function of the protein molecular weight is the basis for the empirical observation that plots of the mobility of the SDS/protein complex in polyacrylamide gel vs the logarithm of the protein molecular weight are linear at a given gel concentration (Fig. 4). This relationship in the SDS-polyacrylamide gel electrophoresis is based on the assumption of a constant and uniform SDS-binding per unit polypeptide chain length (see Sec. I.B. and Fig. 3) (72), but it must be emphasized that this assumption is not satisfied in every case since variations from the expected values has been noted (53,71), e.g. for intrinsic membrane proteins (see Chap. 3) or with proteins with a core of hydrophobic amino acids (34).

D. Conformation of Protein/Surfactant Complexes

It is shown in Sec. I.C. that the retardation coefficient K_R of SDS/protein complexes depends only on the molecular weight of the proteins. In view of the fact that the proteins as such show different conformations, it must

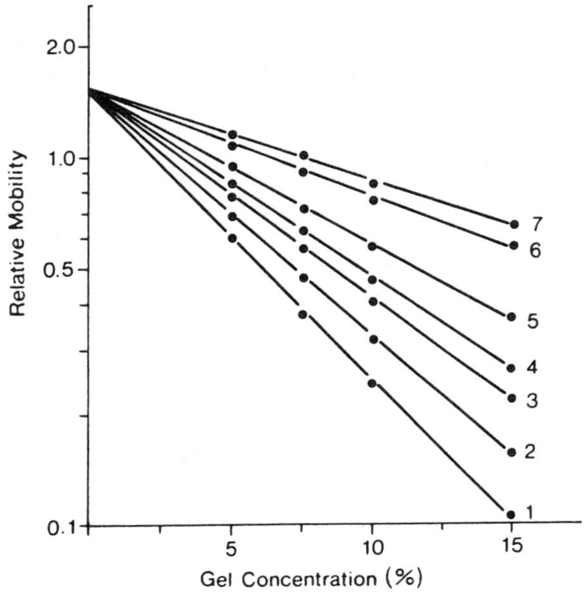

FIG. 5 Electrophoretic mobilities of seven standard proteins on SDS polyacrylamide gels plotted on a logarithmic scale against the gel concentration. Extrapolation of the different lines to zero gel concentration indicates that the different proteins have approximately the same value of "free mobility." The weight ratio of acrylamide to methylenebisacrylamide is 37:1. The acrylamide concentrations are 15, 10, 7.5, and 5%. The numbers refer to the following proteins: 1, catalase; 2, fumarase; 3, aldolase; 4, glyceraldehyde-3-phosphate dehydrogenase; 5, carbonic anhydrase; 6, myoglobin; 7, bacteriophage R17 coat protein. (From Ref. 69.)

be assumed that the conformational differences are destroyed by the binding of SDS (35) and that all SDS/polypeptide complexes show similar structures. Passing through the polyacrylamide gel the frictional drag will depend on complex size and be independent of complex shape (72). The size of SDS/protein complexes hardly varies with the molecular weight of the protein.

An asymmetric SDS/protein complex with a constant minor axis and a major axis varying with the length of the polypeptide would be consistent with this (72). Several models of the SDS/protein complex have been suggested.

Measurements of the intrinsic viscosities of SDS/polypeptide complexes led to the conclusion that proteins assume the shape of rigid prolate ellipsoids in the binding of SDS (42). The prolate ellipsoids possess a constant semiminor axis and a semimajor axis which is proportional to the molecular weight of the protein moiety. The average length of the semiminor axis of the 12 SDS/protein complexes studied was 18 Å in the saturated complex. The semimajor axis was found to be 44.8 Å in Cytochrome c (molecular weight 11,700) and 230.5 Å in serum albumin (molecular weight 69,000).

The shape of the complex was confirmed by gel-filtration experiments (41). Results obtained in free-boundary electrophoresis (71) and from proton magnetic resonance (pmr) studies (73) of SDS/protein complexes led to an alternative model of the SDS/protein complex, the so-called necklace model. In this model the polymer chain is flexible and micellelike clusters of SDS are arranged along the polymer chain. The electrostatic repulsion between SDS clusters with charges of the same sign leads to a statistical elongation of the polypeptide chain, which increases with its molecular weight.

Another model overcomes the differences between the models discussed above (43). In the latter model the SDS/protein complexes are deformable, free-draining prolate ellipsoids. Flexibility of the polypeptide chain in the complex is assumed—as in the last two of the complex models mentioned above (43,71,73)—in a conformational analysis published (74). Based on circular dichroism spectra of SDS/protein complexes the complex is assumed to have an overall conformation which corresponds to a random coil with some helical sections.

The variability of the circular dichroism spectra of the various SDS/protein complexes is explained by the assumption that the cationic amino acids of the polypeptide chain lead to helical substructures with increased probability in the presence of SDS. Anomalous values in molecular weight determinations by SDS-polyacrylamide gel electrophoresis may be explained by this model in the cases of ribonuclease A (44), collagen peptides (75), and histones (76).

E. Binding of Cationic and Nonionic Surfactant to Proteins; Comparison with the Binding of Anionic Surfactants

As shown in previous sections, there are more or less strong interactions between anionic surfactants, particularly SDS, and proteins, which lead to the formation of surfactant/protein complexes.

While the sparingly soluble zein is easily dissolved by anionic surfactants, nonionic surfactants do not cause any solubility changes (Fig. 1). Only few binding studies relating to the binding of nonionic and cationic surfactants are available; therefore, any comparison must necessarily remain incomplete.

It has been shown that Triton X-100 (octylphenolpolyoxyethylene glycol) is bound at high-affinity sites of bovine serum albumin (56,97). However, the cooperative binding, which is characteristic for the binding of alkyl sulfates, alkyl sulfonates, and benzene sulfonates to proteins, is absent in the case of Triton X-100 (54,46).

The most likely explanation is that due to the low critical micelle concentration of Triton X-100 the concentration of monomeric Triton molecules, which is required for cooperative binding, is not reached and therefore both cooperative binding and denaturation of the macromolecule are impossible (56).

Even in micellar solutions hydrophilic proteins—with the exception of albumin—bind only little or no Triton X-100 (97), while lipophilic proteins, i.e., proteins which are associated with considerable amounts of lipids in vivo, are capable of binding considerable amounts of Triton X-100. Presumably, Triton X-100 is bound to hydrophobic segments of the lipoprotein, which are covered by membrane lipids in the native protein (51,97). The interaction of sucrose mono-esters, another type of nonionic surfactants, with bovine serum albumin and ovalbumin were studied in detail (97a). The sucrose esters of caprylic and capric acids were found to bind only at discrete binding sites of native bovine serum albumin and this binding did not induce any detectable change in the protein conformation. Virtually no binding of the surfactants to ovalbumin, which lacks strong affinity sites in its native state (31), was observed. The authors concluded that the interaction of these sucrose esters with the proteins is quite similar to that of Triton X-100. No change in protein structure was also observed after binding of octyl- or decyl glucoside to bovine serum albumin. The average number of surfactant molecules bound increased with increasing hydrocarbon chain length (97b).

In summary, nonionic surfactants generally combine only poorly with water-soluble proteins, but interact strongly with lipophilic proteins, e.g., membrane proteins (97c).

Tetradecyltrimethylammonium chloride, a cationic surfactant, is bound by four high-affinity sites of serum albumin. At higher concentrations there is cooperative binding with both albumin and other proteins. As in the case of SDS, cooperative binding occurs with denaturation of the macromolecule (50). However, the onset of cooperative binding occurs only at a concentration which is 10-fold higher than the corresponding concentration in the case of SDS. The formation of the saturated tetradecyltrimethylammonium/protein complex is prevented by the competing formation of surfactant micelles. Sen et al. described the thermodynamics of the binding of cationic detergents to bovine serum albumin in some detail. They concluded that the binding sites on the protein molecule must be heterogeneous in character so that the binding process is controlled by hydrophobic, and other types of cooperative or noncooperative interactions (27).

Because of the lower critical micelle concentration of the cationic surfactants, denaturation is less pronounced than in the case of SDS. Thus, the inhibitory action of cetyldimethylbenzylammonium chloride on trypsin was found to be reversed by the removal of the surfactant, while the activity loss caused by SDS is irreversible (82). Changes in the helix content of myoglobin were also reversible upon removal of the cationic surfactant (cetylpyridinium chloride), in contrast to findings with SDS (83).

II. EFFECTS OF ANIONIC SURFACTANTS ON ENZYMES

Generally, all proteins are capable of adsorbing anionic surfactants, as discussed in Sec. I.B., and massive cooperative binding is observed, particularly in alkyl sulfonates, alkyl sulfates, and alkyl benzene sulfonates. Massive cooperative binding eventually leads to protein denaturation. Denaturation comprises the disruption of noncovalent, structure-stabilizing bonds with the concomitant loss of secondary and tertiary structure as well as biological activity (85). Therefore, massive cooperative binding of the anionic surfactant will always lead to enzyme inactivation. On the other hand, the specific activity of an enzyme is a sensitive monitor of a possible change of protein conformation induced by the binding of the surfactant to the protein (85a).

The low concentrations that suffice to cause protein and enzyme denaturation are particularly striking in the cases of anionic and cationic surfactants. In contrast to surfactants, structure breakers such as urea and guanidine hydrochloride, compounds that are frequently used for protein denaturation, lead to denaturation only at comparatively high concentrations (8 and 6 mol liter^{-1}, respectively). On the other hand, denaturation by dodecyl sulfate generally occurs at a concentration of 10^{-3}-10^{-4} mol liter^{-1} (15,84), e.g., the enzyme ribonuclease is denatured at a concentration of 8×10^{-3} mol liter^{-1} (85). Dodecyl benzene sulfonate, which is even more strongly hydrophobic than SDS (58), causes serum albumin denaturation even at a concentration of 10^{-5} mol liter^{-1} (86). This means that in these cases enzyme inactivation occurs at concentrations below the critical micelle concentration.

In a recent investigation it was observed that the 6 M guanidinium chloride-denatured enzyme rhodanese could be reactivated by different surfactants (86a). Not all surfactants were useful, no favorable effects in reactivation were found with anionic surfactants, e.g., SDS and sodium cholate, but the cationic surfactant cetyltrimethylammonium bromide (CTAB) and some zwitterionic and nonionic surfactants were effective. Interestingly, at high enough concentration CTAB denatures rhodanese and gives a complex similar to that formed with dodecyl sulfate (86b).

The following mechanisms of enzyme inactivation by surfactants have to be considered (91):

1. Disruption of the quaternary structure of the enzyme when the enzyme protein consists of several subunits
2. Induction of conformational changes in the tertiary or secondary structure of the enzyme protein
3. In the case of membrane-bound enzymes, separation of the enzyme protein from essential membrane lipids
4. Binding at active sites of the enzyme

Ribonuclease is denatured by dodecyl sulfate, as mentioned above. While ribonuclease A is inactivated by sodium dodecyl sulfate at pH 7.3 in aqueous solution, further studies (87,88) have shown that no inactivation occurs in the presence of the cationic surfactants *n*-dodecyltrimethylammonium bromide, *n*-tetradecyltrimethylammonium bromide, *n*-hexadecylpyridinium chloride under comparable conditions. This also indicates that the surfactant/protein interaction is codetermined by the polar group of the surfactant. The polar group of SDS may form strong ionic bonds with the correspondingly charged amino acids of the protein, while such bonds are weakened by steric hindrance due to the methyl groups and the pyridine ring of the cationic surfactants. However, *N*-dodecylamine hydrochloride and *n*-dodecanoic acid again inactivate the enzyme as there is no steric hindrance to the enzyme/surfactant interaction. It is concluded from spectroscopic data that *n*-dodecylamine hydrochloride is also bound to the protein by an ionic interaction which eventually leads to conformational changes, while no conformational changes are induced by the binding of *n*-dodecanoic acid; presumably, the enzyme is inactivated by binding in the vicinity of the active site (87,88).

Glutamate dehydrogenase is inactivated both by anionic and by cationic surfactants (108,209,210). Here, too, a certain amount of steric hindrance due to the structure of the cationic surfactant cetyltrimethylammonium bromide (108) (vide infra) in the adsorption to the protein molecule was observed. With partially purified rat brain protein kinase c, an enzyme found in both membrane and cytosolic fraction of cells it has been found that most commonly used surfactants inhibited the enzyme. Inhibition was found with the anionic surfactants SDS, linear alkylbenzene sulfonate and deoxycholate, with the nonionic surfactants octyl glucoside and Triton X-100 as well as with the cationic surfactant dodecyl trimethylammoniumbromide. An ethoxylated alcohol and digitonin as nonionic surfactants did not inhibit the enzyme whereas an ethoxylated alcohol sulfate inhibited the enzyme activity only slightly (88a).

Several dehydrogenases (glucose-6-phosphate-, lactate-, malate- and isocitrate dehydrogenases) were strongly inhibited by SDS and the long-chain fatty acid palmitoleate (88b). The lactate dehydrogenase isoenzyme LDH-B4 can be inactivated by SDS, but the enzyme could be stabilized to surfactant inactivation by the presence of co-factor or co-factor analogues (88c). Sanford et al. examined the effect of ionic surfactants on the activity of lactate dehydrogenase (LDH) isoenzymes in detail. They found that the cationic surfactants (hexadecylpyridinium bromide and hexadecyltrimethylammonium bromide) rapidly inactivated LDH isoenzymes containing the B unit whereas the anionic surfactants (SDS and sodium decyl sulfate) rapidly inhibited LDH isoenzymes containing the A subunit (88d). They explained the selectivity of the inactivation process by different electrostatic interactions: positively charged cationic surfactants inactivate isoenzymes containing a subunit of net negative charge, and the negatively charged anionic surfactants preferentially inactivate isoenzymes containing a subunit of net positive charge. A similar different action of cationic and anionic surfactants on enzyme activity was found with porcine heart malate dehydrogenase isoenzymes (cytoplasmic or mitochondrial origin). The cationic surfactant hexadecyl trimethylammonium bromide inactivated the cytoplasmic isoenzyme at concentrations that did not effect the mitochondrial isoenzyme. On the other hand, SDS inactivated both enzymes, but the mitochondrial enzyme was significantly more sensitive than its cytoplasmic counterpart. As with the LDH isoenzymes, there seem to be some correlation between the charges of the isoenzymes and the inactivation (88e). Both malate dehydrogenases possessed considerable resistance to the nonionic detergent, Triton X-100.

Enzymes, especially proteases, are used in laundry detergents. In this commercial application the enzymes are used under conditions that are not the most favorable for stability and activity of the enzymes. This is particularly well exemplified by the inclusion of proteases in liquid laundry detergents. The serine proteases, e.g., Subtilisin carlsberg, have interesting stability properties in this respect. They are relative resistant to denaturation by urea, guanidine hydrochloride or SDS (88f). The reasons for the increased stability are not well understood. With aprA-subtilisin, a secreted bacerial serine protease that is 85% homologous to subtilisin BPN′, it was found that the addition of SDS to the enzyme resulted in a slow conversion of the protease from the intact protein to the completely unfolded form. This conversion was accompanied by decreased activity, disruption of tertiary structure, a change in the mobility of the protein in SDS-polyacrylamide gel electrophoresis and an increase in the apparent Stoke radius of the protein. In the presence of 1% SDS for up to 30 min, Subtilisin retained its activity and its original tertiary structure, after 2 hr in 1% SDS at 20°C, 25% of the enzyme was still intact and active (88f). Further investigations are necessary to understand why the subtilisin molecule shows such a high resistence to unfolding in the presence of SDS. Kravetz et al. (1985) determined the effects of nonionic surfactants (ethoxylated alcohols) and anionic surfactants on the stability of proteases and amylases in liquid laundry detergents. The enzymes were more stable when formulated into detergents containing ethoxylated alcohols and alkyl ether sulfates than when formulated with alkyl sulfates or surfactants with sulfonate groups (88g).

A strong pH dependence was observed in the inhibition of urease by SDS (89). While there is complete inactivation of urease by SDS at a concentration of 10^{-3} mol liter^{-1} at pH 5.0 (Fig. 6), the enzyme is fully active at the same concentration of SDS and pH 5.4.

Increasing inactivation with decreasing pH has also been observed with several other enzymes. It may be assumed that low pH values favor the adsorption of negatively charged SDS ions to the enzyme protein, in analogy to findings relating to the adsorption to synthetic polymers (10). All enzymes studied were inhibited at a concentration of or close to 10^{-3} mol liter^{-1} SDS, i.e., below the critical micelle concentration (89). Obviously, only the SDS monomer is adsorbed by the enzymes (see Sec. I.B.). However, all enzymes are not inactivated by SDS at concentrations below the critical micelle concentration. Viscosimetric studies have shown that the enzyme catalase is disrupted into its subunits only at concentrations above 4-5% SDS (above 0.1 mol liter^{-1}) (90). Close to the critical micelle concentration of SDS the enzyme still showed 70% of the initial activity. Catalase is more strongly inactivated by SDS than by sodium laurate. These results have been obtained with preincubation of the enzyme.

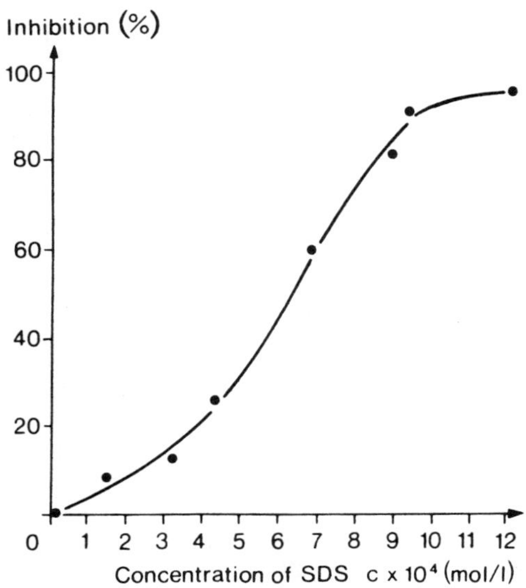

FIG. 6 Comparison of the effect of varying concentrations of SDS on urease at pH 5.0. (From Ref. 89.)

Experiments have shown that enzymes consisting of subunits bind SDS rather slowly. Thus, the binding between SDS and the native tetramer of catalase and the resulting formation of the catalase subunits is a slow process: Under the conditions chosen 29% of the native tetramer were present even after 72 hr. Measurements of enzyme activity indicate that the tetrameric catalase is still fully active.

The binding of SDS to pancreatic lipase has been determined by equilibrium dialysis (99). It was found that massive cooperative binding occurs above the critical micelle concentration (see Fig. 7). In the saturated complex 1.1 g SDS is bound/g protein. It is interesting to note that dodecanoyl taurate, which resembles SDS structurally, does not denature lipase. With chromobacterium lipase, an enzyme with a high tolerance to SDS, at lower concentration than cmc, clusters are formed when SDS interacts with the enzyme. This results in a reduction of the surface tension. However, the activity of the enzyme is retained at this concentration (99a).

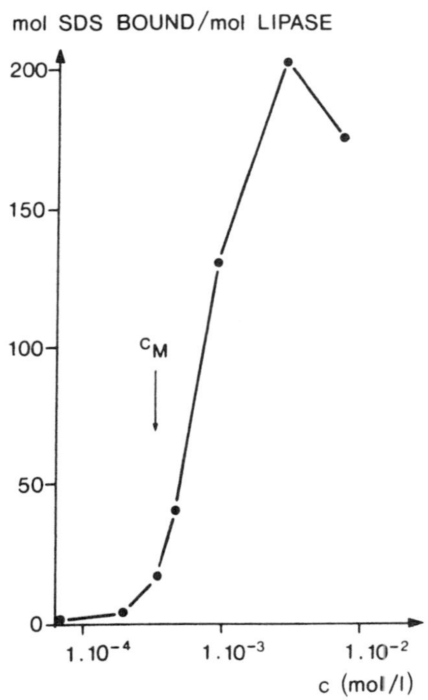

FIG. 7 Binding of SDS to pancreatic lipase (ordinate) vs log concentration of free SDS (abscissa). (From Ref. 99.)

Several enzymes have been found to be extremely resistant against inactivation by SDS. Glucose oxidase is not inactivated within 24 hr even at an SDS concentration of 1%. Papain and pepsin showed 41 and 33% activity, respectively, after 4 hr at the same concentration of SDS. All three enzymes were shown to bind little or no SDS (<0.2 g SDS/g protein) (57). In the case of pepsin the binding of SDS is presumably prevented by the high negative charge of the protein (isoelectric point <1), as binding of SDS is observed in the case of pepsinogen which has an isoelectric point of 3.5. With glucose oxidase its tightly knit structure prevents the initiation of chain unfolding that precedes large-scale SDS binding (57).

As another example, alkaline phophatase isolated from various tissues (92-95,106) is extremely resistant to SDS inactivation. Even after 3 hr at 37°C and at an SDS concentration of 1%, alkaline phosphatase isolated from human liver is found to be fully active (94). The stability of alkaline phosphatase may be explained by the fact that the enzyme is a glycoprotein (94). Glycoproteins are known to bind SDS only to a minor extent (53). Also uncommonly resistant to SDS inactivation is the enzyme superoxide dismutase, which continues to function even in 10 M urea and in 4% SDS (211).

Table 1 summarizes results obtained relating to the activity of various enzymes after incubation with SDS. The table lists those results only that have been obtained by preincubation of the enzyme with SDS. In comparison to results obtained in nonincubation procedures it should be kept in mind that the adsorption of the surfactant onto the enzyme protein may be very slow (57), with the result that nonincubation results are not directly comparable with preincubation results.

Trypsin, which is totally inactivated (Table 1) after 60 min at a concentration of 1% SDS, may be extremely resistant if other conditions are chosen. SDS electrophoresis has shown that trypsin shows proteolytic activity even after 18 hr at a concentration of 1% SDS when a substrate is present (105).

The inhibition mechanism of glutamate dehydrogenase (from bovine liver) has been studied in a series of sodium *n*-alkyl sulfate esters (evennumbered C_6-C_{16}) (108). Glutamate dehydrognease is irreversibly denatured by SDS even at comparatively low concentrations (103) (Table 1). The enzyme is inhibited by all *n*-alkyl sulfate esters studied before the onset of conformational changes of the enzyme protein. The relative potency of the alkyl sulfate esters to inhibit the enzyme is proportional to the number of carbon atoms in the detergent.

From this and the fact that the inhibition is also a function of the lipophilicity of the surfactant (measured as the partition coefficient for the detergent between the phases 1-octanol and water), the authors conclude that hydrophobic bonding interactions are the major force in the combination of the detergents with this enzyme (108).

TABLE 1 Inhibition of Various Enzymes by SDS

Enzyme	Activity, %	Concentration, g SDS/100 ml	Incubation time, min	Temperature, °C	Ref.
Alcohol dehydrogenase	35	1	60	25	57
Aldolase	0	1	60	25	57
Alkaline phosphatase (calfthymus microsomal fraction)	100	1	24 hr	23	95
Alkaline phosphatase (human liver)	100	1	180	37	94
α-Amylase	30	0.15	18 hr	25	100
Catalase	31	1	60	25	57
Glucose oxidase	100	1	24 hr	37	57
Glucose oxidase	0	1	60	60	57
Glucose-6-phosphohydrolase	20	0.1	30	0	102
PP_i-glucose phosphotransferase	15	0.1	30	0	102
Glutamic dehydrogenase	0	0.01	15	25	103
Invertase	98	9.5 mg	24 hr	40	98
Invertase	1	38 mg	24 hr	40	98
Lactic dehydrogenase					
Bovine heart	27	0.5	30	25	101
Chick heart	0	0.5	30	25	101
Pancreatic lipase	50	29 mg	150	25	99
Pancreatic lipase	50	87 mg	2	25	99
Papain	67	1	120	25	57
Pepsin	57	1	120	25	57
Pyrophosphatase	22	0.1	30	0	102
Trypsin	0	1	60	25	57
Tryptophan oxygenase	35	0.004	4	25	104
Xanthine oxidase	2	1	60	25	57

Hydrophobic interactions rather than electrostatic interactions between the surfactant and the enzyme are also held to be responsible for the inhibition of penicillinase by alkyl sulfates and alkylaryl sulfonates (109).

Phospholipase A1 isolated from *E. coli*—a membrane-bound enzymelike phosphatase—has been found to be stable at an SDS concentration of 3% (107). However, testing for enzymatic activity cannot be done in the presence of the surfactant because anionic surfactants are strong inhibitors. Possibly the enzyme is transformed by SDS into a form which undergoes renaturation easily. In inhibition experiments with *n*-alkyl sulfate esters of different chain lengths (even-numbered C_8-C_{14}) maximum inhibition has been observed in the case of C_{14}; the authors conclude that the alkyl sulfates compete with the substrate for hydrophobic regions at the active site.

In addition to the results discussed in Sec. I, studies on the interactions between SDS and Ferricytochrome c (110) have also shown that the binding of surfactants to protein does not only involve hydrophobic interactions. The authors conclude from spectroscopic evidence that SDS molecules react with cationic amino acids (lysyl, arginyl, and histidyl residues) at the protein surface in the first step of SDS/Cytochrome c interaction which results in an unfolding of the protein molecule, which in turn allows hydrophobic side chains of the protein to enter into hydrophobic interactions with the hydrophobic alkyl chain of the surfactant.

An interesting analytic application is the determination of small concentrations (ppm region) of anionic surfactants utilizing lipase (112) and acid phosphatase inhibition (113). Linear and branched alkyl benzene sulfonates as well as dodecyl sulfate may be determined rapidly and quantitatively by this method.

A number of attempts have been made to correlate protein denaturation and enzyme inactivation with the dermatotoxic properties of the surfactants. A general relationship between skin irritation and inhibitor action of anionic surfactants appears to be well established in the case of saccharase, however, there are occasional departures from this rule (98,114). The inhibition of saccharase by linear and branched alkyl benzene sulfonates is appreciably stronger than inhibition by SDS or sodium lauryl diethoxysulfate. As anticipated, SDS inhibited saccharase at a higher degree than the homologous sodium decyl- and octyl sulfates (114a). It has been suggested that the loss of enzymatic activity coincided with micellar concentrations of these anionics and that at least an initial step in denaturation of the enzyme by SDS is ionic. Other enzymes used for a possible correlation of enzyme inactivation with dermatotoxic properties were glucose-6-phosphate dehydrogenase (115b, 115e) and acid phosphatase (115c,115d). With glucose-6-phosphate dehydrogenase it was found that this enzyme was inactivated more strongly by dodecyl sulfate than by the corresponding dodecyl ether sulfate (115b). This finding correlates with the better skin compatibility of the ether sulfates (115b). Imokawa and Mishima investigated the lysosome labilizing effect of surfactants using a combined study of enzyme assay (acid phosphatase) and gel chromatography to establish a relationship of the labilization and skin roughness inducing effect (115c). They found that the labilizing ability of surfactants decreased in the order of cationic, anionic, and nonionic properties of their terminal groups and varies with its alkyl chain length, e.g., SDS had a more pronounced labilizing effect than decyl- or octyl sulfate.

In the case of both anionic and cationic surfactants the concentrations which induced lysosome labilizing effects were below their CMC (critical micelle concentration). The authors suggest that the lysosome labilizing effect of these ionic surfactants is due to an increase of permeability of the lysosome membrane resulting from the adsorption of surfactant molecules rather than their solubilizing action. The denaturation of keratin, albumin, enzymes, and the lysosome membrane by various surfactants has been studied with regard to their skin-irritating action (115,115a). The surfactants have high denaturing potency in the decreasing order of alkyl benzene sulfonate $\geqslant$ alkyl sulfate $>$ α-olefin sulfonate $\geqslant$ sodium alkane sulfonate.

Changes in the enzymatic activities of the following enzymes in the stratum corneum of the skin have been observed after multiple application to the skin of SDS, cetyltrimethylammonium bromide, and Triton X-100 (octylphenolpolyoxyethyleneglycol): acid phosphatase, β-D-glucuronidase, and glucose-6-phosphate dehydrogenase. The smallest change was observed in the case of Triton X-100, while cetyltrimethylammonium bromide led to the strongest enzyme inhibition (116). Human skin was very similar in its reaction to pig skin.

Comparing the effects of anionic, cationic, and nonionic surfactants on isolated enzymes, the following conclusions may be drawn from the studies available. Nonionic surfactants—most of the experimental work has been done with Triton X-100—generally do not cause inactivation or denaturation of enzymes and proteins (88d,112,113,118,119,124,124a). In the case of lipid-dependent enzymes, e.g., enzymes located in membranes, nonionic surfactants can also act as enzyme activators (124a). An occasional enzyme inhibition observed in the case of dopamine β-monooxygenase in the presence of Triton X-100 has been explained by the presence of impurities in the Triton X-100 preparation used (120). Possibly other cases of enzyme inhibition by Triton X-100 (122,123) may be explained by the presence of impurities in the nonionic surfactant. Therefore, nonionic surfactants are frequently employed in the isolation of membrane-bound enzymes and proteins (117). Anionic surfactants, on the other hand, show the strongest inactivating and denaturing potency on enzymes, while cationic surfactants are generally less effective than anionic surfactants (82,88,88b,112,113,124a), but an exception from this rule was found with lactate dehydrogenase isoenzymes (LDH). The greater susceptibility of LDH isoenzymes to complete inactivation by cationic compared with anionic surfactants was explained with the increased instability of the A subunit to cationic surfactants relative to the stability of the B subunit to anionic surfactants (88d). Other factors can also influence the general rule (vide supra) in more complex systems, such as surfactant effects on skin (212).

III. INTERACTION OF SURFACTANTS WITH MEMBRANES

A. Introduction

Because of their amphiphilic nature, surfactants may enter into interactions with the major membrane components, lipids and proteins. For instance, the hemolytic as well as to some extent the bactericidal action of surfactants is caused by the interaction of the surfactant with the erythrocyte membrane and the bacterial membrane, respectively, and has been known for a long time (84,126,127,128).

Surfactants are frequently used for cleaning purposes and the contact with the human skin is unavoidable. Their skin irritancy depends largely on their ability to penetrate the corneal layer and to interact with the Malpighian cells of the epidermis (125).

The membrane is an important part of the cell, the fundamental unit of living tissue. It compartmentalizes biological functions within the cell and separates the cell from its environment. The cell membrane is also essential in regulating ion transport, cell recognition, receptor-mediated processes, energy transduction, and also as a barrier to diffusion (178a). Surfactants play an important role in the elucidation of the complex membrane structure. When applied at sufficiently high concentrations many surfactants will solubilize membranes; the solubilized constituents of the cell membrane may then be characterized by biochemical or biophysical methods (59). In contrast to the usage in the case of the solubilization of sparingly soluble substances in surfactant micelles (147), the term "solubilization" as used in membrane biology only denotes the transformation of the membrane into a water-soluble form with no sediment being separable upon ultracentrifugation (see page 27). A more detailed discussion about criteria for solubility can be found in the paper of Hjelmeland and Chrambach (130a). Under certain conditions the membrane constituents may be used to reconstitute a membrane identical or similar to the initial one after removal of the surfactants (81a,129,130,130b,130c,130f).

B. Composition and Structure of Membranes

Biological membranes contain proteins, glycoproteins, lipids (mainly phospholipids), glycolipids, and ions. According to their function the ratio of the components can vary. Membranes that behave mainly as a barrier (e.g., myelin) contain a relatively high proportion of lipids. The ratio of protein to lipid in the membrane of myelin is 0.23 (131-133), in the human erythrocyte membrane 1.1 (134), in the inner mitochondrial membrane 3.2 (135), and in the membrane of Gram-positive bacteria 3.0 (136). This increase reflects the primarily "functional" rather than "barrier" function of the latter two while the erythrocyte membrane seems to be intermediate.

From the data of the protein and lipid content of the human erythrocyte membrane it is possible to conclude that there is sufficient lipid to cover 80-90% of the membrane surface area with a lipid bilayer (137). These figures strongly favor the presence of a lipid bilayer in this and other membranes (138). The lipid bilayer is arranged in a manner that the hydrophilic head groups of the phospholipids form the inner and outer membrane surfaces and the two hydrophobic tails form the hydrophobic core of the membrane (Fig. 8). The lipid bilayer has a thickness of about 45 Å, whereas the whole membrane measures about 100 Å.

In modification of the concept of the unit membrane model in which the protein was restricted to the surfaces of the lipid bilayer, there exists now strong evidence for the following arrangement of proteins in the membrane (139-141).

Some membrane proteins are readily solubilized by changing the pH of the solution, by increase of the ionic strength, or by chelating agents. These proteins do not penetrate the hydrocarbon core of the lipid bilayer and are called extrinsic or peripheral proteins. Peripheral proteins are soluble in water and in most cases peripheral enzymes do not loose their enzymatic activities by solubilizing.

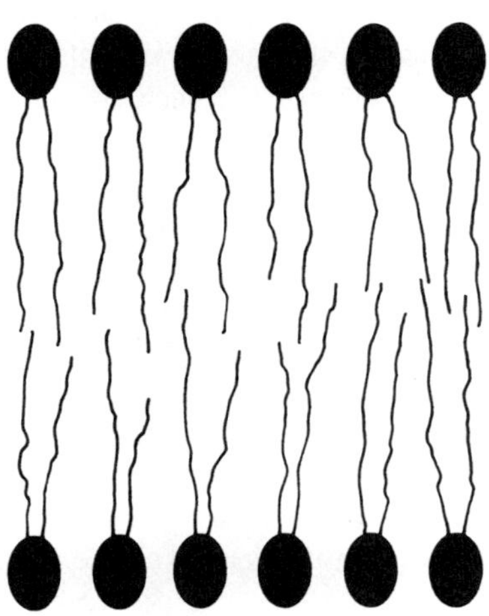

FIG. 8 A phospholipid bilayer: schematic cross-sectional view. The filled circles represent the ionic and polar head groups of the phospholipid molecules, which make contact with water; the wavy lines represent the fatty acid chains. (From Ref. 140.)

Other membrane proteins have substantial contact with the hydrocarbon core of the bilayer. They are embedded in the lipid matrix and can even span the whole membrane. These proteins can only be solubilized from the membrane by destruction of the lipid bilayer (59) and—in the absence of detergents—are insoluble in aqueous media or form multimolecular aggregates. These proteins are called intrinsic or integral proteins. The solubilization of integral membrane proteins is possible by using detergents. In most cases nonionic (141a) or special ionic detergents (sodium deoxycholate) have been used because usual ionic detergents (e.g., SDS) have been shown to denature membrane proteins as well as soluble globular proteins (see Sec. II).

With the aid of modern physical methods (nmr, membrane labeling, electron spin resonance) it could be demonstrated that the membrane lipids diffuse rather freely in the plane of the membrane (142-144), while the motion of the lipids from one side to the other of the membrane (flip-flop) is extremely slow (145). The latereal mobility of the lipids depends upon the proportion of unsaturated fatty acids in the phospholipids of the lipid bilayer and the temperature. In the membranes of mammalian cells a considerable proportion of the lipids in the bilayer are unsaturated with the effect that the melting temperature for the bilayer is below the body temperature. In this fluid state the hydrophobic tails of the phospholipids are able to move in the bilayer.

The lipids in the neighborhood of the intrinsic proteins which are immersed in the hydrophobic core of the bilayer are immobilized by the protein. These "boundary lipids" seem to be necessary for the full activity of intrinsic enzymes.

Besides the phospholipids of the bilayer, the integral proteins, too, show lateral mobility in the plane of most membranes (140,146). These and other observations have led to the proposal that under physiological conditions most membranes exist as a solution of integral proteins in the fluid bilayer (fluid mosaic model; see Fig. 9) (146,153). As a direct consequence, one would expect both lipid and proteins to be randomly distributed within membranes and the lipid phase to exhibit bulk properties in terms of fluidity. In fact, membrane organization in the refined picture is certainly much more complex. It is now well established that lipids and proteins are transversely asymmetrically distributed. Most membrane proteins are not free to diffuse, presumably due to their attachment to cytoskeleton components. Studies on membrane model systems show that lateral motion of lipids strongly depends on their chemical structure and physical state and on the lipid-to-cholesterol and lipid-to-protein ratios. There is increasing evidence suggesting that microheterogenities exist in biological membranes (153a,153b).

Current data suggest that the erythrocyte membrane and other membranes are asymmetric with respect to the two surfaces in the distribution of the

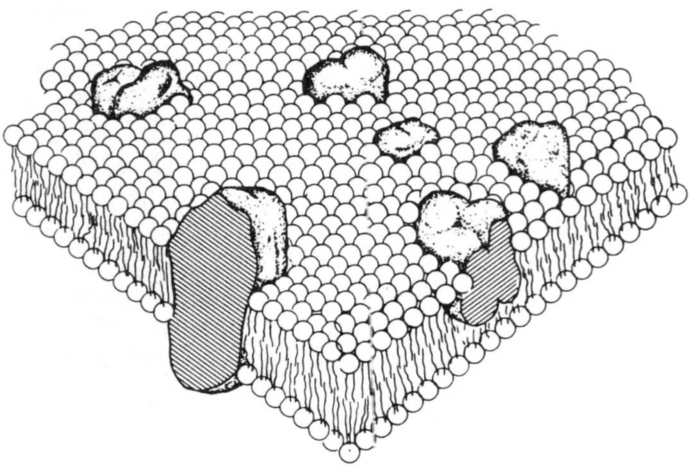

FIG. 9 The lipid-globular protein mosaic model with a lipid matrix (the fluid mosaic model); schematic three-dimensional and cross-sectional views. The solid bodies with stippled surfaces represent the globular integral proteins, which at long range are randomly distributed in the plane of the membrane. At short range, some may form specific aggregates, as shown. (From Ref. 140.)

proteins (138,147,148) and lipids (138). The major phospholipids present at the outer monolayer are phosphatidylcholine and sphingomyelin whereas the inner leaflet contains predominantly phosphatidylethanolamine and phosphatidylserine (149,150,150a). The outer monolayer also seems to contain more cholesterol than the inner one (151). Cholesterol is a major component of many biological membranes and obviously fulfills an important function in them, but it is not known what this function is. The present opinion is that cholesterol condenses and rigidifies membranes without solidifying them. Erythrocyte membranes are rich in cholesterol, the molar ratio of cholesterol/phospholipids is about 0.90 (152). With regard to the hemolytic and bactericidal action of detergents it is interesting to note that bacterial membranes are devoid of sterols (152).

Readers who are interested further in the structure of biological membranes and the erythrocyte membrane should consult the following reviews (140, 146,152,153a,175-178,178a, and 138,179-181, respectively).

C. Interaction of Surfactants with Different Membranes

Because of their amphiphilic nature and the consequent ability to be adsorbed at interfaces, surfactants may interact with biological membranes. The surfactant-induced hemolysis of erythrocytes has been particularly well investigated in the last decades because of the comparatively simple isolation of erythrocytes; therefore, this topic is discussed in a separate section (III.D).

Applied at sufficiently high concentration SDS causes the complete solubilization of many bacterial membranes (91,154,155,156). Complete solubilization of a membrane has been achieved if no sediment is formed upon centrifugation at 100,000 g for 60 min and no membrane structures are discernible by electron microscopy (129). At membrane concentrations of more than 2 mg ml^{-1} about 1.6 mg SDS/mg membrane lipids is required for membrane solubilization (117).

The solubilization of membranes by SDS depends on the membrane structure. Thus, only 75% of the membrane of *Micrococcus lysodeikticus* was dissolved by SDS (154). Generally, anionic surfactants, particularly SDS and cationic surfactants (cetyltrimethylammonium bromide) are more effective than nonionic surfactants (see also Sec. I.E). However, there is the disadvantage that anionic and cationic surfactants may lead to denaturation and, in the case of enzymes, to the loss of biological activity because of their stronger binding to proteins.

Based on this principle of interaction, anionic and more particularly cationic surfactants may show bactericidal and bacteriostatic action even at low concentrations (84,127,157). While anionic surfactants are active only against Gram-positive bacteria, cationic surfactants are active against both types of bacteria. The activity of anionic surfactants increases with decreasing pH value while the opposite behavior is observed in the case of cationic surfactants, which indicates that the first step in the interaction consists of an ionic adsorption to the bacterial cell wall. The consequent membrane damage leads to cytolysis, autolysis, and the eventual inactivation of metabolic processes (157).

Large stretches of dissolved cell membrane are observed immediately after the addition of SDS to *E. coli K12*. While only small amounts of RNA and protein, but no DNA, are released from the cell after incubating for 2 min, larger amounts of RNA, protein, and DNA are released within 45 min without concomitant destruction of the cell (182).

The fatty acid composition of the cell membrane appears to be an important factor determining the sensitivity of the cell toward surfactants. Thus, *E. coli* with a low unsaturated fatty acid content in the phospholipids of the cell membrane is significantly more sensitive against lysis by SDS than *E. coli* with more than 30% unsaturated fatty acids (167).

Severe destructive damage of the cell walls, the cytoplasmatic membrane, and the inner cell structure has been observed after treatment of Yersinia pseudotuberculosis with SDS (158); also, the action of SDS on ciliates led to changes of the cytoplasmatic membrane (166).

Lysis of the single-celled ameba proteus by SDS and dodecyltrimethylammonium bromide (a cationic surfactant) had been observed at concentrations of 5×10^{-3} and 2×10^{-3} mol liter^{-1}, respectively, within 1 hr. While in the case of SDS lysis was not inhibited by the addition of sodium ions, which lower the negative surface charge of the membrane, inhibition was observed in the case of the cationic surfactant. The authors conclude that the first step of the lytic process consists of an ionic reaction of the surfactant with acidic polysaccharides at the cell surface (159).

The effect of SDS and various other surfactants on rat liver plasma membranes have been studied in some detail (106,160,161). Complete solubilization does not occur even at high SDS concentrations (0.8%); the solubilizing action was more pronounced in the case of SDS than in the case of the anionic surfactant deoxycholate. Both surfactants led to the dissociation of the membrane proteins from the phospholipids.

Anionic surfactants (SDS and deoxycholate) are more effective solubilizing agents at higher concentrations than nonionic surfactants (Lubrol, Triton) (161). In the case of deoxycholate, as opposed to SDS, membrane-bound enzymes may be obtained in biologically active form upon solubilization (117). In micellar solutions of deoxycholate, lipophilic proteins bound large amounts of the surfactant (117). In view of their nondenaturing solubilization of membrane-bound proteins, cholic acid and its derivates desoxycholic acid, taurocholic acid, and taurodeoxycholic acid, are frequently used in membrane studies (130a,161a).

In the solubilization of membrane-bound enzymes the inactivation by surfactants generally occurs in the following order: (Lubrol, Triton X-100, Tween 20) < bile salts (cholate, deoxycholate) < dodecyl sulfate and cetyltrimethylammonium (165,213). In the solubilization of erythrocyte membranes conformational changes of the membrane proteins have been observed in the cases of SDS and sodium deoxycholate application, while no such conformational changes were found with Triton X-100 (163). In the case of SDS the onset of conformational changes occurs only at the critical micelle concentration, at which a major portion of the membrane proteins has been dissolved already. Presumably, the proteins are protected against conformational changes by the lipids bound to the dissolved lipophilic proteins; only when the critical micelle concentration has been reached are the lipids removed from the membrane proteins by incorporation into mixed phospholipid/surfactant micelles (163). The mechanisms of solubilization of the various membrane constituents (protein, phospholipids, cholesterol, and individual proteins) are different in the cases of Triton X-100, SDS, and sodium deoxycholate. SDS, for instance, extracts lipids and individual proteins at different SDS concentrations, while Triton X-100 solubilizes most proteins and lipids at a concentration which is identical for all components (164). If the surfactant concentration is increased at constant membrane concentration, three distinct regions of surfactant action can be distinguished (117).

Stage 1. If the surfactant/membrane ratio is low, surfactants may cause permeability changes of the membranes and at higher concentrations cell lysis (e.g., hemolysis) may occur. Solubilization of the membrane does not occur at these low concentrations.

Stage 2. If the surfactant concentration is higher than required for cell lysis, the lamellar structure of the membrane is lost (lamellar-micellar phase transition) and the membrane is solubilized. Mixed micelles, proteins, and lipoprotein complexes are present in the solution. As in stage 1 the monomer form of the surfactant is bound to the membrane (117).

Stage 3. Further increases in the surfactant concentration lead to the removal of the phospholipids from the proteins and adsorption of the surfactant to the hydrophobic regions as they become available. SDS and other surfactants capable of massive cooperative binding generally cause conformational changes with concomitant loss of biological activity, whereas nondenaturing surfactants such as bile salts and nonionic surfactants are effective in maintaining the native conformation of intrinsic membrane proteins (81a,130d). More recently, zwitterionic surfactants, e.g., sulfobetaines, have been used as effective and mild surfactants in membrane research (130a,130e)

Binding data for anionic surfactants, e.g., SDS or deoxycholate to membrane proteins are scarce, since these proteins are generally poorly soluble in aqueous media. Investigations with cytochrome b_5 with the major glycoprotein from the erythrocyte membrane and with the coat protein of bacteriophage f_1 have shown that the binding of deoxycholate and SDS is different from that generally observed for water-soluble proteins in that way that these membrane proteins can interact with surfactant micelles (39). With another intrinsic membrane protein, the major membrane protein of Rhodospirillum rubrum chromatophore, it was found that this protein can bind more than twice the amount of SDS on weight basis than typical water-soluble proteins (167a). It is assumed that with intrinsic membranes the protein-protein interaction is so strong that the solubilization of the protein with SDS is only possible after the critical micelle concentration is attained. Unusual high binding of deoxycholate and SDS was also found with the myelin basic protein, a highly charged, peripheral membrane protein found in larger amounts in the myelin sheath in the central nervous system (30,167b). Here, too, the bulk of the surfactant binding occurred around the cmc of the surfactants. The different surfactant binding properties of membrane proteins compared to water-soluble proteins can be the reason that the common method for determining molecular weight, SDS-polyacrylamide gel electrophoresis, can yield incorrect results for membrane proteins (81a,167c).

More recently, with the easy availability of cultured cells, investigations concerning the cytotoxic effects of surfactants were undertaken (164a-c). In most of these studies, the elucidation of the mechanism of interaction with the cell membrane was not the object, but getting insights into a possible correlation between the local compatibility of the surfactants concerning skin and mucous membrane and their cytotoxic properties. The cytotoxicity of the different classes of surfactants follows the general order of cationic > anionic > nonionic > amphoteric but exceptions from this ranking may occur depending on the structure of the surfactant. It is assumed that with cultured cells at a certain high enough concentration the membrane, too, will be permeable for cell constituents resulting in cytotoxic and possibly irreversible, lytic effects. If leucocytes are incubated with anionic, cationic, or nonionic surfactants at concentrations (1×10^{-4} mol liter^{-1}) below the concentration required to cause lysis (stage 1), it is found that most anionic surfactants stimulate the metabolic processes, while both cationic and nonionic surfactants inhibit the metabolism (168,169). Presumably, ionic bonds are formed between the cell membrane and the anionic surfactants which leads to structural changes of the plasma membrane.

In the case of mitochondria, SDS concentrations below the level required for lytic effects (10^{-6}-10^{-4} mol liter^{-1}) lead to an uncoupling of the oxidative phosphorylation (170) and structural changes. The inhibiting effect increases with the number of carbon atoms in the *n*-alkyl moiety of the *n*-alkyl sulfates. The inactivation of the oxidative functions of submitochondrial preparations by SDS is caused by a general alteration of the membrane structure due to modifications in the surface charge (171). Influences on the membrane fluidity appear to be of minor importance. Many attempts have been made to correlate the toxicity toward isolated cells with the skin irritancy of surfactants (125,172,173).

The influence of surfactants on the potassium permeability of yeast cells and on the alterations of desquamated cells from buccal mucosa have been studied in some detail (125). While surfactant-induced changes in the potassium permeability of yeast cells appear to be irrelevant with regard to the skin irritancy of surfactants, the toxicity toward buccal mucosa may be a model for the skin-damaging effects.

In the incubation of peritoneal mast cells with surfactants a release of histamine is observed (172,173). The amount of histamine released depends both on the polar group of the surfactant molecule and on the length of the alkyl chain. *n*-Alkyl sulfates with 12 or 14 carbon atoms are most effective. Exposure to SDS leads to cell damage due to the cytolytic action of the surfactant, which is clearly seen under the electron microscope.

The ability of surfactants to inhibit the DNA synthesis or to release cytoplasmic proteins due to fibroblast damage has been studied with guinea pig kidney fibroblasts (173). Of the carboxylic acid sodium salts tested (C_8-C_{16}), sodium laurate showed the strongest effects; the effects were increased by the replacement of the polar carboxylic acid head group by the sulfate or the isethionate group. While the mechanism of the initial mediation of the primary irritation phenomenon has not yet been fully elucidated, the authors believe that there is a close relationship between the effects observed and the inflammation of the skin. It should be mentioned that the solubility of zein, a sparingly soluble protein, in the presence of surfactants is well correlated with skin irritancy (11-14) (see also Sec. I.B).

In the determination of the lytic activity and the toxicity of surfactants toward HeLa cell cultures* the biological activity of the surfactants was found to decrease in the following order: sodium dodecyl benzene sulfonate > SDS > alkane sulfonate-hydroxyalkane sulfonate mixture.

Application of the surfactants (20% solutions) to the skin on five consecutive days showed the strongest effects again in the case of sodium dodecyl benzene sulfonate; SDS showed the mildest action (174).

D. Hemolytic Properties of Surfactants

Because of the comparatively simple isolation of erythrocyte membranes, their interaction with surfactants has been studied in considerable detail to probe the architecture of biological membranes (126,183-204,204a). Almost all anionic surfactants are capable of hemolyzing erythrocytes at certain concentrations. Hemolysis occurs when proteins and other macromolecules can pass the membranes and may easily be followed photometrically because of the release of hemoglobin from the cell. Most anionic surfactants will hemolyze erythrocytes even at very low concentrations (about 10^{-5}-10^{-3} mol liter^{-1}).

Hemolysis depends on numerous experimental parameters such as temperature, pH, erythrocyte concentration, surfactant concentration, as well as concentration and nature of ions present. An additional parameter is the animal species from which erythrocytes were isolated. Depending on the species the surfactant concentration required to cause a certain degree of erythrocyte hemolysis may vary (202). Due to these numerous experimental parameters and due to discrepancies in experimental conditions, the lack of data on the quality of the surfactant used, etc., the results of different investigations are very difficult to compare.

---

*Human cervical carcinoma cells which have been cultivated in a tissue culture since 1951.

The limiting concentrations, i.e., the surfactant concentrations that will just cause visible hemolysis after an incubation period of 120 min, are shown graphically in Fig. 10 for carboxylic acid sodium salts and alkyl sulfates in the hemolysis of pigeon erythrocytes (183). Both hemolysis curves show a maximum at C_{14}. Similar hemolysis curves have been observed in the case of cationic surfactants (188).

It is striking that the hemolysis curves of homologous series of cationic and anionic surfactants are similar in shape to the curves of antibacterial action (188).

The hemolytic activity of surfactants is inhibited by the addition of lecithin, cholesterol, and proteins (184,205), presumably because of a decrease in the surfactant monomer concentration due to protein and lipid addition.

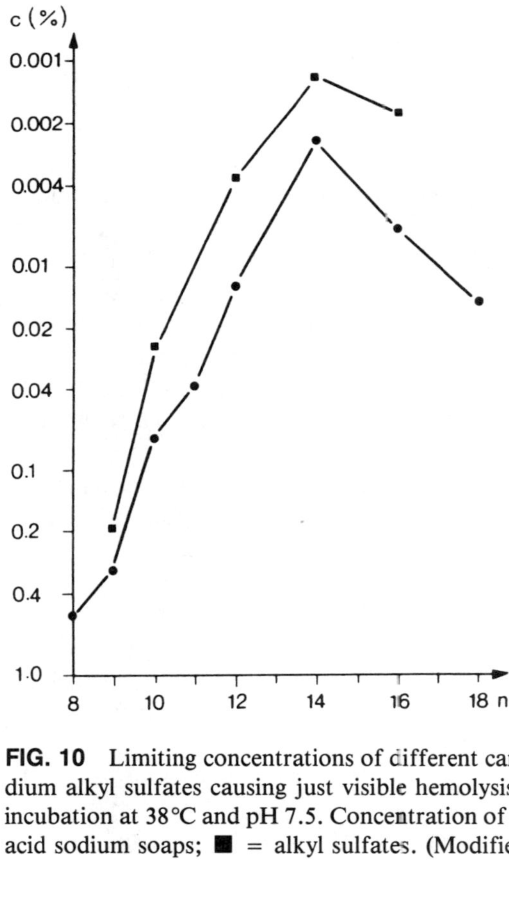

FIG. 10 Limiting concentrations of different carboxylic acid sodium soaps and sodium alkyl sulfates causing just visible hemolysis of pigeon erythrocytes after 2 hr incubation at 38°C and pH 7.5. Concentration of erythrocytes: 1%; ● = carboxylic acid sodium soaps; ■ = alkyl sulfates. (Modified from Ref. 183.)

The hypothesis that decreasing surface tension leads to increasing hemolytic activity of surfactants cannot be maintained in view of the lack of correlation between hemolytic activity and surface activity of surfactants, which has been observed experimentally (206). The decrease of hemolytic activity observed in alkyl sulfates and carboxylic acid soaps with 16 and 18 carbon atoms is presumably caused by an increase in the Krafft temperatures* and a corresponding decrease in the number of free surfactant molecules which are available for reaction with the cell membrane (111,111a).

It is of interest that cationic as well as anionic and nonionic surfactants protect erythrocytes against hypotonic hemolysis at sublytic concentrations, while other typical hemolysins (digitonin, saponin) do not show any erythrocyte stabilization (193,193a,193d). Digitonin and saponin are known to have a high affinity toward cholesterol and presumably cause hemolysis by binding of cholesterol or extraction of cholesterol from the erythrocyte membrane (193b). The stabilization of the membrane against osmotic and mechanical hemolysis due to the binding of surfactants is possibly due to an increase in the size of the membrane surface (117). Anionic surfactants (linear and branched-chain Na dodecylbenzenesulfonates and Na dodecyl sulfate) and nonionic surfactants (Triton X-100 and ethoxylated nonylphenol) had qualitatively similar effects: at lower concentrations (<0.6 and <2 ppm for the anionic and nonionic types) there was no activity; higher concentrations (0.6-90 and 2-150 ppm, resp.) increased the osmotic resistance, but at still higher concentrations this resistance progressively decreased and complete hemolysis occurred at concentrations of > 300 for anionic and > 600 ppm for the nonionic surfactants. Thus, expressed in molarities, a stabilizing action was shown at appr. 10^{-4} M, and both classes of surfactants were hemolytic at appr. 10^{-3} M, suggesting the particular effects to be function of the number of surfactant molecules present in the solution (193b). Alkyltrimethylammoniumsalts (C_{12}, C_{14}, and C_{16}) and SDS induce different shape alterations in rat erythrocytes but it was concluded that the induction of different shapes by cationic and anionic surfactants cannot be due to differences in the distribution of the surfactant molecules within the lipid bilayer of the erythrocyte membrane alone (193d).

In spite of great research efforts, the molecular mechanism of hemolysis is still unknown. When surfactants are used as lytic agents the key physicochemical properties of these compounds in this process seem to be the CMC and the hydrophilic-lipophilic balance as well as the chemical structure, the

---

*The Krafft temperature is the temperature above which the solubility of a surfactant sharply increases. At this temperature the solubility becomes equal to the critical micelle concentration.

nature of the charged head group and the length of the surfactant tail (193c). Hemolysis may be regarded as a process which consists of at least five consecutive reactions: The substance is adsorbed at the membrane surface (1) and penetrates into the membrane (2), where it induces changes in the molecular organization (3), which alter the permeability of the membrane (4) in such a fashion that hemoglobin is released (5) due to a change in the osmotic equilibrium (190,207). In the case of lysolecithin-induced hemolysis the second reaction step—penetration of the substance into the membrane—appears to be the rate-determining step (200).

Binding studies with virus membranes and SDS have shown that the surfactant is bound to a finite number of binding sites at the virus membrane at low SDS concentrations, while massive binding of SDS occurs with the lysis of the membrane (about 5×10^{-4} mol liter^{-1}) (117).

It has been reported that anionic surfactants are bound primarily by the protein of the erythrocyte membrane at physiological pH values (202,204). The binding of surfactants leads to conformational changes of the membrane proteins, which in turn results in a release of hemoglobin from the cell. In the case of cationic surfactants the binding to the phospholipids of the membrane appears to be of major importance (196,199,203,204). It cannot be excluded, however, that membrane lipids are also involved in the anionic surfactant-induced lysis.

It has been concluded from studies of hemolysis kinetics that the cells are destroyed by two different mechanisms operating simultaneously. One of these processes leads to rapid erythrocyte destruction, presumably by reaction of the surfactants with the phospholipids of the membrane. In the other, slow-process lysis occurs by a slow breakdown of a lipoprotein complex (190). This hypothesis explains nicely the observation that erythrocytes show different sensitivity toward one and the same surfactant. Another hypothesis suggests that hemolysis occurs when a critical surface pressure is exceeded as a consequence of the penetration of the surfactant into the lipid layer of the membrane (187). It has been shown that cholesterol is extracted from the membrane in the course of SDS-induced hemolysis (195). In still another hypothesis it is assumed that the lysis of erythrocytes is caused by the formation within the membrane of a transient micelle which is formed by the transformation of the laminar structure of the lipid bilayer under the influence of surfactants (191).

The lytic activity of SDS is higher than that of the corresponding quaternary ammonium salts (186). Among nonionic surfactants the hemolytic activity varies widely: Some sugar derivatives, such as Tween 60 and 80,* do not

*Tween 60 = polyoxyethylene-(20)-sorbitan monostearate. Tween 80 = polyoxyethylene-(20)-sorbitan monooleate.

show any hemolytic activity (208), while other derivatives are strongly active (194). The lytic action of the sugar derivatives seems to be time dependent. No lytic action was found after a short incubation of polysorbate-20, whereas hemolytic action of the compound was demonstrated after some hours of incubation, presumably by influencing the ion permeability of the membrane (214). In the case of polyoxyethylene-derived surfactants the well-known hemolytic activity of these compounds was proposed to be induced mainly by peroxides and free radicals formed by autoxidation (215).

REFERENCES

1. S. Saito, *Kolloid-Z. Z. Polym., 215,* 16(1967).
2. S. Saito, *Kolloid-Z., 154,* 19(1957).
3. S. Saito, *Kolloid-Z., 158,* 120(1958).
4. S. Saito, *Kolloid-Z., 165,* 162(1959).
5. S. Saito, *Kolloid-Z., 168,* 128(1959).
6. M.J. Schwuger and H. Lange, *Proc. Int. Congr. Surface Active Subs., 5th, Barcelona, 2,* 955(1968).
7. H. Lange, *Kolloid-Z. Z. Polym., 243,* 101(1971).
8. M.N. Jones, *J. Colloid. Interfac. Sci., 23,* 36(1967).
9. S. Gravsholt, *Proc. Int. Congr. Surface Active Subs., 5th, Barcelona, 2,* 993 (1968).
10. M.J. Schwuger, *J. Colloid. Interfac. Sci., 43,* 491(1973).
11. J.F. Foster, *J. Phys. Colloid. Chem., 53,* 175(1949).
12. E. Gotte, *Proc. Int. Congr. Surface Active Subs., 4th, Brussels, 3,* 83(1964).
13. E. Gotte, *Aesthet. Med., 15,* 313(1966).
14. M.J. Schwuger, *Kolloid-Z. Z. Polym., 233,* 898(1969).
15. J. Steinhardt and J.A. Reynolds, *Multiple Equilibria in Proteins,* Academic Press, New York, 1969.
16. J.A. Reynolds, J.P. Gallagher, and J. Steinhardt, *Biochemistry, 9,* 1232(1970).
17. M.J. Schwuger, *Kolloid-Z. Z. Polym., 246,* 626(1971).
18. G.J. Puttermann, N.F. Wolejsza, M.A. Wolfram, and K. Laden, *J. Soc. Cosmet. Chem., 28,* 521(1977).
19. U. Zeidler and G. Reese, *Arztl. Kosmetol., 13,* 39(1983).
20. L.D. Rhein, C.R. Robbins, K. Fernee, and R. Cantore, *J. Soc. Cosmet. Chem., 37,* 125(1986).
21. J.C. Blake-Haskins, D. Scala, L.D. Rhein, and C.R. Robbins, *J. Soc. Cosmet. Chem., 37,* 199(1986).
22. M.M. Breuer, *J. Soc. Cosmet. Chem., 30,* 41-64(1979).
23. J.G. Dominguez, J.L. Parra, R.M. Infante, C.M. Pelejero, F. Balaguer, and T. Sastre, *J. Soc. Cosmet. Chem., 28,* 165-182(1977).
24. A. Conrads and H. Zahn, *Int. J. Cosmet. Sci., 9,* 29(1987).
25. J.C. Dohnal and J.E. Garvin, *Biochim. Biophys. Acta, 576,* 393-403(1979).
26. T. Takagi, J. Miyake, and T. Nashima, *Biochim. Biophys. Acta, 626,* 5-14(1980).
27. M. Sen, S.P. Mitra, and D.K. Chattoraj, *Colloids and Surfaces, 2,* 259-275(1981).

28. M.N. Jones, P. Manley, and A. Holt, *Int. J. Biol. Macromol., 6,* 65-68(1984).
29. K.S. Birdi and J. Steinhardt, *Biochim. Biophys. Acta, 534,* 219-227(1978).
30. P.F. Burns, C.W. Campagnoni, I.M. Chaiken, and A.T. Campagnoni, *Biochemistry, 20,* 2463-2469(1981).
31. S.-G. Hjalmarsson, *Biochim. Biophys. Acta, 581,* 210-216(1979).
32. K. Kale, G.C. Krescheck, and G. Vanderkooi, *Biochim. Biophys. Acta, 535,* 334-341(1978).
33. J.A. Reynolds, In *Meth. Enzymology,* Vol. 61, W.B. Jakoby (Ed.), Academic Press, New York, 1979, pp. 58-62.
34. S. Makino and R. Niki, *Biochem. Biophys. Acta, 495,* 99-109(1977).
35. E.G. Brown, *Anal. Biochem., 174,* 337-348(1988).
36. K. Aoki, H. Okabayashi, S. Maezawa, T. Mizuno, N. Murata, and K. Hiramatsu, *Biochim. Biophys. Acta, 703,* 11-16(1983).
37. K. Takeda, M. Shigeta, and K. Aoki, *J. Colloid Interface Sci., 117,* 120(1987).
38. M. Koide, M. Fukuda, K. Ohbu, Y. Watanabe, Y. Hayashi, and T. Takagi, *Anal. Biochem., 164,* 150-155(1987).
39. P.F. Burns and A.T. Campagnoni, *Biochim. Biophys. Acta, 743,* 379(1983).
40. P.H. von Hippel and K.Y. Wong, *J. Biol. Chem., 240,* 3909(1965).
41. W.W. Fish, J.A. Reynolds, and C. Tanford, *J. Biol. Chem., 245,* 5166-5168 (1970).
42. J.A. Reynolds and C. Tanford, *J. Biol. Chem., 245,* 5161-5165(1970).
43. A.K. Wright, M.R. Thompson, and R.L. Miller, *Biochemistry, 14,* 3224-3228 (1975).
44. A.L. Shapiro, E. Vinuela, and J.V. Maizel, Jr., *Biochem. Biophys. Res. Commun., 28,* 815-820(1967).
45. K. Weber and M. Osborn, *J. Biol. Chem., 244,* 4406-4412 (1969).
46. A.K. Dunker and R.R. Rueckert, *J. Biol. Chem., 244,* 5074-5080(1969).
47. M.N. Jones, *Biochem. J., 151,* 109-114(1975).
48. J.A. Reynolds and R.H. Simon, *J. Biol. Chem., 249,* 3937-3940(1974).
49. M.N. Jones, H.A. Skinner, and E. Tipping, *Biochem. J., 147,* 229-234(1975).
50. Y. Nozaki, J.A. Reynolds, and C. Tanford, *J. Biol. Chem., 249,* 4452-4459 (1974).
51. A. Helenius and K. Simons, *J. Biol. Chem., 247,* 3656-3661(1972).
52. J. Steinhardt, J.R. Scott, and K.S. Birdi, *Biochemistry, 16,* 718-725(1977).
53. R. Pitt-Rivers and F.S. Ambesi Impiombato, *Biochem. J., 109,* 825-830(1968).
54. J.A. Reynolds and C. Tanford, *Proc. Nat. Acad. Sci. U.S., 66,* 1002-1007(1970).
55. M.N. Jones and A. Wilkinson, *Biochem. J., 153,* 713-718(1976).
56. S. Makino, J.A. Reynolds, and C. Tanford, *J. Biol. Chem., 248,* 4926-4932 (1973).
57. C.A. Nelson, *J. Biol. Chem., 246,* 3895-3901(1971).
58. H.M. Rendall, *J. Chem. Soc. Faraday Trans., I, 72,* 481-484(1976).
59. C. Tanford and J.A. Reynolds, *Biochim. Biophys. Acta, 457,* 133-170(1976).
60. A.A. Spector, *J. Lipid Res., 16,* 165-179(1975).
61. W.U. Malik and V.P. Saxena, ACS Symp. Ser. 1975, 9 (Colloidal Dispersions Micellar Behav., Pap. Symp. 1974), 299-315; *Chem. Abstr., 83,* 39014(1975).
62. J.A. Reynolds, S. Herbert, and J. Steinhardt, *Biochemistry, 7,* 1357-1361(1968).

63. T. Takagi, K. Tsujii, and K. Shirahama, *J. Biochem., 77,* 939-947(1975).
64. A. Chrambach and D. Rodbard, *Science, 172,* 440-451(1971).
65. K.A. Ferguson, *Metabolism, 13,* 985-1002(1964).
66. D. Rodbard and A. Chrambach, *Anal. Biochem., 40,* 95-134(1971).
67. G.A. Banker and C.W. Cotman, *J. Biol. Chem., 247,* 5856-5861(1972).
68. D.M. Neville, Jr., *J. Biol. Chem., 246,* 6328-6334(1971).
69. K. Weber and M. Osborn, In *The Proteins,* Vol. 1, 3rd, H. Neurath and R.L. Hill (Eds.), Academic Press, New York, London, 1975, pp. 179-223.
70. J.T. Stoklosa and H.W. Latz, *Anal. Biochem., 68,* 358-370(1975).
71. K. Shirahama, K. Tsujii, and T. Takagi, *J. Biochem., 75,* 309-319(1974).
72. A.H. Maddy, *J. Theor. Biol., 62,* 315-326(1976).
73. K. Tsujii and T. Takagi, *J. Biochem., 77,* 511-519 (1975).
74. W.L. Mattice, J.M. Riser, and D.S. Clark, *Biochemistry, 15,* 4264-4272(1976).
75. H. Furthmayr and R. Timpl, *Anal. Biochem., 41,* 510-516(1971).
76. S. Panyim and R. Chalkley, *J. Biol. Chem., 246,* 7557-7560(1971).
77. A. Ray, J.A. Reynolds, H. Polet, and J. Steinhardt, *Biochemistry, 5,* 2606-2616(1966).
78. R.V. Decker and J.F. Foster, *J. Biol. Chem., 242,* 1526-1532(1967).
79. I.M. Klotz and F.M. Walker, *J. Amer. Chem. Soc., 69,* 1609-1612(1947).
80. C. Tanford, *J. Mol. Biol., 67,* 59-74(1972).
81. A. Wishnia and T. Pinder, *Biochemistry, 3,* 1377-1384(1964).
81a. S. Makino, *Adv. Biophys., 12,* 131(1979).
81b. K. Ohbu, N. Jona, N. Miyajima, and N. Mizushima, *J. Jap. Oil Chem. Soc., 29,* 866(1980).
82. K. Nakaya, A. Ushiwata, and Y. Nakamura, *Biochim. Biophys. Acta, 439,* 116-124(1976).
83. K. Nakaya, K. Yamada, M. Onozawa, and Y. Nakamura, *Biochim. Biophys. Acta, 251,* 7-13(1971).
84. F.W. Putnam, *Advan. Prot. Chem., 4,* 79-122(1948).
85. C. Tanford, *Advan. Prot. Chem., 23,* 122-275(1968).
85a. F. Timmermann and A. Domsch, *Parfuem. Kosmet., 68,* 126-134(1987).
86. R.V. Decker and J.F. Foster, *Biochemistry, 5,* 1242-1254(1966).
86a. S. Tandon and P.M. Horowitz, *J. Biol. Chem., 262,* 4486(1987).
86b. P.M. Horowitz and S. Bowman, *Electrophoresis, 7,* 534(1987).
87. M.N. Jones, H.A. Skinner, E. Tipping, and A. Wilkinson, *Biochem. J., 135,* 231-236(1973).
88. C. Blinkhorn and M.N. Jones, *Biochem. J., 135,* 547-549(1973).
88a. T.F. Abidi, C.A. Faaland, D.D. Scala, L.D. Rhein, and J.D. Laskin, *Biochim. Biophys. Acta, 992,* 362-368(1989).
88b. F. Favilli, M. Stio, C. Treves, P. Vanni, and M.T. Vincenzini, *C. R. Soc. Biol., 179,* 307-315(1985).
88c. G. DiSabato and N.O. Kaplan, *J. Biol. Chem., 239,* 438-443(1964).
88d. K.J. Sanford, D.J. Meyer, M.J. Mathison, and J. Figueras, *Biochemistry, 20,* 3207-3214(1981).
88e. K. Smith and T.K. Sundaram, *Biochim. Biophys. Acta, 884,* 109-118(1986).

88f. L.O. Narhi, M. Zukowski, and T. Arakawa, *Arch. of Biochem. Biophys., 261,* 161-169(1988).
88g. L. Kravetz and K.F. Guin, *J. Am. Oil. Chem. Soc., 62,* 943-949(1985).
89. E.D. Wills, *Biochem. J., 57,* 109-120(1954).
90. G. Popescu, M. Marinescu, and I. Brad, *Colloid Polym. Sci., 254,* 55-63 (1976).
91. Z. Ne'eman, I. Kahane, and S. Razin, *Biochim. Biophys. Acta, 249,* 169-176(1971).
92. E.D. Wachsmuth and K. Hiwada, *Biochem. J., 141,* 273-282(1974).
93. I.H. Mather and I.W. Keenan, *FEBS Lett., 44,* 79-82(1974).
94. K.D. Gerbitz, H.J. Kolb, and O.H. Wieland, *Hoppe-Seyler's Z. Physiol. Chem., 358,* 435-446(1977).
95. P.L. Ey and E. Ferber, Biochim. Biophys. Acta, 480, 163-177(1977).
96. H.L. Schrock and R.B. Gennis, *J. Biol. Chem., 252,* 5990-5995(1977).
97. S. Clarke, *J. Biol. Chem., 250,* 5459-5469(1975).
97a. S. Makino, S. Ogimoto, and S. Koga, *Agric. Biol. Chem., 47,* 319-326(1983).
97b. Z. Wasylewski, A. Kozik, *Eur. J. Biochem., 95,* 121-26(1979).
97c. R.D. Morero and G. Weber, *Biochim. Biophys. Acta, 703,* 241-246(1982).
98. H. Wilmsmann, *Fette, Seifen, Anstrichm., 65,* 958-964(1963).
99. B. Borgstrom and J. Donner, *J. Lipid Res., 17,* 491-497(1976).
99a. T. Kawase, W.-S. Kim, T. Fujii, and M. Minagawa, *Chem. Express, 1,* 483-486(1986).
100. Y. Pomeranz, *Biochim. Biophys. Acta, 73,* 105-112(1963).
101. G. di Sabato and N.O. Kaplan, *J. Biol. Chem., 239,* 438-443(1964).
102. R.E. Snoke and R.C. Nordlie, *Biochim. Biophys. Acta, 139,* 190-192(1967).
103. K.S. Rogers and S. C. Yusko, *J. Biol. Chem., 244,* 6690-6695(1969).
104. K. Koike, W.N. Poillon, and P. Feigelson, *J. Biol. Chem., 244,* 3457-3462 (1969).
105. W.H. Porter and J.L. Preston, *Anal. Biochem., 66,* 69-77(1975).
106. R. Sohn and G.V. Marinetti, *Chem. Phys. Lipids, 12,* 17-30(1974).
107. C.J. Scandella and A. Kornberg, *Biochemistry, 10,* 4447-4456(1971).
108. K.S. Rogers and S.C. Yusko, *J. Biol. Chem., 247,* 3671-3676(1972).
109. G.E. Afinogenov and E.F. Panarin, *Antibiotiki (Moscow),* 21, 876-880 (1976); *Chem. Abstr., 86,* 624(1977).
110. R.K. Burkhard and G.E. Stolzenberg, *Biochemistry, 11,* 1672-1677(1972).
111. H. Lange and M. Schwuger, *Kolloid-Z. Z. Polym., 223,* 145-149(1968).
111a. H. Schott, *J. Pharm. Sci., 62,* 341-343(1973).
112. H. Tanaka, K. Tsuji, and K. Konishi, *Anal. Lett., 7,* 425-436(1974).
113. H. Tanaka, Y. Horiuchi, and K. Konishi, *Anal. Biochem., 66,* 489-497(1975).
114. H. Wilmsmann and A. Marks, *Fette, Seifen, Anstrichm., 61,* 965-973(1959).
114a. R. Ernst, *J. Am. Oil. Chem. Soc., 57,* 93-98(1980).
115. G. Imokawa and M. Katsumi, *Yukagaku, 25,* 24-30(1976); *Chem. Abstr., 84,* 100393(1976).
115a. G. Imokawa, K. Sumara, and M. Katsumi, *J. Am. Oil. Chem. Soc., 52,* 484 (1975).
115b. F. Timmermann, personal communication (1986).

115c. G. Imokawa and Y. Mishima, *Contact Dermatitis, 5,* 151-162(1979).
115d. C. Prottey, D. Oliver, and A.C. Coxon, *Int. J. of Cosmet. Sci., 6,* 263-273 (1984).
115e. A. Domsch, *Seifen-Oele-Fette-Wachse, 114,* 209(1988).
116. M. Kermici and G. Aubin, *Labo.-Pharma.-Probl. Techn., 24,* 592-597(1976).
117. A. Helenius and K. Simons, *Biochim. Biophys. Acta, 415,* 29-79(1975).
118. A. Tzagoloff and H.S. Penefsky, In *Methods Enzymology,* Vol. 22, W. Jakoby (Ed.), Academic Press, New York, 1971, pp. 219-230.
119. C. Gloxhuber and W.K. Fischer, Food Cosmet. Toxicol., *6,* 469-477(1968).
120. O. Terland, E. Slinde, T. Skotland, and T. Flatmark, *FEBS Lett., 76,* 86-90(1977).
121. U. Kragh-Hansen and T. Riisom, *Eur. J. Biochem., 70,* 15-23(1976).
122. R.M. Bell, *J. Biol. Chem., 250,* 7147-7152(1975).
123. H.P. Bar and S. Kulshresta, *Can. J. Biochem., 53,* 472-474(1975).
124. G.J. Giotta, *Biochem. Biophys. Res. Commun., 71,* 776-782(1976).
124a. F.E. Beyhl, *IRCS Med. Sci., 14,* 417-418(1986).
125. F.R. Bettley, *Brit. J. Derm., 80,* 635-642(1968).
126. E. Ponder, *Hemolysis and Related Phenomena,* Grune & Stratton, New York, 1948.
127. E.I. Valko, *Ann. N.Y. Acad. Sci., 46,* 451-478(1946).
128. R.D. Hotchkiss, *Ann. N.Y. Acad. Sci., 46,* 479-493(1946).
129. S. Razin, *Biochim. Biophys. Acta, 265,* 241-296(1972).
130. Y. Kagawa, *Biochim. Biophys. Acta, 265,* 297-338(1972).
130a. L.M. Hjelmeland and A. Chrambach, In *Methods Enzymology,* Vol. 104, W.B. Jakoby (Ed.), Academic Press, New York, 1984, pp. 305-318.
130b. J.A. Reynolds and D.R. McCaslin, *Subcell. Biochem., 14,* 1-24(1989).
130c. A. Helenius, E. Fries, and J. Kartenbek, *J. Cell. Biol., 75,* 866-880(1977).
130d. J. van Renswoude and C. Kempf, In *Methods Enzymology,* Vol. 104, W.B. Jakoby (Ed.), Academic Press, New York, 1984, pp. 329-339.
130e. A. Gonenne and R. Ernst, *Anal. Biochem., 87,* 28-38(1978).
130f. R.D. Klausner, J. van Renswoude, and B. Rivnay, In *Methods Enzymology,* Vol. 104, W.B. Jakoby (Ed.), Academic Press, New York, 1984, pp. 340-347.
131. J.A. Lowden, M.A. Moscarello, and R. Morecki, *Can. J. Biochem., 44,* 567-577(1966).
132. J.S. O'Brien and E.L. Sampson, *J. Lipid Res., 6,* 537-544(1965).
133. J. Gagnon, P.R. Finch, D.D. Wood, and M.A. Moscarello, *Biochemistry, 10,* 4756-4763(1971).
134. S.A. Rosenberg and G. Guidotti, *J. Biol. Chem., 243,* 1985-1992(1968).
135. D.F. Parsons and Y. Yano, *Biochim. Biophys. Acta, 135,* 362-364(1967).
136. M.R.J. Salton and J.H. Freer, *Biochim. Biophys. Acta, 107,* 531-538(1965).
137. D.M. Engelman, *Nature, 223,* 1279-1280(1969).
138. G. Guidotti, *Ann. Rev. Biochem., 41,* 731-752(1972).
139. G. Vanderkooi, *Ann. N.Y. Acad. Sci., 195,* 6-15(1972).
140. S.J. Singer and G.L. Nicolson, *Science, 175,* 720-731(1972).
141. D.F.H. Wallach and P.H. Zahler, *Proc. Nat. Acad. Sci. U.S., 56,* 1552-1559(1966).

141a. E.A. Dennis, *Adv. Colloid Interface Sci., 26,* 155-175(1986).
142. E. Sackmann, H. Trauble, H.-J. Galla, and P. Overath, *Biochemistry, 12,* 5360-5369(1973).
143. P. Devaux, C.J. Scandella, and H.M. Mc Connell, *J. Mag. Resonance, 9,* 474-485(1973).
144. A.G. Lee, N.J.M. Birdsall, and J.C. Metcalf, *Biochemistry, 12,* 1650-1659 (1973).
145. R.D. Kornberg and H.M. McConnell, *Biochemistry, 10,* 1111-1120(1971).
146. S.J. Singer, *J. Colloid Interfac. Sci., 58,* 452-458(1977).
147. T. Nakagawa, In *Nonionic Surfactants,* Vol. 1, M. Schick (Ed.), Marcel Dekker, New York, 1967, p. 558.
148. M.S. Bretscher, *Science, 181,* 622-629(1973).
149. A.J. Verkleij, R.F.A. Zwaal, B. Roelofsen, P. Comfurius, D. Kastelijn, and L.L.M. van Deenen, *Biochim. Biophys. Acta, 323,* 178-193(1973).
150. P. Emmelot and R.P. van Hoeven, *Chem. Phys. Lipids, 14,* 236-246(1975).
150a. D. Chapman and P.I. Haris, *Biochem. Soc. Trans., 17,* 951-953(1989).
151. K.A. Fisher, *Proc. Nat. Acad. Sci. U. S., 73,* 173-177(1976).
152. R.A. Demel and B. de Kruyff, *Biochim. Biophys. Acta, 457,* 109-132(1976).
153. S.J. Singer, In *Structure and Function of Biological Membranes,* L. Rothman (Ed.), Academic Press, New York, 1971, p. 145.
153a. J.F. Tocanne, L. Dupou-Cezanne, A. Lopez, and J.-F. Tournier, *FEBS Letters, 257,* 10-16(1989).
153b. C. Richter, *Chem. Phys. Lipids, 44,* 175-189(1987).
154. D.G. Bishop, L. Ruthberg, and B. Samuelsson, *Eur. J. Biochem., 2,* 454-459(1967).
155. D.M. Engelman, T.M. Terry, and H.J. Morowitz, *Biochim. Biophys. Acta, 135,* 381-390(1967).
156. Z. Ne'eman, I. Kahane, J. Kovartovsky, and S. Razin, *Biochim. Biophys. Acta, 266,* 255-268(1972).
157. D.F. Hoelzl Wallach, *The Plasma Membrane: Dynamic Perspective in Genetics and Pathology,* Vol. 18, Heidelberg Science Library, Springer-Verlag, New York, 1972.
158. N. Cherepova, S. Baykousheva, and D. Veljanov, *Zbl. Bakt. Hyg., I. Abt. Orig., A238,* 216-219(1977).
159. J.E. Brewer and L.G.E. Bell, *Nature, 222,* 891-892(1969).
160. W.S. Bont, P. Emmelot, and H. Vaz Dias, *Biochim. Biophys. Acta, 173,* 389-408(1969).
161. J.K. Dunnick, G.V. Marinetti, and P. Greenland, *Biochim. Biophys. Acta, 266,* 684-694(1972).
161a. A.J. Furth, H. Bolton, J. Potter, and J.D. Priddle, *Methods Enzymology,* Vol. 104, W.B. Jakoby (Ed.), Academic Press, New York, 1984, pp. 318-328.
162. D.M. Small, *Fed. Proc., 29,* 1320-1326(1970).
163. F.H. Kirkpatrik and H.E. Sandberg, *Biochim. Biophys. Acta, 298,* 209-218 (1973).
164. F.H. Kirkpatrick, S.E. Gordesky, and G.V. Marinetti, *Biochim. Biophys. Acta, 345,* 154-161(1974).

164a. W. Sterzel, F.G. Bartnik, W. Matthies, W. Kästner, and K. Künstler, *Toxic. in Vitro, 4,* 698-701(1990).
164b. H. Babich and E. Borenfreund, *ATLA, 18,* 129-144(1990).
164c. M. Marinovich, E. Tragni, A. Corsini, and C.L. Galli, *J. Toxicol.-Cut. & Ocular Toxicol., 9,* 169-178(1990).
165. R. Coleman, *Biochim. Biophys. Acta, 300,* 1-30(1973).
166. T. Rockstroh and K. Zapf, *Naturwissenschaften, 54,* 568(1967).
167. Y. Akamatsu, *J. Biochem., 76,* 553-561(1974).
167a. J. Miyake and T. Takagi, *Biochim. Biophys. Acta, 668,* 290-298(1981).
167b. R. Smith and B.J. McDonald, *Biochim. Biophys. Acta, 554,* 133-147(1979).
167c. R.G. Gennis and A. Jonas, *Ann. Rev. Biophys. Bioeng., 6,* 195-238(1977).
168. K. Kakinuma, *Biochim. Biophys. Acta, 348,* 76-85(1974).
169. K. Kakinuma, T. Hatae, and S. Minakami, *J. Biochem., 79,* 795-802(1976).
170. S.C. Yusko, E.S. Higgins, and K.S. Rogers, Proc. Soc. Exp. Biol. Med., 141, 10-13(1972).
171. R.J. Mehlhorn and L. Packer, *Biochim. Biophys. Acta, 423,* 382-397(1976).
172. C. Prottey and T.F.M. Ferguson, *Food Cosmet. Toxicol., 14,* 425-430(1976).
173. T.F.M. Ferguson and C. Prottey, *Food Cosmet. Toxicol., 14,* 431-434(1976).
174. A. Ono, N. Okazaki, R. Akema, Y. Miyamato, H. Wada, H. Ikeda, and M. Kuroda, *Kanagawa-Ken Eisei Kenkyusho Nempo, 23,* 68-69(1973); Chem. Abstr., *85,* 172361(1976).
175. T. Gulik-Krzywicki, *Biochim. Biophys. Acta, 415,* 1-28(1975).
176. J.A. Lucy, *FEBS Lett., 40,* S105-S111(1974).
177. G. Vanderkooi, *Biochim. Biophys. Acta, 344,* 307-345(1974).
178. E.D. Korn, *Science, 153,* 1491-1498(1966).
178a. D. Chapman and D.C. Lee, *Biochem. Soc. Trans., 15 Suppl,* 47S-54S(1987).
179. E.A. Evans and R.M. Hochmuth, *J. Membr. Biol., 30,* 351-362(1977).
180. V.T. Marchesi, H. Furthmayr, and M. Tomita, *Ann. Rev. Biochem., 45,* 667-698(1976).
181. T.L. Steck, *J. Cell. Biol. 62,* 1-19(1974).
182. C.L. Woldringh, *Biochim. Biophys. Acta, 224,* 288-290(1970).
183. F.L. Breusch and H. Bodur, Z. Physiol. Chem., 286, 148-159(1950).
184. E. Ponder, *J. Gen. Physiol., 30,* 15-23(1946).
185. L.H. Love, *J. Cell. Comp. Physiol., 36,* 133-148(1950).
186. S. Ross and A.M. Silverstein, *J. Colloid Sci., 9,* 157-165(1954).
187. B.A. Pethica and J.H. Schulman, *Biochem. J., 53,* 177-185(1953).
188. F.L. Breusch and S. Hersek, Z. Physiol. Chem., 291, 1-6(1952).
189. E. Hutchinson and K.E. Bean, *Arch. Biochem. Biophys., 58,* 81-91(1955).
190. E. Rideal and F.H. Taylor, *Proc. Roy. Soc., Series B., 146,* 225-241(1957).
191. R.W. Bonsall and S. Hunt, *Biochim. Biophys. Acta, 249,* 266-280(1971).
192. J.G.R. Elferink, *Pharm. Weekbl., 112,* 1-10(1977).
193. Y. Ogura and J. Tamura, *Ann. Rept. Inst. Food Microbiol. (Chiba Univ.),* 21, 119-126(1968).
193a. L. Bolis, P. D'Angelo, and G. Delfino, *Boll. Soc. Ital. Biol. Sper., 53,* 1755 (1977).
193b. P. Seeman, *Biochem. Pharmacol., 15,* 1767-1774(1966).

193c. B.Y. Zaslavsky, N.N. Ossipov, S.V. Rogozhin, Y.L. Sebyakin, L.V. Volkova, and R.P. Evstigneeva, *Biochim. Biophys. Acta, 556,* 314-321(1979).
193d. B. Isomaa and G. Paatero, *Biochim. Biophys. Acta, 647,* 211-222(1981).
194. Y. Ogura and J. Tamura, *Ann. Rept. Inst. Food Microbiol. (Chiba Univ.), 20,* 89-94(1967).
195. E. Hutchinson, M.J. Sibley, and C. Abram, *Kolloid-Z. Z. Polym., 225,* 167-175(1968).
196. G.J.M. Hooghwinkel, R.E. de Rooij, and H.R. Dankmeijer, *Acta Physiol. Pharmacol. Neerl., 13,* 304-316(1965).
197. Z. Byczkowska and Z. Paradowski, *Polski Tygodnik Lekarski, 30,* 137-139 (1975).
198. T. Kondo and M. Tomizawa, *J. Pharm. Sci., 57,* 1246-1248(1968).
199. T. Kondo and M. Tomizawa, *J. Pharm. Sci., 58,* 1378-1381(1969).
200. K. Harlos, W.L.C. Vaz, and S. Kovatchev, *FEBS Lett., 77,* 7-10(1977).
201. T. Kondo and M. Tomizawa, *J. Colloid Interfac. Sci., 21,* 224-228(1966).
202. T. Kondo, *Adv. Colloid Interfac. Sci., 6,* 139-172(1976).
203. M. Tomizawa and T. Kondo, *Chem. Pharm. Bull.* (Tokyo), *18,* 2158-2163 (1970).
204. M. Tomizawa and T. Kondo, *Kolloid-Z. Z. Polym., 246,* 694-699(1971).
204a. B.Y. Zaslavsky, N.N. Ossipov, and S.V. Rogozhin, *Biochim. Biophys. Acta, 510,* 151-159(1978).
205. C. Gloxhuber, *Arch. Toxicol, 32,* 245-270(1974).
206. F. Breusch and E. Ulusoy, *Z. Physiol. Chem., 286,* 159-173(1950).
207. F.C. Reman, R.A. Demel, J. de Gier, L.L.M. van Deenen, H. Eibl, and O. Westphal, *Chem. Phys. Lipids, 3,* 221-233(1969).
208. P.H. Elworthy and J.F. Treon, In *Nonionic Surfactants,* M. Schick (Ed.), Marcel Dekker, New York, 1967.
209. M. Nemat-Gorgani and G. Dodd, *Eur. J. Biochem., 74,* 129-137(1977).
210. M. Nemat-Gorgani and G. Dodd, *Eur. J. Biochem., 74,* 139-147(1977).
211. J.L. Abernethy, H.M. Steinman, and R.L. Hill, *J. Biol. Chem., 249,* 7339-7347(1974).
212. G.J. Putterman, N.F. Wolejsza, M.A. Wolfram, and K. Laden, *J. Soc. Cosmet. Chem., 28,* 151-164(1977).
213. N.C. Robinson and R.A. Capaldi, *Biochemistry, 16,* 375-381(1977).
214. W. Pape, U. Pfannenbecker, and U. Hoppe, *Molecular Toxicol., 1,* 525-536(1987).
215. E. Azaz, R, Segal, and I. Milo-Goldzweig, *Biochim. Biophys. Acta, 646,* 444-449(1981).

2

Absorption, Metabolism, and Excretion of Anionic Surfactants

J. GORDON BLACK AND DOUGLAS HOWES Unilever Research, Colworth Laboratory, Bedford, England

I. INTRODUCTION

Traditionally, soaps derived from natural oils and fats have been the classical surfactants used in personal and domestic cleansing. The availability of synthetic surfactants since the 1940s has led to the gradual replacement of traditional soaps in a variety of cleansing products, and has given the technologist a range that can be used to formulate specific products for cleansing fabrics, dishes, carpets, floors, as well as many types of toilet soap bars and shampoos. The physical and chemical properties of surfactants have been reviewed by Jones (1), and their diverse nature can be gauged from such publications as *Surfactants U.K.* by Hollis (2).

The widespread use of surfactants both by industry and in the home means that every individual, certainly in the Western world, is exposed either directly by use of oral hygiene and skin-cleansing products or indirectly from residues on fabrics, on cooking utensils, or by ingestion of contaminated food and water. This low but continuous exposure has led to interest in their potential toxicity, and has been the subject of a review article by Gloxhuber (3).

Anionic surfactants, which represent the most widely used class of surfactant, are characterized by a lipophilic moiety, usually an alkyl or alkylaryl structure, and a hydrophilic moiety or polar head group that is negatively charged. Soaps are the most commonly known member and, in this case, the lipophilic moiety is the fatty acid derived from animal or vegetable fats and oils. The hydrophile of soaps is the ionized carboxylic acid function normally prepared as the sodium salt. The fatty acids are usually straight chains of from 8 to 18 C atoms. In some instances, the fatty acids may be unsaturated, with one or more double bonds in the alkyl chain.

The most common of the synthetic surfactants are the alkyl benzene sulfonates, which come in two forms. The branched alkyl chain type (ABS) was the first used commercially, but its slow rate of biodegradation in sewage treatment plants has led to its gradual replacement by alkyl benzene sulphonates with linear alkyl chains (LAS) that are biodegraded more rapidly than the branched type. Typically, both have 12 C atoms in the alkyl chain, but commercial LAS may contain from 10 to 18 C atoms depending on the source of commercial olefin.

The second most important group is the alkyl sulfates. Within this class are primary sulfates derived from either reduced fatty acids or industrially prepared alcohols that are sulfated and neutralized. Most of the alcohols have straight chains with 8-18 C atoms, but some industrially derived alcohols show a degree of chain branching that is usually α-methyl or α-ethyl. In addition, secondary alkyl sulfates are also commercially available. These are prepared by sulfation of commercial straight-chain olefins and the sulfate becomes distributed randomly along the alkyl chain.

Alkyl sulfonates include primary alkane sulfonates derived from industrial olefins, secondary alkane sulfonates derived from paraffins, and olefin sulfonates that, although relatively minor in terms of total use commercially, may become important in the future. There are, in addition, several minor classes that include sulfated alcohol ethoxylates, isethionic acid esters of fatty acids, phosphate esters, sulfosuccinates, and sarcosinates.

This chapter reviews absorption from the intestine, fate after parenteral administration, and penetration through the skin of members of these main classes of anionic detergents. The absorption of soaps will also be discussed but not their metabolic fate, since these behave as fatty acids, the catabolism of which is widely appreciated.

II. INTESTINAL ABSORPTION, METABOLISM, AND EXCRETION

A. Alkyl Benzene Sulfonates

The first publication on intestinal absorption of alkyl benzene sulfonates is probably that of Havermann and Menke (4), who fed dodecyl benzene [^{35}S]sulfonate to swine. Urine and feces from the pigs were monitored for ^{35}S for up to 8 days after administration and it was found that up to 35% of the administered ^{35}S was excreted in the urine. The feces contained most of the remainder of the administered ^{35}S. There was about 0.5% of the ^{35}S dose remaining in the carcass at 8 days after administration. This was distributed in tissues similarly to the ^{35}S found in tissues after administration of $Na_2{}^{35}SO_4$. The authors concluded that at least 35% of the dose was absorbed from the intestinal tract, but in this early study no chemical identification of fecal or urinary ^{35}S components was performed.

It was not until 1968 that Michael (5) published additional data that confirmed and extended these findings. In this publication, two samples of ^{35}S-labelled alkyl benzene sulfonate were synthesized. One was a linear alkyl benzene sulfonate that had 11, 12, or 13 C atoms in the *n*-alkyl chain and an even distribution of the phenyl sulfonate moiety among the 2-7 positions. For brevity this will be referred to as [^{35}S]LAS. The second was a branched alkyl chain alkyl benzene sulfonate, which had 8 or 9 atoms in the alkyl chain, with 3 or 4 methyl branches that will be referred to as [^{35}S]ABS.

After oral administration of the [^{35}S]LAS to rats at doses of 3, 5, 6, 40, or 250 mg kg^{-1}, 40-57% of the ^{35}S was recovered in the feces. There was no apparent dose-related recovery of ^{35}S in the feces. During 3 days after administration, the urine contained 38-56% of the ^{35}S. The recoveries are similar to those found by Havermann and Menke (4) in swine. Michael (5) extended his studies to examine the route of absorption of the [^{35}S]LAS from the intestinal tract. Oral administration of the [^{35}S]LAS to rats that had been surgically fitted with thoracic duct cannulas resulted in a recovery of only 1.6% of the ^{35}S in the lymph collected during a 42-hr period. It was concluded that LAS absorbed from the intestine was carried via the portal blood system. Other rats were fitted with bile duct cannulas and these studies were extended to a crossover experiment in which the bile from a treated rat entered the intestinal tract of a second rat, the bile of which was collected. In the bile duct-cannulated rats, 46% of the ^{35}S given orally as [^{35}S]LAS was excreted in the urine and 25% in the bile. This indicates that at least 71% of the dose was absorbed from the intestine and that the bile was a significant route of excretion. In the crossover experiment, all of the biliary-excreted ^{35}S was absorbed from the intestine and partially (40%) excreted in the urine, although most (60%) was excreted in the bile of the second rat. Thus, biliary excretion and intestinal resorption of LAS or its metabolites occurs in the rat, and enterohepatic recirculation results in a significant proportion of the ^{35}S being excreted in the urine. By isotope dilution analysis with Na_2SO_4, it was shown that no $^{35}SO_4{}^{2-}$ was present in any of the samples. No unchanged [^{35}S]LAS was detected in the urine or bile, but traces were recovered in the feces. Isotope dilution analysis confirmed that 19% of the fecal ^{35}S was present as [^{35}S]LAS. The nature of the biliary and fecal ^{35}S compounds was not further investigated. Infrared, nmr, and mass spectrometric (MS) studies of the methyl [^{35}S]-esters of the urinary components were performed and compared with spectra of authentic chemical compounds. Using these techniques, the urinary ^{35}S-labeled components were shown to be present as a mixture of sulfophenyl butanoic and sulfophenyl pentanoic acids. Precise quantification of these components was not made but the author concluded that, in the rat, LAS is degraded by ω-oxidation of the alkyl chain, followed by β-oxidation to a series of short-chain sulfophenyl carboxylic acids. This mechanism is similar to that found in bacterial degradation of LAS (6). However, bacteria can degrade LAS further by removal of the sulfonate moiety, enabling degradation of the phenyl ring to occur (7).

The fate of sodium alkyl [^{14}C]benzene sulphonate ([^{14}C]LAS) in rhesus monkey was reported by Cresswell et al. (8). At 5 days after a single oral dose of 30 mg kg^{-1}, 71.2% of the ^{14}C dose was recovered in the urine and 23.1% in the feces. Peak plasma levels were observed at 4 hr after dosing and the biological half-life was 5-6 hours. Similar results were found after doses of 150 and 300 mg kg^{-1} doses. After 7 daily doses of 30 mg kg^{-1}, there was no accumulation of ^{14}C in tissues and the kinetic values remained the same. Urinary metabolites were resolved into at least 5 major components and characterized as more polar than the LAS. No glucuronide or sulfate conjugates were found.

When the LAS isomer 2-sulfophenyl[2-^{14}C]*n*-dodecane ([^{14}C]2-DOBS) was administered orally to rats at a dose of 5 mg kg^{-1}, Rennison et al. (9) recovered 75% of the ^{14}C dose in the urine and 13.3% in the feces within 48 hr of dosing. When the LAS isomer 6-sulfophenyl[6-^{14}C]*n*-dodecane ([^{14}C]6-DOBS) was administered orally to rats at 5 mg kg^{-1} only 18% of the dose was recovered in the urine and 77.9% was in the feces at 48 hr after dosing. In anaesthetised and bile duct-cannulated rats dosed orally with the ^{14}C-labelled DOBS isomers, it was found that most of the ^{14}C was absorbed from the intestine. However, only 6.5% of the [^{14}C]2-DOBS was excreted in the bile at 6 hr after dosing but 83.1% of the [^{14}C]6-DOBS was excreted in the bile. The nature of the ^{14}C in the bile and urine from these rats was not reported by Rennison et al., but analysis of the urinary ^{14}C components from the [^{14}C]2-DOBS by the authors using the Knight-House procedure described by Swisher (10) gave a single radioactive compound which was shown by gas liquid chromatography-mass spectroscopy (GLC-MS) to be 3-methyl-indan-1-one, which can be regarded as a condensation product from butyric acid-3-benzene-*p*-sulphonate, the natural end product of metabolism. The ^{14}C components excreted by the rats dosed with [^{14}C]6-DOBS have been resolved into at least eight components that have not been characterized.

Lay et al. (11) fed a diet containing 1.4 mg of [^{14}C]LAS kg^{-1} diet to rats for 5 weeks. During this time, approximately twice as much ^{14}C was recovered in the feces as in the urine as polar metabolites of LAS. From an estimate of the ingested diet over the 35 days of feeding and the recoveries of ^{14}C in excreta, the authors concluded that bioaccumulation of LAS and its metabolites was occuring. They analysed the ^{14}C levels in tissues on completion of the 35 days feeding but not in animals permitted to eat normal diet for 7 days before sacrificing.

Michael (5) also intubated the branched chain [^{35}S]ABS to rats at doses of 3, 5, 6, or 40 mg kg^{-1}. In these studies, only 6.8-9.5% of the ^{35}S was recovered in the urine and up to 79% in the feces 3 days after dosing. Absorption was shown not to be by the lymphatic system, since only traces of ^{35}S were recovered in the lymph from thoracic duct-cannulated rats. The bile, however, contained 80-90% of the administered ^{35}S and enterohepatic recirculation was demonstrated in a bile-transfer study. Using the same techniques as described for the [^{35}S]LAS above, Michael showed that the metabolites of ABS were also sulfophenyl carboxylic acids, some of which had apparently undergone desaturation as well as partial oxidation. The chain branching apparently interfered with the β-oxidation of the alkyl chain resulting in longer carboxylic acid. As with the LAS, the sulfonate moiety of ABS remained intact.

These studies show that alkyl benzene sulfonates are readily absorbed across the intestinal tract of mammals. The absorbed surfactant is rapidly metabolized after transportation in the portal blood system, by oxidation of the alkyl chain, but no degradation of the polar head group occurs. It is clear that the form of the alkyl chain, i.e., linear or branched, affects the overall metabolism, as does the position of the phenyl sulfonate moiety on the alkyl chain. The data available suggest that the end products are all likely to be sulfophenyl carboxylic acids. The isomers in which the phenyl sulfonate moiety is in the middle of the alkyl chain give rise to a complex mixture of metabolites that may be dicarboxylic acids. It is extremely interesting to note that some of these positional isomers are excreted almost exclusively in the bile, whereas others are excreted almost exclusively in the urine.

B. Alkyl Sulfates

Havermann and Menke (4) reported on the fate of sodium dodecyl [^{35}S]sulfate in swine. After oral administration the ^{35}S was rapidly recovered principally in the urine. At 8 days after dosing, a total recovery of about 90% in the urine and 10% in the feces was made. Traces of ^{35}S remained in the carcasses of the treated pigs and some free $^{35}SO_4^{2-}$ was detected in the urine. The distribution of the ^{35}S in the body was similar to that found after administration of $Na_2{}^{35}SO_4$. The authors concluded that sodium dodecyl sulfate was absorbed extensively from the intestine and that only a small amount of the surfactant was degraded to free sulfate.

Denner et al. (12) investigated the fate of potassium dodecyl [^{35}S]sulfate. Oral administration to rats at doses of 5 mg kg^{-1} resulted in a rapid recovery of [^{35}S] in the urine. Virtually all of the ^{35}S was recovered in the urine during the first 48 hr after treatment and less than 1% of the dose was recovered in the feces during this time. Thus, complete absorption from the intestinal tract had occurred. Whole-body autoradiography at various times after dosing showed that absorption was apparently occurring across the small intestine and that the liver was the main organ of accumulation. Extensive use of paper, thin-layer, ion-exchange, gel-permeation, and gas-liquid chromatographic techniques was made to isolate the ^{35}S-labeled components from the urine. Approximately 20% of the urinary ^{35}S was present as inorganic sulfate. The remainder was present as a single sulfate ester, but not as the parent compound. The ^{35}S-labeled metabolite was more polar than the parent compound and on acid hydrolysis yield $^{35}SO_4^{2-}$ and butyrolactone. This suggested that the metabolite was probably butyric acid-4-sulfate which was synthesized and found to behave identically in several chromatographic systems to the ^{35}S metabolite. From this study it was concluded that the rat is capable of metabolizing *n*-alkyl sulfates by ω-oxidation of the hydrophobe followed by β-oxidation of the resulting carboxylic acid. The β-oxidation of the alkyl chain was not complete since butyric rather than acetic acid sulfate was the main end product. The presence of up to 20% of the ^{35}S in urine as $^{35}SO_4^{2-}$ was thought to be an indication of either sulfatase activity or a capability for at least partial degradation of the whole alkyl chain by β-oxidation.

Ottery et al. (13) studied the fate of [^{35}S] and [^{14}C]butyric acid-4-sulfate by rat liver homogenates. They found that the rate of hydrolysis was pH-dependent and was nonenzymic. In each case the end products were identified as γ-butyrolactone and SO_4^{2-}.

Burke et al. (14) extended these studies to include potassium decyl [^{35}S]sulfate and potassium octadecyl [^{35}S]sulfate. After oral administration of these compounds to rats, all of the ^{35}S was recovered in the urine and the end products of metabolism were similar to those found from rats treated with potassium dodecyl [^{35}S]sulfate, namely, butyric acid-4-sulfate and some $^{35}SO_4^{2-}$. These data suggest that in the rat all *n*-alkyl sulfates with an even number of C atoms in the alkyl chain are readily absorbed across the intestinal wall and metabolized by a common route.

Merits (15) intubated [1-^{14}C]hexadecyl sulfate and hexadecyl [^{35}S]sulfate as their erythromycin salts into rats and dogs at a dose of 14.4 mg kg^{-1} and gave by capsule to humans at 1.4 mg kg^{-1}. The results from the rats were essentially similar to that obtained by Denner et al. (12) after administration of potassium dodecyl [^{35}S]sulfate. However, some 3-7% of the ^{14}C was expired at $^{14}CO_2$. In the dog the recoveries of ^{35}S after oral administration of hexadecyl [^{35}S]sulfate were somewhat different in that 37% was recovered in the feces and 52% was recovered in the urine. With the [1-^{14}C]hexadecyl sulfate similar recoveries to the rats were found with one dog, but in a second dog 79% of the ^{14}C was recovered in the urine and only 12% in the feces. From one human volunteer, 80% of the ^{14}C was recovered in the urine and 7% in the feces up to 72 hr after dosing with [1-^{14}C]hexadecyl sulfate. In a second human volunteer, 72% of the dose was recovered in the feces and 20% in the urine up to 118 hr after ingestion.

Detailed chromatographic examination of the urinary ^{14}C-labeled compounds from the animals dosed with [1-^{14}C]hexadecyl sulfate showed that the main metabolite was butyric acid-4-sulfate from both species. In the dog, however, about 20% of the urinary ^{14}C was present as the sulfate ester of glycolic acid. In man, only 5% of the ^{14}C in the urine was present as the glycolic acid ester. In addition to these two sulfate esters, inorganic $^{35}SO_4^{2-}$ was detected in the urine of rat and dog. From the animals dosed with the [1-^{14}C]hexadecyl sulfate, small amounts (2-7%) of $^{14}CO_2$ were expired. In the dog, approximately 10% of the urinary ^{14}C was recovered as γ-butyrolactone and, in the rat, 3% of the urinary ^{14}C was identified as γ-butyrolactone. Small proportions of unchanged surfactant were also recovered in the urine from dog and rat. These findings extend those of Denner et al. (12) and Burke et al. (14) to show that alkyl sulfates with even numbers of C atoms in the alkyl chain are readily absorbed from the intestine and are metabolized by dog and man in a similar manner to the rat. The further β-oxidation of the principle metabolite, butyric acid-4-sulfate, apparently occurred to an extent in dog and in man so that one of the products of metabolism was acetic acid-2-sulfate. The presence of [^{14}C]-γ-butyrolactone in the urine of dog and man indicated that desulfation of the metabolite occurred and the presence of $^{14}CO_2$ expired by the rat and dog indicated that either small amounts of the parent surfactant were desulfated or that the desulfated metabolite was degraded in these species.

Studies conducted by Maggs (16) with octan-2-sulfate in the rat showed that while the stereospecific isomers were absorbed, distributed, and excreted similarly there was a suggestion that the ω oxidation step was more rapid for the D-(+) isomer than for the L-(−) isomer.

Work in our own laboratory has confirmed most of these findings (Howes, unpublished). However, it is worth noting that the hydrolysis of butyric acid-4-sulfate occurs spontaneously (13) and examination of fresh urine samples is to be recommended when investigating sulfate ester metabolites.

Tovell et al. (17) showed that goldfish rapidly absorbed sodium [1-^{14}C]-dodecyl sulfate across the gill and metabolized the surfactant by a route similar to that in the rat.

Burke et al. (18) published data on the fate of an *n*-alkyl isomer with an odd number of C atoms in the alkyl chain. This isomer, *n*-undecyl sulfate, was prepared with ^{35}S as the potassium salt. Rats were intubated with aqueous solutions at 5 mg kg^{-1} and approximately 80% of the administered ^{35}S was recovered in the urine during 48 hr after dosing. Fecal ^{35}S levels accounted for approximately 10% of the administered ^{35}S during this time. At 48 hr, between 0.02 and 7.6% of the dose was recovered in the carcasses of these rats. Chromatographic techniques similar to those reported by Denner et al. (12) were employed to examine the nature of the ^{35}S compounds in the urine from the treated rats. From these investigations it was found that less than 5% of the urinary ^{35}S was present as $^{35}SO_4^{2-}$, but there were three ^{35}S-labeled components with lower Rf values on TLC than the parent undecyl sulfate. The urine did not contain undecyl [^{35}S]sulfate. There was one major metabolite and two minor ones although the authors did not quantify the proportions. The major metabolite co-chromatographed with propionic acid-3-sulfate on several TLC systems, and on hydrolysis with HCl all of the ^{35}S was recovered as inorganic $^{35}SO_4^{2-}$. No positive identification of the resulting alkyl moiety was made, although the authors concluded that the metabolite was propionic acid-3-sulfate. The minor metabolites were not identified but the authors suggested that one was valeric acid-5-sulfate.

Unpublished data from our own laboratory have shown that rats intubated with sodium [1-^{14}C]pentadecyl sulfate at 5 mg kg^{-1} excreted up to 90% of the administered ^{14}C in the urine within 72 hr, and that 3-5% was expired as $^{14}CO_2$. We have identified the alkyl moiety of the main metabolite, obtained after acid hydrolysis and methylation, as 3-hydroxy-propionic acid (as the methyl ester) by GLC-MS techniques.

Webborn et al. (19) examined the fate of tridecan sulfate isomers after oral administration to rats. The 1, 3, 5, and 7 isomers were studied and their overall disposition was similar being excreted mainly in the urine as carboxylic acids. However, the 5 and 7 isomers produced conjugated metabolites that were excreted in the bile.

The fate of nonan-5-sulfate after oral administration to the rat was studied by Bains et al. (20) and was found to be almost exclusively and rapidly excreted in the urine. Metabolism was extensive but parent material was detected in the urine with two main metabolites that were identified as nonan-1-ol-5-sulfate and nonanoate-5-sulfate.

As mentioned in the introduction, some alkyl sulfates contain branched chain hydrophobes. To our knowledge the only published data on this type of surfactant is that of Knaak et al. (21), who examined the fate of sodium 2-ethyl [1-^{14}C]hexyl sulfate ([^{14}C]EHS) and sodium 2-ethyl hexyl [^{35}S]sulfate in the rat. This sulfate ester, with the chain branching on the 2 carbon of the alkyl chain, is typical of branched chain alkyl sulfate isomers and is probably present in many synthetic alkyl sulfates.

After oral administration of [^{14}C]EHS to rats at a dose of 650 mg kg^{-1}, 77.5% of the administered ^{14}C was removed in the urine during 4 days after dosing. During this time 6.6% of the ^{14}C was recovered in the feces and 7.1% in the expired air as $^{14}CO_2$. Less than 0.1% of the ^{14}C remained in the carcasses of the rats at 4 days after dosing. After administration of [^{35}S]EHS at the same dose level, 80.4% of the ^{35}S dose was recovered in the urine, 2.4% in the feces, and less than 0.1% was recovered in the carcass at 4 days after dosing.

The nature of the ^{35}S-labeled compounds excreted in the urine was examined by chemical and chromatographic means. It was found that 35.4% of the urinary ^{35}S was probably inorganic sulfate and the remainder was ether sulfate at the time of analysis. Ion exchange chromatography of the ^{14}C-labeled urinary components on columns of DEAE cellulose resolved the rat urine into four ^{14}C-containing fractions. These fractions were examined by GLC after ether extraction or acid hydrolysis and acetylation, and the retention times compared with standard compounds. Infrared and nmr spectra of the peaks were also used to characterize some of the metabolites. From these data, it was concluded that the four peaks were unchanged 2-ethyl

[1-^{14}C]hexyl sulfate (60%), 2-ethylhexanoyl glucuronide (5%), 2-ethyl-2,3-dihydroxy-hexanoic acid, and 2-ethylhexanol. These findings are not in keeping with the ω, β-oxidation degradation of the *n*-alkyl sulfate surfactants.

Unpublished results on the fate of the sodium 2-ethyl [3-^{14}C]decyl sulfate after oral administration to rats at doses of 10 mg kg^{-1} showed that within 4 days after administration, approximately 90% of the ^{14}C is recovered in the urine, about 5% in the feces, and about 5% in expired $^{14}CO_2$. The urine contained one metabolite which contributed >90% of the urinary ^{14}C. Acid hydrolysis of the urine gave an end product, readily extractable into $CHCl_3$, which co-chromatographed with 2-ethylbutyrolactone on several GLC column systems. Mass spectral analysis of the ^{14}C-containing peak confirmed that the end product of metabolism was 3-ethylbutyric acid-4-sulfate.

All available data would suggest a common mode of degradation in mammals of alkyl sulfates, based upon ω, β-oxidation of the alkyl chain. As with normal lipid metabolism, branching of the alkyl chain will affect the degradation.

Stewart et al. (22) recently presented data that showed that rat liver microsomes readily oxidised sodium [1-^{14}C]dodecyl sulfate in vitro to give ω and ω-1 hydroxylated SDS. The prefeeding of rats with SDS, phenobarbitone, or clofibric acid caused an increase in the rate of metabolism of SDS by the microsomes isolated from the livers of these rats. In addition to the hydroxylated SDS, the microsomal preparations also produced ω-carboxy-SDS but β oxidation fo the alkyl chain was not found.

It appears that oxidation can only proceed to within three to four C atoms of the polar head group before either steric hindrance prevents further enzymic attack or the polarity of the metabolite allows withdrawal from the enzyme site. It is clear that absorption of alkyl sulfate is efficient from the intestinal tract of mammals and through the gill epithelia of fish.

C. Alkyl Sulfonates

There are little published data on alkyl sulfonates, which include alkane sulfonates and olefin sulfonates. Taylor (23) studied the fate in rats of sodium [1-^{14}C]dodecyl sulfonate, sodium dodecyl [^{35}S]sulfonate, sodium [1-^{14}C]hexadecyl sulfonate, and sodium [1-^{14}C]undecyl sulfonate. Male and female MRC hooded rats or Colworth Wistar rats were dosed orally with the alkyl sulfonate isomers at doses of 5 mg kg^{-1}. Urine, feces, and expired air were monitored for 48 hr after intubation for ^{14}C or ^{35}S. With [1-^{14}C]dodecyl sulfonate, the urine contained 95.7% of the dose to male rats and 87.4% of the dose to female rats, while in the feces 4.7% was recovered from the males and 7.7% from the females. These differences were not significant and indicate an efficient absorption from the intestinal tract. No expired $^{14}CO_2$ was detected.

The results with the ^{35}S-labeled sodium dodecyl sulfonate were essentially the same and no differences were noted between the two strains of rats. The urine contained ^{14}C-labeled metabolites, which were examined by thin-layer, paper, and ion-exchange chromatography and by paper electrophoresis. These results showed the presence of a single ^{14}C component which was more polar than the parent compound. Using techniques similar to those described by Denner et al. (12) for the purification of butyric acid-4-sulfate, the metabolite from [1-^{14}C]dodecyl sulfonate was methylated and identified as butyric acid-4-sulfonate using radio GLC-MS techniques.

After intubation of sodium [1-^{14}C]hexadecyl sulfonate at 5 mg kg^{-1} of an aqueous suspension, 65.1% of the ^{14}C was recovered in the urine and 39.4% in the feces of male rats. From female rats, 43.1% was recovered in the urine and 42.8% in the feces. No expired $^{14}CO_2$ was recovered from any of the animals. Analysis of urinary and fecal ^{14}C components showed that the urinary metabolite was butyric acid-4-sulfonate and the fecal ^{14}C was unchanged surfactant. Thus, even after administration of this alkyl sulfonate isomer, which has a very low solubility, a considerable proportion was absorbed from the intestine. Taylor deduced from this study that after absorption from the intestine the rat metabolized the even-chained alkyl sulfonate isomers to a common metabolite without desulfonation of the alkyl chain.

After intubation of sodium [1-^{14}C]undecyl sulfonate to male and female rats, the ^{14}C recoveries from males were 77.1% in the urine and 11.3% in the feces. The females excreted 75.4% in the urine and 22.3% in the feces. Chromatographic examination of the urinary ^{14}C showed that two ^{14}C metabolites were excreted, a major metabolite ($>75\%$) and a minor metabolite ($<25\%$), but no parent [1-^{14}C]undecyl sulfonate. The metabolites were identified as propionic acid-3-sulfonate and pentanoic acid-5-sulfonate, respectively, and indicate the ω, β-oxidation mechanism of degradation in the rat, similar to that for the alkyl sulfates.

Whole-body autoradiograms of rats intubated with the ^{14}C-labeled surfactants were made at 0.25, 0.5, 1, and 4 hr after administration. These showed the rapid removal of radioactivity from the intestinal tract and accumulation in the livers of the rats. At 2 hr, radioactivity was observed in the kidney tubules and bladders.

D. Sulfated Alcohol Ethoxylates

McDermott et al. (24) presented data on the fate of some isotopically labeled alkyl ethoxysulfates. Two model surfactants were synthesized from hexadecanol which was ethoxylated with 3 or 9 mol of ethylene oxide and the resulting ethoxylates were sulfated. These alcohol ethoxysulfates were labeled with ^{14}C incorporated into the alkyl or ethylene oxide moieties or with ^{35}S in the sulfate moiety. In man and rat, an oral dose of [1-^{14}C]hexadecyl tritheoxysulfate gave a recovery of 65% in the urine, 15% in the feces, and 11% as $^{14}CO_2$, indicating extensive absorption and biotransformation. In rat and man, an oral dose of [1-^{14}C]hexadecylnona-ethoxysulfate resulted in an 85% recovery of ^{14}C in the feces and 5% in each of the urine and expired $^{14}CO_2$. Oral administration of hexadecyl [^{14}C]triethoxysulfate or hexadecyl triethoxy [^{35}S] sulfate resulted in similar isotope recoveries of 62% in the urine, 26% in the feces, and none in the expired air, indicating little metabolism of the sulfate and ethoxylate moieties. The authors state that the urinary ^{14}C is present as a mixture of metabolites but 75% of the ^{14}C in the urine after administration of the [1-^{14}C]hexadecyl triethoxysulfate or hexadecyl [^{14}C]triethoxysulfate is present as the triethoxysulfate derivative of glycolic acid, i.e., acetic acid-2-triethoxysulfate.

In this study, it is not clear whether the isotopically labeled compounds were pure or whether they were mixtures of different proportions of ethylene oxide (e.g., 2, 3, or 4 ethylene oxide). An interesting point not discussed by the authors is the production of $^{14}CO_2$ from the alkyl ^{14}C-labeled ethoxysulfate. It is clear from this study that significant absorption from the intestine of the triethoxysulfates occurs readily and that the degradation of the surfactants is by ω,β-oxidation of the alcohol hydrophobe component.

Taylor (23) examined the fate of two pure isomers of alkyl ethoxysulfates. These were sodium [1-^{14}C]undecyl triethoxysulfate and sodium [1-^{14}C]dodecyl triethoxysulfate, which were administered to rats by oral intubation at doses of about 8 mg kg^{-1}. During the 48 hr after dosing male rats with the [1-^{14}C]undecyl triethoxysulfate, 66% of the dose was recovered in the urine, 4% in the feces, and 20% in expired air. Recoveries from female rats were 45, 5, and 35%, respectively. These results indicate that, although the urine is the major route of excretion, metabolism of the alkyl moiety to $^{14}CO_2$ occurred, especially in female rats.

The recoveries from male rats dosed with the [1-^{14}C]dodecyl triethoxysulfate over 48 hr were 95% in the urine, 7.5% in the feces, and 1% in expired air. From the female rats the ^{14}C recovers were 86% in the urine, 11% in the feces, and 2% as expired $^{14}CO_2$. These data indicate a rapid absorption from the intestinal tract. Female rats appear to be capable of metabolizing the odd-numbered carbon alkyl chain to a greater extent than male rats, although both sexes exhibited some degradation of the whole alcohol chain as shown by the formation of $^{14}CO_2$.

Whole-body autoradiograms of rats dosed at 20 mg kg^{-1} with these two compounds showed the removal of surfactant from the small intestine, accumulation in the liver, and subsequent excretion via the kidney. Although a general low level of ^{14}C distribution was observed throughout the body, accumulation of ^{14}C was not recorded in any tissue other than liver, kidney, and bladder.

Urinary metabolites were partially purified by column and thin layer chromatographic techniques similar to those described by Denner et al. (12). The urine from the [1-^{14}C]undecyl triethoxysulfate-treated rats contained three bands that accounted for 21, 75, and 4%, respectively, of the ^{14}C and from the [1-^{14}C]dodecyl triethoxysulfate four bands that accounted for 87, 10, 2, and 1%, respectively, of the ^{14}C. All of the metabolites were more polar than the parent surfactant and were retained by anion exchange resin. Metabolites were hydrolysed with acid to remove the sulfate moiety and the trimethylsilyl (TMS) derivatives were formed from the freeze-dried hydrolysate. From the partially purified urine from the [1-^{14}C]undecyl triethoxysulfate-treated rats, one major [^{14}C]TMS derivative was found on radio-GLC. Mass spectrometry of this peak failed to give a conclusive structure but suggested that the peak was the TMS derivative of propionic acid-3-triethoxylate. From the [1-^{14}C]dodecyl triethoxysulfate-treated rat's urine, four ^{14}C peaks were detected on radio-GLC. Only the major metabolite afforded sufficient amounts for meaningful mass spectra that were in keeping with the structure being the TMS derivative of acetic acid-2-triethoxylate. From these data it was concluded that the alkyl ethoxysulfates follow the same overall metabolic route as the other surfactants, i.e., ω,β-oxidation of the alkyl chain but no degradation of the polar head moiety.

It is interesting to note that β-oxidation of the alkyl chain is more complete in the alkyl ethoxysulfate surfactants than with the alkyl sulfates, which may be indicative of less steric hindrance afforded by the sulfate moiety due to the intervening ethylene oxide groups.

In summary, it would appear that these anionic surfactants are readily absorbed from the intestinal tract in contrast to the statement by Chasseaud (25) and are transported to the liver probably via the hepatic portal blood supply. Involvement of the lymphatic system is minimal. Metabolism in the liver is rapid, as evidenced by the short duration of surfactants and their metabolities in the body, and complete, since the parent surfactant is rarely recovered in the urine. For most anionic surfactants, the urine is the main route of excretion of the metabolites, although the biliary route is an important one for some isomers of the alkyl benzene sulfonate type. The data also indicate that surfactants do not become localized in any of the tissues of the body other than those associated with the detoxification mechanism and excretory process. The catabolism by ω and β-oxidation of the alkyl chains of anionic surfactants would appear to be common to all the major types of surfactants, although the results on branched-chain alkyl sulfates of Knaak (21) appear to be at variance with the breakdown of the straight-chain alkyl sulfates. It is interesting that β-oxidation stops at about the fourth carbon of the alkyl chain. This may be connected with the relative size of the SO_4 moiety, which might sterically hinder the enzyme system, or with the "shell" of water that aggregates around the polar head group of anionic surfactants, preventing further enzymic degradation. This shell extends along the alkyl chain of *n*-alkyl sulfates to about the fourth carbon atom (26). It may, however, be related to the change in physical properties due to the shortening of the alkyl chain and the introduction of the carboxyl function which will increase the water solubility and decrease the surface activity of the molecule.

III. PARENTERAL ADMINISTRATION AND DISPOSITION

A. Alkyl Benzene Sulfonates

Howes (27) published data on the recoveries of ^{14}C after intraperitoneal (i.p.) and subcutaneous (s.c.) injection to rats of sodium 1-sulfophenyl [1-^{14}C]dodecane, a model isomer of linear alkyl benzene sulfonate, at a dose of 5 mg kg^{-1}. In the 24 hr immediately after injection, 78% of the ^{14}C was recovered in the urine, 1.5% in the feces, and the remainder in the carcasses. No $^{14}CO_2$ was expired and there were no significant differences between the i.p.- and s.c.-treated rats. Other unpublished data (Howes) showed that ^{14}C was present in the urine as a single ^{14}C compound, which after desulfonation gave a single ^{14}C-containing peak on GLC. This peak was identified by GLC-MS as tetralone (3,4-dihydro-1-(2H)-naphthalenone), which is a dehydration product of desulfonated butyric acid-4-(benzene-*p*-sulfonate). No parent dodecyl-1-benzene sulfonate was detected in the urine.

The studies by Cresswell (8) examined the fate of [^{14}C]LAS s.c. administered to rhesus monkey. After a single dose of 1mg.kg^{-1}, 64% of the dose was recovered in the urine and 11% in the feces at five days after dosing. After s.c. doses of 0.1, 0.5 and 1.0 mg.kg^{-1} peak plasma concentrations, at 2-4 hr after dosing were proportional to the dose being 0.16, 0.72, and 1.13 μg.ml^{-1}, respectively. The half-life was approximately 8 hr, and no bioaccumulation was observed.

Rennison et al. (9) administered the [^{14}C]2 and 6 sulfophenyl dodecane isomers intravenously (i.v.) and intradermally (i.d.) to rats at a dose level of 1 mg.kg^{-1}. The biliary excretion of the 6-isomer and the urinary excretion of the 2-isomer was confirmed. The bioavailability of the i.d. administered ^{14}C was less than after the i.v. administration as shown by the reduced ^{14}C excretion after the i.d. dose.

When Lay et al. (11) administered [14C]LAS by i.p. injection to rats at a dose of 2.3 mg.kg^{-1}, 84.7% of the dose was recovered in the 24 hr after dosing. Slightly more 14C was recovered in the urine than the feces as polar metabolites.

No parent surfactant was excreted after oral or parenteral administrations. We have concluded from these studies that absorption and metabolism of alkyl benzene sulfonates occurs readily in the body but the disposition and overall metabolism is isomer dependent.

B. Alkyl Sulfates

Denner et al. (12) examined the fate of potassium dodecyl [^{35}S]sulfate after oral, i.p. or i.v. administrations. The fate of the dodecyl [^{35}S]sulfate was not affected by the route of administration and no parent dodecyl [^{35}S]sulfate was eliminated in the urine. After an i.v. administration of 5 mg kg^{-1} to rats which had been surgically prepared with bile duct and ureter cannules, virtually all of the administered ^{35}S was recovered in the urine during 12 hr after injection. Qualitative examination of the urine indicated that butyric acid-4-sulfate was the main urinary component but significant quantities of $^{35}SO_4^{2-}$ were present. The levels of $^{35}SO_4^{2-}$ were apparently greater in the urine of male rats than female rats. During the 12 hr of collection, only 1.4% of the administered ^{35}S was recovered in the bile. Results obtained from rats i.p. injected with dodecyl [^{35}S]sulfate were virtually identical to those from rats dosed by intubation.

Merits (15) administered sodium hexadecyl [^{35}S]sulfate at 2.9 mg kg^{-1} and the trimethylammonium salt of [1-^{14}C]hexadecyl sulfate at 14.4 mg kg^{-1} i.v. to dogs. Over a 72-hr collection period about 85% of the administered ^{35}S was recovered in the urine from the dogs treated with hexadecyl [^{35}S]sulfate and 81% of the ^{14}C from the dog treated with the trimethylammonium salt of [1-^{14}C]hexadecyl sulfate. In both cases, only 3 or 4% of the isotope was recovered in the feces. Plasma levels of the hexadecyl [^{35}S]sulfate were seen to fall very rapidly with a half-life of less than 15 min and at 6 hr the plasma level of hexadecyl sulfate was 0.02 μg ml^{-1}. Similar studies showed that the plasma levels of the trimethylammonium [1-^{14}C]hexadecyl sulfate after oral administration to dogs of 14.4 mg kg^{-1} were at a maximum level at 1 hr after administration. Analysis of urinary ^{14}C-labeled compounds obtained after i.v. administration of trimethylammonium [1-^{14}C]hexadecyl sulfate to the dog were almost identical to that obtained after oral administration. In both cases, some 7% of the dose was recovered as parent surfactant, 55% as butyric acid-4-sulfate, 11% as butyrolactone, and 20% as glycolic acid sulfate.

The studies of Burke et al. (14) included i.p. and i.v. administration to rats of potassium decyl [^{35}S]sulfate and potassium octadecyl [^{35}S]sulfate at doses of 1 or 5 mg kg^{-1}. After these administrations urine and either bile or feces were collected over 6 or 48 hr. During 48 hr after i.p. administration of decyl [^{35}S]sulfate, approximately 90% of the ^{35}S was recovered in the urine. During 6 hr after i.v. administration, approximately 70% was recovered in the urine. Chromatographic examination of the urinary ^{35}S showed that there was no parent decyl [^{35}S]sulfate present and that about 85% was present as butyric acid-4-sulfate and 15% as $^{35}SO_4^{2-}$. There were no significant differences between the male and female rats in each group. Administrations to rats of the octadecyl [^{35}S]sulfate gave very similar findings for the route of excretion and the pattern of ^{35}S urinary metabolites. Again no parent octadecyl [^{35}S]sulfate was recovered in the urine of i.v.- or i.p.-dosed rats.

Maggs et al. (16) administered the D-(+) and L-(−) isomers of [^{14}C] octan-2-sulfate by i.v. injection to rats. The rapid and extensive urinary excretion was similar for both isomers, and metabolism was the same as found after oral administration. These findings together with those of Burke et al. (14) extend those of Denner et al. (12) and show that by oral or parenteral administration *n*-alkyl sulfates with chains of even numbers of carbon atoms are absorbed, metabolized, and excreted in a manner that is independent of the mode of administration.

Later, Burke et al. (18) investigated the fate of potassium undecyl sulfate that was administered to rats at doses of 5 mg kg^{-1} by i.p. and i.v. routes. The pattern of excretion was identical from both modes of administration, although note was made of a slightly higher proportion of the ^{35}S dose recovered in the bile from female rats after i.v. administration. The rate of excretion of ^{35}S in the urine was higher after i.v. administration than after either i.p. or oral administration.

The i.v. administration of potassium [^{14}C]nonan-5-sulfate (KDS) to rats, reported by Bains et al. (20), resulted in virtually all of the ^{14}C being recovered in the urine. Some of the parent KDS was recovered in the urine but most of the ^{14}C was present as metabolites. Similarly the [^{14}C]tridecan sulfate isomers administered i.v. to rats by Webborn et al. (19) clearly showed the urine was the main route of excretion but the 5 and 7 isomers were found to be excreted in the bile to a greater extent than the 1 and 3 isomers. However this did not result in increased fecal output of ^{14}C for the 5 isomer and only a slight increase for the 7 isomer. This situation is somewhat similar to that of LAS where the 6 isomer is more extensively excreted in the bile than the 2 isomer.

Howes (27) administered sodium [1-^{14}C]dodecyl sulfate to rats by s.c. and i.p. injections at a dose of 18 mg kg^{-1}. No differences in the recoveries of ^{14}C in the excreta were noted during the 24 hr after dosing. The recoveries were 77% of the ^{14}C in the urine, 1.5% in the expired $^{14}CO_2$, 2.7% in the feces, and 15% in the carcass. No chromatographic examinations of the ^{14}C compounds in the excreta were conducted, but other data have shown that the rat urinary ^{14}C was present exclusively as butyric acid-4-sulfate (17).

Unpublished data from our laboratories have shown that butyric acid-4-sulfate is the end product of metabolism after i.p., s.c., or i.v. injection into the rat and no parent [1-^{14}C]dodecyl sulfate is excreted in the urine.

Thus, there is little difference in the nature of the end products of metabolism of *n*-alkyl sulfates given parenterally or by oral administration. The rate of excretion is more rapid after i.v. injection than after i.p. injection or oral administrations. There appears to be no published data on the fate of either branched chain alkyl sulfate or secondary alkyl sulfates administered by a parenteral route. Our own unpublished data on the fate of an oral, i.p. or s.c. dose of the branched-chain surfactant, sodium 2-ethyl [3-^{14}C]decyl sulfate, have shown that there are no major differences in the end products of metabolism between the three routes of administration and only slight differences in the rate of urinary excretion of ^{14}C.

C. Alkyl Sulfonates

Taylor (23) studied the fate in rats of sodium dodecyl sulfonate, sodium hexadecyl sulfonate, and sodium undecyl sulfonate labeled either with ^{35}S or with 1-^{14}C in the alkyl chain, and given i.v. and i.p. in doses of about 5 mg kg^{-1} in each case. Essentially, there were no significant differences in the rate or route of excretion for any of these surfactants between male and female rats. There were, however, differences due to the mode of administration. Apparently the rate of excretion after i.p. injection was more rapid than when given orally or intravenously. It should be noted however that the rats dosed intravenously were anesthetized and were fitted with bile duct and ureter cannulas that might have affected urine production. There were only slight differences in the rate of excretion of ^{14}C from rats after i.p. administrations of [1-^{14}C]dodecyl sulfonate, [1-^{14}C]undecyl sulfonate, and [1-^{14}C]hexadecyl sulfonate.

Chromatographic analysis of the urinary ^{14}C showed that from oral, i.v. or i.p. administrations of either the [1-^{14}C]dodecyl sulfonate or the [1-^{14}C]hexadecyl sulfonate a single metabolite was produced and no parent surfactant was excreted. Chromatography of the urinary ^{14}C components from the [1-^{14}C]undecyl sulfonate treated rats showed essentially the same result from oral, i.p. or i.v. administration, except that after i.v. administration, the 2-hr urine sample contained the major metabolite and in addition trace amounts of unchanged [1-^{14}C]undecyl sulfonate as well as a second metabolite that was less polar than the main metabolite. Taylor (23) explained this "breakthrough" of parent compound into the urine during the first 2 hr after i.v. administration as due to initial overloading of the degradative mechanism(s). It is interesting that this effect was not observed after similar administration of either of the even-chain alkyl sulfonates.

D. Sulfated Alcohol Ethoxylates

Although McDermott et al. (24) reported on the fate ^{14}C-labeled alkyl ethoxysulfates by intubation into rats and humans, they did not report on the fate of these compounds after parenteral administration. Taylor (23), however, investigated the fate of sodium [1-^{14}C]undecyl triethoxysulfate after i.p. and i.v. administration to rats at a dose of 5 mg kg^{-1}. During the 48 hr after i.p. administration of [1-^{14}C]dodecyl triethoxy sulfate, about 65% of the ^{14}C was recovered in the urine, 33% in the feces, and 2.5% as expired CO_2. There was no significant differences in the recoveries from males and females. Similar proportions of the ^{14}C dose were recovered after i.p. administration of the [1-^{14}C]undecyl triethoxysulfate, but from these rats recoveries of ^{14}C as $^{14}CO_2$ were significantly higher accounting for 12.5-19.3% of the dose. The urine was still the major route of excretion.

Anesthetized rats with bile duct and ureter cannulas were injected i.v. with these two ^{14}C-labeled alkyl ethoxysulfates. The urine contained up to 75% of the administered ^{14}C and only small proportions of the dose were recovered in the bile during 6 hr after injection. With [1-^{14}C]undecyl triethoxysulfate, more ^{14}C was recovered in the bile from female rats than from male rats, but both males and females expired about 10% of the dose. The recoveries of $^{14}CO_2$ after intravenous administration of [1-^{14}C]dodecyl triethoxysulfate were not recorded.

Thin layer chromatography of the urinary ^{14}C components and fecal ^{14}C compounds that were extracted with methanol showed a number of ^{14}C-labeled compounds. Urine contained up to four components that were only partially characterized. No parent surfactant was detected in any of the urine samples and the range of metabolites was similar to that recovered after oral intubation of the same surfactants. Fecel ^{14}C extracts from the rats injected intraperitoneally contained both parent surfactant and metabolites although the proportions of these were not quantified.

The studies by Taylor (23) suggest that in rats there is no significant difference in the fate of alkyl ethoxysulfates given orally or by i.p. or i.v. injection.

E. Miscellaneous Surfactants

There are few published data on the fate of anionic surfactants administered to animals by parenteral routes. Howes (27) administered sodium [1-^{14}C]dodecanoylisethionate to rats by i.p. injection. During the 24 hr after administration recoveries of ^{14}C were 80% as $^{14}CO_2$, 2.7% in the urine, and 1.7% in the feces. Approximately 12% of the administered dose remained in the carcass at 24 hr after dosing. These findings indicate the cleavage of the ester linkage between the isethionic acid moiety and the [1-^{14}C]dodecanoic acid.

However, no additional studies to elucidate the mechanism of this cleavage were reported. It may be that any such ester linkage in the structure of anionic surfactants is amenable to esterase action.

In summary, these data suggest that alkyl benzene sulfonates, alkyl sulfates, alkyl sulfonates, and alkyl ether sulfates are rapidly metabolized and excreted via the same routes as orally administered surfactants. The findings by Taylor (23), Knaak (21), and Merits (15) that traces of unchanged surfactant were present in the urine of rats after i.v. administration may indicate that the metabolic system can be overloaded, resulting in a breakthrough of unchanged surfactant into the urine. Metabolites that have been identified suggest a common path of catabolism. The efficiency and rapidity with which the surfactants were metabolized by naive rats suggests that the enzyme systems involved require no induction as might be expected for systems normally employed in endogenous lipid metabolism.

V. PERCUTANEOUS ABSORPTION IN VITRO

A. Absorption of Anionic Detergents

Many factors combined to encourage investigators to use isolated tissue for studies of skin absorption. Chemical methods with good sensitivity for the analysis of detergents were lacking and the supply of pure detergents was limited. The increasing availability of radioactively labeled detergents and a supply of human skin from autopsies made this combination a popular one for studies of the penetration of detergents into skin.

Using isolated human abdominal skin, Blank and Gould (28) compared the penetration of sodium [1-^{14}C]dodecanoate with that of sodium [1-^{14}C]dodecyl sulfate. The concentration of the test solutions was 0.005 M, which is below the critical micelle concentration (CMC) of each, and the exposure time was 18-24 hr at room temperature (20-23°C). Tape strippings were taken and their radioactive content was measured. Chemical analysis of surfactants was also used but was limited by a low sensitivity of the method and by the strong binding of the anionics to protein. The route of absorption into the excised skin was also examined by autoradiography.

The counting of radioactivity in tape strippings showed that most of the anionic surfactants in the skin were recovered in the first two strippings of the corneum. The data for sodium dodecanoate showed appreciable radioactivity in the epidermis and dermis whereas with sodium dodecyl sulfate little or none was recovered from the dermis.

Autoradiography of skin after exposure to sodium [1-^{14}C]dodecanoate and to sodium [1-^{14}C]dodecyl sulfate and from which the stratum corneum had been removed by tape stripping supported the difference between the ability of sodium dodecanoate and sodium dodecyl sulfate to penetrate into the epidermis and dermis. Fogging of the emulsion occurred in the epidermis and dermis of the skin exposed to sodium [1-^{14}C]dodecanoate, less so in the sebaceous glands, and to a still lesser extent in the sweat glands. Blood vessels also showed fogging of the emulsion. The authors cautiously pointed out that the presence of shadows in the emulsion associated with the glands of the skin did not necessarily indicate a route of entry of the sodium [1-^{14}C]dodecanoate by the respective ducts but might indicate only an accumulation of the soap, as the autoradiograms were prepared after a single prolonged exposure of the skin to the soap. In contrast, autoradiograms prepared from skin treated with sodium [1-^{14}C]dodecyl sulfate showed little fogging of the emulsion over the epidermis or dermis, and none associated with the sebaceous and sweat glands or hair follicles.

Aqueous solutions of sodium dedecanoate and sodium dodecyl sulfate have different pH values, and in order to control this variable, Blank and Gould (28) investigated the penetration into skin of these radioactively labeled anionic detergents from buffered solutions. The pH values of the buffered solutions varied within the range 9.3-12.7 and the duration of contact with skin was 18 or 21 hr, respectively. With both solutions, the epidermis appeared damaged when the pH was 11 or higher, making it impossible to use the stripping method. Quite small amounts of each surfactant were present in the dermis of skins exposed to solutions buffered to pH 11 or below, but above this pH much greater amounts of both were found in the dermis. Differences in the ability of sodium dodecanoate to penetrate more easily than sodium dodecyl sulfate, as found with unbuffered solutions, were not consistent, nor was the penetration of sodium dodecanoate consistent from unbuffered and buffered solutions.

In a later paper, Blank and Gould (29) investigated the amount of penetration into excised human skin of sodium dodecanoate from dilute, buffered solutions of pH 7.5-12 and compared the data with the degree of erythema that developed in human skin in situ. They showed first that the pH of unbuffered 0.005 M sodium dodecanoate decreased by 1.5 pH units during 5 hr contact with isolated skin, but buffered sodium dodecanoate decreased by less than 0.2 pH units during 20-24 hr contact with excised skin. Moreover, they found that penetration into the dermis was minimal between pH 8.5 and 10.5 but increased below pH 8.5 and above pH 10.5. When excised skin was pretreated overnight with a dilute solution of sodium phosphate (pH 12.1), then exposed to a buffered solution of sodium dodecanoate (pH 10.0) that previously showed virtually no penetration into excised skin, readily measurable amounts of sodium dodecanoate, based on radioactivity measurement, were found in the dermis. Furthermore, pretreatment with buffer of pH 7.5 without sodium dodecanoate followed by exposure to sodium dodecanoate buffered at pH 9.5 gave minimal values of sodium dodecanoate in the dermis. The authors concluded that the high alkalinity damaged the skin barrier to penetration.

The reason for the high penetration of sodium dodecanoate into excised skin from solutions buffered in the range pH 7.5-8.5 was explained by a study of the relationship between penetration and the diethyl ether-water partition coefficients. The partition coefficient was high at pH 8.1 and decreased in a linear fashion to a low value at pH 9.0 and above. Thus the observed high penetration in the pH range of 7.5-8.5 was believed to be due to an increased lipid solubility.

Application of 0.005 M buffered solutions of sodium dodecanoate to human skin in situ showed that during 6 hr contact erythema and pruritis developed with solutions of pH 7.5, in agreement with the high penetration observed in vitro, but no erythema or pruritus was recorded when the pH of the applied solution was 9.5, at which pH virtually no penetration had been found in vitro.

In a further study, Blank and Gould (30) compared the absorption into the dermis of isolated human abdominal skin of a homologous series of pure alkyl sulfates, ranging from octyl to octadecyl sulfate, and a series of pure alkyl benzene sulfonates. Two radioactive sulfur-containing detergents, sodium dodecyl sulfate and alkyl benzene sulfate, of undefined mixed homologues were also examined. The concentration of the nonradioactive surfactants was measured spectrophotometrically after extraction from the separated dermis. Following extended contact (18-20 hr) of the isolated skin with 0.005 M aqueous solutions (or 60% (v/v) aqueous acetone solutions to solubilize the hexadecyl and octadecyl sulfates) less than 15 n mol cm^{-2}, the sensitivity limit of the method, of the various surfactants, had penetrated into the dermis, i.e., a penetration of less than approximately 5 μg cm^{-2}. Various methods of decreasing the binding of surfactants failed to show that penetration then occurred. Removal of the upper layers of the stratum corneum did not increase the penetration of dodecyl sulfate into the dermis, but complete tape stripping of the corneum or exhaustive extraction of skin lipids with organic solvents did increase the penetration of dodecyl sulfate into the dermis. The authors concluded that the barrier to penetration seemed to be in the lower stratum corneum.

In a study of radioactivity labeled potassium octadecenoate through isolated human abdominal skin, Bettley (31) found that for both 0.4 M (13% w/v) and 0.2 M (6.4% w/v) potassium [^{131}I]octadecenoate, penetration was 0.42-1.83% on day 4, and thereafter increased considerably over 24-26 days indicating progressive damage by the pure soap to the epidermis.

Later, Bettley (32) used the potassium salts of octanoic, dodecanoic, and hexadecanoic acids labeled with ^{14}C. Human abdominal skin was obtained post mortem, and the epidermis separated by heating and mounted in a diffusion cell to which the soap solution at 0.04 M was added, while the dermal side of the skin was in contact with distilled water. The radioactivity of samples taken from each chamber daily for 7 days was determined. It was found that a total of 24.4% of the original quantity of potassium dodecanoate penetrated through the isolated human epidermis, whereas only 5.3% of potassium octanoate and only 0.15% of potassium hexadecanoate penetrated. The irritancy of these potassium soaps was assessed after a 48-hr patch test on the upper back of 28 human volunteers, of whom 18 reacted to potassium dodecanoate, while only 2 responded to potassium hexadecanoate and 2 to potassium octanoate. Bettley concluded that the ability of potassium dodecanoate to penetrate skin most easily was compatible with its greater power of irritancy.

Using whole guinea pig skin mounted in a continuous-flow diffusion cell with a collecting chamber of small volume to reduce dilution of the perfusate, Scala et al. (33) studied the percutaneous absorption of 0.03 M (or 1%) sodium tetrapropylene benzene sulfonate at pH 7. Initially penetration was low, until about 4 hr when small amounts of radioactivity were just detectable in the perfusate. Over a 20-hr period the diffusion curve was nonlinear showing that the barrier properties of the skin were decreasing during contact with the detergent. When separated guinea pig dermis was placed in the diffusion cell, penetration was observed earlier and to a much greater extent. In this case, the diffusion was linear after 6-hr contact, showing that a steady state had been reached. It was not possible to use separated guinea pig epidermis in this system, but a comparison of the perfusion through whole skin with that through separated dermis showed that the epidermis provided the main barrier to absorption through skin. The authors concluded that, because of the time lag before radioactivity from sodium tetrapropylene benzene [^{35}S]sulfonate appeared in the perfusate of isolated skin, little surfactant would be absorbed through skin under normal use conditions. They also pointed out the importance of duration of contact of the surfactant with skin and of the concentration of the penetrant in diffusion through skin.

Absorption of potassium dodecyl [^{35}S]sulfate from aqueous 0.0016 M solutions through excised human abdominal skin mounted in diffusion cells was studied by Emery and Dugard (34). No detergent appeared in the diffusate when whole skin or the isolated horny layer were used, but appreciable amounts were present when stripped skin was tested, again confirming that the barrier properties of skin reside in the stratum corneum. It is interesting to note that the absorption of the potassium dodecyl [^{35}S]sulfate in dimethyl sulfoxide was reduced in contrast to the usual effort of promotion of absorption assigned to this aprotic solvent (35).

Fiker et al. (36) investigated the extent of retention and absorption by human and mouse skin in vitro of sodium dodecyl sulfate using the methylene blue method of determination. Interestingly, they found that cellophane appeared to be a good model for human skin. At 1 g liter^{-1} sodium dodecyl sulfate at a pH of 6, 9, or 12, the authors recorded a maximum absorption of 13 μg cm^{-2} and a retention of 166 μg cm^{-2} for human skin at pH 12. Most of the absorption occurred within the first 2 hr and extending the duration of contact to 24 hr did not increase the absorption significantly. Human skin and depilated mouse skin were very similar in their behaviour.

Howes (27) investigated the penetration through isolated rat skin and human epidermis of five homologous pure soaps and three pure synthetic detergents, all of which were radioactive. The dermal side of the skin was bathed with saline which was stired magnetically. At 24 hr, for rat skin, and at 48 hr for human epidermis, the epidermal surface was rinsed and the rinse water kept for measurements of radioactivity. An autopsy of the skin and human epidermis was taken for determination of radioactivity.

In rat skin, sodium octadecanoate did not penetrate during the 24 hr of the experiment and sodium hexadecanoate was only detectable at 24 hr. The sodium decanoate, dodecanoate, and tetradecanoate all showed the same low penetration at 2 hr, a moderate penetration at 6 hr, and a larger penetration at 24 hr. None of the three pure synthetic detergents, sodium dodecyl sulfate, sodium dodecanoyl isethionate, and sodium dodecyl-1-benzene-*p*-sulfonate, gave measurable penetration during 24 hr. At the end of the experiment, rinsing of the rat skin removed 60-70% of the applied soaps and isethionate and 30-40% was retained by the skin. For dodecyl sulfate and dodecyl benzene sulfate the rinse water contained only 30% of the applied radioactivity and 70% was retained by the skin.

In human epidermis, a similar response to the soaps was recorded, except that the sodium dodecanoate gave greater penetration at 6 and 24 hr than in rat skin, and sodium decanoate gave a greater penetration at 24 hr than in rat skin. With the synthetic detergents, no dodecyl benzene sulfonate penetrated during 48 hr, but both dodecyl sulfate and dodecanoyl isethionate showed progressively increased penetration over 48 hr, which indicated a loss of barrier properties. All samples of human epidermis showed a degree of swelling at 48 hr, which was most notable with the dodecyl sulfate treatment. Retention of radioactivity by human epidermis was greatest with dodecyl sulfate treatment, after which approximately 75% was not removed by rinsing. Some 30-50% of dodecanoyl isethionate and dodecyl benzene sulfonate was retained by human epidermis after rinsing.

The difficulties in measuring the absorption of detergents per se through isolated skin led many investigators to study the effect of detergents on the permeability of the skin to other compounds. The selection of such markers was usually based on their availability in radioactive form and their relevance to clinical studies or commercial products.

B. Effect on Absorption of Water

Bettley and Donoghue (37) used heat-separated human epidermis from abdominal skin in a Perspex cell and measured daily for 20-30 days the decrease in weight due to loss of water through the epidermis under recorded conditions of temperature and humidity. A 1% solution of a standard hospital toilet soap bar increased the water loss to a maximum of 25-30 mg day^{-1} at 5-7 days, compared to the water control rate of 5-10 mg daily. At 5% soap, the rate of water loss increased steadily to about 70 mg day^{-1} at day 14 and replacement of the soap solution by water caused a return to the control level in 3-4 days. A similar increase in the rate of water loss was recorded with 5% solution of potassium hexadecanoate. The commercial detergent Teepol (secondaryalkyl sulfates) caused an increase in the rate of water loss of the order of 15 mg day^{-1}, an effect which is intermediary between the water (control) and soap response.

In describing a new improved brass cell for measurement of water diffusion through whole rat skin, Isherwood (38) recorded the effect of 1, 2, and 3% (w/v) potassium hexadecanoate. The results showed a linear increase in the mean rate per day, during 4 days of the water diffusion rate, from 0.13 mg cm^{-2} hr^{-1}, the control rate, to 0.20, 0.41, and 0.64 mg cm^{-2} hr^{-1} for 1, 2, and 3% solutions, respectively.

In a diffusion cell with a continuously circulating fixed volume of physiological saline in which the accumulation of radioactivity was recorded, Sprott (39) investigated the effect of chain length (C_8-C_{18}) of the sodium salts of fatty acids and sodium *n*-alkyl sulfates upon the permeability to tritiated water of isolated rat whole skin. In this study, the skin was exposed to solutions of the detergent and tritiated water. At 0.50 mM and pH 10.8, to ensure solubilization of C_{16} and C_{18} soaps, the C_8, C_{10}, C_{12}, and C_{14} soaps increased the penetration of tritiated water. Of the alkyl sulfates at 0.50 mM, C_{12}, C_{14}, and C_{16} increased the penetration of tritiated water whereas the effects of C_8, C_{10} and C_{18} were of doubtful significance. The C_{12} and C_{14} soaps and alkyl sulfates gave similar increases in the absolute penetration rate. It is interesting that the lower members of the soap series had a greater effect on the penetration rate than the intermediary ones, while the higher members were without effect. The more moderate effect on the penetration rate of tritiated water of the alkyl sulfates appeared to be greater at longer chain lengths. When the pH of the solutions of the alkyl sulfates was raised by buffers from 6.1 to 9.1 and 10.8, the effect on the permeability of tritiated water was reduced due to an increase in the control value caused by the more alkaline pH itself.

Scheuplein and Ross (40) investigated the effect of soap and detergent on the permeability of tritiated water through isolated human abdominal epidermis. Sodium dodecanoate at 1% (w/v) caused a large increase in the permeability to tritiated water after an initial lag period of approximately 5 hr. The permeability curve increased exponentially indicating that damage by the soap to the epidermis was continuous. When the concentration of sodium dodecanoate was increased to 5% (w/v), the permeability of the isolated epidermis to tritiated water rose more quickly after a reduced lag phase. These investigators also observed that the epidermis regained some of its barrier properties when the soap solutions were removed, but that use of the more concentrated solutions caused permanent damage. At 5% (w/v), Scheuplein and Ross (40) found that sodium dodecanoate had a greater and earlier effect on the permeability of isolated human epidermis to tritiated water than had sodium dodecyl sulfate. Raising the pH of the sodium dodecyl sulfate to that of sodium dodecanoate (pH 10) had little effect on increasing the permeability of tritiated water by sodium dodecyl sulfate. These authors concluded from additional studies with sodium sulfate and acetic acid that the dodecanoate and dodecyl sulfate were the effective anions of soap and alkyl sulfate, respectively. Scheuplein and Ross believe that the mechanism of the effect of sodium dodecanoate on the stratum corneum is due to a reversible α-β-conversion of keratin resulting from a denaturation by the soap, which facilitates the diffusion of water through the more porous membranes.

C. Effect on Absorption of Electrolytes

Using his diffusion chamber, Bettley (41) investigated the effect of various surface active agents including soaps of different chain lengths, sodium dodecyl sulfate, and alkyl benzene sulfonate, all at 0.04 M, on the penetration of sodium ions through isolated human epidermis after 1 and 2 weeks contact. Compared to sodium tetradecanoate and sodium hexadecanoate, both of which were similar to the control, sodium dodecanoate exerted a striking effect on the penetration of sodium ions. Sodium dodecyl sulfate and sodium dodecyl benzene sulfonate were approximately equal in their effects and only about half as potent as sodium dodecanoate. Neither pH nor surface tension appeared to be directly related to the increase in permeability. Sodium lauroyl sarcosinate had no effect, sodium sulfosuccinate lauric monoisopropanolamine polyglycol had a slight effect on the permeability to potassium ions, whereas sodium sulfosuccinate lauric monoethanolamine had marked effects. The effect of sodium sulfosuccinate lauryl polyglycol on potassium ions were approximately equal to that of sodium dodecylsulfate and sodium dodecyl benzene sulfonate on sodium ions, whereas sodium sulfosuccinate lauric monoethanolamide had a relatively greater effect on potassium ions than had sodium lauroyl sarcosinate and sodium dodecyl benzene sulfonate on sodium ions. The effects of these various anionic detergents could only be partly related to their ability to denature protein of isolated human epidermal stratum corneum (42).

D. Effect on Absorption of Organic Compounds

Bettley (41) used sodium salicylate as a marker for soap damage to human epidermis. The penetration of sodium salicylate through isolated abdominal epidermis was measured chemically before and after addition of 0.034 M (1% w/v) potassium hexadecanoate. During the first week, the average penetration of sodium salicylate in the control cell was 82.5 μg or approximately 21 $\mu g\ cm^{-2}\ day^{-1}$, and at the end of the second week penetration was essentially the same. However, in presence of the soap, penetration of sodium salicylate increased from the control level of 82.5 to 675 μg at the end of the first week, to 1200 μg after 2 weeks, and to 1430 μg after 3 weeks. Thus the action of the soap was to increase progressively the penetration of the salicylate.

Bettley (41) showed that the permeability of the epidermis from human abdominal skin to glucose (200 mg % w/v) was rarely above the sensitivity of the method. In the presence of 1% (w/v) potassium hexadecanoate the permeability of the epidermis to glucose increased during the test period of 3 weeks, although inspection of the data suggests that there is no statistical significance in penetration in the presence of soap at 1, 2, or 3 weeks, as Bettley (41) found with sodium salicylate.

Sprott (43) investigated the effects of four pure *n*-alkyl sulfates on the penetration of [^{35}S]thiourea and [^{14}C]urea through isolated whole rat skin. Only dodecyl sulfate gave an increase in the recorded absolute penetration rate of [^{35}S]thiourea.

Scala et al. (33) investigated the effect of sodium tetrapropylene-benzene sulfonate or soap on the absorption of [^{14}C]nicotinic acid and [^{14}C]thiourea through isolated guinea pig whole skin. In the absence of the detergent, the absorption of [^{14}C]nicotinic acid had a lag time of 5-6 hr, after which time absorption was low but linear. Preexposure of the skin to 0.5% tetrapropylenebenzene sulfonate for 1 hr did not significantly increase the penetration of nicotinic acid, but preexposure for 3 hr did. In both instances of preexposure, the subsequent penetration of nicotinic acid was linear with respect to time. However, when the nicotinic acid was placed in the solution of tetrapropylenebenzene sulfonate or soap the penetration became nonlinear with a quantitatively similar change in rate to that observed for the penetration of tetrapropylenebenzene sulfonate and soap themselves. Like nicotinic acid, thiourea showed a linear but low rate of penetration which became nonlinear in the presence of tetrapropylbenzene sulfonate.

The effect of sodium dodecyl sulfate on the absorption of chloramphenicol through isolated hairless mouse skin was reported by Aguiar and Weiner (44). They showed that absorption through the skin of hairless mice was time-dependent and that there was no lag time in absorption, which the authors attributed to the thinness of the stratum corneum. Absorption was proportional to temperature, from which the authors calculated an apparent activation energy of 15,900 cal mol^{-1}. Addition of 0.2% sodium dodecyl sulfate caused the permeability of the skin to chloramphenicol to double, whereas addition of 0.4% sodium dodecyl sulfate had only a slightly greater effect than the 0.2%. At a concentration of sodium dodecyl sulfate (0.02%) below the CMC, the effect of the surfactant was to reduce the penetration of chloramphenicol through isolated hairless mouse skin. The authors suggest that below the CMC the surfactant may complex with the drug and that at the high concentrations the surfactant may damage the stratum corneum.

V. PERCUTANEOUS ABSORPTION IN VIVO

A. Absorption of Anionic Detergents

Interest in measuring absorption through skin in vivo arose as it became obvious that, despite the attractions of an isolated skin system, in vitro measurements of absorption of substances such as anionic detergents, which bind strongly to skin, did have some shortcomings. A comparison of the penetration of sodium dodecyl [^{35}S]sulfate into living and excised guinea pig skin was made by Choman (45). Clipped guinea pig skin was treated with sodium dodecyl [^{35}S]sulfate for 6 min without rinsing and the treated area excised. Sections of the frozen skin were cut parallel to the surface of the skin from the dermal side to avoid the transfer of radioactivity from an area of high deposition to one of low deposition which would occur in the traditional method of obtaining transverse sections of skin. Four hours after sacrifice other areas of the dorsal skin were excised and considered to be nonliving. Two were placed in a static stainless steel diffusion chamber and sodium dodecyl [^{35}S]sulfate placed on the epidermal surface. The distilled water in contact with the dermal side was monitored for radioactivity by evaporating samples on a glass plate, taken at various time and exposing the residues to X-ray film. Other areas of the nonliving skin were treated with sodium dodecyl [^{35}S]sulfate, and after 60 min autoradiograms were prepared from skin sections.

The results showed that in vivo skin absorption occurred to a depth of some 800 μm whereas in the nonliving skin absorption was recorded to only some 250 μm. The position of radioactivity on the autoradiograms of the living skin coincided with the position of the pilosebaceous appendages. No detectable radioactivity was recorded at any time up to 24 hr from the dermal perfusate in the diffusion cell. The author concluded that skin in vivo was more permeable to sodium dodecyl sulfate than skin in vitro, but by the nature of this technique he was unable to quantify his data.

The first direct study of percutaneous absorption of alkyl sulfates in vivo was reported by Sprott (39). Anesthetized rats were prewashed with a 1% (w/v) aqueous solution of sodium dodecyl sulfate or sodium hexadecyl sulfate, rinsed and dried. Immediately a solution of the appropriate sodium *n*-alkyl [^{35}S]sulfate was applied to the pretreated skin for 10 min, after which the skin was rinsed and dried, but not protected with a patch, and the rats placed in individual metabolism cages. The excretion in the separated urine of radioactivity was measured daily for 9 days. The results showed that approximately 40% of the applied radioactivity was recovered from the rats treated with sodium dodecyl sulfate, whereas some 20% was recovered in the urine from the rats treated with sodium hexadecyl sulfate. In both cases the rate of excretion was rapid, with approximately half of the total recovery being excreted in the first day. When the duration of contact of sodium hexadecyl sulfate was increased from 15 to 30 min, the excretion of radioactivity in the urine was doubled and, significantly, absorption could be demonstrated in the short time of 15 min by the presence of radioactivity in the blood. Other types of surfactants were able to be absorbed through skin. The α-sulfonated fatty acids and alkyl carboxylic acid esters of isethionic acid were stated to be absorbed through skin, although experimental data were not provided.

Penetration into rat skin of sodium hexadecyl sulfate was also reported following exposure of the skin for 10, 30, and 60 min to hexadecyl [^{35}S]sulfate. After application and without rinsing, the treated skin was excised and separated by heat into the epidermis and dermis. The frozen dermis was then cut into 25-μm sections in the plane parallel to the skin surface. Radioactivity of the separated epidermis and individual section of the dermis was measured. The amount of hexadecyl sulfate on the epidermis was constant during 60 min. The dermis contained small but significant amounts of radioactivity (approximately 0.01% of the amount in the epidermis), which decreased with depth into the skin although the total radioactivity in the 150 μm of dermis seemed to remain constant for the various exposure times.

Studies of percutaneous absorption in rats of a series of homologous soaps from saturated chain lengths C_{10}-C_{18} was reported by Howes (27). A model soap system (30 mM, approximately 1% w/v) in which the five pure Na soaps were soluble at 37°C was used. Separate solutions were prepared containing in turn each of the 1-^{14}C soaps (6 mM), which was applied to a defined area of the dorsal skin of anesthetized rats. After 15 min contact, the skin was rinsed and dried and a protective, nonocclusive patch was fitted to the rats to prevent ingestion of the radioactive soap by grooming of the treated area. Other rats were fitted with restraining collars to allow prewashing with a nonradioactive model soap solution before applications of the various radioactive soap solutions. The extent of percutaneous absorption was calculated from the recovery of radioactivity in the expired breath, urine, feces, and carcass after excision of the treated area of skin, at 6 hr after application. For the sodium salts of the five n-1-^{14}C fatty acids, the order of absorption through rat skin was $C_{12} > C_{10} = C_{14} > C_{16} = C_{18}$ when there was no pretreatment of the skin. The absorption of C_{12} soap at 6 mM (0.13% w/v) was 0.675 μg cm^{-2} during the 6 hr of the experiment and the absorption for the C_{16} soap was approximately 10 times less, emphasizing the importance of chain length in absorption. After three prewashes, the order of penetration was $C_{12} = C_{10} > C_{14} > C_{16} > C_{18}$, for which only the quantities of both C_{12} and C_{10} soaps had increased significantly to approximately 1.2 μg cm^{-2}.

Howes (27) also investigated the percutaneous absorption of three synthetic detergents in the rat. In this instance, aqueous solutions of the relatively insoluble sodium [1-^{14}C]dodecyl-1-benzene-*p*-sulfonate (3 mM), sodium [1-^{14}C]dodecyl sulfate (24 mM), or sodium [1,^{14}C]dodecanoyl isethionate (25 mM) were applied and the localization by autoradiography of the detergents in skin as well as quantities absorbed through skin were studied. There was heavy deposition of the synthetic detergents on the surface of the skin and in the upper hair follicles, but only sodium dodecyl sulfate was seen to penetrate into the lower hair follicles and into the dermis.

The absorption through skin of dodecyl benzene sulfonate at 3 mM was below the sensitivity of detection (<0.1 μg cm^{-2} 24 hr^{-1}) as was the absorption of the dodecanoyl isethionate from a 25 mM solution (0.3 μg cm^{-2} 24 hr^{-1}), whereas at 25 mM (0.7%) the penetration of sodium dodecyl sulfate was 0.26 μg cm^{-2} 24 hr^{-1}. When dodecanoyl isethionate of higher specific activity, thus conferring much greater sensitivity to the detection of radioactivity, was used, a low penetration, 0.09 μg cm^{-2} in 24 hr, was recorded by application of a 20 mM aqueous solution (approximately 0.3% w/v). Thus, for a constant chain length of C_{12}, various head groups strongly influenced the ability of the surfactant to be absorbed through skin.

A comparative study of the absorption of pure ^{14}C-labeled detergents through guinea pig skin was reported by Prottey and Ferguson (46). Different head groups were substituted on the hydrophobe, which remained constant at 12 carbon atoms. Aqueous solutions of 5 mM were applied to a defined area of clipped guinea pig skin for 10 min, then the skin was rinsed, dried, and protected with a nonocclusive patch for 24 hr. Absorption through skin was calculated from the amount of radioactivity excreted in the expired breath, urine, and feces together with that remaining in the carcass after removal of the treated area of skin, and corrected for the disposition of a dose of each surfactant given intraperitoneally.

The nature of the head group was found to be very important in determining the extent of percutaneous absorption. Of the anionics examined, sodium laurate was the best penetrant at 7.32 $\mu g\ cm^{-2}$ in 24 hr. Sodium dodecyl triethoxysulfate and sodium dodecanoyl isethionate were intermediary with an absorption of 1.16 and 0.63 $\mu g\ cm^{-2}$, respectively whereas sodium dodecyl sulfate and sodium dodecyl sulfonate gave the least absorption of 0.13 and 0.09 $\mu g\ cm^{-2}$, respectively.

Another comparative study of the absorption of pure ^{14}C-labelled alcohol sulfates, alcohol ether sulfates, and alcohol ethoxylates of varying chain length and degree of ethoxylation through rat skin was reported by Black and Howes (47). The data showed that penetration was low for the alcohol sulfates and alcohol ether sulfates, but greater for the alcohol ethoxylates. For each anionic, penetration was proportional to the applied concentration and the number of applications, but increased more gradually with extended duration of contact.

By comparing the amount of radioactivity excreted together with that remaining in the main organs after application of [^{14}C]α-olefin sulphonate to the intact skin of rats without wiping off or after removal at 0.5 hour or 1.5 hour after application, Minegishi et al. (48) concluded that percutaneous absorption was almost finished by 1.5 hr after application. When the stratum corneum was removed, penetration of a 0.02% solution of the anionic surfactant increased to nearly 50% compared with 0.63% from treatment to intact skin of a 0.2% detergent solution.

B. Effect on Absorption of Water

Sprott (39) used tritiated water to assess the effect of toilet soap bars on the permeability of rat skin. Groups of rats were clipped and washed for 5 min with 4% (w/v) solutions of the toilet soaps, one of which was a conventional bar containing the sodium salts of the fatty acids from saponified tallow and palm kernel oil. The other toilet bar contained a high proportion of sulfonated fatty acids and fatty acid esters of isethionic acid. After rinsing, one group of rats from each treatment had tritiated water applied to the skin. The rats were sacrificed at 5 min after application of tritiated water and samples of heart blood withdrawn, weighed, and freeze-dried. The water from the cold traps of the lyophylizer was counted to determine its content of tritium. Both groups of rats washed with solutions of the toilet bars showed a significantly increased tritium content in the blood, but the differences between the conventional soap bar and the toilet bar containing synthetic detergents was not statistically significant.

Other groups were rinsed after application of the tritiated water, placed in metabolism cages, and separated urine was collected for 72 hr. There was a tendency for the urine of the rats washed with solutions of the toilet bars to contain greater amounts of radioactivity than the control group, but only the effect of the conventional soap was statistically greater.

C. Effect on Absorption of Electrolytes

There have been many reports on the way in which surface-active agents enhance the absorption of electrolytes, particularly those known to elicit a contact dermatitis or delayed hypersensitization, such as nickel and chromate salts. Nilzen and Wikstrom (49) showed that, while neither nickel sulfate or sodium dodecyl sulfate alone would cause an eczematous response, the combination did produce a contact dermatitis. Similarly, Kvornung and Svendsen (50) reported that secondary alkyl sulfate (Teepol) facilitated the sensitization response to nickel chloride and potassium dichromate in susceptible patients. In human patch tests, these authors were able to show a skin response to the allergen in a more dilute solution when the challenge solution contained Teepol. Vinson and Choman (51) showed that sodium dodecyl sulfate and sodium dodecyl benzene sulfonate applied to the skin of guinea pigs in a two-stage multiple-application test produced an eczematous response to nickel sulfate and they observed the deep penetration of nickel into the skin. The application of soap or the other test anionic detergents Igepon A and Igepon T did not elicit a nickel response under the conditions of the test. Lindberg et al. (52) used the technique of energy dispersive X-ray microanalysis to investigate the effect of SLS on the penetration of nickel through the stratum corneum of the guinea pig. The authors found in the SDS treated skin a greater nickel penetration which they believed might be due to an increase in the "water channel" volume of the stratum corneum.

Friberg et al. (53) investigated the effect of soap and alkylaryl sulfonate on the penetration of mercuric chloride and methyl mercury dicyandiamide, both radioactively labeled, through guinea pig abdominal skin by disappearance measurements. Pretreatment of the skin by painting with a soft brush for 2 min twice daily with 1% aqueous solution of soap did not affect the absorption of either of the two mercury compounds. However, a 1% aqueous solution of the alkylaryl sulfonate did show a tendency to increase the absorption of the methyl mercury dicyandiamide. Radioactive mercury was present in the kidneys, liver, and blood of the guinea pigs. When the mercuric chloride or methyl mercury dicyandiamide were dissolved in the soap or alkylaryl sulfonate solution and applied to the dorsal skin of the guinea pigs, Skog and Wahlberg (54) showed that, as Friberg et al. (53) had reported, only the alkylaryl sulfonate affected the absorption of the methyl mercury dicyandiamide. There was no difference between the absorption through guinea pig dorsal or abdominal skin. However, the disappearance technique is not sufficiently sensitive to quantify the degree of absorption.

In comparison of the absorption of sodium chloride and mercuric chloride in soap or alkylaryl sulfonate through normal or stripped guinea pig skin, Wahlberg (55) reported that absolute absorption was related to concentration of the applied salt solution and was greater through stripped skin.

In similar disappearance measurements with sodium [^{51}Cr]chromate in solutions of 0.1, 1.0, and 10% sodium dodecyl sulfate, Wahlberg (56) showed that at 0.239 M sodium chromate in 10% sodium dodecyl sulfate, there was a significantly increased absorption compared to the salt in distilled water.

The effect of surfactants upon the absorption of radioactively labeled sodium [^{131}I]iodide, chosen to represent highly polar, water-soluble substances, and also because of the ease with which its rapid localization can be measured by the scanning in vitro of the thyroid, was reported by Sprott (39). Rats were washed with solutions of two toilet bars of different composition followed by application of [^{131}I]iodide. Autoradiography of sections of the thyroid from rats killed 15 min after application of radioactive marker showed that the silver grain density was greater in the rat washed with the toilet bar containing the sodium salts of the fatty acid esters of isethionic acid and sodium dodecyl benzene sulfonate. In a second identically washed group of rats, the thyroid [^{131}I]iodine radioactivity rose rapidly to a level that was greater in animals washed with the nonsoap toilet bar compared with animals washed with the soap bar. Water washed rats showed a lower level of accumulation of ^{131}I in the thyroid. While it was clear that both toilet bars increased the permeability of skin in this single experiment, the statistical significance of differences between soaps and synthetic detergents was unresolved.

D. Effect on Absorption of Organic Compounds

Studies into the response of human, guinea pig, and rabbit skin to 2,4-dinitrochlorobenzene (DNCB), a known irritant and sensitizer, after preexposure of the skin with up to 2% (w/v) aqueous solutions of soap and alkyl benzene sulfonate were reported by Skog (57). In humans and guinea pigs, preexposure to the surfactants tended to increase the response to DNCB, particularly when the subjects had been sensitized to DNCB. In rabbits no gross differences in the response to DNCB were observed between pretreated and nonpretreated skin.

The bacteriostatic compound 3,5,3′,4′-tetrachlorosalicylanilide (T_4CS) in radioactive form was applied in 75% (w/v) aqueous 2-methoxyethanol to rats, the skin of which had been prewashed with water or 1% (w/v) sodium dodecyl sulfate (39). Excess of the T_4CS was removed after 5 min contact with the skin and the radioactively remaining on the skin and excreted in the urine on five consecutive days was measured. Washing of the rats with detergents seemed to increase the capacity of the stratum corneum and epidermis to bind T_4CS and to promote its absorption through skin compared to the water-washed (control) rats, although the difference was small.

The treatment of guinea pigs with aqueous solutions of soap (5% w/v) or alkyl ether sulfate (0.045 N) for 1 min had little effect on the permeability to SARIN, an organophosphorus cholinesterase inhibitor (58). A significantly greater effect on permeability was recorded when the duration of contact was increased to 5 and 30 min. The progressive damage to skin with time was not altered by rinsing off the surface-active agents with water, indicating that their effect was not reversible. A slow recovery of the barrier properties was observed (59), which might be related to the regeneration of nondenatured proteins by the epidermis. It would have been interesting to use this well-defined system (60) to study the effect of surfactants on the skin barrier properties in relation to concentration of surfactant and to compare the differences between various chemical structures of detergents of the anionic class.

That anionic surfactants have different effects on the skin barrier properties was shown in the studies of Black et al. (61) with pyridine-2-thione-*N*-oxide, also called zinc pyrithione (ZnPT). The absorption of Zn[^{3}H]PT through rat skin washed with shampoos containing different detergent actives was greatest when the active was sodium dodecyl sulfate, and least when triethanolamine lauryl sulfate was used. Sodium lauryl ether sulfate gave an intermediary absorption of Zn[^{3}H]PT. The substituted pyridine, Octopirox, was applied to rat skin in a diluted shampoo base containing sodium lauryl ether sulfate (62). Use of occlusive protective patches greatly increased the skin penetration of Octopirox, as did increases in both duration of contact with the skin and concentration of the Octopirox in the shampoo. Blood levels after topical application without rinsing and with occlusion were greatly reduced when rinsing and nonocclusive protective patches were used.

Application of the hair dye, 2-nitro-*p*-phenylenediamine in a hair colorant product base to clipped rat skin gave a high penetration if the skin was occluded and not rinsed (63). Penetration was much reduced after rinsing and protection of the skin by nonocclusive patches and was also shown to be related to the duration of contact before rinsing and to the applied concentration.

The effect of sodium lauryl sulfate on the percutaneous absorption of butanol as a microemulsion through guinea pig skin was studied by Boman et al. (64) by sampling arterial blood at regular intervals through a catheter. For different concentrations in sodium dodecyl sulfate and water there were marked increases in the blood concentration of butanol compared with the levels after application of neat solvent which the authors believed might be due to denaturation of keratin and hydration of the stratum corneum by sodium dodecyl sulfate, thus facilitating skin absorption of butanol.

Enhancing the penetration of the drug naloxone through isolated human skin by using a variety of compounds including sodium lauryl sulfate sodium laurate or sodium oleate, at 10% in polyethylene glycol 400 was studied by Aungst et al. (65). They found a flux of 3.8 ± 0.5 for sodium oleate and 4.6 ± 0.9 for SLS, while sodium laurate gave a flux of 7.3 ± 0.8. All these figures were much less than found for fatty acids of various chain lengths. Vehicles other than PEG 400 were also examined. In propylene glycol and isopropanol, sodium lauryl sulfate approximately doubled naloxone flux compared with the vehicles alone, while there was no effect when in PEG 400. In a mineral oil vehicle with sodium lauryl sulfate, however the naloxone flux increased markedly.

It is clear that absorption of anionic detergents through skin is low and that determination by in vitro and in vivo methods yield results that are not directly comparable, probably due mainly to the strong binding of detergents to protein. The effects on the skin barrier are also progressive, so that increased duration of exposure will increase the absorption of the detergent. Higher concentrations will also promote absorption. While synthetic anionic detergents in aqueous solution have a pH near to neutrality, the alkalinity per se of aqueous soap solutions or of built detergent products, can reduce the barrier properties of skin. The effect of detergents on the skin barrier also allows better absorption of other compounds. This is important in predicting the skin's response to soap and detergent products, toiletries, and cosmetics, and also in the extrapolation to possible systemic toxic effects.

The low absorption through the skin of animals and the greater impermeability of human skin (66,67) means that only small amounts of anionic detergent are likely to be absorbed by humans, even when the total body surface is exposed. Confirmation of this low absorption in humans requires the development of specific and sensitive techniques beyond the scope of those presently available, independent of radiotracer methodology, for the determination of the parent anionic detergent or metabolite(s) in blood and urine.

Although anionic detergents are well absorbed from the intestine, in contrast to the statement by Chasseaud (25), the expected intake from residues is small (see Chap. 8). The low oral toxicity of anionic detergents (3) related to the low intestinal intake and poor percutaneous absorption suggests that the possibility of systemic toxic effects in humans is extremely unlikely.

REFERENCES

1. K. Jones, In *Basic Organic Chemistry* (Part 5), J.M. Tedder (Ed.), Wiley, London, 1974.
2. G.L. Hollis (Ed.), *Surfactants U.K., 1st, Tergo-Data,* Darlington, England, 1976.
3. C. Gloxhuber, *Arch. Toxicol., 32,* 245-270(1974).
4. V.H. Havermann and K.H. Menke, *Sonderheft Zeitschr. Landwirlsch. Forsch., 9,* 140-146(1957).
5. W.R. Michael, *Toxicol. Appl. Pharmacol., 12,* 473-485(1968).
6. L.G. Scharpf, R.D. Swisher, R.A. Kimerle, and W.E. Gledhill, *Household Personal Products Inds., 3,* 44-49(1976).
7. A.J. Willetts and R.B. Cain, *Biochem. J., 129,* 389-402(1972).
8. D.G. Cresswell, G.A. Baldock, L.F. Chasseaud, and D.R. Hawkins, *Toxicology, 11,* 5-17(1978).
9. P.M. Rennison, G.M. Powell, A.H. Olavesen, D. Howes, and J.G. Black, *Biochem. Soc. Trans., 15,* 456(1987).
10. R.D. Swisher, *Int. Water Pollution Control Fed., 35,* 877-881(1963).
11. J.D. Lay, W. Kein, and F. Korte, *Toxicology Letters, 17,* 187-192(1983).
12. W.H.B. Denner, A.H. Olavesen, G.M. Powell, and K.S. Dodgson, *Biochem. J., 111,* 43-51 (1969).
13. J. Ottery, A.H. Olavesen, and K.S. Dodgson, *Life Sci., 9*(2), 1335-1340(1970).
14. B. Burke, A.H. Olavesen, C.G. Curtis, and G.M. Powell, *Xenobiotica, 5,* 573-584(1975).
15. I. Merits, *Biochem. J., 148,* 219-228(1975).
16. J.L. Maggs, G.M. Powell, K.S. Dodgson, D. Howes, J.G. Black, and A.H. Olavesen, *Xenobiotica, 12,* 611-616(1982).
17. P.W.A. Tovell, D. Howes, and C.S. Newsome, *Toxicology, 4,* 17-29(1975).
18. B. Burke, A.H. Olavesen, C.G. Curtis, and G.M. Powell, *Xenobiotica, 6,* 667-678(1976).
19. P.J.H. Webborn, A.H. Olavesen, G.H. Powell, C.G. Curtis, D. Howes, and J.G. Black, *Biochem. Soc. Trans., 13,* 891 (1985).
20. S.K. Bains, A.H. Olavesen, J.G. Black, D. Howes, C.G. Curtis, and G.H. Powell, *Xenobiotica, 17,* 709-723(1987).
21. J.B. Knaak, S.J. Kozbelt, and L.J. Sullivan, *Toxicol. Appl. Pharmacol., 8,* 369-379(1966).
22. F.P. Stewart, M.N. Pieters, J.C. Wood, J.G. Black, and D. Howes, *European Society of Biochemical Pharmacology*, 11th European Workshop on Drug Metabolism, 1988.
23. A.J. Taylor, Ph.D. thesis, University of Wales, Cardiff (1975).
24. J.A. McDermott, D.H. Hughes, and P.M. Quinlin, *Toxicol. Appl. Pharmacol., 33,* 145-151(1975).

25. L.F. Chasseaud, In *Foreign Compound Metabolism in Mammals,* Vol. 1, The Chemical Society, 1970, p. 15.
26. J. Clifford and B.A. Pethica, *Trans. Faraday Soc., 60,* 1483-149(1964).
27. D. Howes, *J. Soc. Cosmet. Chem., 26,* 47-63(1975).
28. I.H. Blank and E. Gould, *J. Invest. Dermatol., 33,* 327-336(1959).
29. I.H. Blank and E. Gould, *J. Invest. Dermatol., 37,* 485-488(1961).
30. I.H. Blank and E. Gould, *J. Invest. Dermatol., 37,* 311-315(1961).
31. F.R. Bettley, *Brit. J. Dermatol., 73,* 448-454(1961).
32. F.R. Bettley, *Brit. J. Dermatol., 75,* 113-116(1963).
33. J. Scala, D.E. McOsker, and H.H. Reller, *J. Invest. Dermatol., 50,* 372-379 (1968).
34. G. Emery and P.H. Dugard, *Brit. J. Dermatol., 81 Supp.,* 4, 63-68(1969).
35. R.B. Stoughton and W. Fritsch, *Arch. Dermatol., 90,* 512-517(1964).
36. S. Fiker, V. Janeckova, and V. Kacovska, *Acta FAc. Med., Univ. Brun., 41* (*1*), 177-180(1972).
37. F.R. Bettley and E. Donoghue, *Nature, 185,* 17-20(1960).
38. P.A. Isherwood, *J. Invest. Dermatol., 40,* 143-145(1963).
39. W.E. Sprott, *Trans. St. John's Hosp. Dermatol. Soc., 51,* (*2*), 56-71(1965).
40. R. Scheuplein and L. Ross, *J. Soc. Cosmet. Chem., 21,* 853-873(1970).
41. F.R. Bettley, *Brit. J. Dermatol., 77,* 98-100(1965).
42. D.C.F. Wood and F.R. Bettley, *Brit. J. Dermatol., 84,* 320-325(1971).
43. W.E. Sprott, In *De Structura et Functione Stratorum Epidermidis S.D. Barrierrae,* G. Lejhanec and P. Hybasek (Eds.), Facultas Medica Universitatis Brunensis, Czechoslovakia, 1965, pp. 219-232.
44. A.J. Aguiar and M.A. Weiner, *J. Pharm. Sci., 58,* 210-215(1969).
45. B.R. Choman, *J. Soc. Cosmet. Chem., 11,* 138-145(1960).
46. C. Prottey and T. Ferguson, *J. Soc. Cosmet. Chem., 26,* 29-46(1975).
47. J.G. Black and D. Howes, *J. Soc. Cosmet. Chem., 30,* 157-165(1979).
48. K-I Minegishi, M. Osawa, and T. Yamaha, *Chem. Pharm. Bull.* (*Tokyo*)*, 25,* 821-825(1977).
49. A. Nilzen and K. Wikstrom, *Acta Dermatol. Venereol., 35,* 292-299(1955).
50. S.A. Kvornung and I.B. Svendsen, *J. Invest. Dermatol., 26,* 421-426(1956).
51. L.J. Vinson and B.R. Choman, *J. Soc. Cosmet. Chem., 11,* 127-137(1960).
52. M. Lindberg, S. Sagstrom, G.M. Roomans, and B. Forslind, *Scanning Microscopy, 3,* 221-224(1989).
53. L. Friberg, E. Skog, and J.E. Wahlberg, *Acta Dermatol. Venereol., 41,* 40-52 (1961).
54. E. Skog and J. Wahlberg, *Acta Dermatol. Venereol., 42,* 17-20(1962).
55. J.E. Wahlberg, *Acta Dermatol. Venereol., 45,* 335-343(1965).
56. J.E. Wahlberg, *Acta Dermatol. Venereol., 48,* 549-555(1968).
57. E. Skog, *Acta Dermatol. Venereol., 38,* 1-14(1958).
58. T. Fredricksson, *Acta Dermatol. Venereol., 49,* 55-58(1969).
59. T. Fredricksson, *Acta Dermatol. Venereol., 49,* 481-483(1969).
60. T. Fredricksson, *Acta Dermatol. Venereol., 43,* 91-101(1963).
61. J.G. Black, D. Howes, and T. Rutherford, *Actualites de Dermocosmetologie, VII,* 127-142, Centre European de Dermocosmetologie, 69 Lyon (1975).

62. J.G. Black and V.B. Kamat, *Fd. Chem. Toxicol., 26,* 53-58(1988).
63. D. Howes and J.G. Black, *Inter. J. Cosmet. Sci., 5,* 215-226(1983).
64. A. Boman, I. Blute, P. Fernstrom, J. Carlfors, and L. Rydhag, *Contact Dermatitis, 21,* 92-104(1989).
65. B.J. Aungst, N.J. Rogers, and E. Shefter, *Inter. J. Pharmaceutics, 33,* 225-234(1986).
66. M.J. Bartek, J.A. La Budde, and H.I. Maibach, *J. Invest. Dermatol., 58,* 114-123(1972).
67. D. Howes and J.G. Black, *Toxicology, 6,* 67-76(1976).

3

Acute, Subacute, and Chronic Toxicity Data on Anionics

MATJAŽ S. POTOKAR Henkel KGaA, Düsseldorf, Germany

I. INTRODUCTION

Of the synthetic surfactants found in modern cleaning products and cosmetics, anionic surfactants are most common. On the basis of environmental protection, two aspects must be considered in view of the widespread use of these surfactants:

1. The most important aspect of environmental protection is human protection. Therefore, toxicological safety for humans has the highest priority. Contact with surfactants becomes more frequent and less avoidable

with increasing population. Therefore, it is an important duty of industry to ensure careful toxicological evaluation before marketing surfactants.

In this respect, both acute and chronic toxicity data are important. The acute toxicity is of interest particularly in the cases of ingestion, e.g., of cleaning products by children. On the other hand, long-term studies are of interest with respect to the unavoidable intake of trace amounts which occurs as a consequence of the application of cleaning and rinsing agents as well as other household care products. Also, there will always be a small intake with the drinking water.

2. The other relevant aspect of environmental toxicology has to do with the surfactant toxicity toward living organisms found in rivers and other waters.

   In this context, fish toxicity has to be considered. However, we do not include this aspect in our chapter as it has to do with the ecological evaluation of the surfactants.

The purpose of this chapter is to present a quick survey of the data relating to the acute and chronic toxicities of various surfactant classes.

In the study of the toxicological properties of surfactants it has to be kept in mind that surfactants are not chemically pure substances but rather products which are mixtures of individual chemical compounds, the composition of which may vary depending on both the production process and the starting materials used. Strictly speaking, the toxicological data obtained with such mixtures are valid only for the mixture studied; however, these data will also hold approximately for products which are chemically closely related.

II. ACUTE TOXICITY

Table 1 summarizes the acute toxicity data of anionic surfactants. Acute toxicity provides information that is used for the toxicological knowledge of a chemical in order to provide data relevant to the protection of human health and to meet the requirements of regulatory authorities. The route of exposure may influence the toxic response and, thus, its selection is of paramount importance in designing studies. Parenteral routes such as intravenous (i.v.) injection are relevant only for pharmaceuticals, whereas oral, inhalation, and dermal routes are more relevant for industrial chemicals. In individual cases it is important to have acute toxicity data from several routes of exposure, since absorption may change significantly with the route of administration of a compound. The LD_{50}-value itself depends on a large number of environmental and experimental conditions, and there are many factors known to influence the outcome of an acute toxicity test. Even when experiments are repeated under standardized conditions, there may be considerable variation in the resulting LD_{50}.

TABLE 1 Acute Toxicity Data

A. Carboxylic Acid Salts (Soaps) $R\text{-}COO^-Na^+$

Preparation	Species	Application[a]	LD_{50}, mg kg^{-1}	Ref.
Soap (U.S.P.)	Rat	i.p.	980	14
Soap (Ivory)	Rat	Oral	>16,000	49
Sodium myristinate	Rat	Oral	>10,000	23
Sodium palmitate	Rat	Oral	>10,000	

[a]Type of application: Oral, via stomach; i.p., injection into the peritoneal cavity; i.v., into vein by injection, dermal, on the skin s.c., under the skin.

B. *n*-Alkyl Sulfates[a] $R—O—SO_3^-Na^+$

Surfactant	Species	Application	LD_{50}, mg kg^{-1}	Ref.
Alkyl sulfate	Rat	Oral	2730	49
n = 8	Rat	Oral	3200	19
n = 10	Rat	Oral	1950	
n = 12	Rat	Oral	2640	
n = 14	Rat	Oral	>3500	
n = 16	Rat	Oral	>3000	
n = 18	Rat	Oral	3000	
n = 8	Mouse	i.p.	400	
n = 10	Mouse	i.p.	290	
n = 12	Mouse	i.p.	250	
n = 14	Mouse	i.p.	240	
n = 16	Mouse	i.p.	260	
n = 18	Mouse	i.p.	480	
n = 6	Mouse	Oral	6400	23
n = 8	Mouse	Oral	2900	
n = 10	Mouse	Oral	2200	
n = 12	Mouse	Oral	2700	
n = 14	Mouse	Oral	3000	
n = 16	Mouse	Oral	>8000	
n = 18	Mouse	Oral	>8000	
n = 12-15	Rat	Oral	1413	55
Sodium dodecyl sulfate	Rat	Oral	1200	40
	Rat	i.p.	210	14
	Rat	i.v.	118	11
	Rat	Oral	1288	55
(Duponol)	Rabbit	Dermal	580	10
	Rabbit	i.v.	121.2(= LD_{100})	
	Rat	Dermal	2000(= LD_{100})	

TABLE 1 *(continues)*

TABLE 1 (Continued)

Surfactant	Species	Application	LD_{50}, mg kg^{-1}	Ref.
Sodium dodecyl sulfate				
(Duponol)	Guinea pig	Dermal	>1200	10
	Guinea pig	Dermal	2000(= LD_{100})	
Sodium dodecyl sulfate				
(technical grade)	Rat	Oral	1000	56
(Texapon K 12)	Rat	Oral	1700	58
(Texapon Z) n = 12-14	Mouse	Oral	2010	
(Texapon V) n = 12-16	Mouse	Oral	2300	
Ammonium dodecyl sulfate				
(Texapon A400) 34%	Rat	Oral	>5000	
Lithium dodecyl sulfate				
(Texapon LLS) 30%	Rat	Oral	5160	
Ammonium alkyl sulfate				
(Dobanol 23) base	Rat	Oral	1000-2000	7
coconut alcohol base	Rat	Oral	1000-2000	
Monoethanolamine alkyl sulfate				
(Dobanol 23) base	Rat	Oral	1000-2000	
coconut alcohol base	Rat	Oral	2000-3000	
Monoethanolamine dodecyl				
sulfate (Texapon MLS) 33%	Rat	Oral	5620	58
Diethanolamine dodecyl sulfate				
(Texapon DLS) 31%	Rat	Oral	8730	
Triethanolamine dodecyl sulfate				
(Texapon TH) 48%	Rat	Oral	7940	
Triethanolamine dodecyl sulfate	Rat	Oral	2750	40
Triethanolamine alkyl sulfates				
(Dobanol 23)				
(Dobanol 23) base	Rat	Oral	1000-2000	7
coconut alcohol base	Rat	Oral	1000-2000	

[a]n = number of carbon atoms in *n*-alkyl residue.

C. *n*-Alkyl Sulfates with Branched Hydrophobic Residue (Tergitol Brands)

Surfactant	Species	Application	LD_{50}, mg kg^{-1}	Ref.
Sodium-2-ethylhexyl sulfate				
(Tergitol 08), 40%	Rat	Oral	10300	49
	Guinea pig	Oral	3800	
	Rat	s.c.	4730	50
	Rat	i.p.	320	
	Rabbit	Dermal	6540	

TABLE 1 (Continued)

Surfactant	Species	Application	LD_{50}, mg kg^{-1}	Ref.
Sodium-2-ethylhexyl sulfate (Tergitol 08), 40% (continued)	Rat	Inhalation	0.1% Survived for 7 hr	
Sodium-7-ethyl-2-methyl-undecyl-4 sulfate (Tergitol 04) 25%	Rat	Oral	5000	49
	Guinea pig	Oral	2600	
Sodium-3,9-diethyl-tridecyl-6-sulfate (Tergitol 07) 25%	Rat	Oral	5700	
	Guinea pig	Oral	1700	
(Tergitol 08) 40%	Rabbit	i.v.	>100	22
	Rabbit	i.p.	>250	
	Rabbit	Oral	>250	

D. Alkylpolyglycolether Sulfates R—$(OCH_2CH_2)_m$—O-SO_3Na^{+}[a]

Surfactant	Species	Application	LD_{50}, mg kg^{-1}	Ref.
Sodium alkyl ether sulfate m = 2				
(Dobanol 23) base	Rat	Oral	1000-2000	7
coconut alcohol base	Rat	Oral	2000	
Sodium alkyl ether sulfate m = 3 synthetic base	Rat	Oral	2138	55
Sodium dodecyl ether sulfate (Texapon N70) 70%)				
m = 1	Rat	Oral	4500	58
m = 3	Rat	Oral	1820	54
m = 3	Rat	Oral	1995	55
Sodium dodecyl myristic ether sulfate (Texapon K14S)	Mouse	Oral	6020	58
Ammonium dodecyl ether sulfate 24%	Mouse	Oral	>10000	
Magnesium dodecyl ether sulfate (Texapon MS) 30%	Mouse	Oral	>10000	
Monoethanolamine dodecyl ether sulfate ca. 28%	Mouse	Oral	>10000	
Sodium octyl phenolether sulfate m = 1	Rat	Oral	3700-5400	15

[a]m = number of OCH_2CH_2 groups in the molecule.

TABLE 1 (*continues*)

TABLE 1 (Continued)

E. Alkyl Benzene Sulfonates (ABS) with Branched (B) and Unbranched (L) Hydrophobic Residue R—(C_6H_4)—$SO_3^-Na^+$

Surfactant	Species	Application	LD_{50}, mg kg^{-1}	Ref.
B-ABS	Rat	Oral	1220	5
	Rat	Oral	2500	9
n = 6-12[a]	Mouse	i.p.	95	12
n = 12-18	Mouse	i.p.	133	
B-ABS	Rat	Oral	2200	13
	Rat	Oral	2300	21
	Mouse	Oral	1620	38
	Mouse	s.c.	350	
	Mouse	i.p.	120	
	Mouse	i.v.	121	
	Rat	Oral	428-700[b]	34
	Mouse	Oral	1330-2090[b]	
	Mouse	Oral	1400	39
	Rat	Oral	1400[c]	40
n = 10-14	Rat	Oral	520	42
B-ABS	Mouse	Oral	1700	58
n = 10	Mouse	i.p.	90	
n = 11	Mouse	i.p.	110	
n = 12	Mouse	i.p.	103	
n = 13	Mouse	i.p.	110	
B-ABS	Rat	Oral	1400	49
	Rat	Oral	2200-3200[d]	51
	Mouse	Oral	4600[d]	
	Rat	Oral	1500	56
	Mouse	Oral	2800	
L-Sodium dodecyl benzene sulfonate n = 12 (10-14)	Rat	Oral	1260	5
Sodium dodecyl benzene sulfonate (Santomerse 3)	Mouse	Oral	2000	29
	Mouse	i.v.	105	
Sodium dodecyl benzene sulfonate (D-40)	Rat	Oral	2300	13
	Mouse	Oral	2010	28
	Hamster	Oral	1131	
	Rat	Oral	2320	
	Rabbit	Oral	1730	
Sodium decyl benzene sulfonate (Santomerse D)	Mouse	Oral	2100	29
	Mouse	i.v.	115	
	Mouse	Oral	2000	56

TABLE 1 (Continued)

Surfactant	Species	Application	LD_{50}, mg kg^{-1}	Ref.
L-ABS	Rat	Oral	404-1820[b]	29
	Mouse	Oral	1575-1950[b]	
n = 10-14	Rat	Oral	650	42
L-ABS	Mouse	Oral	2300	52
(Santomerse 3)	Rabbit	i.v.	>100	22
	Rabbit	i.p.	>250	
	Rabbit	Oral	>250	
(Santomerse D)	Rabbit	i.v.	>100	
	Rabbit	i.p.	>250	
	Rabbit	Oral	>250	

[a]n = number of carbon atoms in hydrophobic alkyl residue.
[b]Dependent on age, weight, and breed.
[c]Triethanolammonium salt.
[d]20% active content.

F. Other Anionic Surfactants

Surfactant	Species	Application	LD_{50}, mg kg^{-1}	Ref.
Fatty acid condensation products				
Oleonic acid methyl tauride				
(Igepon T)	Mouse	Oral	6630	29
	Mouse	i.v.	350	
	Mouse	Oral	6300	56
	Rat	Oral	>4000	
Sulfosuccinic acid esters				
Dioctylsulfosuccinate				
(Aerosol O.T.)	Mouse	Oral	4800	29
	Mouse	i.v.	60	
	Rat	Oral	1900	40
	Mouse	Oral	1500	3
	Guinea pig	Oral	650	36
	Horse	Oral	<650(=LD)	
	Rabbit	i.v.	>100	22
		i.p.	>250	
		Oral	>250	
n-Alkyl sulfonates				
n-Alkyl sulfonate	Rat	Oral	2700	13
	Mouse	Oral	1440	58
n = 12-17	Rat	Oral	3000	21

TABLE 1 *(continues)*

TABLE 1 (Continued)

Surfactant	Species	Application	LD_{50}, mg kg^{-1}	Ref.
n-Alkyl sulfonates (continued)				
n = 8	Mouse	Oral	2600	24
n = 10	Mouse	Oral	2200	
n = 12	Mouse	Oral	4200	
n = 14	Mouse	Oral	>16000	
n-α-Olefin sulfonates				
n = 15-18	Mouse	Oral	3260	38
	Mouse	s.c.	1660	
	Mouse	i.p.	304	
	Mouse	i.v.	94	
n = 14-16	Mouse	Oral	2900	58
n = 15-18	Mouse	Oral	2930	
n = 16-18	Mouse	Oral	3350	
(Hostapur)	Mouse	Oral	2720	
Various				
Sodium dodecyl glyceryl ether sulfonate	Rat	Oral	1820	54
Sodium isopropyl naphtalene sulfonate	Mouse	i.v.	120	29
(Aerosol O.S.)	Mouse	Oral	1900	
Sodium monobutylphenyl phenol monosulfonate	Mouse	i.v.	180	
(Areskap 100)	Mouse	Oral	3800	
Sodium monobutyldiphenyl monosulfonate	Mouse	i.v.	250	
(Aresket 300)	Mouse	Oral	3500	
Sodium dibutylphenylphenol disulfonate (Aresklene 400)	Mouse	i.v.	200	
	Mouse	Oral	2200	
Sodium polyalkylaryl sulfonate	Mouse	Oral	2570	28
(Oronit wetting agent)	Hamster	Oral	1000	
	Rat	Oral	1620	
	Young rat	Oral	2320	
Sodium lauryl sulfoacetate	Rat	Oral	700	12d
	Rat	i.p.	250	14
Sodium stearoyl-2-lactylate	Rat	Oral	25000	36a
Calcium stearoyl-2-lactylate	Rat	Oral	25000	
Sodium capryl lactylate	Rat	Oral	5840	
Sodium lauroyl lactylate	Rat	Oral	6810	

TABLE 1 (Continued)

Surfactant	Species	Application	LD_{50}, mg kg^{-1}	Ref.
Various (continued)				
Sodium isostearoyl-2-lactylate	Rat	Oral	6100	
Triethanolamine salt of lauroyl myristoyl lactylic acid	Rat	Oral	5000	
Isostearyl lactate	Rat	Oral	7500	

The acute oral systemic toxicity varies widely from harmful materials to relatively innocuous compounds. Very little has been reported concerning systemic short-term studies in an attempt to correlate structure with toxicity. Furthermore, the study protocols and approaches reported for the determination of toxicity have varied considerably among different laboratories, and many of experimental details are not available. In most cases the surfactant was applied orally, in view of the likelihood of oral intake. In numerous cases more than one animal species was used in these studies; in most cases LD_{50} values above 1 g kg^{-1} have been obtained, which shows that the acute toxicity is generally low.

This generalization holds for anionic surfactants of widely different nature. There are only few acute toxicity studies relating to parenteral application. The data obtained show much higher acute toxicities than the ones found upon oral administration. It may be concluded that there is either no significant absorption of the surfactants from the intestinal tract or there is a rapid conversion to less toxic metabolites. The hemolysis is a most characteristic consequence of the presence of the unchanged substances in the blood vessels (see Chap. 1). A number of deadly abortion attempts have been described in the older literature; in these cases, soaps—which belong to the group of anionic surfactants—were injected into the uterus. As the soaps entered the circulatory system they caused severe hemolysis, which was fatal in many cases.

While the group of anionic surfactants shows little toxicity upon oral administration, a certain dependency of the toxicity on the chain length of the alkyl group has been observed in the homologous series of alkyl sulfates and alkyl benzene sulfonates. It was found that compounds with alkyl chains of 10-12 carbon atoms are more highly toxic than compounds with shorter or longer alkyl chains. To our knowledge, this study is the only one in this field.

The finding that compounds with 10-12 carbon atoms in the aklyl chain are less well tolerated than compounds with shorter or longer chains is reflected also in studies relating to the local irritancy.

Because of its biological degradability, linear alkyl benzene sulfonate has now replaced the branched chain tetrapropylene benzene sulfonate, and it was found that the acute oral toxicities of the linear and the branched chain product hardly differ. In some cases the test animals were dissected and local reactions in the intestinal tract were observed.

III. SUBACUTE AND CHRONIC TOXICITY

Animal studies, especially long-term toxicity studies, are sometimes referred to as "safety experiments" to establish "human safety." The use of these terms can be misleading. Animal studies are intended to provide, so far as possible, a specificity of the compound toxicity profile and not to prove that no effects occur. Therefore, all types of long-term toxicity studies are conducted with different dose levels and corresponding control. The high dose level should provoke symptoms of systemic toxicity, the lowest one should show no compound-related findings at a multiple proposed human exposure. If there is no evidence of toxicity, even at high-dose level of the test substance, this does not necessarily mean that a compound can be regarded as "generally safe for humans." In the most cases it may only mean that it is being poorly or not being absorbed at all through physiological barriers (e.g., skin, gastrointestinal tract). In such cases the hazard assessment can only be stated for the specific routes of possible human exposure (e.g., dermal, oral).

Otherwise the data obtained from animal studies of different species may identify toxicity at narrow safety margins that suggest unacceptable risk for humans at proposed use conditions of chemicals.

Thus, safety evaluation in a frame of concluding remarks should be considered in association with other relevant chapters of this work.

The toxicity of surfactants upon repeated intake is of interest mainly in view of the unavoidable intake of traces; traces of surfactants remaining on dishes have to be considered as must contamination of the drinking water by surfactant-containing sewage and application as wetting agents in toothpastes and many other fields of use (see Chap. 8).

Rats are most widely used in long-term studies of anionic surfactant toxicity; however, other animal species, such as mice, guinea pigs, rabbits, dogs, and pigs, are also used. A few substances have even been tested with humans, generally in the context of therapeutic measures.

The doses administered ranged from a few milligrams to toxic and lethal amounts. In most cases, the surfactant was applied with the food; however, dosing in gavage or as admixture to the drinking water was used occasionally. The duration varied between a few weeks and 2 years.

The data obtained with the individual substances are given in Table 2. It is seen that none of the substances gave rise to severe and specific toxic symptoms or severe organ injury. In most cases, the low-grade effects related to a reduced increase in body weight during the experiment, which presumably has to do with reduced food intake or utilization. These effects may be related to the finding of certain irritation in the mucous membrane of the intestinal tract, which have been observed in animals which had received high doses of surfactants. In some cases, low-grade changes were found in liver and kidneys, depending on the surfactant dose received.

IV. ACUTE TOXICITY DATA

See Table 1.

V. SUBACUTE AND CHRONIC TOXICITY DATA

See Table 2.

VI. CLINICAL FINDINGS AND PATHOLOGICAL ANATOMICAL CHANGES OBSERVED IN ANIMAL EXPERIMENTS

The following paragraphs discuss pathohistological and clinical observations that were omitted from the summary of publications on acute and chronic toxicity studies. Observations of this kind are of interest with regard to the ingestion of single large doses of surfactants which occur, e.g., in the cases of attempted suicide or in accidents. The amounts of surfactants which are unavoidably ingested over long periods of time is many powers of 10 below the doses required to cause organic changes, as shown by animal experiments.

The presence of 8% sodium lauryl sulfate in the food is lethal within 2 weeks as shown by Fitzhugh and Nelson (16) with rats as test animals. Dissection showed intestinal inflammation and distention of the digestive system. The authors also studied the toxicological effects of large doses of sodium dioctyl sulfosuccinate (DSS) (addition of 2, 4, and 8% to the food). While the growth was inhibited by the surfactant at the two lower doses during the course of the 4-month experiments, the test animals died of gastrointestinal symptoms within 1 week, when 8% of the surfactant was supplied with the diet.

Similar observations were made in the case of rats which had received 4% sodium alkylaryl sulfonate with the diet. With the exception of diarrhea, gastrointestinal irritation, and intestinal distention, no systemic actions were found.

TABLE 2 Subacute and Chronic Toxicity Data

A. Carboxylic Acid Salts (Soaps) $R\text{-}COO^-Na^+$

Surfactant	Species	Dose and application	Exp. duration	Findings	Ref.
Soap (U.S.P.)	Rat	2500, -5000 and 10000 ppm i.w.[a]	120 day	No reaction	14
	Rat	50000 ppm i.w.	120 day	20% of test animals died after 30 days, reduced food intake, reduced weight gain	
Soap (Ivory)	Rat	200 mg kg^{-1} day		No reaction	49

[a]Parts per million = mg/liter water.

B. *n*-Alkyl Sulfates $R\text{-}O\text{-}SO_3Na^+$

Surfactant	Species	Dose and application	Exp. duration	Findings	Ref.
Sodium dodecyl sulfate	Rat	1000 ppm i.w.	120 day	Minimal growth inhibition	14
	Rat	2500 ppm i.w.	140 day	Minimal irritation of alimentary tract	
	Rat	5000 ppm i.w.	140 day	About 50% of test animals dead after 50 days	
	Rat	10000 ppm i.w.	160 day	About 60% of test animals dead after 36 days	
	Rat	50000 and 100000 ppm i.w.	5 day	Absolutely lethal following inflammation of alimentary tract	
	Rat	30 mg day^{-1} oral	160 day	No reaction	14
	Rat	>100 mg kg^{-1} day^{-1} oral	8 wk	Liver damage	26

	Rat	60 mg kg^{-1} day^{-1} oral	5 wk	No reaction	
	Rat	40, -200, and 1000 ppm i.d.[a]	90 day	No reaction	55
	Rat	5000 ppm i.d.	90 day	Increased liver and kidney weights	
	Dog	135 mg kg^{-1} day^{-1} oral	10 mo	No reaction	17
	Mouse	1000 ppm i.d.	3 mo	No reaction	37
	Guinea pig	2000 ppm i.w.	190 day	No reaction	56
	Man	0.1 g kg^{-1} day^{-1} or 6000-9000 mg day^{-1} oral	25 day 38 day	No reaction	33
	Rat	2500, -5000, and 10000 ppm i.d.	2 yr	No reaction	16
	Rat	20000 ppm i.d.	4 mo	No reaction	
	Rat	40000 ppm i.d.	4 mo	Growth inhibition	
	Rat	80000 ppm i.d.	2 wk	Lethal following diarrhea and abdominal distention	
(Duponol)	Rabbit	0.01, -0.1-, and 0.25 of LD_{50} dermal 1 × 50/day	90 day	Dose-dependent local reactions but no cumulative effects	10
Triethanolamin dodecyl sulfate (Texapon T 42) 42%	Rat	70, 250, 750 mg kg^{-1} day^{-1} oral	28 day	Middle-dose local irritation of fore stomach. High-dose irritation of fore stomach, ulceration, increased leucocytes	58
Sodium alkyl sulfate	Dog	2 g day^{-1} oral	10 mo	No reaction	17
n = 12, n = 16, n = 18,	Man	1000 mg day^{-1} oral	8 wk	No reaction	
Oleylic mixture	Man	1000 mg day^{-1} oral	30 day	No reaction	

TABLE 2 *(continues)*

TABLE 2 (Continued)

Surfactant	Species	Dose and application	Exp. duration	Findings	Ref.
Sodium hexadecyl sulfate	Man	3600 mg day^{-1} oral	15-80 day	No reactions, except for vague discomfort of several test persons	46
Sodium alkyl sulfate n = 12-15	Rat	40, -200, -1000 ppm i.d.	90 day	No reaction	55
	Rat	5000 ppm i.d.	90 day	Increased liver, spleen, and kidney weights	

[a]In diet.

C. *n*-Alkyl Sulfates with Branched Hydrophobic Residue (Tergitol Brands) $R\text{-}O\text{-}SO_3Na^+$

Surfactant	Species	Dose and application	Exp. duration	Findings	Ref.
Sodium 2 ethylhexyl sulfate	Rat	2500 and 5000 ppm i.w.	30 day	No reaction	49
(Tergitol 08) 40%	Rat	1000 ppm i.w.	30 day	Hist. kidney damage	
	Rat	20000 ppm i.w.	30 day	Additional albuminuria	
	Rat	40000 ppm i.w.	30 day	Same + decreased intake of water	
	Rabbit	100 mg day^{-1} oral	20 day	Irritation of alimentary tract	30
	Guinea pig	Inhalation			
		8 hr day^{-1} 0.1%	6 day	No reaction	
		0.1%	6 day	Minimal irritation of the lungs	49
Sodium 7 ethyl 2 methyl undecyl 4 sulfate (Tergitol 04) 25%	Rat	25 mg kg^{-1} day^{-1} oral	30 day	No reaction	
	Rat	Higher doses	30 day	Kidney damage	

Sodium 3, 9 diethyl tri-decyl 6 sulfate (Tergitol 07) 25%	Rat	65 mg kg^{-1} day^{-1} oral	30 day	No reaction	
	Rat	Higher doses	30 day	Kidney damage	
(Tergitol 08) 40%	Rat	100, -500, -2500 ppm i.d.	90 day	No reaction	50
	Rat	12500 ppm i.d.	90 day	Liver and kidney damage	
	Rat	100, -400, -1600 and 6400 ppm i.d.	2 yr	No reaction	
	Dog	400, -1600, -6400 ppm i.d.	2 yr	No reaction	

D. Alkylpolyglycolether Sulfates $R\text{-}(OCH_2CH_2)_m\text{-}O\text{-}SO_3^-Na^+$

Surfactant	Species	Dose and application	Exp. duration	Findings	Ref.
Sodium dodecylether sulfate m = 3	Rat	40, -120, -1000 ppm i.d.	90 day	No reaction	55
	Rat	5000 ppm i.d.	90 day	Increased kidney weights	
Sodium alkylether sulfate m = 3	Rat	40, -200, -1000 ppm i.d.	90 day	No reaction	
	Rat	5000 ppm i.d.	90 day	Increased liver weights	
Sodium dodecylether sulfate m = 3	Rat	1000, and 5000 ppm i.d.	2 yr	No reaction	54
	Mouse	5% solution 2 ×/week dermal	105 wk	No reaction	

E. Alkyl benzene sulfonates (ABS) with Branched (B) and Unbranched (L) Hydrophobic Residue $R\text{-}(C_6H_4)\text{-}SO_3^-Na^+$

Surfactant	Species	Dose and application	Exp. duration	Findings	Ref.
ABS	Sucking pig	2000 ppm i.d.	39 day	No reaction	2
B-ABS (Marlon TP 42)	Rat	100 ppm i.w.	100 wk	No reaction	6

TABLE 2 (*continues*)

TABLE 2 (Continued)

Surfactant	Species	Dose and application	Exp. duration	Findings	Ref.
ABS	Rat	100 mg kg^{-1} day^{-1} oral	not reported	No reaction	9
	Dog	1000 mg kg^{-1} day^{-1} oral	6 mo	No reaction	18
	Dog	3000-4000 mg kg^{-1} day^{-1} oral	Repeated application	Gastrointestinal symptoms, death	
	Rat	5000 ppm i.d.	65 day	No reaction	
	Rat	0.1 of LD_{50} day^{-1}	45 day	Increased liver, kidney, and adrenal weights	21
	Rat	700 mg kg^{-1} day^{-1}	2 mo	General symptoms	45
	Guinea pig	2000 ppm i.w.	180 day	No reaction	56
	Mouse	Dermal 10%	101 wk	No reaction	47
	Rabbit	10% 2×/week	115 wk	No reaction	47
	Rat	30 mg kg^{-1} day^{-1}	6 mo	No reaction	39
	Pig	2600 ppm i.d.	not reported	No reaction	44
	Mouse	5-, 50-, 500-, and 5000 ppm i.d.	13 wk	No reaction	48
	Mouse	0.02 g kg^{-1} single dose	6 mo observation	No reaction	
	Mouse	0.05-, 0.10-, 0.15-, and 0.20 g kg^{-1} single dose	6 mo observation	Growth inhibition 4 wk to 6 mo following application	
	Mouse	20 mg kg^{-1} day^{-1} oral	4 wk	84% dead[a]	48
	Mouse	200 mg kg^{-1} day^{-1} oral	4 wk	80% dead[a]	
	Mouse	300 mg kg^{-1} day^{-1} oral	4 wk	80% dead[a]	
	Mouse	400 mg kg^{-1} day^{-1} oral	4 wk	80% dead[a]	
	Mouse	1000 mg kg^{-1} day^{-1} oral	2 wk	100% dead[a]	
	Rat	1000- and 5000 ppm i.d.	13 wk	No reaction[b]	51

L-Sodium dodecyl benzene sulfonate n = 12(10-14) (Marlon BW 2043)	Rat	100 ppm i.w.	100 wk	No reaction	6
L-ABS n = 10-14	Rat	200, -1000, and 5000 ppm i.d.	2 yr	No reaction	8
n = 6 + n = 10 + n = 12, Alkyl (Santomerse E)	Rat	200, -1000-, and 5000 ppm i.d.	90 day	No reaction	32
	Dog	200, -1000-, and 5000 ppm i.d.	90 day	No reaction	
n = 12(9-15)	Rat	50 mg kg^{-1} day^{-1} i.d.	12 wk	No reaction	42
	Rat	250 mg kg^{-1} day^{-1} i.d.	12 wk	Increased liver weights	
Sodium alkylaryl sulfonate	Rat	5000- and 10000 ppm i.d.	4 mo	No reaction	16
	Rat	20000 ppm i.d.	4 mo	Growth inhibition	
	Rat	40000 ppm i.d.	1 wk	Deaths after diarrhea and abdominal distention	
	Man	100 mg day^{-1} oral	4 mo	No reaction	18
(D-40)	Rat	0.1, -0.25-, and 0.5 of LD_{50} day^{-1}	45 day	No reaction	28
	Rabbit	Dermal 20 mg kg^{-1} day^{-1}	30 day	No reaction	
	Rat	1, -10-, and 20 ppm i.d.	6 mo	No reaction	
	Rat	Dermal 70 mg kg^{-1} day^{-1}	30 day	No reaction	
(Oronit wetting agent)	Rat	0.1, -0.25-, and 0.5 of LD_{50} day^{-1}	45 day	No reaction	
	Rabbit	Dermal 20 mg kg^{-1} day^{-1}	30 day	No reaction	
	Rat	Dermal 70 mg kg^{-1} day^{-1}	30 day	No reaction	
	Rat	1, -10-, and 20 ppm i.d.	6 mo	No reaction	
L-ABS	Rat	100, -500-, and 1000 ppm i.w.	2 yr	No reaction	52

TABLE 2 *(continues)*

TABLE 2 (Continued)

Surfactant	Species	Dose and application	Exp. duration	Findings	Ref.
L-ABS (continued)	Rat	700 ppm i.d.	6 mo	No reaction	57
	Rat	2000 ppm i.d.	6 mo	Increased cecal weights, slight kidney damage	
	Rat	6000 ppm i.d.	6 mo	Minimal loss of weight, increased cecal weights, elevated alkaline phosphatase, reduced total protein (blood), kidney damage	
	Rat	18000 i.d.	6 mo	Diarrhea, loss of weight, marked histological changes of cecum, liver, kidney	
B-ABS (Santomerse 85)	Dog	200- and 1000 ppm i.d.	2 yr	No reaction	32
	Dog	5000 ppm i.d.	2 yr	Signs of liver damage	
n = 12(11-13) (Santomerse 3)	Rat	200, -1000-, and 2000 ppm i.d.	2 yr	No reaction	43
	Rat	10000 ppm i.d.	33 day	Growth inhibition	
	Rat	20000 ppm i.d.	33 day	Marked growth inhibition	
B-ABS	Rat	1000- and 5000 ppm i.d.	2 yr	No reaction	53
	Rat	500 ppm i.w.	2 yr	No reaction	
	Rabbit	100 mg day^{-1}	20 day	Irritation of gastrointestinal tract	30
	Rat	700 ppm i.d.	6 mo	No reaction	57
	Rat	2000 ppm i.d.	6 mo	Increased cecal weights, slight kidney damage	

	Rat	6000 ppm i.d.	6 mo	Slight loss of weight, serum changes, kidney damage	
	Rat	18000 ppm i.d.	6 mo	General symptoms, marked cecal damage, marked liver and kidney damage	
n = 12(10-14)	Rat	50 mg kg^{-1} day^{-1} i.d.	12 wk	No reaction	42
	Rat	250 mg kg^{-1} day^{-1} i.d.	12 wk	Increased cecal weights	
ABS + dodecyl sulfate mixture	Pig	9000 mg day^{-1}	206 day	No reaction	27
Sodium dodecyl benzene sulfonate	Rat	250 mg kg^{-1} day^{-1} i.d.	4 wk	Increased liver weights, elevated fasting blood sugar	1

F. Other Anionic Surfactants

Surfactant	Species	Dose and application	Exp. duration	Findings	Ref.
Olefin sulfonates					
n-α-Olefin sulfonate n = 14-18	Rat	1000, -2500-, and 5000 ppm i.d.	2 yr	No reaction	31
Fatty acid condensation products					
Oleic acid methyl tauride (Igepon T)	Rat	20000 ppm i.d.	4 mo	No reaction	16
	Rat	40000 ppm i.d.	4 mo	Growth inhibition	
	Rat	80000 ppm i.d.	4 mo	Deaths	
	Rabbit	10 mg/animal i.v.	?	No reaction	20
	Guinea pig	5 mg/animal i.v.	?	No reaction	

TABLE 2 *(continues)*

TABLE 2 (Continued)

Surfactant	Species	Dose and application	Exp. duration	Findings	Ref.
Fatty acid condensation products (continued)					
Oleic acid methyl tauride	Guinea pig	2000 ppm i.w.	180 day	No reaction	56
fatty acid isethionate (Hostapon KA)	Guinea pig	0.1 g kg^{-1} day^{-1} oral	4 wk	No proven toxic reactions	4
	Mouse				
Sulfosuccinic acid esters					
Diethylhexyl sulfo-succinate	Rat	200-870 mg kg^{-1} day^{-1} oral	6 mo	No reaction	3
	Rat	Higher doses, oral		Diarrhea	
	Rabbit	500 mg kg^{-1} day^{-1} oral	6 mo	Deaths	
	Dog	100- and 250 mg kg^{-1} day^{-1} oral	6 mo	No reaction	
	Dog	Higher doses, oral		Immediate vomiting	
	Monkey	125 mg kg^{-1} day^{-1} oral	6 mo	No reaction	
	Monkey	Higher doses, oral		Immediate vomiting	
	Rat	2500- and 5000 ppm i.d.	2 yr	No reaction	16
	Rat	10000 ppm i.d.	2 yr	Growth inhibition	
	Rat	20000 ppm i.d.	4 mo	Growth inhibition	
	Rat	40000 ppm i.d.	4 mo	Minimal weight increase, deaths	
	Rat	80000 ppm i.d.	1 wk	100% mortality after gastrointestinal symptoms	
	Rabbit	250 mg day^{-1} oral	5 mo	No reaction	30

	Mouse	4 mg day^{-1} oral	7 mo	No reaction	35
	Guinea pig	2000 ppm i.w.	180 day	No reaction	56
Various					
Sodium dodecyl glycerylether sulfonate	Mouse	Dermal, 5% 2×/week	105 wk	No reaction	54
	Rat	1000- and 50000 ppm i.d.	2 yr	No reaction	
Sodium alkyl sulfonate	Rat	0.1 of LD_{50} day^{-1} oral	45 day	No reaction	21
Sodium lauryl sulfoacetate	Pig	Dermal 0.7% 2 ×/day	30 day	No compound related reactions	52c
	Rat	50, 200, 800 mg kg^{-1} day^{-1} oral	28 day	High dose, reduction of body weight gain, middle and high dose reduced food consumption	52a
	Rat	75, 250, 750 mg kg^{-1} day^{-1} oral	13 wk	Middle and high dose gastritis, erosions of the stomach, gavage or "stress" related	52b
Fatty acid methyl ester sulfonate (Texin ES 68)	Rat	100, 300, 900 mg kg^{-1} day^{-1} oral	90 day	Middle dose—the lowest lethal dose. Damage of fore stomach, hematopoetic system, liver. High dose lethal after ematiation, high grade damage of alimentary tract	58
Tallow fatty alcohol sulfate (Sulfopon T 55)	Rat	100, 300, 900 mg kg^{-1} day^{-1} oral	90 day	High dose, damage of liver, kidney, thymus, and fore stomach	58

[a]Toxic myocardial changes: hyperplasia of capillary endothelium; brain: necrotic lesions; liver: histocyte proliferation; lungs: lymphohistocyte infiltration.
[b]20% active content.

The hepatotoxic action of pure sodium dodecyl sulfate has been studied by Rauen et al. (66) with rats as test animals and intraperitoneal application. Glutamate oxalic acetate transaminase (GOT), glutamate pyruvate transaminase (GPT), and sorbitol dehydrogenase (SDH) were used as the parameters, as well as an analysis of free carboxylic acids.

Increases in serum activity taken as criteria, liver damage occurs very soon after intraperitoneal administration of 16.1 mg dodecyl sulfate per animal (Fig. 1). The increase in GOT is significant 60 min after the injection, while significant increases in GPT and SDH occur later: significant increases of all three parameters are observed 4 hr after the application.

Figure 2 shows the changes in serum enzyme activities and indicates that the changes observed are largely reversible within 24 hr. The enzyme activities found in animals pretreated with barbital were found to be of the same orders of magnitude at various time intervals after application.

If the single intraperitoneal dose is increased from 16.1 mg per animal to 32.2 mg sodium dodecyl sulfate per animal, the period required for a total reversal of the serum activities of GOT, GPT, and SDH is increased to about 48 hr after application, as shown in Fig. 3. Also, the free carboxylic acids present in the serum as determined by vapor phase chromatography indicate metabolic disturbance in the liver, the strongest deviation from normal values being found 8 or 24 hr after administration, respectively.

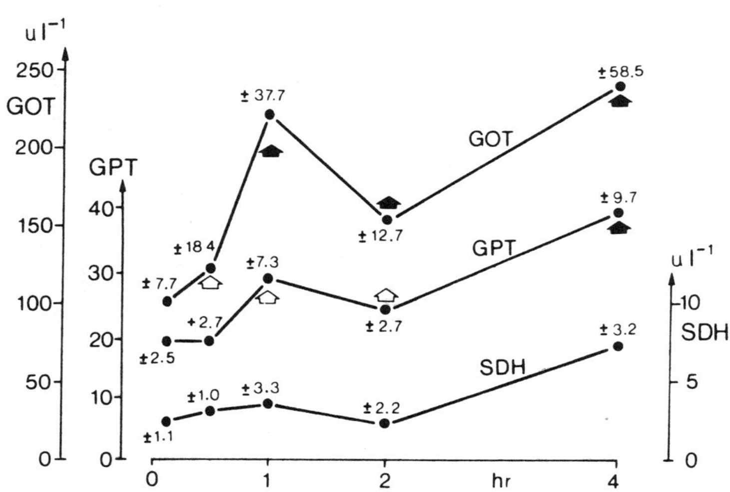

FIGURE 1 Acute hepatotoxic action as a function of time: 16.1 mg Na-dodecyl sulfate/animal 0-4 hr. Open arrows, $P < 0.05$; closed arrows, $P < 0.01$. (Modified from Ref. 66.)

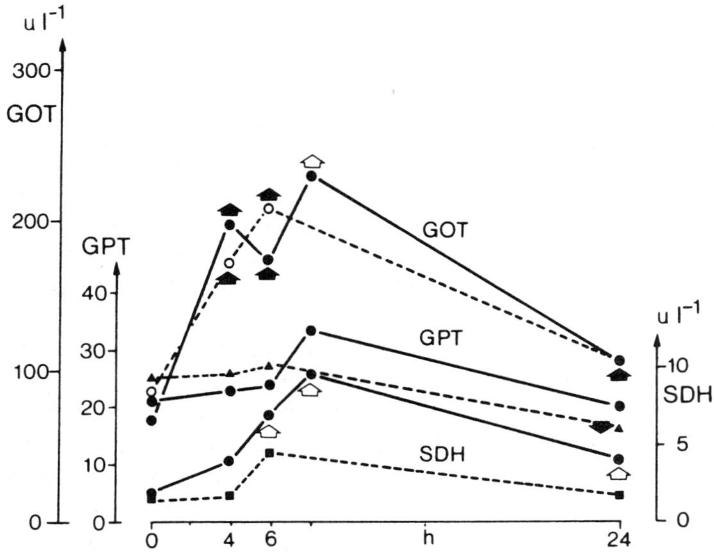

FIGURE 2 Acute hepatotoxic action as a function of time: 16.1 mg Na-dodecyl sulfate/animal 0-24 hr. Open arrows, $P < 0.05$; closed arrows, $P < 0.01$; closed circles, animals with pretreatment; open circles, closed triangles, and closed squares, pretreatment with barbital. (Modified from Ref. 66.)

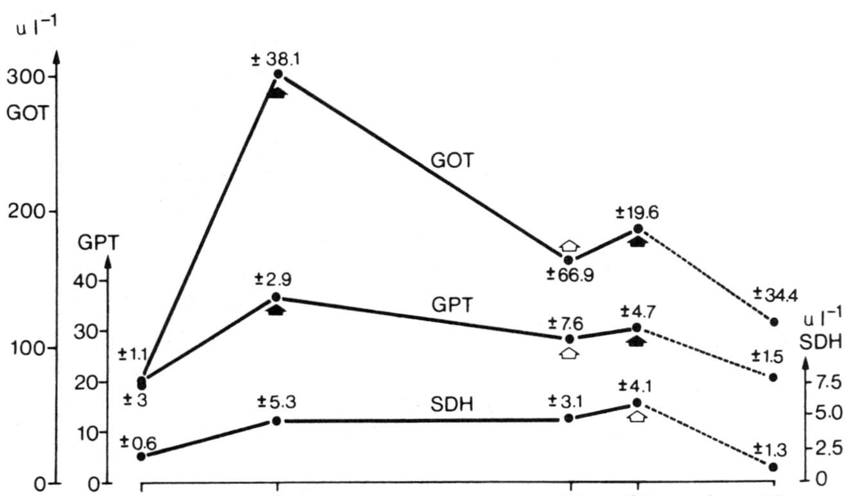

FIGURE 3 Acute hepatotoxic action as a function of time: 32.2 mg Na-dodecyl sulfate/animal, 0-48 hr. Open arrows, $P < 0.05$; closed arrows, $P < 0.01$. (Modified from Ref. 66.)

The influence of sodium dodecyl sulfate and sodium tetradecyl sulfate on the epithelial cells of the small intestine of mice at concentrations of 3×10^{-2} and 3×10^{-3} mol/liter has been studied by Ruska (67). Both substances behaved similarly. Cell elongation, swelling of mitochondria, and a slight widening of the interior of the endoplasmatic reticulum in the Golgi complexes and a fine granulation of the nuclei were observed. Nucleus and cytoplasma structures were retained in finest dispersion. Also, the endoplasmatic reticulum remained unchanged. These findings may be attributed to an increased solubility of the surfactants at interfaces, to the removal of an impermeable membrane, or to increased permeability.

Yoneyama et al. (70) have discussed the influence of alkyl benzene sulfonate on liver and kidney function. The substance was administered to mice and rats as an addition to the drinking water (0.07, 0.02, and 0.6%) and to the diet (0.6 and 1.8%). During the duration of the experiment (9 months) a decrease in liver weights was observed. Also, there were reductions in the activities of liver enzymes, such as glucose-6-phosphate dehydrogenase, lactate dehydrogenase, and glutamate pyruvate transaminase. Finally, the activities of kidney glucose-6-phosphatase and Na-, K-ATP ase were reduced.

A number of cumulative toxic effects were observed by Yoneyama et al. (69) in their 6-month experiments with branched chain sulfonate (ABS) and linear alkyl benzene sulfonate (LAS). Diarrhea, reduced weight increases, and damage to internal organs such as cecum, liver, and kidneys were observed in the test animals which had been fed with 1.8% LAS or ABS in the diet.

The administration of 0.6% of LAS or ABS in the diet led to a slight growth retardation, increases in cecal weights, increased activity of alkaline phosphatase, decreases in the total amount of protein in the blood, and histological kidney changes. Increased cecal weights and slight kidney damage (histological changes) were also observed when LAS and ABS were administered at a level of 0.2% in the diet. The kidney damage, which depended on the amount of surfactant administered, consisted of atrophy of the glomeruli and also a swelling and histiolysis of the glomeruli. The tubuli showed cloudy swelling, vacuolation, and deposition of lipofuscin. Henle loops showed a disposition of PAS-positive material. Also, there was a dilation of the distal tubules, destruction of the uriniferous tubules, and edema of the stroma as well as proliferation into the interstitial connective tissue. The liver was engorged and there was a loss of basoplasma of the liver cells. The mucosal epithelium of the colon was reduced.

The changes observed by Tiba (52) in the internal organs of rats upon 2-year exposure to 0.01, 0.05, and 0.1% LAS in the diet occasionally included pneumonia, local vacuolar degeneration of the cardiac muscle, small round-cell infiltrates in the liver parenchyma, small necrotic lesions in the liver, engorgement of the renal cortex, and small delineated round-cell infiltrates in the region of the glomeruli. These findings are much more consistent with the general histological observations in infected animals and are probably not related to cumulative toxicity of LAS.

The test animals (rats and mice) were also examined histologically in the study of the acute toxicities of L-ABS and B-ABS of Kobayashi et al. (34). Mainly local and dose-dependent changes in the stomach were observed, such as vasodilation and congestion in the submucosa, as well as degeneration of the mucosal epithelium. Similar changes were observed in the small intestine. Flattening of the mucosal epithelium and congestion were observed in the colon. Of the parenchymatous organs only the liver showed a reaction with dilation of the sinusoid spaces and congestion.

Some evidence of liver degeneration was found in dogs fed a dietmixture of 125 mg sodium dodecyl benzene sulfonate (branched) kg^{-1} body weight day^{-1} for 2 years while no toxicity was reported in animals administered 25 mg kg^{-1} day^{-1} (no further details available (32). There was apparently no toxicity in dogs fed 125 mg sodium-ABS (branched) kg^{-1} day^{-1} for 90 days (no further details available) (32). In rhesus monkeys an oral administration of 300 mg ABS (linear C_{10}-C_{13} mixture) kg^{-1} day^{-1} applied simultaneously for 28 days with a subcutaneous dose of 1 mg kg^{-1} day^{-1} produced vomiting and diarrhea. Clinical symptoms were weaker at a combination of oral dose of 150 mg kg^{-1} day^{-1} applied together with a subcutaneous dose of 0.5 mg kg^{-1} day^{-1}. The oral dose of 30 mg kg^{-1}, combined with subcutaneous injection of 0.1 mg kg^{-1} day^{-1} revealed no clinical effects or cellular injury (27a).

Dermal application of about 400 mg ABS (linear, average C 11.7) kg^{-1} for 21 days reduced body weight gain in groups of 20 rats after open exposure for 30 min day^{-1}. There were no observable effects with 100 mg kg^{-1} day^{-1} (12a).

Toxic effects of synthetic LAS in the chicken were reported by Hoshino Sadao et al. (29a). The admixture of the compound did not affect body weight, food intake, organ weights, and sperm concentration at 10-100 ppm, but caused increased GOT in serum (10-15 weeks) as well as total cholesterol level (15 weeks). Serum triglyceride were at week 10 decreased. When LAS admixture was raised to 1500 ppm up to the end of the study (28 weeks) significant decreases in water intake, body weight, testes weight, and sperm concentration in seminal fluid were observed. Furthermore, in this high concentration the cokerels showed significant decrease of liver enzymes and alkaline phosphatase and a significant increase of serum triglyceride level. It was concluded that intake of LAS at 80 mg kg^{-1} would seriously inhibit growth and sexual maturation in the male chicken.

TABLE 3 Histopathology of Tissues of Guinea Pigs Receiving Oral 0.7 g kg^{-1} DSS

	6[a]	21[a]	3[a]	15[a]	30[a]	1[a]
Time of death[c]	2.8	2.8	3.8	4.5	4.6	5.4
Generalized vascular engorgement	+	+	+	+	+	+
Gastric hyperemia and necrosis	+	+	+	+	+	+
Spleen contraction and loss of pmns[d]	+	+	+	+	+	+

[a]Died.
[b]Euthanatized.
[c]In hours following administration of DSS.
[d]Polymorphonuclear leucocytes.

According to an abstract (no further details given) of a Hungarian, approximately 50 mg dodecyl benzene sulfonate kg^{-1} day^{-1} reduced some pancreatic enzymes, among them particularly trypsin and lipase in rats (70a).

After inhalation exposure to 3.59 mg/m³ of dodecyl benzene sulfonate in the workplace atmosphere, blood changes were apparently induced in humans (no further details available) (68a).

Intravenous application of 50 mg keryl benzene sulfonate resulted in minimal respiratory trouble and certain minor changes in blood pressure in the case of cats as the test animals. Major amounts, e.g., 100 mg kg^{-1} lead to slight bradycardia, relaxation of the smooth musculature of the uterus, contractions of the nicticating membrane, and marked respiratory and circulatory problems. Paralysis of the smooth musculature of the gastrointestinal tract was achieved at a concentration of 0.047% keryl benzene sulfonate (61).

In the context of pharmaceutical applications of DSS, Moffat et al. (36) have also carried out toxicological studies in order to determine the proper therapeutic dose levels.

After the application (in water, stomach sound) of 0.65 or 1.0 g DSS kg^{-1}, horses showed symptoms of discomfort within 1 hr, and increased cardiac and respiratory frequency within 3 hr followed by watery diarrhea, abdominal pain, cumulation of intestinal juice, cardiac malfunction, and death within 14-72 hr. Hematological examination showed increased hemoglobin levels as well as an increase in the corpuscular blood elements with a concomitant slight decrease in serum chloride. Doses of up to 200 mg DSS kg^{-1} were recommended as a laxative in horses.

Doses of 0.5-0.98 g DSS kg^{-1} led to growth stagnation and inactivity in guinea pigs. Death is generally observed 4-7 hr after application. The animals show muscular atony, dyspnea, and convulsions prior to death. Increases in the amounts of corpuscular cell elements in the blood—correlated with the

Animal									
20[a]	29[a]	2[b]	23[b]	7[b]	5[b]	4[b]	16[b]	19[b]	24[b]
6.2	10.3	12.5	25.1	47.9	48.3	48.3	95.4	95.7	97.1
+	+	±	–	–	–	–	–	–	–
+	+	+	+	–	–	–	–	–	–
+	+	±	±	–	–	–	±	±	–

death rate—are observed after application of 0.1-0.9 g kg^{-1}, while increased amounts of lymphocytes are observed upon application of 0.7 or 0.9 g DSS kg^{-1}.

Dehydration and hemoconcentration were observed upon dissection. Engorgement of various organs, emphysema of the lungs, and edema were observed in the horses. An accumulation of bile was observed in the liver, and a slight degeneration of the proximal tubuli and medullary hemorrhages were found in the kidneys. Bleeding in the cortex of the suprarenal glands was also observed. Those animals which survived 24 hr after the application showed a loss of the intestinal epithelium with villous contraction. Necrosis of the stomach mucosa as well as generalized changes of the blood vessels and massive infestation of the spleen with polymorphonuclear leucocytes were observed in guinea pigs which had died after the application of 0.5 g kg^{-1}.

The results are summarized in Tables 3 and 4. According to the authors, these findings indicate that further work is required for the clearance of DSS in therapeutic applications.

In further animal experiments lethal doses lead to reduced activity in mice accompanied by muscular weakness and diarrhea (8a,35a).

Oral doses of 500-3000 mg DSS kg^{-1} also induced diarrhea in mice but 250 mg kg^{-1} were tolerated without clinical symptoms (47a). In hamsters, 125 or 250 mg kg^{-1} given oral increased fluid accumulation in small intestine (24a). Tablets containing 30 mg DSS kg^{-1} day^{-1} administered to 8 dogs over a period of 1 year did not affect body weight gain, blood- and biochemistry parameters, urinalysis, or the weight and microscopic examination of a comprehensive range of organs (8a).

Single installation of DSS into trachea of dogs of an aerosol containing 15 mg kg^{-1}, as a 1% solution in ethanol and saline induced a decrease of

TABLE 4 Histopathology of Tissues of Horses Receiving Oral DSS

	Animal						
	1	2	3	4	5	6	7
Age	Mature	3.5 yr	4 yr	8 yr	3 yr	Mature	Mature
Sex	MC[a]	MC[a]	MC[a]	M[b]	MC[a]	M[b]	F[c]
DSS administered[d]	1.0	1.0	1.0	1.0	1.0	0.65	0.65
Time of death[e]	14	15	29.5	24	72	28	44.5
Adrenal cortical hemorrhage	+	+	NE[f]	NE[f]	+	+	+
Generalized vascular engorgement	+	+	+	+	+	+	+
Small intestine:							
Denudation	–	NE[f]	+	+	+	+	+
Villous atrophy	–	NE[f]	+	+	+	+	+
Submucosal edema and congestion	+	NE[f]	+	+	+	+	+

[a]Male castrate
[b]Male
[c]Female
[d]In grams per kilogram
[e]In hours following administration of DSS
[f]Not examined

lung volume, partial lung collapse, and fluid accumulation. After 4 hr, changes in lung function were still detectable, microscopically no organ injury could be seen (37a). An intratracheal instilled dose of 50 mg DSS kg^{-1} still caused fluid accumulation in the lung (6a).

Infants given oral 50 mg DSS kg^{-1} tolerated the dose without clinical symptoms (71). Oral doses of 40, 80, or 200 mg DSS (adult?) revealed weak effects in the stomach lining as a result of a lowering of the transmucosal electrical potential difference (12b).

DSS was given orally in divided doses of 50-300 mg day^{-1} as a laxative for adults and at 5 mg kg^{-1} for children (31a,35b). Oral administration, which may continue for several weeks, was tolerated without any adverse effects or clinical symptoms (12c,14a). In a further oral administration study, diarrhea occured in seven of 471 volunteers treated with unspecified doses of DSS (51a), although 200 mg day^{-1} given orally for an average of 26 days to 34 nonconstipated persons revealed no effect on stool excretion (24b).

Periodic clinical examinations failed to reveal any DSS related symptoms in workmen who handled the compound daily over periods of years (numbers, levels of exposure, and extent of examination are not described) (3).

Berenson and Temple (60) contacted esophageal and intestinal mucosa with 10 and 20% preparations of sodium lauryl sulfate using a catheter or direct operative introduction into the stomach. Edema of the esophageal mucosa as well as congestion and loss of the superficial epithelial cells were observed within 15 min after the application of the 10% preparation by stomach sound. In the stomach, this treatment led to hydropic degeneration, loss of superficial mucosal cells, vascular congestion, edema of the submucosa, as well as occasional focal ulceration. More pronounced changes, such as extensive edema of the submucosa as well as destruction and erosion of the esophageal and gastric mucosa, were observed after the application of the 20% preparation of sodium lauryl sulfate.

These findings suggest that complications due to perforations in the digestive tract are probably not to be expected after the ingestion of sodium lauryl sulfate.

Octyl sulfate, decyl sulfate, dodecyl sulfate, and tetradecyl sulfate, applied intravenously to cats at levels of 4.6, 5.2, 5.7, and 6.3 mg min^{-1} showed different effects on pulmonary hypertension according to Schumacher et al. (68). While slight increases in the peripheral circulatory resistance and decreases in the respiratory minute volume as well as pulmonary hypertension were observed upon administration of 5.2 mg min^{-1} of decyl sulfate or 5.7 mg min^{-1} of dodecyl sulfate, both octyl sulfate and tetradecyl sulfate were found to be inactive or only weakly active with respect to the above mentioned parameters. Pulmonary hypertension in cats is to be regarded as a symptom of direct action on the smooth musculature of the blood vessels and the bronchi. Blood sugar levels were not affected in these experiments.

The acute toxicity of sodium lauryl sulfate upon intravenous administration has been studied in depth by Cascorbi et al. (11) using dogs and rabbits (intraabdominal) as test animals. The following histological changes in lungs, kidneys, and livers were observed upon 3-hr infusion of 10 mg/100 ml of the substance in an emulsion:

Lungs: slight to pronounced atelectasis, acute congestion, pneumonia
Kidneys: cloudy swelling, one case of acute pyelonephritis
Liver: cloudy swelling, necrotic cells, acute congestion

Similar observations were made with rabbits upon intravenous administration of sodium lauryl sulfate at a level of 10 mg/100 ml (5 × 30 min).

In a 90-day test, Smyth et al. (50) found that 1.25% sodium-2-ethyl-hexyl sulfate applied with the diet gave rise to diffuse cloudy swelling of the proximal kidney tubuli in rats. Also, a cloudy swelling of the central parts of the hepatic cords and increased fatty infiltration of the liver cells were observed.

Sodium alkyl sulfate added to the drinking water at a concentration of 0.5-500 mg liter^{-1} leads to a decrease in the phagocytic activity of the neutrophile leucocytes in blood after 1-8 months, according to Linkov et al. (65).

A detergent based on sodium lauryl sulfate was fed to dogs at a dose of 0.5 mg kg^{-1} day^{-1} for a period of 110 days by Bartorelli et al. (59). The kidneys were found to be the target organs with histomorphological changes relating to glomerular reactions, such as ectasies of the glomerular structures and hydropic degeneration of the glomerular epithelium, changes of Bowman capsule, and to vasodystrophic arterial changes in the renal cortex as well as to intertubular capillary hyperemia.

Peritoneal inflammation was observed in rats upon repeated intraperitoneal injection of sodium-2-ethylhexyl sulfosuccinate at a level of 0.01 kg week^{-1} for an experimental duration of 12 weeks (64). The increased permeability of pulmonary vessels which led to a swelling of the connective tissue was among the most pronounced histological findings.

The dermal application of acutely toxic doses (580 mg kg^{-1}) of sodium lauryl sulfate to rabbits led to irritation and tremor, tonic-clonic spasma, and respiratory collapse, according to Carson and Oser (10). Apparently these phenomena depend on the attainment of a critical concentration in the blood.

Death was observed within 4 weeks upon repeated and daily oral applications of 0.02-1 g alkyl benzene sulfonate to white mice (48). The dead animals showed reactive hyperplasia of the capillary endothelium in various organs, particularly in the brain, heart, lungs, and liver, where histological changes were also observed as necrotic lesions and histiocyte proliferation.

The potential danger of inhalation upon contact with anionic surfactants has also been considered. Therefore, Hall (62) has studied the action of sodium lauryl sulfate (Duponol C), sodium alkylaryl sulfonate (Aerosol OS), sodium-2-ethylhexyl sulfate (Tergitol 08), and dialkyl sulfosuccinate (Aerosol MA) upon repeated inhalation with guinea pigs as test animals. The concentrations used were 0.1, 0.5, and 1% and the duration of the exposure was 8 hr day^{-1} on 6 consecutive days. Trachea, bronchi, and lung tissue were examined histologically. All substances gave rise to a dose-dependent dyspnea of the test animals with the most severe effects in the 1% groups.

Slight to moderate respiratory symptoms were observed in the 0.5% groups, while the 0.1% aerosols were tolerated without toxic effects. Three of the four test animals which had received 1% Duponol C died after short exposure. The other surfactants applied at the 1% level did not cause any deaths, but led to dyspnea and lethargy. None of the test animals died upon inhalation of the anionic surfactants applied at the 0.1 and 0.5% levels. The onset of the respiratory symptoms was rapid and they persisted for 1-3 hr after exposure. The observations are summarized in Table 5.

With the exception of a harder consistency of the lungs upon inhalation of surfactants at the 1% level, there were no findings in the dissections. However, dose-dependent changes in the lung tissue could be observed histologically. There was a pronounced segregation of polymorphonuclear cells and macrophages, as well as interstitial fibroblast proliferation, degeneration and desquamation of the pulmonary epithelium, and a proliferation of the alveolar epithelium, which could be observed in the lungs exposed at the 1% level (exception Tergitol 08). Large alveolar regions were filled with inflammatory material.

The guinea pigs exposed to the 0.5% solutions showed only weak histological reactions, while 0.1% solutions were tolerated also without histological findings. In the case of Tergitol 08 only minimal changes were observed in the lungs even at the 1% level.

Acyl lactylates are a relatively new group of anionic surface-active compounds prepared from naturally occurring raw materials. Murphy et al. (36a) recommended this group of substances for cosmetic applications as the order of primary irritation is very low, and median oral lethal values are in excess of 5000 mg kg^{-1}. There are no published data on long-term systemic toxicity studies.

VII. CONCLUDING REMARKS

Although anionic surfactants can be absorbed through the gastrointestinal tract, following ingestion the results of systemic acute toxicity studies of a number of important substances indicate a low order of systemic toxicity.

TABLE 5 Concentration of Wetting Agents and Results of Experiment

	Concentration of wetting agent, %											
	Duponol C			Aerosol OS			Aerosol MA			Tergitol 08		
Result	0.1	0.5	1.0	0.1	0.5	1.0	0.1	0.5	1.0	0.1	0.5	1.0
Number of animals	4	4	6	4	4	4	4	4	4	2	2	2
Number of deaths	0	0	4	0	0	0	0	0	0	0	0	0
Degree of dyspnea	0	2+	4+	0	2+	4+	0	2+	4+	0	1+	2+

These findings agree with clinical observations. The major dose-dependent effect of anionic surfactants is restricted to local damage of mucous membranes of the gastrointestinal tract. High doses lead to vomitus and diarrhea.

The possibility of chronic toxicity has been studied by intense investigations in different species. Results obtained from animal experiments and exposures of up to 2 years revealed, even at high-dose levels, no risks attributable to possible intake under user conditions. Some long-term investigations which have been conducted with human volunteers lead to favorable results. No serious side effects have been observed and significant accumulation of this compound class in the organism has never been detected.

Based on the available data and taking into account the possible oral intake (see Chap. 8) anionic surfactants can be regarded as nonhazardous for human health as currently used in household products.

REFERENCES

1. M. Antal, *Z. Ernährungsw., 10,* 35-39(1970).
2. W.M. Beeson, T.W. Perry, and P.J. Reynolds, *J. Animal Sci., 12,* 619-622 (1953); *Chem. Abstr., 48,* 3492b(1954).
3. A.E. Benaglia, E.J. Robinson, E. Utley, and M.A. Cleverdon, *J. Ind. Hyg. Toxicol., 25,* 175-180(1943).
4. J. Borneff, *Arch. Hyg. Bakteriol., 141,* 578-595(1957).
5. G. Bornmann and A. Loeser, *Fette, Seifen, Anstrichm., 63,* 938-940(1961).
6. G. Bornmann and A. Loeser, *Fette, Seifen, Anstrichm., 65,* 818-824(1963).

6a. C.E. Brandenberg et al., *J. Appl. Physiol., 60,* 253(1986).

7. V.K.H. Brown and C.M.C. Muir, *Tenside, 7,* 137-139(1970).
8. E.V. Buehler, E.A. Newman, and W.R. King, *Toxicol. Appl. Pharmacol., 18,* 83-91(1971).

8a. M.T. Case et al., *Drug Chem. Toxicol., 1,* 89(1977).

9. I. Cabejszek, M. Rybak, and B. Styczynski, *Roczniki Panstwowego Zakladu Hig., 14*(4), 307-316(1963); Chem. Abstr., 60, 6121c(1964).
10. St. Carson and B.L. Oser, *Soc. Cosmet. Chem., 15,* 137-147(1964).
11. H.F. Cascorbi, F.G. Rudo, and G.G. Lu, *J. Pharm. Sci., 52,* 803-805(1963).
12. T. Chruściel, M. Chruściel, Z. Kleinrok, and Z. Herman, *Med. Pracy., 10,* 235-246(1959); Chem Abstr., 13425(1960).

12a. I.W. Daly et al., *Fd. Cosmet. Toxicol., 18,* 55(1980).

12b. K.M. Cochran et al., *Gut, 18,* A422(1977).

12c. G.S. Clayden, *Lancet, ii,* 787(1978).

12d. Cosmetic, Toiletry and Fragrance Association (CTFA) (September 5, 1978). Submission of data by CTFA. Unpublished acute oral, dermal, ocular, and mucous membrane testing of a bath product containing 50% sodium lauryl sulfoacetate (2-18-1). [Published in *J. Am. Toxicol., 6,* 261-277(1987).]

13. S.M. Drachev, A.A. Bylinkina, and V.F. Garshenin, *Sb. Vysoke Skoly Chem.-Technol. Praze. Technol. Vody, 8*(1), 161-166(1965); *Chem. Abstr., 64,* 4789h (1966).
14. S. Epstein, A.H. Throndson, W. Dock, and M.I. Tainter, *J. Amer. Dental Ass., 26*(9), 1461-1471(1939).
14a. A.M. Fain et al., *Sth. Med. J., 71,* 677(1978).
15. J.K. Finnegan and J.B. Dienna, *Proc. Sci. Sec. Toilet Goods Ass., 20,* 16-19 (1953).
16. O.G. Fitzhugh and A.A. Nelson, *J. Amer. Pharm. Ass., 37*(1), 29-32(1948).
17. S.J. Fogelson and D.E. Shock, *Arch. Int. Med., 73,* 212-216(1944).
18. S. Freeman, M.W. Burrill, T.-W. Li, and A.C. Ivy, *Gastroenterology, 4,* 332-343(1945).
19. L.E. Gale and P.M. Scott, *J. Amer. Pharmac. Ass.* (*Scient. Ed.*), *42,* 283-287(1953).
20. M. Garcia de Jalon and P. Garcia de Jalon, *Farmacoterap. Actual.* (*Madrid*), *3,* 833(1946).
21. V.F. Garshenin, *Maslob.-Zhir. Prom., 29*(10), 20-23(1963); *Chem. Abstr., 60,* 6617h(1964).
22. L. Gershenfeld and B. Witlin, *Amer. J. Pharm., 113,* 215(1941).
23. Chr. Gloxhuber, *Fette, Seifen, Anstrichm., 74,* 49-57(1972).
24. Chr. Gloxhuber, *Arch. Toxicol., 32,* 245-270(1974).
24a. G.W. Gullikson et al., *J. Pharmac. Exp. Ther., 220,* 236(1982).
24b. J. Goodman et al., *J. Chron. Dis., 29,* 59(1976).
25. G.C. Hall, *Amer. J. Dis. Child., 80,* 408-412(1950).
26. E.H. Hatton, L.S. Fosdick, and J. Calandra, *J. Dental Res., 19,* 87-92(1940).
27. H. Havermann and H. Rutter, *Arch. Tierernähr. Beih., 5,* 60-76(1954).
27a. R. Heywood et al., *Toxicology, 11,* 245(1978).
28. C.H. Hine, H.H. Anderson, I. Finley, and J.G. Moore, *J. Amer. Pharm. Ass., 42*(8), 489-495(1953).
29. S.H. Hopper, H.R. Hulpieu, and V.V. Cole, *J. Amer. Pharm. Ass., 38*(8), 428-432(1949).
29a. Hoshino, Sadao, Wakita, Masaaki, et al., *Nippon Kakin Gakkaishi, 18,* 51-59 (1981).
30. W.C. Hueper, *Arch. Pathol., 38,* 381-391(1944).
31. B. Hunter and H.G. Benson, *Toxicology, 5,* 359-370(1976).
31a. C.M. Hyland, J.D. Foran, *Practitioner, 200,* 698(cited in Martindale, 1982).
32. J.H. Kay, F.E. Kohn, and J.C. Calandra, *Toxicol. Appl. Pharmacol., 7(6),* 812-818(1965).
33. J.B. Kirsner and R.A. Wolff, *Gastroenterology, 2,* 93-101(1944).
34. H. Kobayashi, H. Ichikawa, T. Fujii, N. Yano, T. Konno, K. Hiraga, H. Nakamura, Y. Watanabe, and S. Mimura, *Tokyo Toritsu Eisei Kenkyusho Kenkyu Nempo, 24,* 397-408(1972).
35. E. Lorenz, M.B. Shimkin, and H.L. Stewart, *J. Nat. Cancer Inst., 1,* 335-360 (1940).
35a. L. Lindholm, N. Svedmyr, *Acta Pharmac. Tox., 15,* 373(1959).

35b. Martindale, *The Extra Pharmacopoeia,* 28th Ed., The Pharmaceutical Press, London, 1982, p. 1440.

36. R.E. Moffatt, L.L. Kramer, D. Lerner, and R. Jones, *Can. J. Comp. Med., 39,* 434-441(1975).

36a. L.J. Murphy et al., *Drug & Cosmetic Industry, 5,* 35(1978).

37. J.A. Nissim, Nature, *187,* 308-310(1960); *Chem. Abstr., 56,* 1968d(1962).

37a. G.F. Nieman and C.E. Brendenberg, *J. Appl. Physiol., 58,* 129(1985).

38. Y. Ogura and J. Tamura, *A.R. Inst. Food Microbiol.* (*Chiba Univ.*), *20,* 89-94(1967).

39. K. Okahara, Y. Kurayuki, M. Iseki, S. Taniguchi, and A. Yamada, *Shokuhin Eiseigaku Zasshi, 4,* 15-31(1963); Chem. Abstr., *59,* 14478b(1963).

40. K.J. Olson, R.W. Dupree, E.T. Plomer, and V.K. Rowe, *J. Soc. Cosmet. Chem., 13,* 469-479(1962).

41. D.L. Opdyke, F.H. Snyder, and H.L. Rubenkoenig, *Toxicol. Appl. Pharmacol., 6,* 141-146(1964).

42. B.L. Oser and K. Morgareidge, *Toxicol. Appl. Pharmacol., 7,* 819-825(1965).

43. O.E. Paynter and R.J. Weir, *Toxicol. Appl. Pharmacol., 2*(6), 641-648(1960).

44. T.W. Perry, W.M. Beeson, and B.W. Vosteen, *J. Animal Sci., 12,* 310-315 (1953); *Chem. Abstr., 47,* 12544g(1953).

45. B. Polak, A. Preis, M. Dulhős, M. Beranova, O. Bilek, and B. Pohl, *36th Int. Congr. Industr. Chem., Brussels* (1967).

46. J.F. Prudden, *Gastroenterology, 15,* 426-437(1950).

47. U. Saffiotti, P. Shibik, and D.L. Opdyke, *Toxicol. Appl. Pharmacol., 4,* 763-769(1962).

47a. E. Sakurai et al., *Folia Pharmac. Jap., 76,* 293(1980).

48. J. Sanz Ibanez and A.T. Jiminez-Castellanos, *Arch. Inst. Farmacol. Exptl., 16*(2), 107-113(1964); Chem. Abstr., *63,* 12212(1965).

49. H.F. Smyth, Jr., J. Seaton, and L. Fischer, *J. Ind. Hyg. Toxicol., 23,* 478-483(1941).

50. H.F. Smyth, Jr., C.P. Carpenter, C.S. Weil, and J.M. King, *Toxic. Appl. Pharmacol., 17,* 53(1970).

51. F.H. Snyder, D.L. Opdyke, J.F. Griffith, and H.L. Rubenkoenig, *Toxicol. Appl. Pharmacol., 6,* 133-140(1964).

51a. R.B. Stewart and L.E. Cluff, *Am. J. Dig. Dis., 19,* 1(1974).

52. S. Tiba, *J. Food Hyg. Soc.* (Jap), *13,* 509-515(1972).

52a. Toxicol. Laboratories Limited (August 1985). Submission of unpublished data by CTFA. Twenty-eight day oral range finding study in the rat. Report Reference No. SUS/1/C. Test Material: 910-74. Test Sponsor: Stepan Company. [Published in *J. Am. Coll. Toxicol., 6(3),* 261-277(1987).]

52b. Toxicol. Laboratories Limited (March 1986). Submission of unpublished data by CTFA. Ninety-day oral toxicity study in the rat. Report Reference No. SUS/2/C. Test Material 910-74. Test Sponsor: Stepan Company. [Published in *J. Am. Coll. Toxicol., 6(3),* 261-277(1987).]

52c. T. P. S. Inc. (October 20, 1978). Submission of data by CTFA. Unpublished subacute dermal toxicity studies in swine on an acne wash containing Sodium

Lauryl Sulfoacetate (2-18-9). [Published in *J. Am. Coll. Toxicol., 6(3),* 261-277(1987).]

53. T.W. Tusing, O.E. Paynter, and D.L. Opdyke, *Toxicol. Appl. Pharmacol., 2,* 464-473(1960).
54. T.W. Tusing, O.E. Paynter, D.L. Opdyke, and F.H. Snyder, *Toxicol. Appl. Pharmacol., 4,* 402-409(1962).
55. A.I.T. Walker, V.K.H. Brown, L.W. Ferrigan, R.G. Pickering, and D.A. Williams, *Food Cosmet. Toxicol., 5,* 763-769(1967).
56. G. Woodard and H.O. Calvery, *Proc. Sci. Sect. Toilet Goods Assn., No. 3,* 1-4(1945).
57. M. Yoneyama, T. Fujii, M. Ikawa, H. Shiba, Y. Sakamoto, N. Yano, H. Kobayashi, H. Ichikawa, and K. Hiraga, *Tokyo Toritsu Eisei Kenkyusho Kenkyu Nempo, 24,* 409-440(1972).
58. Hemkel KGaA,, unpublished results.
59. A. Bartorelli, P. Capua, and A. La Comba, *Arch. Sci. Med., 131*(3), 139-145 (1974).
60. M.M. Berenson and A. R. Temple, *Clin. Toxicol., 8*(4), 399-404(1975).
61. T.L. Chrusciel, Z. Kleinrock, M. Chrusciel, R. Brus, W. Janiec, and H. Trzeciak, *Med. Pracy., 17*(2), 104-108(1966).
62. G.C. Hall, *Amer. J. Dis. Child., 80,* 408-412(1950).
63. V.V. Ivanov, E.A. Ievleva, and L.I. Sych, *Vestn. Dermatol. Venerol., 9,* 62-66(1975).
64. S. Kosmider, M. Wieczorek, J. Szymaniec, M. Zielinski, H. Machalska, *Bromatol. Chem. Toksykol., 7*(3), 375-383(1974); *Chem. Abstr., 82,* 52320s(1975).
65. Yu. V. Linkov, M.M. Avkhimenko, and N.N. Litvinov, *Gig. Vody Sanit. Okhr. Vodoemoy,* 72-76(1973); *Chem. Abstr., 82,* 68972(1975).
66. H.M. Rauen, H. Schriewer, O. Kopp, and U. Peters, *Arzneim.-Forsch., 23,* 134-136(1973).
67. C. Ruska, *Z. Zellforsch., 53,*867-878(1961).
68. K.-A. Schumacher, H.G. Classen, P. Marquardt, and L. Späth, *Arzneim.-Forsch., 22,* 1489-1491(1972).
68a. A. Sawinsky et al., *Egészsegtudomány, 28,* 105(1984).
69. M. Yoneyama, T. Fujii, M. Ikawa, H. Shiba, Y. Sakamoto, N. Yano, H. Kobayashi, H. Ichikawa, and K. Hiraga, *Tokyo Toritsu Eisei Kenkyusho Kenkyu Nempo, 24,* 409-440(1973).
70. M. Yoneyama, Y. Mabuchi, M. Ikawa, H. Kobayashi, and H. Ichikawa, *Tokyo Toritsu Eisei Kenkyusho Nempo,* 27-32, 105-112(1976); *Chem. Abstr., 86,* 151209b(1977).
70a. J.N.A. Zsinka, *Egészsegtudomány, 18,* 9(1974).
71. J.L. Wilson and D.G. Dickinson, *J. Am. Med. Ass., 158,* 261(1955).

4

Observations on Tolerance in Cases of Ingestion

URSULA OBERDISSE, AXEL HAHN, AND SIGRID RITTER *Berlin Poison Control Center, Berlin, Germany*

I. INTRODUCTION

Household washing and cleaning products almost never contain anionic surfactants without also containing other additives. Such products are mixtures of organic and inorganic compounds, which include a wide variety of synthetic detergents (anionic, nonionic, cationic, amphoteric surfactants), builders (sodiumcarbonate, -silicate, pentasodiumtrisphosphate, sodiumaluminiumsilicate), bleaches (sodiumperborate, -percarbonate), and others. An outline of the formulation of household and washing care products is summarized in Table 1.

II. EPIDEMIOLOGY

Because they are used on a large scale, washing and cleaning products are accessible to children in practically all households. Therefore, cases of ingestion

TABLE 1 Outline Formulation of Household and Washing Care Products

Ingredients	Heavy-duty detergents			Specialty detergents without phosphate					
					For woolens				
	Powdered, without phosphate, %	Powdered, with phosphate, %	Liquid, without phosphate, %	Powdered, %	Powdered, %	Liquid, %	For curtains, powdered, %	Manual dish-washing, liquid, %	General purpose cleaner, liquid, %
Anionic surfactants	5-10	5-10	10-15	10-20	10-20	10-20	0-10	10-40	2-10
Nonionic surfactants	0-10	2-10	10-15	2-5	1-5	2-5	2-7	0-10	2-10
Sodium phosphate	—	10-35	—	—	—	—	—	—	—
Zeolithe	20-25	0-20	—	10-35	10-25	—	15-25	—	—
Sodium carbonate	5-20	0-10	—	—	—	—	5-15	—	—
Sodium perborate	10-25	10-25	—	—	—	—	0-15	—	—
Aluminum silicate	—	—	—	0-25	0-20	—	—	—	—
Sodium silicate	2-6	2-6	—	—	—	—	2-6	—	—
Sodium sulfate	2-15	2-15	—	20-50	20-50	—	10-35	—	—
Alcohols	—	—	6-12	—	—	0-10	—	0-6	0-10
Soaps, silicone oil	0.5-5	0.5-5	10-15	1-5	0-5	0-10	1-5	—	—
Urea	—	—	—	—	—	—	—	0-10	0-10
Tetrasodiumdiphosphate	—	—	—	—	—	—	—	—	0-10

Some products may contain as further ingredients: citric acid 0-6%, polycarboxylic acids 0-4%, EDTA 0-5%, cellulose ethers 0-3%, enzymes 0-1.5%, optical brighteners 0.1-0.3%, fragrances up to 1%, coloring agents $<1\%$, stabilizers 0.5%, preservatives 0-0.1%, sodium chloride 0-6%.

occur frequently, usually among small children, particularly between 1 and 4 years of age (2) and, in our experience, with the highest risk between the 1st and 2nd year (Fig. 4). In the literature of recent years there are no precise epidemiological data related to cases of detergent ingestion. Therefore, we refer to the 1988 Annual Report of the American Association of Poison Control Centers (AAPCC) National Data Collection System, in which 137,240 cases of ingestion of cleaning products are summarized. This number is about 10% of a total of 1,368,748 human exposures reported by 64 participating Poison Control Centers. Only in 29,218 cases were washing and cleaning products, based on anionic/nonionic surfactants, involved (10). This is about 2% of cases reported. The contribution of different cleaning products to these cases is summarized in Figure 1.

The vast majority of poison exposures was accidental (about 98%); suicidal intention was existent in only about 1%. As shown in Fig. 2, most cases of accidental exposure involved children under 6 years of age. The majority (about 70%) of all cases reported to poison control centers showed none or only minor symptoms (Fig. 3). From the 17 poison control centers of the Federal Republic of Germany, no comparable data are available. In 1989, the Berlin Poison Control Center registered 44,772 poison calls from the Federal Republic of Germany. In a random sample of 10,414 inquiries taken between January 1 and March 31, 1989, we noted 697 cases (about 7%) of ingestions with cleaning products, based mainly on anionic and nonionic surfactants. In comparison with these data, our records from 1963 to 1975 (56,711 inquiries) show a frequency of ingestions of these substances of about 5%, and in 99% of the exposures none or only minor symptoms were exhibited (Table 2).

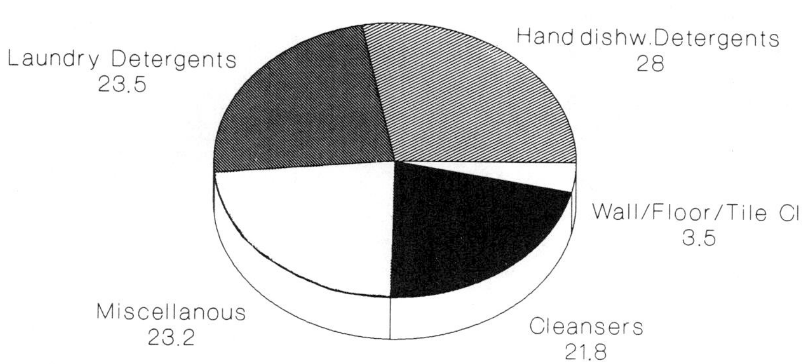

FIGURE 1 Contribution (in %) of the different cleaning products to 29,218 cases of ingestions (AAPCC, 1988).

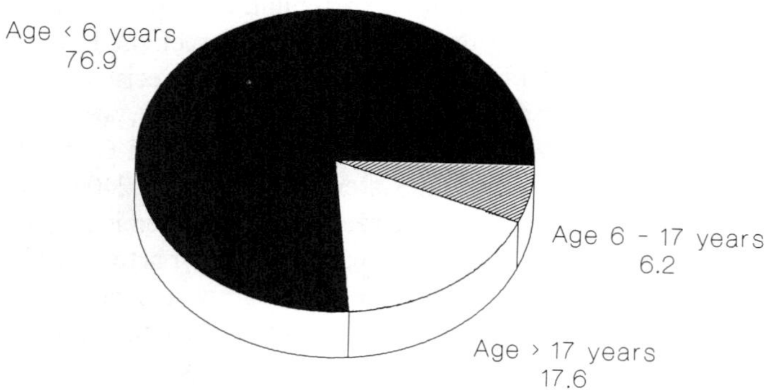

FIGURE 2 The distribution of age (in %) of the 29,218 cases reported by the AAPCC in 1988.

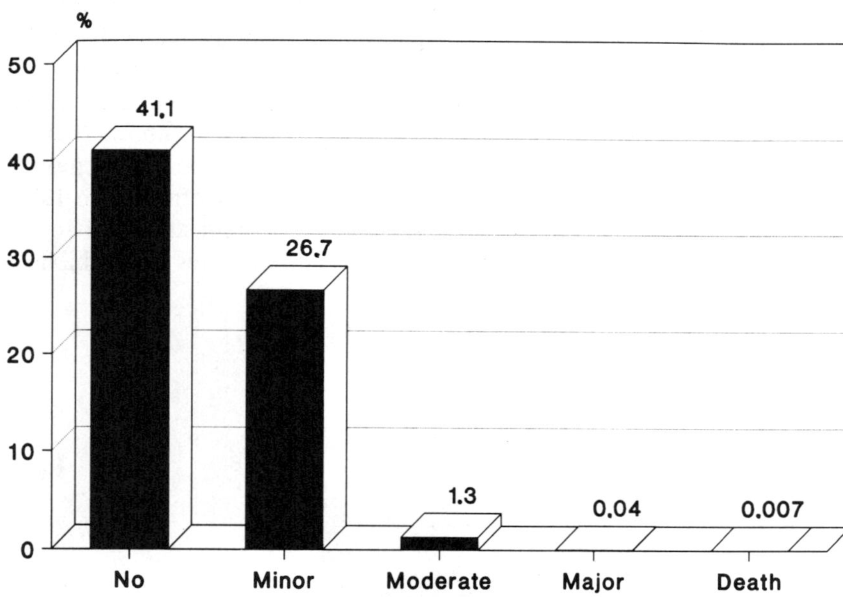

FIGURE 3 The classification of the 29,218 cases reported by the AAPCC in 1988 according to the severity of symptoms.

TABLE 2 Severity of Symptoms in 2942 Accidental Ingestions of Household Products Based Mainly on Anionic/Nonionic Surfactants of 56,711 Inquiries to the Berlin Poison Control Center Between 1963 and 1975

	Symptoms				
Type of product	None	Minor	Moderate	Severe	Total
Dishwashing detergents (manual)	1164	375			1539
Special detergents for delicate fabrics	56	17			73
Heavy-duty detergents	232	69			301
Other cleaning products	30	3			33
Household cleaners	745	228	21	2	996
Total	2227	692	21	2	2942

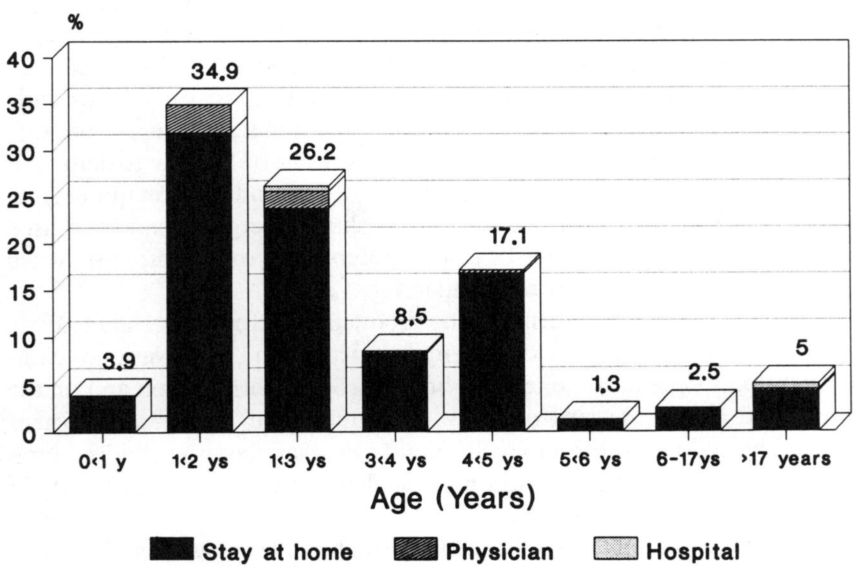

FIGURE 4 The distribution of age in 697 cases with ingestions of detergents and the management recommendation by the Berlin Poison Control Center (Random sample 1/1/90-3/31/90).

The analysis of our present data from 1989 revealed that about 93% of the cases of ingestion pertained to the group of children under 6 years and only 5% to adults (over 17 years). In our statistical data, we make no distinction between symptoms of different severity, whether minor, moderate, or major, or of fatal outcome, respectively. Instead, we use, as an indirect parameter, the management recommendations given by our center (Fig. 4). In 94% of cases of ingestion with anionic/nonionic surfactants, the patient could remain at home because we did not anticipate severe symptoms. In about 5% of all cases, we advised consultation with a practitioner and, only in 1% of cases, we counseled admission of the patient to hospital.

III. CLINICAL ASPECTS

Because the most common washing and cleaning products used in households contain not only anionic surfactants but usually a mixture of these substances with other ingredients of differing concentrations, we can only state clinical and toxicological experience obtained with these mixtures.

The degree of systemic toxicity caused by anionic surfactants is considered by different authors to be relatively small. Velvart (15) has stated that, for adults, the oral ingestion of anionic surfactants up to 1.0 g/kg of body weight is relatively harmless, and Dreisbach (4) estimates the maximum safe amount for children to be 0.1-1.0 g/kg. As a result of the low toxicity and the emetic properties of liquid detergents, Giesker and Trautmann (11) reported on 15 cases in which liquid handwashing detergents had been used as an emetic in cases of ingestion with potentially toxic substances when Syrup of Ipecacuanhae was not available.

Most of the other ingredients in the products affect not systemic toxicity but, rather, the intensity of skin and mucosa irritation. The severity of clinical symptoms depends on one hand on the product formulation and on the other hand on the ingested quantity. The amount of powdered detergent ingested by children is especially difficult to estimate because children typically ingest only a small proportion of the detergent due to the difficulty in handling of the powdered form.

The ingested quantity of liquid products can be better judged and is usually seldom more than two mouthfulls, where the quantity of a mouthfull is about 0.3-1.0 ml/kg of body weight.

The symptoms observed after ingestion of cleaning and washing products (in our experience and in the literature) differ between powder and liquid preparation. The topical effects of all detergents may range from mild irritation of the mucosa of the mouth, esophagus and stomach to corrosivelike effects seen only with powder preparations (1,4,5,9,13,15). The mucosa-irritating effect of liquid detergents (pH 7-9) is considered to be lower than that of granules because the builders give the powder detergents more alkalinity (pH 9.5-10.5, in solution!).

The most common symptoms after ingestion are nausea or vomiting shortly after the ingestion; salivation is sometimes described. Further symptoms are diarrhea, occasional abdominal discomfort, and intestinal distension. If surfactant-containing powders or aerosols are inhaled or if surfactant-containing fluids are aspirated, there is the risk of edema of the lungs and suffocation as a consequence of the formation of foam in the pulmonary system. In some of our cases, we found coughing but without evidence of pulmonary aspiration as checked by auscultation and X-ray examination. Aspiration or inhalation of powders or liquids is feared mostly because of possible pulmonary complications.

Serious symptoms caused by detergents reported in the literature were due mostly to aspirations or when the substances were ingested with suicidal intention.

1. The report by Krepler et al. (8) concerns permeability changes as a consequence of the aspiration of a surfactant solution. There is supporting evidence from animal experiments that the degree of resorption of the solution via the intestinal tract is considered to be less than via the lungs.

A 16 month old boy fell into a detergent solution (0.5% Sunil). During clinical observation, serious deterioration of his overall condition was observed after stomach lavage. There was a high-grade edema of the face and the entire body was edematous. The child was not responsive and cyanotic and his breathing was irregular and groaning. The pulse was weak and frequent, whereas cardiac action was normal. Moist rales were occasionally heard over the lungs. There was a further rapid deterioration with unconciousness attributed to acute cerebral edema. With suitable therapy the case was kept under control. This report does not allow any conclusion as to which extent the symptoms described could be attributed to the anionic and nonionic surfactants present in the Sunil detergent or to other ingredients such as condensed sodium phosphates, sodiumsilicate, sodiumperborate, sodiumsulfate, or optical brighteners.

2. A similar case has been described by von Clarmann (3), a case of poisoning by the inhalation/aspiration of a bubble bath. In this case edema of the lungs was also observed with nonhomogeneous spotty shadows over both lungs. In addition, there was a transient increase of the transaminases (SGOT) up to a maximum of 144 U/ml. The state of acute lung edema as well as the bronchopneumonia was brought under control after 36 hr of treatment. The possibility of an aspiration of surfactants (including anionic surfactants) should be kept in mind when therapeutic measures are taken.

3. Other effects on mucous membranes, such as those observed with detergent enemas ranging from mild proctitis to gangrene in the rectal area (12,14), have not been reported in cases of accidental ingestions; we also have not observed effects of this kind on the gastrointestinal tract among the cases we studied.

4. Velvart (15) described the attempted suicide by a man with 39 g of a detergent powder that was dissolved in 200 ml of water. The symptoms observed included vomiting, hyperemia of the oropharynx, and also a pseudomembraneous circular lesion of the upper part of the esophagus.

IV. THERAPEUTIC RECOMMENDATIONS

In cases of accidental ingestion of detergent only small amounts of tea or water should be swallowed immediately for the purpose of dilution, with the subsequent addition of a carminative (Dimeticon) within one hr. If excessive vomiting or severe diarrhea occur (which is very seldom), the water and electrolyte status has to be maintained and symptomatic treatment may be indicated.

Gastric emptying is required only if large amounts have been swallowed, such as in suicidal attempts where spontaneous vomiting did not occur. In such a case the administration of dimeticon dissolved in water must be given before gastric emptying in order to minimize the risk of foam aspiration. *In our experience these cases are very rare.*

If the patient has clinical signs of inhalation or aspiration, observation and the treatment of pulmonary symptoms are necessary. In case of eye contact, irrigation should be performed immediately until the substance is completely removed. After contact with preparations of high alkalinity, the eye should be rinsed continuously for about 15 min as first aid and then the patient transferred to an ophthalmologist immediately. In case of skin contact the product should to be removed with only water.

V. CONCLUDING REMARKS

Detergent ingestions with anionic surfactants are generally harmless if small quantities are swallowed. Therefore, we would urgently warn against overtherapy. On the other hand, these substances should not be considered to be totally without risk, particularly because there is the possibility of their aspiration into the respiratory tract.

REFERENCES

1. J.M. Arena and R.H. Drew (Eds.), *Poisoning. Toxicology, Symptoms, Treatments,* 5th ed., C.C. Thomas, Springfield, Ill, 1986, pp. 681-693.

2. J.C. Calandra and O.E. Fancher, *Cleaning Products and Their Accidental Ingestion,* The Soap and Detergent Association, New York, 1976.
3. H.v. Clarmann, Vergiftung durch Inhalation bzw. Aspiration eines Schaumbades, *Fortschr. Med., 83,* 137-138(1965).
4. R.H. Dreisbach and W.O. Robertson, *Handbook of Poisoning,* 12th ed., Appleton & Lange, Norwalk/Los Altos, 1987.
5. M.J. Ellenhorn and D.G. Barceloux, *Medical Toxicology, Diagnosis and Treatment of Human Poisoning*, Elsevier,, New York.
6. C. Gloxhuber, Vergiftungen mit Reinigungs- und Spülmitteln, *Med. Heute, 17,* 232-235(1968).
7. C. Gloxhuber, Über die Behandlung nach Einnahme von Wasch- und Reinigungsmitteln, *Med. Welt, 19,* 351-357(1968).
8. P. Krepler, J. Eichler, and W. Kovac, Studium einer akuten Permeabilitätsstörung nach Aspiration einer Detergentienlösung, *Wien. Klin. Wschr., 75,* 787-795(1963).
9. E.G. Krienke, K.E.v. Mühlendahl, and U. Oberdisse (Eds.), *Vergiftungen im Kindesalter,* Enke, Stuttgart, 1986, pp. 202-207.
10. T.L. Litovitz, B.F. Schmitz, and K.C. Holm, 1988 Annual report of the American Association of Poison Control Centers National Data Collecting System, *Amer. J. Emergency Med., 7,* 495-545(1989).
11. H.C. Mofenson, J. Greensher, and T.R. Caraccio, Ingestions considered nontoxic, *Emergency Med. Clin. North Amer., 2,* 159-174(1984).
12. B.F. Pike, P.J. Phillippi, and E.H. Lawson, Soap colitis, *New Engl. J. Med., 285,* 217-218(1971).
13. A.R. Temple, Bleach, soaps and detergents. Clinical management of poisoning and drug overdosage. In: L.M. Haddad and J.F. Winchester (Eds.), Saunders, Philadelphia, 1983.
14. P.B. Toffler and J.M. Barry, Colonic mucosal slough following detergent enemas, *Amer. J. Gastroenterol., 58,* 638-640(1972).
15. J. Velvart, Vergiftungen durch Haushaltsprodukte, *Dt. Apotheker Z., 129,* 509-515(1989).

5

Local Tolerance (Animal Tests): Mucous Membranes and Skin

WERNER KÄSTNER Henkel KGaA, Düsseldorf, Germany

I. INTRODUCTION

Anionic wetting agents (anionic surfactants) belong to the larger group of surface-active agents (surfactants). They may also be classified as detergents, synthetic detergents (syndets), wetting agents, or surfactants. All surfactants possess the ability to reduce the surface tension of water and to solubilize water-insoluble materials. These properties are related to the effects on the mucous membranes and the skin.

The anionic surfactants are widely used in washing, cleansing, and rinsing agents in the home and in industry. They are not only important for the cleaning of surfaces and objects, especially textiles, but also come into contact with the skin and hair of humans and animals, and are used in the region of the mucous membranes. Various cosmetic products contain them as solubilizers or stabilizers in aqueous systems, which are additional routes by means of which the products come into contact with the human skin.

In the case of certain cosmetics which contain anionic surfactants, contact with the mucous membranes is intentional. When cleaning surfaces and objects with products designed for such applications, skin contact can seldom be avoided. Contact with the mucous membranes can take place intentionally (intimate wash lotions, teeth and mouth care products) or can be accidental. Therefore, in order to protect the consumer from harmful effects, it is necessary to subject the raw materials and finished products to toxicological and dermatological examinations. The toxicological and dermatological tests must be so planned that they give information concerning effects which could occur in case of accident or misadventure (results concerning such studies are described elsewhere).

Results concerning local compatibility of anionic surfactants obtained in animal experiments based on published investigations are presented in this chapter; in certain individual cases, however, our own unpublished results are given.

Because the products containing anionic surfactants are applied to various regions of human body, the local tolerance must address itself to the problem of compatibility to the skin and to the mucous membranes found at the body orifices. Whereas the reactions taking place on mucous membranes are almost exclusively of a primary nature, those taking place on skin are primary, i.e., skin changes occur a short time after application, or secondary, i.e., effects occur after a longer time of exposure or after repeated applications. Especially interesting among the secondary irritations are those resulting from the formation of specific antibodies. Substances which cause the formation of antibodies are also known as sensitizing agents. As a number of anionic surfactants are used in both household and cosmetic products which come into daily contact with certain regions of the body (skin or mucous membranes), it is of utmost importance to ensure that the anionic surfactants in such products do not possess sensitizing properties. Because of the importance of recognizing sensitizing substances, the subject is treated in Sec. IV under the general topic of skin irritation.

Anionic surfactants can reveal irritant behavior when brought into contact with living tissue, which is dependent on chemical structure of the surfactant, its concentration, duration of contact, and other factors. (Concerning soaps and alkyl sulfates, see Ref. 1.) Table 1 lists some of the fields of application of anionic surfactants, together with their concentration in formulated products and their concentrations for recommended use. That contact with the undiluted product is always a possibility must be taken into consideration when testing the product.

A. Concentrations for Use of Anionic Surfactants

It is evident from the data in Table 1 that the anionic surfactants are applied in wide concentrations; thus it is not surprising that some authors have examined diluted product for their local compatibility, while others have examined the undiluted materials, depending on the region of application in question. Many investigators have determined the highest concentration tolerated without reaction, whereby the time factor (time of exposure) plays an important role when comparing results. It was found that the frequency and intensity of testing of a substance were generally dependent on the practical importance of the substance. This is the reason that some alkyl sulfates and sulfonates have been tested repeatedly and others only once or twice.

TABLE 1 Products, Concentrations, Frequency, and Duration of Contact with Human Skin

Product	Conc. product, %	Conc. application, %	Contact with human skin[a] A	B	C
Household					
Heavy-duty	10-30	0.04-0.12		+	
Light-duty	18-40	0.09-0.40		+	
Special-duty	5-30	0.03-0.30		+	
Presoak	5-20	0.03-0.20	+		
Washing paste	30-50	15-25	+		
Dishwashing					
Liquid hand dishwashing	12-40	0.01-0.05			+
Powdered dishwashing	20-40	10-20			+
Hard surface cleanser	3-15	1-8		+	
Cosmetic					
Shampoo	5-20	0.5-2	+		
Bubble bath	10-30	0.001-0.003		+	
Shower bath	10-30	1-3	+		
Hair dyeing	up to 5	2.5		+	
Tooth paste	2	2			+
Soap, solid	90-95	8			+
Soap, liquid	10-30	5-15			+
Shaving soap, solid	90	up to 5		+	
Shaving cream	60	up to 5		+	
Hand cleanser	5-50	up to 125			+
Syndet soaps	60	6			+
Intimate care	20	up to 5			+
Skin cream and sun protection cream	up to 5	up to 5		+	
Cold cream	5-10	2.5-5	+		
Miscellaneous					
Hand disinfectants	up to 10	up to 5			+
Car shampoo	15	0.15	+		

[a]A = brief or occasional contact; B = long or permanent contact; C = brief repeated daily contact.

B. List of Trade Names of the Products Cited

As the anionic surfactants examined are often technical products that cannot always be exactly defined chemically, the literature often refers to a trade name rather than a chemical description. Frequently, but not always, information concerning the chemical composition is given. Table 2 lists all trade names mentioned in the following three sections by the various substance

TABLE 2 Trade Names

Trade name	Carbon chain length	Cation	AS%[b] or form	
Alkyl sulfates[a]				
Duponol C	—	Sodium	—	Powder
Duponol ME dry	—	Sodium	—	Powder
Tergitol 08	2-ethylhexyl	Sodium	40	
Tergitol 4	Branched C_{14}	Sodium	28	
Tergitol 7	Branched C_{17}	Sodium	27	
Sipon LS	—	Sodium		Liquid
Sulfopon T-55	C_{16}-C_{18}	Sodium	55	
Texapon K-12	C_{12} (pure)	Sodium	90	
Texapon L-100	C_{12} (pure)	Sodium	99	
Texapon Z high conc.	C_{12}-C_{18}	Sodium	90	
Texapon V high conc.	C_{12}-C_{16}	Sodium	90	
Texapon MGS	C_{12}-C_{16}	Magnesium	30	
Wetanol	—	Sodium	—	—
Texapon A-400	C_{8}-C_{14}	Ammonium	34	
Texapon ALS	C_{12}-C_{18}	Ammonium	30	
Sipon L-20	—	Ammonium	—	—
Texapon MLS	C_{12}-C_{14}	MEA[c]	33	
Texapon DLS	C_{12}-C_{14}	DEA[c]	30-32	
Sipon LT	—	TEA[c]	40	
Duponol WAT	—	TEA	—	
Texapon TH	C_{8}-C_{14}	TEA	47	
Texapon T-42	C_{12}-C_{14}	TEA	42	
Alkyl ether sulfates[a]				
Texapon N-40 (2EO)	C_{12}-C_{14}	Sodium	27-28	
Texapon N-25 (2EO)	C_{12}-C_{14}	Sodium	27-28	
Triton X-30	Arylalkyl	Sodium	—	
Texapon K-14S special (3EO)	C_{12}-C_{14}	Sodium	30	
Sipon ES	—	Sodium	27	
Texapon ASV (7EO)	C_{12}-C_{18}	Sodium and magnesium	28-30	

Trade name	Chemical structure[d]
Diverse alkyl sulfates	
Triton X-30	Sodium salt of arylalkyl ether sulfate
Triton X-135	An alkylated monosodium benzene sulfate
Ultra-wet 30-DS	An alkylated monosodium benzene sulfate

TABLE 2 *(continues)*

TABLE 2 (Continued)

Trade name	Chemical structure[d]
Sulframin DR	Sodium salt of an hydroxyalkylamidoalcohol sulfate
Sulframin DT	Sodium salt of alkylaminosulfate
Sulframin LW	Sodium salt of alkylaminosulfate
Triton X-301	Sodium alkylaryl polyether sulfate
Alkylaryl sulfonates and other alkyl sulfonates	
Areskap 100	Monobutylphenyl phenol sodium monosulfonate
Aresket 300	Monobutyl diphenyl sodium monosulfonate
Aresklene 400	Dibutylphenyl phenol sodium disulfonate
Santomerse 3	Dodecyl benzene sodium sulfonate
Santomerse D	Decyl benzene sodium sulfonate
Aerosol OS	Isopropyl naphthalene sodium sulfonate
Nacconol NRSF	Sodium alkylaryl sulfonate
NAS	A purified mixture of alkylaryl sodium sulfonates
D-40	Alkylaryl sulfonate (Mol wt 352)
OWA (Oronite Wetting Agents)	Polyalkylaryl sulfonate (mol wt 375)
Ultrawet 60L	Triethanolamine salt of alkylaryl sulfonate
Triton X-200	Alkylaryl polyether sulfonate
Entsufon	Sodium octyl phenoxyethoxyethyl ether sulfonate
E-4161	Sodium salt succinic acid half-ester sulfonate
Diverse Anionics	
Igepon A	Sodium lauryl isethionate
Igepon T	Oleic acid methyl taurate
Igepon TN-74	Sodium *N*-methyl-*N*-acyl laurate
Aerosol O.T.	Dioctyl ester of sodium sulfosuccinate
Sulfatex	Sodium and potassium salts of a sulfated fatty alcohol amide

[a]Alkyl implies lauryl, or other alkyl chain lengths, which are noted when known.
[b]AS% = active substances, noted when known.
[c]MEA = monoethanolamine; DEA = diethanolamine; TEA = triethanolamine.
[d]Description or the declaration of the authors were used.

classes. The chemical compositions are also given, as many as are available in the literature. In further discussion, preference is given to chemical formulas, if known, over trade names.

Because of the greater sensitivity of the mucous membranes and the serious consequences of severe mucous membrane damage, e.g., in the eye in comparison to skin damage, first mucosa compatibility, then skin compatibility, and then sensitizing properties of anionic surfactants are described.

See the list at the end of this chapter for abbreviations used throughout the text.

II. TESTING OF MUCOSA COMPATIBILITY

A. General Remarks

1. Scope of test

The evaluation of mucosa compatibility is of special importance for anionic surfactants because a number of cosmetics when used can or should come into contact with the eyes (or other mucous membranes, e.g., bubble and shower baths, shampoos, toothpastes, intimate care products). Furthermore, surfactants are present in many household products that can have accidental eye contact (washing,, rinsing, and cleansing products). The testing of mucosa compatibility of surfactants must therefore be carried out with the following goals:

To allow marketing of only those products that do not cause damage to the mucous membranes when used correctly

To make predictions concerning likely effects of accidental contact of the mucous membranes with the surfactant, together with the first-aid recommendations

2. Test methods and systems of scoring

The common test animal for testing the mucous membrane tolerance on the eye is the albino rabbit. A definite quantity of substance, either undiluted or as a solution, is instilled into the conjunctival sac of one eye, and either left there or rinsed out with water after a certain time. The right eye is usually treated; the left eye generally remains untreated and serves as a control. The eyes were inspected for inflammatory symptoms on the cornea, the iris, and bulbar and palpebral conjunctiva, at fixed time intervals. Numerical scores are assigned to lesions observed according to a standard scoring system.

The examination and evaluation of the results follow several procedures. The method of Draize et al. (2) is frequently used. This method has been adopted by the U.S. government (3-5). A more frequently used method has been described by Neuman and Aalbers (6). These methods have been subject to criticism (7-9). A uniform evaluation of substances by different laboratories presents difficulties (7,10,11), as well as an objective conclusion (6,7,12). Therefore studies emanating from different authors can be compared only with difficulty. However, official test procedures for the classification of substances, including mucous membrane tolerance, are being developed by various government agencies. It may be expected that the difficulty of comparing results from different laboratories will be problematic.

According to the regulations of the Food and Drug Administration (FDA) (3), based on the procedure developed by Draize et al. (2) and Draize (13), the mucosa toxicity of a product is determined using the eyes of nine albino rabbits (Draize test). One-tenth of a milliliter of the test substance is instilled into the conjunctival sac of the right eye, while the other eye remains untreated and serves as a control. The treated eyes of three rabbits remain unwashed. The six remaining animals are divided into two equal groups. Two or four seconds after instillation of test substance, both the treated and untreated eyes of these three rabbits of each group are rinsed out with 20 ml of lukewarm water. The injuries are assessed either macroscopically or by means of a hand slit-lamp, or of a binocular loupe or a special microscope 24, 48, 72, 96, and 168 hr after application of the substance. The evaluation is carried out according to Scheme 1. The points are summarized as shown in the scheme.

The original Draize method for the assessment of potential hazards of chemicals and products to the eyes was adopted and modified by various regulatory bodies. With the publication of the OECD Test Guideline No. 405 (13a) a harmonization of the different existing eye-testing methods was achieved (Dose level: 0.1 ml or 100 mg undiluted liquids, solids, and pastes; Number of animals: at least 3; Contact time: no rinsing, wash out after 24 hr if appropriate; Observation time: duration not fixed, but should be long enough to determine (ir)reversibility of effects, max. 21 days; Scoring: at observation time 1, 24, 48, 72 hr; Score responses on following scale: cornea 0-4, iris 0-2, conjunctiva redness 0-3, conjunctiva chemosis 0-4). The intensity of the irritant response in different regions of the anterior eye and the level of the response during the first 3 days after exposure are the basis of classification according to the Council Directive 61/548/EEC (13b). When the mean intensity of irritancy exceeds defined levels, two risk phrases are identified (R 36: irritating to eyes, and R 41: risk of serious damage to eyes). One of the critical points of this method is the application volume. The instillation of 0.1 ml of test material into the rabbit eye leads to an overestimation of the human hazard. The volume of 0.1 ml greatly exceeds both the capacity of the rabbit eye and the quantity of material which may enter the human eye accidentally.

A series of studies has been carried out to evaluate methods of modifying existing ocular test protocols, with the aim of reducing the number of animals used, and lessening the severity of the response with the minimum effect on the predictive value of the tests. Initial studies showed that when the dose administered is reduced from 100 to 10 μl, the severity of the response is reduced but the rank order of severity of a series of test compounds is unchanged (13c,13d).

Griffith et al. (13e) compared the human hazard reported in Grant (13f) to the response in rabbit eye tests of different chemicals that included surfactants. They came to the conclusion that a dose volume of 0.01 ml more accurately predicted human hazard. The extensive surveys of consumer and employee eye response to a variety of detergent products and the comparison

SCHEME 1 Scale for Scoring Ocular Lesions

Cornea	
Opacity degree of density (area most dense taken for reading)	
No opacity	0
Scattered or diffuse area, details of iris clearly visible	1
Easily discernible translucent areas, details of iris slightly obscured	2
Opalescent areas, no details of iris visible, size of pupil barely discernible	3
Opaque, iris invisible	4
Area of cornea involved	
One-quarter (or less), but not zero	1
Greater than one-quarter, but less than half	2
Greater than half, but less than three-quarters	3
Greater than three-quarters, up to whole area	4
A × B × 5 Total maximum = 80	
Iris	
Values	
Normal	0
Folds above normal, congestion, swelling, circumcorneal injection (any or all of these or combination of any thereof), iris still reacting to light (sluggish reaction is positive)	1
No reaction to light, hemorrhage, gross destruction (any or all of these)	2
A × 5 Total maximum = 10	
Conjunctiva	
Redness (refers to palpebral conjunctiva only)	
Vessels normal	0
Vessels definitely injected above normal	1
More diffuse, deeper crimson red, individual vessels not easily discernible	2
Diffuse beefy red	3
Chemosis	
No swelling	0
Any swelling above normal (includes nictitating membrane)	1
Obvious swelling with partial eversion of lids	2
Swelling with lids about half-closed	3
Swelling with lids about half-closed to completely closed	4
Discharge	
No discharge	0
Any amount different from normal (does not include small amounts observed in inner canthus of normal animals)	1
Discharge with moistening of the lids and hairs just adjacent to lids	2
Discharge with moistening of the lids and hairs, and considerable area around the eye	3
Score (A + B + C) × 2 Total maximum = 20	

Maximal total score = 110 points for the eye is the sum of all scores obtained for the cornea, iris, and conjunctiva.
Source: Ref. 2.

of results from rabbit tests using 0.01 ml of the test material with the findings by the results of an eye study with human volunteers led to the same conclusion (13g-i).

Most of the literature on the low-volume approach regarding detergent-type products appears to result in a more accurate prediction of the human response under normal conditions of exposure. However, in the absence of supportive human data, it is not possible to extend this conclusion in a broader way to a wider range of materials (13j).

A suggested method by which the number of rabbits used in ocular irritancy evaluation might be reduced is to pre-test material for its primary skin irritation so that further ocular testing might not be required.

A literature survey (1968-1982) was conducted by Williams (13c,13d) to find all DuPont compounds that were severe primary skin irritants and had been tested also for ocular irritancy. Of the 60 materials that met both of these criteria, 39 were severe eye irritants, 15 were nonirritants to minimal-irritants, and 6 were moderate irritants. Therefore, severe primary skin irritation was predictive of a similar degree of ocular response for only 65% of the materials tested. The author came to the conclusion that the results obtained from primary skin irritation tests probably do not provide a reliable index of ocular irritation potential.

The irritant effects on skin and eyes of 145 detergents products were compared by Seabough et al. (13d1). All 145 detergents were rabbit eye irritants; 82% of those were either moderate or severe. Almost half (47%) of the detergents were either primary skin irritants or corrosives. These studies show that results of primary skin-irritancy testing do not give a good prediction for eye irritating potential.

Gilman et al. (13d2) demonstrated by a comparison of the Primary Irritation Index and the ocular irritation scores on petrochemicals, neat product ingredients, household, and cosmetic products that the hypothesis that a chemical or product found to be irritating to animal skin need not to be tested in the Draize Eye Irritancy Test but simply assumed to be eye irritant is often invalid. Other investigators (13d3) published similar results.

In the last 15 years some test methods have been developed to test the irritating potential of chemicals without using animals. These in vitro methods can be classified into three groups: cell cultures, organ cultures, and other procedures.

The model system that most closely resembles the in vivo test is the isolated eye maintained in a perfusion system (13k-m). Another suggested in vitro model is that of the isolated, spontaneously contracting ileum (13m,n).

Bell et al. (13n1) made a proposal how to test shampoos for eye irritation. Detergent concentrations are the principal factor influencing findings in the rabbit eye and an active matter content of 2.5-3% (usually representing a

20% aqueous dilution of shampoo) gives the best discrimination between formulations whilst producing no signs of distress in the animals. The authors recommend a test concentration of 20% without subsequent rinsing for typical anionic shampoos. A control formulation of equivalent concentration, known to be well tolerated under marked conditions should be tested together with the new formulations. For preliminary screening, an in vitro test using freshly isolated buccal mucosa cells from human volunteers may be useful. Irritancy is assessed by the proportion of cells showing loss of visible nuclei when examined by phase-contrast microscopy.

The most important of these methods in predicting in vivo ocular irritancy are the cytotoxicity tests and the test on the chorionallantoic membrane of fertilized incubated hen's eggs (HET-CAM-Test).

Various cell cultures have been utilized to measure the effects of known in vivo eye irritants on their morphology and biochemical characteristics. In any approach of this nature, one is always faced with the assumption that it must be more informative to use cell cultures derived from ocular tissue. But numerous experiments showed that rabbit corneal epithelial and endothelial cells gave similar results found with other types of cells such as Balb/c 3T3 murine fibroblasts, HepG2 human hepatoma cells, human epithelioid cell lines (HeLa, HEp2), Chinese hamster V79 lung fibroblasts (13o,13p). The investigators found that, although the cell lines and primary cultures differed in absolute sensitivity, the overall ranking of the various toxicants was similar and that the results correlated well with in vivo data when the in vitro parameters measured included trypan blue exclusion, morphological assessment, total cell protein, and uridine or neutral red uptake (13q-v).

In connection with testing of anionic surfactants for their mucous membrane irritancy, only the Chorionallantoic Membrane Test will be described shortly, because this test method has a good chance for predicting the ocular-irritant potential.

The test material in a suitable form (solution), is applied to the chorioallantoic membrane of fertilized hen's eggs, which have been incubated for 10-14 days. The immediate and subsequent effects on the membrane are noted, which include: hyperaemia and haemorrhage, lysis of the blood vessels, lesion size, and coagulation of the protein in and around the blood vessels. The speed of appearance and the degree of change are the scores for estimation of the irritating potential. This method was used for testing surfactants and the results were compared with the results of the OECD Test Guideline No. 405. A good correlation was observed (13w).

In contrast to these findings, Parish (13l) and Price et al. 13x) tested other chemicals and came to the conclusion that there was no correlation between the intensity of the chick chorionallantoic membrane (CAM) response and the severity of in vivo eye responses.

In an attempt to develop a test method not dependent on the subjective experience of the testing personnel, and in order to enable general comparisons to be made, Neumann and Aalbers (6) altered the method already described. This standard procedures uses 2% aqueous sodium lauryl sulfate as a standard substance. This substance is regarded as possessing an intermediate degree of irritation and would be used as the standard against which the other products would be compared. The experimental animal to be used is the albino rabbit; 0.05 ml of the test solution is instilled in one or both eyes of the animals. By reducing the application volume from 0.1 ml to 0.01 ml (low-volume method) in the rabbit eye test (Draize eye test), a closer correlation with the reactions of human eyes (human eye data) will be achieved (13e,13g-i,13y).

An objective procedure for quantitating eye irritation base upon changes of corneal thickness was published by Kennah et al. (13z).

The characteristic irritation level for each hour is arrived at by addition of the numerical scores for redness, chemosis, and discharge of the conjunctiva with subsequent multiplication of these values by the corresponding number of hours. By addition of these hourly values, the ocular lesion (irritating index) is obtained as an average of 6-10 eyes. The irritating index of the standard product, sodium lauryl sulfate (specially purified), is given as 10. The concentrations of the substances to be tested should be so chosen that the reactions have disappeared within 6-10 hr, as such degrees of irritation render optimum reproducibility. In this way, excessive lesions and overshadowing of the primary irritations by secondary effects, such as microbial effects, are avoided. The scoring of ocular lesions deviates from that of Draize (2) (see Scheme 2).

B. Changes of Mucous Membranes through Influence of Anionic Surfactants

Because of the absence of keratin in the mucous membranes, the epithelial tissues are very sensitive. The phenomena caused by anionic surfactants are not very different macroscopically from those generated by other chemicals. However, they are less characteristic than those caused by acids or bases or substances, which have a tanning effect analogous to that of formaldehyde. The destruction of tissues by highly concentrated anionics surfactants resembles more a colliquation necrosis than a coagulation necrosis. Although the surfactants, in analogy with acids, have a proteindenaturing effect resulting in coagulation, they are also capable of dissolving difficult water-soluble proteins.

In vitro tests carried out on excised beef corneas with various anionic surfactants using a concentration of 0.1 M (approximately 3%) in water have

SCHEME 2 Scale for Scoring Ocular Lesions

Redness of conjunctiva	
Normal aspect	0
Vessels slightly injected	½
Slight enlargement of vessels, slight general redness	1
Stronger enlargement of vessels and increase of blood vessels, general redness	2
Individual blood vessels no longer detectable, deep red color	3
Chemosis	
Normal aspect	0
Just a little visible swelling	½
Slight glassy swelling	1
Stronger swelling, a little turbid, slight lifting of eyelid	2
Strong opaque/milky swelling, eyelid partially closed because of swelling	3
Whitish secretion	
No secretion	0
Some small, whitish flocks	1
Several individual flocks of secretion	2
Coherent mass of secretion	3

Source: Ref. 6.

shown that the substances do not loosen the contact between the corneal epithelium and the stroma, but appear to attack the epithelial cells themselves, gradually transforming the epithelium into a slimy mass. This takes place progressively, commencing at the surface and extending to the basal layer (14). This action on the epithelial cells may be analogous to the hemolytic action of the anionic surfactants on blood cells (15,15a).

1. Gross visible symptoms

The effects of anionic surfactants on mucous membranes depend on chemical structure, concentration, and duration of contact. Inflammatory symptoms such as redness, chemosis, and pain are observed. However, the eye possesses various types of tissue, which react differently on account of variations in blood supply.

Generally, the first symptom observed after application of the test substance to the eye of the experimental animal is pain or discomfort. In the case of rabbits, a blepharospasm is seen, lasting from seconds to minutes, depending on the irritating activity of the substance. The conjunctival mucous membranes initially exhibit an increased injection or expansion of the blood vessels, which results in damage of the cell walls. A more intensive irritation causes diffuse redness (erythema) where differences are seen from light to dark red color, facilitating a differentiation concerning the degree of irritation. In addition, an edema is observed, which may lead to a swelling of the eyelid (chemosis). This swelling can result in a total closure of the eye. A further criterion for evaluation is the secretion formation. Whereas the human eye reacts to the smallest irritation by producing tears, which rinse the product out of the conjunctival sac, or at least strongly dilute it, such a secretion from the rabbit eye is of secondary importance. The secretion proceeds slowly, and is viscous or slimy, and in later stages even purulent. This has caused several investigators to conclude that the rabbit eye is more sensitive than the human eye (9,16-18) and the animal methods overestimate the human response to accidental eye exposure (13e,13g-i,13y). Therefore it has been suggested that other test animals should be included when testing for mucous membrane compatibility is necessary (17-19,19a). Extreme irritation can result in the appearance of necrosis of the conjunctiva, sklera, or cornea.

A slight irritation of the cornea results in changes which cannot be seen macroscopically. More intense irritations cause a corneal opacity or dulling of corneal luster. The degree and extent of clouding as well as its reversibility or irreversibility are a measure of the intensity of the irritation. Extensive destruction of the corneal epithelium may lead to ulcerations of the cornea, with the result of blood vessels growing into the cornea (keratitis), with subsequent formation of a leucoma.

The visible processes taking place on the iris concern the expansion of blood vessels, the resultant tissue swelling, and the absence of the pupillary reflex. According to Aalbers and Newmann (6), irritant activity after instillation of diluted anionic surfactants, e.g., 2% sodium lauryl sulfate, commences 10-30 min after application and is visible by reddening of the conjunctiva.

2. Microscopic visible symptoms

Slight corneal lesions may be made visible using the fluorescein method. Sites of epithelial damage retain the fluorescein despite rinsing with water, so that the corneal stippling are visible using UV light. More severe damage, resulting in damage to the whole corneal epithelium, results in the observation of large areas of fluorescence. Aids to the judgment of corneal damage are the slit-lamp and special microscopes. Occasionally, the rabbits are sacrificed after a certain time following the application and histological sections of the eyes are prepared and microscopically examined.

Burton (7) and Kennah et al. (13z) describe a method for objective measurement of eye irritation. The use of Draize's scheme, especially for the assessment of corneal lesions, results in subjective values, which often differ widely. Therefore, these authors prefer the measurement of corneal thickness, which correlates well with the results of the Draize test. The authors report good correlation between transparency (grade of opacity), thickness, and hydration of the cornea.

Measurements are carried out at specific time intervals after application with a special slit-lamp. (The normal corneal thickness amounts to 0.346 ± 0.02 mm for rabbits weighing 1300-2000 g and 0.382 ± 0.017 mm for rabbits weighing 2000-2700 g.) Experiments concerned with the K^+/Na^+ exchange on the cornea in connection with the action of surfactants have also been carried out (20).

Weltman et al. (21) used ophthalmoscopic, photographic, and histological methods to assess the changes in the rabbit eye. These studies were especially useful in revealing erosions, loss of stratified epithelial layers of cornea, and the leucocytic infiltration in the substantia propria. The animals were sacrificed at different intervals after treatment, and histological sections were prepared. By this method the rate of damage and regeneration could be easily followed in correlation of the time after applications.

Kambe (22), using light microscopic, scanning, and transmission electron microscopic techniques, managed to detect fine-tissue changes resulting from the contact with anionic surfactants. These are:

Numberous microvilli between the individual cells at the outer surface of the examined cornea.

Extrafoliations of the epithelial surface at the cornea, which had the appearance of a mosaic.

Enlargement of the intercellular spaces, which were highly joined by desmosomes.

The layers of cells forming the corneal epithelium were markedly decreased and irregularly wrinkled, stereoscopically. The epithelial cell layers showed markedly high electron density.

3. Tests with different mucous membranes

As previously mentioned, the predominant part of mucous membrane tolerance tests have been carried out on rabbit eyes. Draize et al. (2) described a method using the penile mucosa of male rabbits: 0.2 ml of the test substances is applied to the penile mucosa, and the intensity of reaction is evaluated according to a point system at regular time intervals.

Anionic surfactants (linear alkyl benzene sulfonate, sodium alkyl ethoxy(3EO) sulfate, α-olefin sulfonate) were also applied to the vaginal mucosa of dogs in order to test the compatibility of the vaginal tissue (22a). The irritative effect of topically applied sodium lauryl sulfate and other chemicals in the buccal mucosa were investigated in a rat model (22b). The evaluation of the irritations was a semiquantitative one of cellular infiltrates, which were performed in routine histologic preparations.

Several surfactants have also been examined on the mucous membranes of the human mouth and cheek (23,24). The following two procedures are used: Either a specific surfactant in a definite concentration is applied to the mucous membranes and then a smear is prepared so that changes of the mucous membrane cells could be examined microscopically (24), or fresh mucous membrane cells are removed, mixed in vitro with the surfactant, and, after a certain contact time, the changes of these cells (shrinkage, pyknose of the cell nucleus) examined microscopically (23). In addition, local changes of the intact mucous membranes (roughness, stinging of tongue and lips, painful lesions of mucous membranes) (24) as well as disturbances of the sense of taste (24,25) are tested, with surfactants to be formulated into toothpastes or mouthwashes.

Berenson and Temple (26) examined the effect of detergents on the mucous membranes of the esophagus and stomach of cats. Two techniques were used, an irrigation and a tampon technique. According to the concentration used, mild changes (surface cell desquamation), medium-cell desquamation associated with edema, congestion and inflammation, or severe reactions (extensive mucosal disruption with ulceration and necrosis) were observed. The degree of irritation was scored on a quadripartite scale.

The compatibility of surfactants with the upper respiratory tract was investigated by Ciuchta (27) using guinea pigs, mice, and rabbits. Because sensations such as stinging or burning cannot be recorded for animals, the release of a reflex, such as inhibition of respiration, was used to assess the irritation.

Ciuchta and Dodd (27a) assessed a relatively potential of different surfactants on the mouse upper respiratory tract and with the mouse writing test and compared the test results of these both test methods with the test results of the Draize eye and skin test.

C. Dependence of Irritation on Structure of Substances, Carbon Chain Length, pH Value of Applied Solutions, and Duration of Contact

In addition to the chemical structure of the surfactant, the concentration plays a dominant role in irritation. Likewise, the duration of contact is an important factor. Only the first seconds of contact should have the decisive influence (16) and only immediate rinsing markedly decreased tissue damage (27b). In contrast, the amount of applied solution is of secondary importance since the conjunctival sac has only a limited capacity. Most authors apply a volume of 0.1 ml. Scholz (9) considers this too much, however, and recommends 0.05 ml. The same suggestion made Arthur et al. (27c).

By reducing the volume from 0.1 to 0.01 ml applied to the rabbit eye, a better prediction for reactions on human eyes in case of accidental contact is achieved (13e,13g-i,13x).

The irritation of surfactants differs widely. Draize and Kelley (28) report a "maximum tolerated concentration" for cationic surfactants of 0.5-1.0%, for anionic surfactants 10-20%, and for nonionic surfactants 100%. Anionic surfactants and soaps are regarded by Hazleton (29) as moderate irritants, cationic surfactants as strong, and nonionic surfactants as mild irritants. Other authors (15,29a) have reached similar conclusions.

Among the anionic surfactants, the *n*-alkyl sulfonates and dodecyl benzene sulfonates (DBS) are the most severe irritants for the rabbit eye, whereas the protein fatty acid condensates and actyl lactylates, a new group of anionic surfactants possess the weakest mucous membrane irritant properties (30,30a). Oba et al. (31) compared the irritant activity of α-olefin sulfonates (AOS), linear alkyl sulfonates (lASF), and tetrapropylene benzene sulfonate (TPBS), and found thereby that lASF caused the strongest response on mucous membranes, followed by TPBS and AOS.

Kästner and Frosch (15a) compared 16 anionic surfactants from four different groups (alkyl sulfates (AS), alkyl sulfonates (ASF), alkyl ether sulfates (AES) and others (1 sulfosuccinate and 1 fatty acid protein condensate) and found that the alkyl sulfates and sulfonates caused markedly stronger reactions than the alkyl ether sulfates and the others.

Within a group, e.g., nAS or the AOS, those with a middle carbon chain length (C_{12}-C_{14}) are the strongest irritants, whereas shorter and especially longer carbon chains are less irritating. According to Daweke (32), it is not important whether concentrations of the same weight percent or isomolar concentrations are compared. The sole exception was octadecyl sulfate, where differences in the irritation between the two concentration types were observed.

Panzer and Nehmsmann (32a) found in their studies about structure/property relationship of a series of anionic/amphoteric blends that eye irritation decreases as alkyl anionic chain length and/or ethoxylation degree increase.

A clear correlation between the increase of ethoxylation and the diminution of irritancy in the rabbit eyes were observed in our own investigations (34).

In contrast, Iimory (33) does not confirm a dependence of mucous membrane-irritating properties on the chain length with nAOS and with synthetic AS in the range of C_{10}-C_{18}.

Grant (15) reports the maximum irritation for the dodecyl sulfate among the *n*-alkyl sulfates.

The results of our own investigations (34) with the sodium salts of nAS and nASF of defined chain length in the range of C_6-C_{18} and C_8-C_{18}, respectively, indicate that surfactants with C_{10}-C_{14} chain lengths are clearly more irritating than those with shorter or longer chain lengths. The C_{12} compound was not different than the C_{10} and C_{14} substances with regard to its irritating properties. The 1% surfactant solutions caused irritations only in the conjunctiva. The cornea and iris did not show any macroscopically visible changes. As evident from Table 3, the conjunctival reactions disappeared in all cases within 2 days.

In addition to chemical structure and carbon chain length the pH value of the solution influences the irritating action of surfactants. This is also so for soaps, which can exhibit pH values up to 12.

Tronnier and Schneider (35) investigated fatty acid methyl taurate and fatty acid protein condensate (mol wt 750) on the rabbit eye, using the method of Neumann and Aalbers, taking special account of the pH values of the solutions. These ranged between 5.0 and 9.0. For comparison, a 2% sodium lauryl sulfate solution of pH 7 was used. An improvement in tolerance with increasing pH value was observed with an optimum value at pH 8, higher values resulting in a subsequent decrease of compatibility.

According to the method described by Draize (2), from the nine treated rabbits the conjunctival sac of six rabbits have to be rinsed within 2 or 4 sec after application. This time interval is to be adhered to experimentally but appears to be too short under practical circumstances (8). In the case of mis-

TABLE 3 Mean Reactions of Conjunctiva in Percent of Maximal Possible Reactions (Scored According to Scheme 1) after Application of 50 μl of 1% Anionic Surfactant-Solution, No Irrigation, Four Rabbits per Group

	Reaction after (hr)				
Substance	2	6	24	48	72
Sodium hexyl sulfate	0	2.5	0	0	0
Sodium octyl sulfate	2.5	5.0	2.5	0	0
Sodium decyl sulfate	32.5	27.5	7.5	0	0
Sodium dodecyl sulfate	27.5	27.5	2.5	0	0
Sodium tetradecyl sulfate	27.5	25.0	5	0	0
Sodium hexadecyl sulfate	12.5	10.0	0	0	0
Sodium ocatadecyl sulfate	12.5	7.5	0	0	0
Sodium salt of					
n-Octyl sulfonate	0	0	0	0	0
n-Decyl sulfonate	10.0	17.5	2.5	0	0
n-Dodecyl sulfonate	30.0	17.5	0	0	0
n-Tetradecyl sulfonate	30.0	25.0	7.5	5	0
n-Hexadecyl sulfonate	0	0	0	0	0
n-Octadecyl sulfonate	0	0	0	0	0

adventure, i.e., by the inadvertent entry of a small amount of product into the eye, undoubtedly a longer interval will elapse before the damaged eye can be rinsed, i.e., before a wash bottle or a water tap can be reached. Therefore, several authors (16,34) have variously extended the contact time up to 3 min after instillation into the conjunctival sac. It has been shown that rinsing must occur within the first 30 sec after application in order to attain a significant effect in comparison to the unrinsed eye. After longer time intervals the damage is so marked that rinsing has no effect. It was also examined as to whether the amount of rinsing water had an effect.

York et al. (35a) used for the assessment of eye irritancy of shampoos an in vitro test (enucleated rabbit eyes) and found a strong correlation between the intensity of corneal effects (corneal swelling and opacification) and contact time. They tested the shampoos undiluated and in a 10% dilution. The contact time was 10 and 30 sec, and 2 and 5 min.

Davies et al. (16) carried out tests with sodium lauryl sulfate (SLS), as with this substance much practical experience was available. They applied 0.1 ml of a 10% SLS solution to the rabbit eye. The rinsing out was either omitted or carried out 4-120 sec after the application with 20-100 ml water at 37°C. The scoring was performed after 1 and 4 hr and 1, 2, 3, 4, 7, 14, 21, 28, and 35 days later, or as long as no further reaction was noticeable. The observed reactions were scored numerically according to the system recommended in the Code of Federal Regulations (see Scheme 3), which conforms extensively with the scoring scheme published by Draize et al. (2). The authors were able to demonstrate that rinsing must take place in less than 20 sec to prevent corneal opacity or a dulling or normal corneal luster. If rinsing is carried out after 30 sec, it is without major influence on the irritating action. Similar responses are seen when rinsing is carried out after 120 sec or if the rinsing stage is completely left out. The faster the rinsing, the faster the symptoms disappear. The amount of water used for rinsing, 20 or 100 ml, did not have a significant influence on the symptoms.

Olson et al. (36) tested the effect of 1, 5, and 25% aqueous solutions of various anionics for their mucous membrane-irritating properties on the rabbit eye. In one group of rabbits the solutions remained in the eye. The treated eyes of another group were thoroughly rinsed with flowing water 30 sec after instillation of the solution in the conjunctival sac for 2 min. The rinsing procedure prevented the formation of severe corneal, i.e., permanent damage in the case of the 25% concentration. Such severe damage occurred when using the same substances in the same concentration (25%), but without rinsing. Intensive rinsing also showed an effect in the case of the 5% concentration.

Kambe (22) also found that a marked effect by rinsing is only achieved at higher concentrations. After application of 10% alkyl benzoic sulfonic

SCHEME 3 Grades for Ocular Lesions

Cornea	
No ulceration or opacity	0
Scattered or diffuse areas of opacity (other than slight dulling of normal luster), details of iris clearly visible	[1]
Easily discernible translucent areas, details of iris slightly obscured	2
Nacreous areas, no details or iris visible, size of pupil barely discernible	3
Completed corneal opacity, iris not discernible	4
Iris	
Normal	0
Markedly deepened folds, congestion, swelling, moderate circumcorneal injection (any of these or combinations of any thereof)	[1]
No reaction to light,, hemorrhage, gross destruction (any or all of these)	2
Conjunctiva	
Redness (refers to palpebral and bulbar conjunctiva excluding cornea and iris)	
Vessels normal	0
Some vessels definitely injected	1
Diffuse, crimson red, individual vessels not easily discernible	[2]
Diffuse beefy red	3
Chemosis	
No swelling	0
Any swelling above normal (including nictitating membrane)	1
Obvious swelling with partial eversion of lids	[2]
Swelling with lids about half-closed	3
Swelling with lids more than half-closed	4

Figures in brackets indicate lowest grades considered positive under Title 16, Section 1500.42 of the Code of Federal Regulations.
Source: Ref. 16.

acid sodium salt (SABS), a mild chemosis and redness of the conjunctiva was seen 48 hr after application in the eye which had not been rinsed. In addition needle-size pointed remainders of fluorescein were observed on the cornea. In contrast, the eyes that had been rinsed did not exhibit any symptoms. The differences were even more marked by using a 20% SABS concentration. While the unrinsed eyes showed needle-size fluorescence of the cornea and moderate redness and chemosis of the conjunctiva, these symptoms were completely absent in the rinsed eyes exposed to 20% SABS solution before irrigation. The results cited appear to show that the interval between instillation and rinsing is more important than the intensity of rinsing under the chosen test conditions (rabbit eye). In case of an immediate beginning of lacrimation after instillation into the eye the irritant activity can be considerably reduced (human eye). The critical interval for development of corneal opacity is, according to Davies et al. (16), within an exposure time of 4-10 sec. Therefore, these authors recommend that rinsing of the eye after accidental contact with a surfactant should take place within 10 sec, in order to avoid corneal lesions.

D. Dependence of Irritation on Concentration

In addition to the parameters employed to assess the intensity of irritation mentioned above, concentration is significant in evaluating the damage to mucous membranes. There is a good relation between concentration and irritation. If a surfactant possesses a relatively high degree of irritating action but is used only in low concentrations, so that either no irritation or very little results, then this product is just as suitable for use as an anionic, which is only a weak irritant, but must be used at high concentrations, i.e., low concentrations of a strongly irritating anionic surfactant have the same effect as high concentrations of a weakly irritating surfactant.

In the case of products destined for use on or near the mucous membranes, the concentration which can be tolerated by the mucous membranes (threshold or limiting concentration) is of special importance. The concentration dependence of the irritant effect on mucous membrane of the eyes has been investigated by several authors.

Matsuura et al. (29a) showed a clear dependence of irritation on concentration with the anionic surfactant sodium lauryl sulfate (SLS) in concentrations of 1, 5, 10, and 25%. The 1% solution did not cause opacity of the cornea. The higher concentrations tested caused opacity. The rabbit eyes were more sensitive to SLS than were the eyes of guinea pigs.

Neumann and Aalbers (6) examined the dependence of irritation on concentration of SLS and observed a linear relation. The threshold concentration for SLS under the conditions used was below 0.5%.

Olson et al. (36) observed on a series of surfactants tested a slight irritation, even with the lowest examined concentration of 1%.

Oba et al. (31) tested α-olefin sulfonate (AOS) and tetrapropylene benzene sulfonate (TPBS) on the rabbit eye and found that they caused virtually no reactions at 0.05% concentration. However, the concentration of linear alkyl sulfonate (lASF) had to be lowered to 0.01% before the absence of reactions was established.

Serrano et al. (27b) tested lauryl sulfates (LS) and lauryl ether sulfates (LES) (salts of ammonium, lithium, sodium, magnesium, and triethanolamine) and dioctylsulfosuccinate sodium salt in 5 concentrations (1.25, 2.5, 5, 10, and 20%), each without rinsing, and rinsing 2 and 4 sec after application. All concentrations caused more or less strong eye irritations. The concentration of 1.25% was in no case the limit concentration of irritancy.

Iimori et al. (33) observed higher limiting concentrations such as 0.1% for IASF, 0.5% for AOS of varying chain lengths (AOS mix), 0.1% for synthetic alkyl sulfate (AS), 0.5% for pure AOS of defined chain lengths, and 0.1% for pure synthetic AS of defined chain length.

According to Brown and Muir (37), the threshold concentrations, which did not lead to any response, of various alcohol sulfates are 0.1%. Experiments carried out by Draize et al. (2) proved that dioctyl sodium sulfosuccinate could not be tolerated without irritation at a concentration of 0.5% with the human eye.

The following concentrations of surfactants which do not cause visible gross lesions in the eye mucosa are given by Martin et al. (38): tri-ethanolamine alkyl sulfate (Sipon LT) = 1.5%; sodium-*n*-methyl-*n*-acyl laurate (Igepon TN 74) = 1.1%; and the sodium salt of an alkyl aminosulfate (Sulframin LW) = 2.5%.

Lawrence et al. (38a) used the chorionallantoic membrane (CAM) of chick embryo in order to test nine chemicals including SLS with known irritancy potential for the eye in vivo. Each was tested at four different concentrations. For SLS a concentration of 0.1% was found as the threshold concentration for irritant effect on CAM, and concentrations of 0.1 and 0.01 caused minimal effects in the Draize eye test.

According to Ginsburg and Robson (39), the rabbit eye tolerance of dioctyl ester of sodium sulfosuccinate (Aerosol OT) solution is below 1%. No irritation but rather a blepharospasm appearing immediately after the instillation of the solutions was observed with 0.1% concentrations of Aerosol OT, sodium isopropylnaphthalene sulfonate (Aerosol OS), sodium alkyl sulfate (branched C_{14}) (Tergitol 4), sodium alkyl sulfate (branched C_{17}) (Tergitol 7), sodium alkyl (2-ethylhexyl) sulfate (Tergitol 08), and sodium alkyl sulfate (Duponol ME dry) (40).

Lehman and Draize (41) give tolerance levels of various anionic surfactants which appear to be very high in comparison to those threshold concentrations reported by other authors (see Table 4). Threshold concentrations, i.e., highest concentrations tested, which are tolerated by the eyes without any irritation (A) and lowest concentrations tested that still cause mucous membrane irritation (B) have been given by various authors (15,22, 29a,34,42-44) (Table 5).

E. Improvement of Mucous Membrane Compatibility by Additives or by Mixtures with Other Surfactants

Experimental studies and practical use have shown that specific additives improve the mucous membrane compatibility of anionic surfactants or, in other words, a reduction of irritating properties can be achieved by additives.

TABLE 4 Limiting Concentrations[a] Tested on the Eyes

Surfactant	A, %	Ref.	B, %	Ref.
Soap	4	22		
n-Alkyl sulfates				
Sodium alkyl sulfates				
Coconut alcohol-based	0.1	37	1.0	37
Dobanol 23-based	0.1	37	1.0	37
Higher aliphatic alcohol sulfate derived from technical lauryl alcohol				
(Duponol ME dry)	0.1[b]	40		
(Duponol ME)			0.5	15
Texapon V high. conc. (C_{12}-C_{16}; 90% AS[c])			0.5	34
Texapon Z high conc. (C_{12}-C_{18}; >90% AS)			1.0	34
Natrium alkyl sulfate (synthetic alcohol base)	0.1	33	0.5	33
Sodium alkyl sulfate (C_8-C_{18})			0.5	15
Sodium decyl sulfate (C_{10})	0.1	33	0.5	33
Sodium dodecyl sulfate (C_{12})	0.1	33	0.5	6
			1.0	36
			0.5	33
			0.5	15
			0.1	39
Sodium lauryl sulfate (Sipon LS)	20.0	41		
Texapon K-12 (C_{12}; 90% AS, L-100 (C_{12}; 99% AS)			0.5	34
Sodium tetradecyl sulfate (C_{14})	0.1	33	0.5	33
Sodium cetyl sulfate (C_{16})	0.1	33	0.5	33
Sodium octadecyl sulfate (C_{18})	0.1	33	0.5	33
Modified sodium salt of sulfonated fatty alcohol (Wetanol)	25.0	41		
Ammonium alkyl sulfates				
Coconut alcohol-based	0.1	37	1.0	37
Dobanol 23-based	0.1	37	1.0	37
Ammonium lauryl sulfate				
(Sipon L-20)	30.0	41		
(Texapon A-400; C_8-C_{14}, 33-35% AS)			0.5	34
Monoethanolamine alkyl sulfates				
Coconut alcohol-based	1.0	37	25.0	37
Dobanol 23-based	1.0	37	25.0	37
Texapon MLS (C_{12}-C_{14}; 33% AS)			0.5	34
Triethanolamine alkyl sulfates				
Coconut alcohol-based	1.0	37	25.0	37
Dobanol 23-based	1.0	37	25.0	37
Texapon TH (C_8-C_{14}; 47% AS)			0.5	34

TABLE 4 Continued

Surfactant	A, %	Ref.	B, %	Ref.
n-Alkyl sulfates (continued)				
Triethanolaminolauryl sulfate				
Sipon LT	1.15[b]	38		
	40.0	41		
Triethanolaminododecyl sulfate			1.0	36
			0.5	15
Alkyl sulfates with branched hydrophobic groups				
Sodium-2-ethyl-hexyl sulfate (Tergitol 08)	0.1	40		
Sodium-7-ethyl-2-methyl-undecyl-4-sulfate (Tergitol 4)	0.1	40		
Sodium-3,9-diethyl-tridecyl-6-sulfate (Tergitol 7)	0.1	40		
Alkyl polyglycol ether sulfates				
Sodium alcohol (2EO) ethoxy sulfate				
Coconut alcohol-based	0.1	37	1.0	37
Dobanol 23-based	0.1	37	1.0	37
Sodium dodecyldiglycol ether sulfate (pure)	0.5	9		
Sodium dodecylmyristyl ether sulfates				
Texapon N-25 (C_{12}-C_{14}, 28% AS) (2EO)	0.1	34	0.5	34
Texapon N-40 (C_{12}-C_{14}, 28% AS) (2EO)	0.5	34	1.0	34
Texapon K-14S special (30% AS) (3EO)	0.5	34	1.0	34
Sodium magnesium alkyl (C_{12}-C_{18}) ether (7EO) sulfate (Texapon ASV)	1.0	34	5.0	34
Sodium salt of arylalkyl ether sulfate (Triton X-30)			0.5	15
			1.0	42
Sodium alkylaryl polyether sulfate (Triton X-301)	10.0	41		
Diverse alkyl sulfates				
Sodium salt of an alkyl aminosulfate (Sulframin LW)	2.5[b]	38		
	10.0	41		
Sodium salt of an hydroxyalkyl amido alcohol sulfate (Sulframin DR)	25.0	41		
An alkylated monosodium benzene sulfate (Ultrawet 30-DS)			0.5	15
Alkyl sulfonates				
Alkyl benzene sulfonates				
Sodium alkyl benzoic sulfo acid	4.0	22	10.0	22

TABLE 4 Continued

Surfactant	A, %	Ref.	B, %	Ref.
Alkyl sulfonates (continued)				
Tetrapropylenbenzene sulfonate	0.05	31	0.1	31
Decyl benzene sodium sulfonate (Santomerse D)			0.5	15
			1.0	43
Dodecyl benzene sodium sulfonate (Santomerse 3)			1.0	43
Dodecyl benzene sulfonate triethanolamine salt			1.0	36
Sodium alkylaryl sulfonate (Nacconol NRSF)	5.0	41	0.5	15
Alkylaryl polyether sulfonate (Triton X-200)	28.0	41		
Linear alkyl sulfonates				
Linear alkylate sulfonates	0.01	31	0.05	31
	0.1	33	0.5	33
α-Olefin sulfonate	0.05	31	0.1	31
	0.5	33	1.0	33
AOS C_{10}-C_{18}	0.5	33	1.0	33
Diverse sulfonates				
Isopropyl naphthalene sodium sulfonate (Aerosol OS)	0.1	40	0.5	15
	1.0	44	2.0	44
Monobutylphenylphenol sodium monosulfonate (Areskap 100)			1.0	43
Monobutyl diphenyl sodium monosulfonate (Aresket 300)			1.0	43
Dibutylphenyl phenol sodium disulfonate (Aresklene 400)			1.0	43
Sodium dodecylated diphenyl ether disulfonate			1.0	36
Dodecylated diphenyl ether disulfonate magnesium salt			1.0	36
Triethanolamine salt of dodecylated diphenyl ether disulfonate			1.0	36
Sodium dodecylated diphenyl ether monochlorosulfonate			1.0	36
Sodium octyl phenoxyethoxyethyl ether sulfonate (Entsufon)			0.5	15

TABLE 4 Continued

Surfactant	A, %	Ref.	B, %	Ref.
Diverse anionic surfactants				
Diotyl ester of sodium sulfosuccinate				
(Aerosol OT)			0.5	15
			0.5	2
	0.1	40	1.0	39
			1.0	43
Sodium lauryl sulfoacetate			3.0	45a
Calcium stearoyl-2-lactylate	100	30a		
Sodium capryl/lauroyl lactylate	10	30a		
Oleic acid methyl taurate				
(Igepon T)			1.0	43
Sodium *N*-methyl-*N*-acyl laurate				
(Igepon TN-74)	1.1[b]	38		
	10.0	41		

[a]Highest concentrations tested that were tolerated without any irritation with the rabbit eyes (A), and lowest concentrations tested which still caused mucous membrane irritations (B) after a single application without rinsing.
[b]No gross lesions.
[c]% AS = active substances in %, noted when known.
Note: To avoid errors and ensure agreement with the cited publications, the nomenclature follows the original literature.

The same effect was observed with mixtures of various surfactant groups. In particular, addition of proteins, protein hydrolysates, or the less irritating protein fatty acid condensates were all especially mentioned in this context. The question as to whether these additives which reduce the irritating properties reduce simultaneously the washing activity of surfactants and whether a reciprocal relationship between washing effectiveness and compatibility exists cannot be discussed in this context.

Meinicke (45) studied commercial fatty alcohol sulfates and alkylaryl sulfonates at higher concentrations in the rabbit eye and compared the irritant activity with the reactions caused by fatty acid protein condensates. In addition, he examined mixtures of these three anionic surfactants. The following products were investigated:

A. Commercial alkylaryl sulfonate (35% solution)
B. Commercial fatty alcohol sulfate (35% solution)
C. Commercial protein fatty acid condensate (35% solution)
D. Mixture of the concentrates A, B, and C
E. Mixture of equal parts of concentrates A and C
F. Mixture of equal parts of concentrates B and C
G. Mixed product of equal parts of concentrate A and a protein containing protein hydrolysate of approximately 35% concentration.

TABLE 5 Tests of Mucous Membrane Tolerance of Surfactant Mixtures on Rabbit Eyes[a]

		Results 1 hr after application			Results 24 hr after application			Results 48 hr after application		
Product	Conc.	Conj.	Cornea	Iris	Conj.	Cornea	Iris	Conj.	Cornea	Iris
A	5 + 10%	+ +/+ + +[b]	–	–	+	–	–	–	–	–
	20%	+ + + +	+ + +	+ +	+ + + +	+ +	+ +	+ + + +	+ +	+ +[c]
B	5 + 10%	+ + +	–	–	+	–	–	–	–	–
	20%	+ + +	+	+	+ + +	+ +	+ +	+ +	+ +	+[d]
C	5 + 10%	–	–	–	–	–	–	–	–	–
	20%	–	–	–	–	–	–	–	–	–
D	5 + 10%	–	–	–	–	–	–	–	–	–
	20%	+ +	+	–	+ +	–	–	+	–	–
E	5 + 10%	+	–	–	–	–	–	–	–	–
	20%	+	–	–	+ + +	–	–	+ + +	–	–
F	5 + 10%	+ +	–	–	+	–	–	–	–	–
	20%	+ + +	+	+	+ + +	+	+	+ +	–	–
G	5 + 10%	+	–	–	–	–	–	–	–	–
	20%	+	–	–	+	–	–	–	–	–

[A]Instilled volume 0.1 ml, contact time 4 sec, pH values of solutions 5.4 [according to Ref. 45].
[b]Each + corresponds to a grade of the scoring from Scheme 4.
[c]Conjunctival tissue: up to 17 days + + + +, after 28 days –; cornea: up to 39 days + +; iris: up to 14 days + +.
[d]Conjunctival tissue: up to 5 days + +, after 5 days –; cornea and iris: after 3 days –.

The damages caused were assessed according to Scheme 4. The results of the tests are summarized in Table 5. The assessments were carried out 1, 24, and 48 hr after application. It is evident that the combination of the stronger irritants A and B with the product C, which had good compatible properties, resulted in a marked improvement of the mucous membrane-irritating properties. Likewise the addition of protein hydrolysate also lessened the irritating action of anionic surfactants.

Russel and Hoch (10) tested two shampoos which consisted essentially of triethanolamine lauryl sulfate (20%) for their mucous membrane compatibility on rabbit eyes using the method of Draize. The shampoos had similar compositions and differed only by the addition of 3% lanolin alcohol ether and 1% lanolin oil. Similar investigations following Draize's method were carried out in various other laboratories. It was established that the addition of lanolin alcohol ether and lanolin oil improved the mucous membrane-irritating properties of the triethanolamine lauryl sulfate. This finding could not be confirmed by Weltman et al. (21).

F. Local Anesthesia in Eye Mucosa

Since a series of nonionic surfactants exhibits marked local anesthetic properties, it appeared mandatory to examine also anionic surfactants for such effects. This effect requires attention with applications of surfactants where a contact with the mucous membranes, e.g., the conjunctival tissues cannot be avoided (shampoos) and thus possible irritations would not be noticed in time.

SCHEME 4 Scale for Scoring Ocular Lesions

Conjunctiva	
Intensity I	Slight redness. Vessels excessively injected, slight swelling of conjunctive tissues including membrana nictitans
Intensity II	Diffused red areas, extended vessels, marked swelling and partial turning of eyelid, slight secretion, wetting of eyelids and hairs
Intensity III	Dark red discoloration of conjunctiva tissues. Vessels nondistinguishable, strong swelling. Eyelids half-closed, strong secretion. Hair wetted within a circumference of 1 cm
Intensity IV	Redness comparable to intensity III. Eyelids almost completely closed, strong secretion and wetting within a large circumference around eyes
Cornea	
Intensity I	Dispersed or extended areas, a little turbidity, iris visible
Intensity II	Slight turbidity, iris slightly clouded
Intensity III	Opalescence of wide areas, iris not completely visible
Intensity IV	Cornea totally nontransparent, iris invisible
Iris	
Intensity I	Response to light positive, vessels injected
Intensity II	No reaction to light

Source: Ref. 45.

Martin et al. (38) investigated sodium *N*-methyl-*N*-acyl laurate, triethanolamine salt of lauryl alcohol sulfate, and sodium salt of an alkyl aminosulfate for their local anesthetic properties using the rabbit eye according to the method of Regnier (46) and observed no local anesthetic properties. They found, with concentrations that produce no significant macroscopically visible lesions in the eye mucosa, that the tolerance for pain (threshold of pain) can be increased.

Five commercial bath and face soaps and four hair shampoos were tested by Harris et al. (47) for their local anesthetic properties on human and rabbit cornea. The concentrations applied to the human and rabbit eye were 10% (soaps), undiluted, 50% and 25% (shampoos). All four soaps and the shampoos caused a corneal anesthesia in the rabbit eye (up to 4 hr). In the human eye, no difference between the treated and the untreated eyes was found. The animal experiments gave results which were comparable to the anesthesia produced by a 0.5% tetracaine solution.

The irritating effects of the soaps and shampoos which were applied in 10% (soap), undiluted, 50% and 25% concentrations (shampoos) in the rabbit or human eye were unfortunately not described. However, one cannot assume that these concentrations are tolerated without any reaction by the mucous membranes of the eyes.

G. Irritation of Mucous Membranes by Specific Anionic Surfactants

As described in Section I.A, some anionic surfactants have been investigated frequently, others only once. At first the results of tests with the oldest anionic surfactants, i.e., the soaps, are presented, followed by those with sulfates and sulfonates, and finally those that cannot be classified into these three groups.

1. Soaps

Soaps were used in ancient times for many purposes (see Fig. 1) (48), and compatibility problems certainly played a role because the products then in use did not have the purity of today's products.

Hazleton (29) classified soaps, together with other anionic surfactants, as moderate irritants, in contrast to the cationic surfactants that were classified as severe irritants and the nonionics that he classified as mild irritants. According to Grant (15), the only soaps that lead to permanent corneal damage are those that contain much free alkali and have a pH value of more than 12.

Feldman et al. (49) examined the irritation of rabbit eye mucous membranes by soap powders; 0.1 ml (!) was instilled in the conjunctival sac and the eyelids were closed for a few seconds. Rinsing was not done. The soap

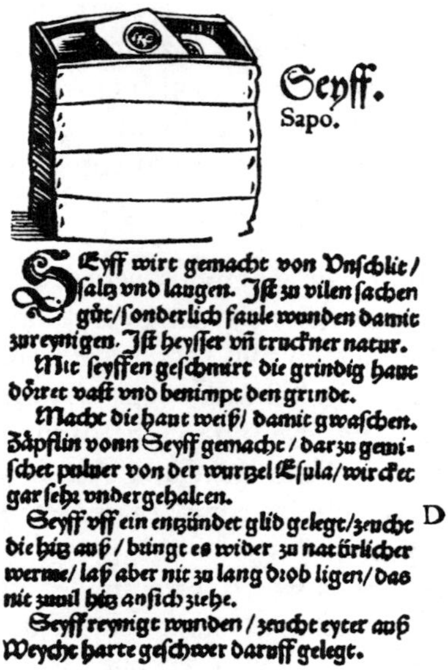

FIG. 1 *From Hortus Sanitatis, IV* book. Printed and completed at Strassburg by Mathias Apiario in 1536.

powder caused conjunctivitis, chemosis, and corneal opacity within 24 hr. However, the corneal opacity largely disappeared after 72 hr. Histological changes of the eyes were not detectable 14 days after treatment.

Opdyke et al. (50) reported that pure soaps (3 mg per application) caused only slight irritation in the form of temporary conjunctivitis and slight transient corneal haziness, which did not persist.

Griffith et al. (13e) carried out dose-response studies with sodium coconut oil soap powder, 10% sodium coconut oil soap and a granulated soap product and placed 0.01, 0.03 and 0.10 ml directly on the cornea of one eye of three rabbits each. The eyes were not irrigated. At 0.01 ml slight reactions occurred, which disappeared within 1-4 days. After the application of 0.03 and 0.10 ml, moderate reactions were observed, which disappeared between 2 and 7 days (0.03 ml), respectively, and between 4 and 14 days (0.1 ml). The 10% solution was less irritating than the undiluted soap.

Carpenter and Smyth (12) examined the effect of soaps for their ability to damage the rabbit cornea. Two hours after examination of the cornea for intactness (with 5% fluorescein solution), 0.005 ml of a 5 or 15% soap solution (granulated white soap) was placed at the center of the cornea, the eyelids then being prevented from closing for 1 min. The assessment of reactions was made 18 or 24 hr after application. The 5% soap solution caused reactions below 5 points, the 15% solutions over 5 points, with a maximum in the scale of 20 points (see Scheme 5).

SCHEME 5 System for Numerical Scoring of Injury to the Rabbit Eye 24 hr after Application of a Material

Symptom	Points	Maximum
Visible before fluorescein staining		
Cornea dull	2	
Cornea opaque, less than half of area	4	
Cornea opaque, more than half of area	6	6
Keratoconus		6
Iritis, slight internal congestion	1	
Iritis, marked internal congestion	2	2
Visible after fluorescein staining		
(If necrosis is diffuse, assign points corresponding to half observed area)		
Necrosis on less than 5% of cornea	1	
Necrosis on 5-12%	2	
Necrosis on 13-37%	3	
Necrosis on 38-62%	4	
Necrosis on 63-87%	5	
Necrosis on 88-100%	6	6
Total		20

Source: Ref. 12.

Burton (7), who assessed the irritation of the mucous membranes by scoring the swelling of the cornea, studied the irritating activity of a 10% toilet soap. The swelling resulting from the soap application had completely disappeared after 2 days. The 10% Ivory soap is being used as a standard by the Food and Drug Administration according to Harris (47) as a model for the absence of mucous membrane irritation of the eye (51).

Shadduck et al. (51a) tested the cytotoxicity of six surfactants, including coconut oil soap and compared the results with Federal Hazardous Substance Act (FSHA) eye irritations ranking. According to the FHSA, coconut soap is classified as "mild," sodium dodecylsulfate as "moderate." The same differentiation was found with the cytotoxicity test (chromium release) with rabbit corneal epithelial cells.

Ferguson and Prottey (51b) examined in vitro a series of highly pure surfactants for their capacity to inhibit DNA synthesis in guinea-pig kidney fibroblasts, to release histamine from isolated rat peritoneal mast cells, and to effect the cytolytic release of cytoplasmic proteins from guinea-pig kidney fibroblasts. The surfactants studied included a homologous series of carboxylic acid sodium soaps of chain length C_8-C_{16}. Sodium laurate (chain length C_{12}) was more potent at inhibiting DNA synthesis, at releasing histamine, and at lysing fibroblasts than either the higher or lower homologoues.

A different ranking between soap and SLS is published by Dossou and Vinsonneau (51c), who used a special test method for testing the mucous membrane compatibility. A characteristic body position after intraperitoneale application of the test solution assumed by male and female mice was used for estimation the relative discomfort and stinging potential of different types of surfactants. Soap was a little more irritating than SLS.

Beckley et al. (19) compared the irritating activity of a 5% soap solution on the human eye with the reactions on the rabbit eye and rhesus monkey eye. The assessment of the eye damage was carried out by means of a slit-lamp biomicroscope and fluorescein and also according to the commonly used Draize visual observation method. In all three species 0.1 ml of solution was instilled into the eyes, which were not rinsed afterward. This was done in order to simulate accidental instillation into the eye of humans with limited mobility a situation that would permit maximum damage. Before application of the soap solution, the cornea was examined for uninjurence using fluorescein solution. The reaction was examined after 1 and 6 hr, and then daily; in the case of human subjects after 15 min, 1, 4, 6, and 24 hr after application. At these times after application, the coloration of the eyes was carried out with fluorescein, the human subjects having this operation carried out at least once. The test subjects could rinse the eyes after 1-2 min after the instillation. The intensity of the reactions was graded into three stages such as "accept," "accept with caution," "probably injurious to human eyes" (see Scheme 6). After application of a 5% soap solution the following reactions in the eyes were observed.

SCHEME 6 Proposed Eye Scoring Scheme Based on Use of a Slit-Lamp Biomicroscope and Fluorescein

Site	"Accept"	"Accept with caution"	"Probably injurious to human eyes"
Conjunctiva	Hyperemia without chemosis	Chemosis, less than 1 mm at the limbus	Chemosis, greater than 1 mm at the limbus
Cornea	Staining, corneal strippling without confluence at 24 hr[a]	Confluence of staining at 24-48 hr[b]	Staining with infiltration or edema
Anterior chamber	0	0	Flare (visibility of slit beam). Rubeosis of iris[c]

[a]Corneal stippling: Multiple discrete punctate irregularities in the corneal epithelial layer which retain fluorescein.
[b]Confluence: Uniform zones of fluorescein retention larger than 1 mm in diameter.
[c]Flare: Tyndall effect in a beam traversing the aqueous humor.
Source: Ref. 19.

Rabbit eye: Redness of the conjunctival tissues (degree 1-2) up to 48 hr following instillation. Chemosis and discharge (degree 1) up to 24 hr after treatment.

Monkey eye: Redness of the conjunctival tissues with four of six monkeys. In two cases reaction did not disappear after 48 hr. No chemosis, no discharge.

The biomicroscopic examination showed

Rabbit eye: Conjunctivitis, which had healed after 72 hr. Corneal stippling and stain retention up to 72 hr after treatment. Lacrimation, no loss of epithelium.

Monkey eye: After treatment conjunctivitis up to 24 hr, corneal stippling and stain retention did not disappear with three out of the six monkeys after 96 hr. Epithelial loss with three of six monkeys no longer visible after 72 hr.

In the human volunteers, instillation of the soap solution resulted in immediate lacrimation and complaints of irritation. The conjunctiva became hyperemic in 15 min and remained so for 6 hr. In one subject punctate superficial loss of the epithelium occurred in 15 min but disappeared by 6 hr. The examination demonstrated the usefulness of the use of primates for the testing of mucous membrane compatibility of substances which can easily penetrate the human eye. Whereas rabbit eyes did not show any corneal epithelial defects after contact with 5% soap solution, these were observed on the human and the monkey cornea. Such lesions could be detected only by using of the slit-lamp biomicroscope. If one uses for the 5% soap solution the scoring of rabbit's eyes, this soap solution would be classified as "accept." However, if one uses the results of monkey eyes, it is classified as "accept with caution" and this applies also to the irritation of a 5% soap solution on the human eyes.

In contrast to the findings of Beckley et al. (19), Kambe (22) saw no irritations on the mucous membranes of rabbit eyes after instillation of 0.1 ml of a 4% soap solution into the eyes. The conjunctival tissue showed no change and the examination with the slit-lamp also failed to reveal any corneal lesions.

As evident from Table 6, which summarizes the reported tests with soap regarding its mucous membrane compatibility, the authors have arrived at markedly different results, especially with soap powders. This may be attributed to a differing residual content of alkali and differences in carbon chain lengths of the fatty acids. Furthermore, the amount of applied soap was not uniform.

Since the described experiments were carried out without rinsing of the conjunctival sac after instillation of soap or soap solution and still only resulted

TABLE 6 Summary of Reactions on Mucous Membranes of Rabbit Eyes after Instillation of Soap Powders or Soap Solutions

Conc. %	Observed reactions	Time to disappearance of reactions	Ref.
100	Irreversible corneal damages caused with soap of pH value >12	—	15
100	Conjunctivitis, chemosis, opacity of corneal luster	72 hr	49
100	Transient conjunctivitis, mild corneal damage	120 hr	50
15	>25% of the maximal possible reaction	a	12
5	<25% of the maximal possible reaction	a	12
10-20	Moderate irritations	a	29
10	Corneal swelling (edema)	48 hr	7
10	Ivory soap no reactions on mucous membranes	—	51
5	Corneal stippling (rabbits), loss of epithelium, and slight to moderate conjunctivitis (monkeys)	48 hr 48 hr	19
4	Cornea: no change; conjunctiva: no change	—	22

[a]No information.

in temporary reactions in the rabbit eye, it may be expected that accidental contact of the human eye with soap or soap powder followed by rapid rinsing of the conjunctival sac does not cause severe reactions and that the caused reactions disappear quickly.

2. Alkyl sulfates and alkyl ether sulfates

The alkyl sulfates (AS) or fatty alcohol sulfates are powerful surface active substances and are presently among the most important groups of synthetic detergents. They are derived either from natural oils and fats, particularly coconut oil, or from petroleum. Whereas the fatty alcohol sulfates derived from animal and vegetable fats and oils only have an even number in the carbon chain, the synthetic AS contain also odd-numbered carbon chains, which may be of interest with respect to their mucous membrane compatibility. Further, the technical products are generally mixtures of substances with different carbon chain lengths, which is also important for their compatibility.

Although AS concentrations of 10% or more are primary eye irritants in rabbits, a concentration of 1% produces little or no ocular irritation. Concentrated AS samples are primary skin irritants in rabbits and guinea pigs but are nonirritating to laboratory animals at a 0.1% concentration (51d).

Among the alkyl sulfates, the compound sodium lauryl sulfate (SLS) was most thoroughly investigated for its mucous membrane irritation. SLS was often chosen as standard or model compound to assess the irritation of other anionic surfactants. Thus, Neumann and Aalbers (6) developed a scale for scoring using the testing method described in Sec. II.A.2, which is being referred to again at this point. Davies et al. (16) have also used 10% SLS solution as a model substance for investigation concerning the importance of rinsing. The reason was that the most extensive experience was available for this alkyl sulfate. The results published by Modde et al. (52) and Tronnier et al. (53) included sodium lauryl sulfate. Likewise Meinicke (45) used a technical fatty alcohol sulfate containing 35% active substance for his comparative studies.

Table 7 summarizes the experimental results concerning the testing of SLS solutions for their mucous membrane compatibility on rabbit eyes. The results are comparable only in so far as the identical test method and scoring scheme has been used. Other AS and alkyl ether sulfates (AES) will be presented only in connection with the description of the test method when necessary. In most cases, results may only be compared with each other when they are described in the same publication.

Olson et al. (36) tested, besides a series of sulfonates, sodium and triethanolamine lauryl sulfate. Since these tests were concerned mainly with sulfonates, the methods and scoring scheme are described in Sec. II.G.3. Two drops of 1, 5, or 25% preparations of the two AS cited above were instilled in the rabbit eye. Rinsing of the eyes 30 sec after application was carried out with a certain proportion of the animals. No significant differences in tolerance between the two salts were observed.

Instillation of 0.1 ml of 2, 10, or 20% aqueous solutions of the sodium, ammonium, or triethanolamine salts of C_{12} AS into rabbits' eyes produced mild irritation at the 2% level and moderate to severe irritation that persisted at 7 days at the higher surfactant concentration (27a). The relative irritancy of the three samples could not be determined since they were fairly well grouped together.

Several anionic surfactants (sodium lauryl polyether(12)sulfate = SLES, sodium coco methyl tauride = SCMT, triethanolamine lauryl sulfate = TEALS, ammonium lauryl sulfate = ALS and SLS were investigated for their irritancy on mucous membranes with different animal-irritancy test methods whose criteria include sensory response, pain/discomfort, or tissue damage. The test results of these tests (upper respiratory tract irritancy test

TABLE 7 Assessment of Irritating Properties of Sodium Lauryl Sulfate on Mucous Membranes of Rabbit Eyes[a]

Conc., %AS[b]	Irritation index according to Ref. 6 (see Scheme 2)	Irritation on corneal and conjunctival tissues scored by Scheme 7	Ref.
4.0	28.1		6
2.0	14.4		6
0.5	3.6		6
10.0		Corneal opacity until 4.5 days Conjunctival redness until 9 days	16
25.0		Grade 4.0 (Scheme 7)	36
5.0		Grade 3.5 (Scheme 7)	36
1.0		Grade 2.0 (Scheme 7)	36
2.0	10.0		52
0.05 M	10.0		52
2.0 pH = 5	12		35
2.0 pH = 6	11.5		35
2.0 pH = 7	10.0		35
2.0 pH = 8	10.4		35
2.0 pH = 9	11.5		35
2.0	10.0		53
0.5	6.8		53

[a]Eyes were not rinsed after instillation.
[b]AS = washing active substance.

and mouse writing test) were compared with the results of the Draize eye irritancy test. Data of both tests indicated that SLES was the least irritating and SLS, ALS, and TEALS were the most irritating substance. The data of the Draize eye test came to the same differentiation (27a).

An interesting comparative investigation was carried out by Daweke (32) with a homologous series of alkyl sulfates comprising the sodium salts of *n*-octyl sulfate, *n*-decyl sulfate, *n*-dodecyl sulfate, *n*-tetradecyl sulfate, *n*-hexadecyl sulfate, and *n*-octadecyl sulfate. Experiments were carried out and scored according to the method of Draize (2), and 2.5% and 86.5 mM solutions, respectively, have been investigated (0.1 ml application volume, 18 or 24 animals per solution). These applied concentrations of the substance did not lead to damage of the cornea or iris. The average irritation symptoms of the conjunctiva tissues are summarized in Table 8. As evident from this table, maximum irritation occurs at a chain length of C_{12}. This applies to both concentrations. However, in the case of the isomolar concentration, *n*-octadecyl sulfate (C_{18}) possessed a similar irritant effect to that of *n*-dodecyl sulfate, which, it is suggested, could have a connection with the

TABLE 8 Mean ($\bar{x}$) and Standard Deviations of Single Values (s) of Irritations on Conjunctival Tissues of Albino Rabbits of Sodium Alkyl Sulfates with Carbon Chain Length C_8-C_{18}

Carbon chain length	2.5% by weight				86.5 mM (isomolar)			
	n[a]	$\bar{x}$	s	V[b]	n[a]	$\bar{x}$	s	V[b]
8	18	5.9	2.9	0.49	18	3.2	1.7	0.53
10	18	16.5	4.5	0.27	18	14.7	3.3	0.22
12	18	25.7	5.2	0.20	18	25.7	5.2	0.20
14	18	23.1	7.3	0.32	24	21.1	7.6	0.36
16	18	21.4	8.4	0.39	24	20.4	8.2	0.40
18	18	14.9	5.1	0.34	18	25.4	7.5	0.29

[a]n = number of tested eyes.
[b]V = coefficient of variation (s:$\bar{x}$).
Source: Ref. 32.

method of solution preparation. The coefficients of variation given show that the smallest variation of the individual values was found with the most irritating substance, i.e., dodecyl sulfate.

Iimory et al. (33) investigated, among others, sulfates of synthetic alcohols and sulfates of natural alcohols with definite chain lengths between C_{10} and C_{18} on rabbit eyes. The reactions were scored according to the scheme of Draize (see Scheme 1). The results obtained with 0.1, 0.5, 1, and 5% concentrations are given in Table 9.

As evident from Table 9, with a concentration of 0.1% no reactions were found. However, a 0.5% concentration caused mild reactions, which decayed completely within 48 hr. The reaction arising from the 1% solutions, in the case of the sulfates with definite chain length, had also disappeared completely within 48 hr. The synthetic sulfate, in contrast, caused stronger irritations. The 5% solutions produced symptoms noticeably more severe, and up to 5 days was required for some of them to disappear (see Table 10). It is noteworthy that Iimory et al. (33) did not find a dependence of irritating activity on chain length of the alkyl sulfates.

In contrast to these results, Prottey and Ferguson (53b) found a dependency between the irritating potential and the length of the alkyl chain, with the lauryl and myristyl moieties being the most potent. The test were carried out with rat peritoneal mast cells, which released the histamine stored in the cell granules when incubated with pure surfactants.

Brown and Muir (37) compared the mucous membrane compatibility of five various anionic surfactants derived from, respectively, coconut alcohol, and synthetic alochol Dobanol 23, but otherwise identical. Coconut alcohol

TABLE 9 Effects of Alkyl Sulfates on Rabbit Eye Mucosa[a]

Preparation	Time post instillation, hr	Conc., % 0.1	0.5	1	5
Alkyl sulfate from natural alcohols C_{10}	24	0	2	4	30
	48	0	0	0	9
Alkyl sulfate from natural alcohols C_{12}	24	0	2	4	30
	48	0	0	0	12
Alkyl sulfate from natural alcohols C_{14}	24	0	2	4	21
	48	0	0	0	12
Alkyl sulfate from natural alcohols C_{16}	24	0	2	4	30
	48	0	0	0	16
Alkyl sulfate from natural alcohols C_{18}	24	0	2	4	30
	48	0	0	0	16
Sulfate of synthetic alcohol (syn. AS)	24	0	8	10	46
	48	0	0	2	23

[a]Method of Draize (3). The figures listed under "concentration" indicate the degree of irritation; in the least favorable case a value of 110 may result (33).

consists mainly of lauryl and myristyl alcohol. Dobanol 23 is a similar alcohol of technical manufacture which, in addition to even-numbered carbon chains, also contains odd-numbered carbon chains. The test method used was that described in the *U.S. Federal Register* (4) (concentrations tested: 0.1, 1.0, and 25%). The test results are presented in Table 11. The results show the following order of irritating activity (the first sulfates have the slightest, the

TABLE 10 Evaluation of Rabbit Eye Irritation by Draize Method (Theoretical Total Maximum = 110)

Sample (conc. 5%)	Time after instillation, hr 24	48	72	96	120
syn. AS[a]	46	23	9	4	0
C_{10} AS[b]	30	9	6	2	0
C_{12} AS	30	12	6	2	0
C_{14} AS	21	12	4	0	0
C_{16} AS	30	16	4	2	0
C_{18} AS	30	16	4	2	0

[a]Sulfate of synthetic alcohol.
[b]Sulfates of natural alcohol.
Source: Ref. 33.

TABLE 11 Eye Irritation Reactions to Aqueous Solutions of Surfactants

Product	Eye response		
	0.1% w/v active matter	1% w/v active matter	25% w/v active matter
Monoethanolamine alcohol sulphate	No reaction	Conjunctivitis (duration up to 48 hr)	Severe conjunctivitis (duration up to 72 hr)
Triethanolamine alcohol sulphate	No reaction	Conjunctivitis (duration up to 48 hr)	Severe conjunctivitis (duration up to 72 hr)
Ammonium alcohol sulphate	No reaction	Mild conjunctivitis (duration up to 48 hr)	Severe conjunctivitis and some corneal opacity (duration more than 72 hr). Eye lids very swollen and closed for 24 hr. No permanent damage to the eyes.
Sodium alcohol sulphate	No reaction	Mild conjunctivitis (duration up to 48 hr)	Severe conjunctivitis (duration more than 72 hr). No permanent damage to the eyes.
Sodium alcohol (2 EO) ethoxy sulphate	No reaction	Conjunctivitis (duration up to 72 hr)	Severe conjunctivitis (duration more than 7 days); some corneal opacity (duration 7 days); lids very swollen and closed for 24 hr. No permanent damage to the eyes.

Source: Ref. 37.

last ones the strongest irritating properties): monoethanolamine alcohol sulfate $=$ triethanolamine alcohol sulfate $<$ sodium alcohol sulfate $\triangle$ ammonium alcohol sulfate $<$ sodium alcohol (2EO) ethoxy sulfate.

In general, the alkyl ether sulfates have less irritant potential on the eyes than the alkyl sulfates with equal carbon chain length (15a). But undiluted AES were determined to be positive eye irritants and were classified as corrosive in tests done according to the Draize procedure. Eye irritation studies in rabbits with several undiluted AES (e.g., $C_{10\text{-}12}$ AE_3S, $C_{12\text{-}13}$ AE_3S, $C_{12\text{-}13}$ AE_6S, $C_{14\text{-}16}$ $AE_{11}S$) resulted in extensively corneal damage, inflammation and haemorhage of the iris and maximal conjunctival irritation with no significant improvement seen over a 7-day period (22a).

Alkyl ether sulfates based on fatty alcohols cause moderate, but reversible, irritations in the rabbit eye, when they are applied in a concentration of about 30% active ingredients without irrigation (53c). Most investigators regard them as less irritating than alkyl sulfates with the same alcoholic fatty base.

In contrast to other authors, Brown and Muir (37) found that the ethoxylated compound caused stronger irritations than the other tested anionic surfactants without EO molecules, and Adam and Neumann (53d) did not find a significant difference in animal experiments (Draize-test on rabbit eyes) between anionic surfactants of the type alkyl ether sulfate and alkyl sulfate.

But Ciuchta and Dodd (27a) found that AES with a high number of EO-molecules are less irritating. Instillation of 0.1 ml of 2, 10 or 20% aqueous solution of C_{12} AES into rabbits' eyes produced minimal eye irritation. Irritancy scores were similar in magnitude for all three concentrations (15 of a possible 110 points), with no signs of irritation evident by 72 hr. Leopold (40) investigated various anionic surfactants which are suitable for use in eye medicaments to increase the penetration in order to ascertain their mucous membrane irritation. The author assumed that the eyes to be treated are frequently damaged. Therefore, he tested the tolerance on eyes with corneal abrasions as well as on eyes with intact corneas. Among the anionic surfactants tested were sodium-7-ethyl-2-methyl-undecyl-4-sulfate (Tergitol 4), sodium-3,9-diethyltridecyl-6-sulfate (Tergitol 7), sodium-2-ethylhexyl sulfate (Tergitol 08), and a higher aliphatic alcohol sulfate derived from technical lauryl alcohol (Duponol ME dry). The tested concentrations were 0.1, 0.25, 0.5, 1, and 100%; isotonic solution of sodium cloride was used as solvent and the pH values of the resultant solutions ranged from 6 to 7.5. The following applications were carried out:

1. Single instillation of two drops
2. Repeated instillation four times daily (with the exception of the undiluted products) for a period of 3 days

The intact eyes were tested using all test concentrations, whereas the eyes with corneal abrasions were only tested with 0.1 and 0.5% solutions. The following observations were described with all compounds.

1. Eyes with intact cornea (single application):
   - 0.1%: Blepharospasm immediately after instillation, but failed to produce any signs of corneal involvement or conjunctival injection.
   - 0.25%: Slight conjunctival hyperemia.
   - 0.5%: Conjunctival hyperemia, edema, loosening of the conjunctival epithelium, and minute areas of corneal staining. These changes disappeared entirely within 24 hr. Tergitol 08 produced the smallest changes.
   - 1.0%: Blepharospasm, conjunctival hyperemia, edema, loosening of conjunctival epithelium, superficial corneal haziness, and areas of corneal staining. All these changes disappeared within 24 hr. Tergitol 08 produced less conjunctival changes than the other products did, and no corneal changes.
   - Undiluted substances: Tergitol 4, 7, and 08 and the powder of Duponol ME dry produced marked conjunctival injection, edema, and considerable sloughing of the conjunctival epithelium and staining of one-third of the cornea within the first few minutes. Superficial punctate areas and corneal haziness occurred in all eyes. With most eyes, these phenomena had disappeared by the end of one week, but a few eyes showed a slight corneal haze at the end of 2 weeks. No significant differences were observed among the five sulfates tested. No changes were seen in the iris, lens, or anterior chamber of the eyes.
2. Eyes with intact cornea (repeated application):
   - 0.1 and 0.25%: After each instillation considerable blepharospasm occurred, which lasted about 4 min. Only mild conjunctival injection was observed, which disappeared entirely within 24 hr after instillations were discontinued.
   - 0.5%: Produced pronounced conjunctival edema, sloughing of epithelium, and hyperemia; a mucoid discharge also developed. Corneal staining and superficial haziness occurred. All eyes cleared within 48 hr after instillations were stopped. (Tergitol 08 caused the mildest reactions among the substances tested.)
   - 1%: Findings analogous to those at 0.5% concentration but more severe and did not fully disappear in all eyes until 72 hr. (Tergitol 08 produced the cornea changes in this concentration only after 7 days of such treatment.)

3. Application four times within 24 hr on eyes with abraded cornea:
   The treatment with the 0.1 and 0.5% alkyl sulfate solutions led to a retardation of the corneal epithelium regeneration. This retardation was marked when a solution containing 0.5% of the surfactants was used. Also, the 0.1% concentration still retarded epithelium regeneration. The compound with the least influence on the regeneration of the epithelium was Tergitol 08.

These experiments demonstrated that the anionic surfactants tested lead to irritation of mucous membranes, which are dependent on concentration, but even their use in undiluted form caused only superficial damages.

Sodium lauryl diglycol ether sulfate (pure) was tested in 0.5, 1.0, 2.0, and 8% concentrations on rabbit eyes by Scholz (9) (applied volume 0.05 ml, observations up to 96 hr). The irritations were concentration-dependent. With 2% concentration the reactions were minimal and 0.5% did not result in any adverse reactions at all. The 8% solution only caused reversible damage (concerning comparable experiments with coconut fatty acid sodium sarcoside and coconut fatty acid sodium methyl taurate; see page 197). The same substance was tested by Tronnier et al. (53), according to the method of Neumann and Aalbers (6), using a 2% concentration. An SLS rating of 8-9 was found; thus the substance may be classified as more compatible than lauryl sulfate.

Serrano et al. (27b) studied the ocular tolerance of surfactants used in champoos. Among others, nine anionics (different salts of lauryl sulfates and lauryl ether sulfates) were tested in a concentration row from 1.25 to 20%. Only the 20% concentration caused strong irritations, which did not disappear totally within 7 days. In the higher concentrations, marked differences between sulfates and ether sulfates could not be stated, but between the different concentrations.

According to the FHSA classification, no significant differences in the eye-irritation ranking between sodium dodecyl sulfate (SDS) and alkyl ethoxylated sulfates (AES) with one resp. three EO exist. But with the cytotoxicity test method, a significant difference between SDS and AES was found. SDS was markedly less irritating than AES and AES with one EO was less irritating than AES with three EOs (51a).

Some technical alkyl sulfates and alkyl ether sulfates (Texapon types) were investigated in our own comparative experiments (34) for their mucous membrane irritation under standardized conditions on rabbit eyes. All rabbits (New Zealand White rabbits) originated from the same breed, were kept separately, were fed with standard food, and were male. The Texapon types examined (A400, ASV, K12, K14S special, L100, MLS, N25, N40, TH, V highly conc., Z highly conc), were instilled in 20, 10, 5, 1, and 0.5% concentrations based on active substance; 0.1 ml of the solutions was dropped into the conjunctival sac of the right eyes of eight rabbits (20 and 10% concentrations) or four rabbits (5, 1, and 0.5% concentrations) using a dropping pipette. The eyes of four rabbits of each group treated with the 10 and 20% solutions were rinsed 10 sec after application with 60 ml tap water at body temperature, whereas the eyes of the other four rabbits of these groups were left unrinsed. In the case of the low concentrations, 0.5 and 1%, the left eyes of the rabbits were treated with 0.1 ml of a 0.9% sodium chloride solution, for comparative purposes (labeled with a C in Table 12c). At 2, 6, 24, 48, and 72 hr as well as 4, 7, 10, 12, 18, and 21 days after the application or until the irritations had disappeared the inflammatory symptoms of cornea, iris, and conjunctival tissues were scored according to the scheme of Draize et al. (2) (see Scheme 1). If the reactions had not subsided in 21 days after the instillation, the damage on the mucous membranes were regarded as irreversible and the rabbits were sacrificed under anesthesia.

The scores were averaged for four rabbits with application of identical concentration and contact time. These mean values were expressed as percent of the maximal possible reactions. These percent values are summarized in Tables 12a-c.

Table 12a lists those anionic surfactants and their concentrations that, among the 11 Texapon types tested, had led to corneal damage. The Texapon types K14S special, ASV, and MLS did not cause corneal damage, even at a concentration of 20% and without rinsing. The Texapon types A400, N25, N40, and TH led to more or less severe damage only in high concentrations and with unrinsed conjunctival sac. The Texapon types K12, V highly conc., and Z highly conc. caused corneal irritations, even when applied at 5% concentration. Irreversible corneal damage was observed with 20% solutions of Texapon types L100, TH, V highly conc., and Z highly conc. and with 10% solutions of Texapon types V highly conc. and Z highly conc., but only if the conjunctival sac of treated eyes was left unrinsed.

Those Texapon types and their concentrations which caused changes in the iris are listed in Table 12b. The severity of the reactions presents a picture similar to that found for the corneal damage. The Texapon types K14S special, ASV, MLS, and TH did not cause irritations of the iris. Irreversible damages were observed only with a single animal in each of the groups treated with 20% solutions of L100 and Z highly conc. With both animals no irrigation of the eyes had been carried out.

Table 12c lists the conjunctival irritations, expressed as before, as percentage of the maximum possible reactions. The Texapon types K14S special and ASV caused the least and almost identical conjunctival reactions, followed by the Texapon types MLS and TH. Whereas the alkyl ether sulfates occur in the middle of the tolerance range, the Texapon types K12 and L100 (pure dodecyl sulfates) caused the most severe conjunctival irritations.

TABLE 12a Average Corneal Reactions in Percent of the Maximal Possible Reactions[a]

Texapon[b]	Conc.,[c] %	Time of exposure	Hours after instillation					Number of animals free of reactions on day					21st day
			2	6	24	48	72	4	7	10	12	18	
A-400	20	No rinse	1.6	4.7	6.2	9.4	4.7	1				3	
	10	No rinse	0	0	1.6	1.6	1.6	3		1			
K-12	20	No rinse	3.1	3.1	7.8	3.1	1.6	3	1				
	20	10 sec	4.7	4.7	4.7	3.1	0						
	10	No rinse	0	1.6	1.6	1.6	0						
	10	10 sec	1.6	1.6	0	0	0						
	5	No rinse	1.6	1.6	0	0	0						
L-100	20	No rinse	10.9	10.9	14.1	7.8	7.8	2	1				1 (25%)[d]
	20	10 sec	3.1	9.4	4.7	4.7	3.1	2				1	1 (6%)
	10	No rinse	3.1	4.7	3.1	3.1	0						
	10	10 sec	1.6	1.6	1.6	0	0						
N-25	20	No rinse	4.7	4.7	6.2	6.2	4.7	1				3	

N-40	20	No rinse	1.6	0.9	6.2	4.7	3.1	2	2	
	20	10 sec	0	3.1	1.6	0	0			
TH	20	No rinse	3.1	6.2	14.1	21.9	15.6		3	1 (19%)
	10	No rinse	0	1.6	1.6	0	0			
V highly conc.	20	Rinse	3.1	6.2	12.5	23.4	17.2	1		3 (20%)[d]
	20	10 sec	0	0	1.6	0	0			
	10	No rinse	3.1	4.7	12.5	15.6	10.9		3	1 (12.5%)
	5	No rinse	1.6	1.6	4.7	6.2	0			
Z highly conc.	20	No rinse	3.1	3.1	12.5	25	29.7	1		3 (62.5%)
	20	10 sec	1.6	4.7	0	0	0			
	10	No rinse	0	4.7	15.6	15.6	10.9	2	1	1 (12.5%)
	5	No rinse	0	4.7	4.7	1.6	0			

[a]Assessment according to Scheme 1 (2). Maximal irritations 80 points = 100%.
[b]Chemical structure, see Table 12c.
[c]Concentration related to active substance (AS).
[d]One or three animals showed reactions (percent of maximal possible reaction) on the 21st day and these reactions are supposed irreversible.

TABLE 12b Average Iris Reactions in Percent of Maximal Possible Reactions[a]

Texapon[b]	Conc.,[c] %	Time of exposure	Hours after instillation					Number of animals free of reactions on day					
			2	6	24	48	72	4	7	10	12	18	21st day
A-400	20	No rinse	12.5	37.5	37.5	37.5	25.0	2				2	
K-12	20	No rinse	0	12.5	25	25	0						
	20	10 sec	0	25	25	12.5	0						
L-100	20	No rinse	12.5	12.5	25	25	12.5	3					1 (50%)[d]
	20	10 sec	0	0	12.5	25	25	2				2	
	10	No rinse	12.5	12.5	12.5	12.5	0						
N-25	20	No rinse	0	25	50	25	25	2				2	
N-40	20	No rinse	12.5	50	37.5	25	0						
	20	10 sec	0	25	0	0	0						
V highly	20	No rinse	0	0	12.5	12.5	25	2	1	1			
conc.	10	No rinse	0	0	12.5	25	12.5	3			1		
	5	No rinse	0	0	12.5	0	0						
Z highly	20	No rinse	0	0	0	12.5	25	1	2				1 (100%)[d]
conc.	10	No rinse	0	0	0	12.5	12.5	3	1				

[a]Assessment according to Scheme 1 (2). Maximal reactions 10 points = 100%.
[b]Chemical structure, see Table 12c.
[c]Concentration related to active substance (AS).
[d]One animal showed reactions (50 or 100% of maximal possible reaction) on the 21st day and these reactions are supposed as irreversible.

TABLE 12c Conjunctival Reactions in Percent of the Maximal Possible Reactions[a]

			Hours after instillation					Number of animals free of reactions on day					
Texapon[b]	Conc.,[c] %	Time of exposure	2	6	24	48	72	4	7	10	12	18	21st day
A-400													
Ammonium alkyl	20	No rinse	65.0	52.5	55.0	47.5	30.0			4			
(C_8-C_{14}) sulfate	20	10 sec	37.5	42.5	25.0	15.0	5.0	3	1				
(34% AS)	10	No rinse	50.0	52.5	52.5	30.0	15.0	1		3			
	10	No rinse	57.5	52.5	27.5	10.0	2.5	3	1				
	5	No rinse	55	55	35	5	0						
	1	No rinse	20	5	0	0	0						
	C[c]	No rinse	0	0	0	0	0						
	0.5	No rinse	10	2.5	0	0	0						
	C	No rinse	0	0	0	0	0						
ASV													
Sodium and	20	No rinse	55	45	25	10	0						
magnesium alkyl	20	10 sec	25	15	2.5	0	0						
(C_{12}-C_{18}) ether	10	No rinse	27.5	27.5	7.5	2.5	0						
(7EO) sulfate	10	10 sec	20	15	0	0	0						
(29% AS)	5	No rinse	12.5	7.5	0	0	0						
	1	No rinse	0	0	0	0	0						
	C	No rinse	0	0	0	0	0						
K-12													
Sodium dodecyl	20	No rinse	62.5	62.5	45	45	30		3	1			
(C_{12}) sulfate	20	10 sec	60	57.5	42.5	32.5	12.5	2	1	1			
(90% AS)	10	No rinse	52.5	60	47.5	22.5	7.5	3	1				
	10	10 sec	50	60	37.5	17.5	2.5	3	1				

TABLE 12c (Continued)

Texapon[b]	Conc.,[c] %	Time of exposure	Hours after instillation 2	6	24	48	72	Number of animals free of reactions on day 4	7	10	12	18	21st day
K-12	5	No rinse	50	55	25	12.5	0						
(continued)	1	No rinse	22.5	10	0	0	0						
	C[c]	No rinse	0	0	0	0	0						
	0.5	No rinse	12.5	2.5	0	0	0						
	C	No rinse	0	0	0	0	0						
K-14S Special													
Sodium dodecyl-	20	No rinse	45	32.5	10	0	0						
myristyl ether	20	10 sec	30	25	2.5	0	0						
(3EO) sulfate	10	No rinse	32.5	22.5	5	0	0						
(80% AS)	10	10 sec	17.5	17.5	2.5	0	0						
	5	No rinse	17.5	5	0	0	0						
	1	No rinse	12.5	7.5	0	0	0						
	C[c]	No rinse	0	0	0	0	0						
	0.5	No rinse	2.5	0	0	0	0						
	C	No rinse	0	0	0	0	0						
L-100													
Sodium dodecyl	20	No rinse	60	57.5	55	45	35		3				1 (40%)[d]
(C_{12}) sulfate	20	10 sec	65	57.5	42.5	32.5	25	1	2			1	
(99% AS)	10	No rinse	65	60	50	32.5	17.5	1	3				
	10	10 sec	62.5	62.5	32.5	15	5	3	1				
	5	No rinse	52.5	50	20	7.5	5	3	1				
	1	No rinse	15	15	0	0	0						
	C[c]	No rinse	0	0	0	0	0						
	0.5	No rinse	15	5	0	0	0						
	C	No rinse	0	0	0	0	0						

MLS											
Monoethanol-	20	No rinse	55	55	15	2.5	0				
amine dodecyl-	20	10 sec	20	27.5	12.5	5	0				
myristyl sulfate	10	No rinse	45	42.5	5	0	0				
(33% AS)	10	10 sec	25	12.5	0	0	0				
	5	No rinse	27.5	17.5	0	0	0				
	1	No rinse	22.5	2.5	0	0	0				
	C[c]	No rinse	0	0	0	0	0				
	0.5	No rinse	17.5	2.5	0	0	0				
	C	No rinse	0	0	0	0	0				
N-25											
Sodium dodecyl-	20	No rinse	67.5	67.5	50	40	27.5				4
myristyl ether	20	10 sec	55	60	35	15	2.5	3	1		
(2 EO) sulfate	10	No rinse	62.5	57.5	60	37.5	17.5	1	3		
(28% AS)	10	10 sec	37.5	32.5	5	0	0				
	5	No rinse	52.5	50	12.5	0	0				
	1	No rinse	20	7.5	0	0	0				
	C[c]	No rinse	0	0	0	0	0				
	0.5	No rinse	10	2.5	0	0	0				
	C	No rinse	0	0	0	0	0				
N-40											
Sodium dodecyl-	20	No rinse	62.5	62.5	55	52.5	30			4	
myristyl ether	20	10 sec	45	42.5	17.5	17.5	5	3	1		
(2EO) sulfate	10	No rinse	57.5	50	50	32.5	10	3	1		
(28% AS)	10	10 sec	47.5	35	25	15	5	3	1		
	5	No rinse	42.5	47.5	25	7.5	0				
	1	No rinse	15	5	0	0	0				
	C[c]	No rinse	0	0	0	0	0				
	0.5	No rinse	0	2.5	0	0	0				
	C	No rinse	0	0	0	0	0				

TABLE 12c (Continued)

Texapon[b]	Conc.,[c] %	Time of exposure	Hours after instillation 2	6	24	48	72	Number of animals free of reactions on day 4	7	10	12	18	21st day
TH													
Triethanol--	20	No rinse	52.5	55	45	25	17.5		1	1	2		
amine alkyl	20	10 sec	40	45	22.5	5	0						
(C_8-C_{14}) sulfate	10	No rinse	47.5	62.5	37.5	10	2.5	3	1				
(47% AS)	10	10 sec	32.5	37.5	15	2.5	0						
	5	No rinse	37.5	32.5	0	0	0						
	1	No rinse	10	2.5	0	0	0						
	C[c]	No rinse	0	0	0	0	0						
	0.5	No rinse	10	2.5	0	0	0						
	C	No rinse	0	0	0	0	0						
V highly conc.													
Sodium alkyl	20	No rinse	55	57.5	52.5	45	32.5		1			2	1[e]
(C_{12}-C_{16}) sulfate	20	10 sec	30	35	20	12.5	10	3	1				
(90% AS)	10	No rinse	67.5	62.5	57.5	45	30		3			1	
	10	10 sec	47.5	47.5	27.5	12.5	0						

	5	No rinse	60	67.5	27.5	5	2.5	3	1		
	1	No rinse	7.5	7.5	0	0	0				
	C[c]	No rinse	0	0	0	0	0				
	0.5	No rinse	5	2.5	0	0	0				
	C	No rinse	0	0	0	0	0				
Z highly conc.											
Sodium alkyl	20	No rinse	62.5	65	65	42.5	37.5		1	1	2 (35%)[d]
(C_{12}-C_{18}) sulfate	20	10 sec	42.5	50	25	10	2.5	3	1		
(90% AS)	10	No rinse	55	52.5	45	37.5	25		3	1	
	10	10 sec	27.5	27.5	12.5	7.5	5	3	1		
	5	No rinse	60	60	35	20	7.5	3	1		
	1	No rinse	5	5	0	0	0				
	C[c]	No rinse	0	0	0	0	0				
	0.5	No rinse	2.5	0	0	0	0				
	C	No rinse	0	0	0	0	0				

[a]Assessment according to scheme 1 (2). Maximal reactions 20 points = 100%.
[b]Concentration related to active substance (AS).
[c]C = the left eyes served as a control. Into these eyes were instilled a volume of 0.1 ml physiological sodium chloride solution (0.9%).
[d]One or two animals showed reactions (percent of maximal possible reaction) on the 21st day and these reactions are supposed irreversible.
[e]Reactions had disappeared on 21st day.

The intensity of the conjunctival reactions show a clear dependence on contact time with the mucous membranes and on the concentration of the applied products. Texapon ASV at low concentrations proved to be the most compatible product. One percent concentrations were tolerated on rabbit eyes without any irritation. The 0.5% concentrations of the Texapon types N40, K14S special, and Z highly conc. caused practically no mucous membrane irritations on the rabbit eye and thus the 0.5% concentration of these Texapon types may be considered as the threshold concentration (see Table 4). The remaining Texapon types tested caused clear irritations at 0.5% concentration, even though they rapidly subsided.

If all the reactions on all mucous membranes (cornea, iris, and conjunctiva) over the entire scope of concentrations tested are taken into consideration together with the irritations observed after the third day, including irreversible eye changes caused by some of the Texapon types, then the following order is found. The sequence starts with the most compatible product and ends with the most irritating product. K14S special $\leqslant$ ASV $<$ MLS $<$ TH $<$ N40 $\leqslant$ N25 $<$ K12 $<$ A400 $<$ Z highly conc. $<$ V highly conc. $<$ L100. In contrast, on the basis of the results obtained at only the low concentrations 5, 1, and 0.5%, the following sequence of tolerance commencing with the most compatible substance is found: ASV $<$ K14S special $<$ MLS $\leqslant$ TH $<$ N40 $<$ N25 $<$ A400 $\leqslant$ K12 $<$ L100 $<$ Z highly conc. $<$ V highly conc.

Comparative assessment of the alkyl sulfates and the alkyl ether sulfates, wich consideration of specific cations, has shown that the sodium salts of AES are less irritating to mucous membranes than the sodium salts of AS. These findings correspond well to those observed by Finkstein (53a), who reported that a 9 M or more EO-ethoxylated sodium laureth sulfate system is practically nonirritating to eyes at use levels in shampoos, i.e., in 7-8% concentrations. But the Texapon types MLS and TH (mono- and triethanolamine alkyl sulfates, respectively) proved to be less irritating as well; Texapon A400 (ammonium alkyl sulfate) proved to be more irritating. The sodium salts of AS caused the most intensive irritations of mucous membranes, but the pure dodecyl sulfates (Texapon K12 and L100) caused in no case stronger irritations than the AS with varying carbon chain lengths (see the results of Texapon V highly conc. and Z highly conc.).

Table 13 gives a survey of all the alkyl sulfates examined, with the exception of sodium lauryl sulfate and those alkyl sulfates that are tested in our own laboratories (Cf. Tables 7 and 12a-c). As cited before, comparisons among results of various authors are not permissible.

Up to now, the tests on mucous membrane compatibility of AS have been confined to the reactions caused to the mucous membranes of the eyes because the rabbit eye is the predominantly used model for the testing of substances for such properties. However, it should be stated that studies concerned with the irritant action of AS upon other mucous membranes have also been carried out. These experiments are discussed in Sec. II.B.3.

Oral and vaginal mucous membranes show empirically higher tolerance to surfactants than the extremely sensitive mucous membranes of the eye. It could be demonstrated that toothpastes, containing up to 2% anionic surfactants, do not cause any damage of oral mucous membranes during continuous use.

A 1% aqueous solution of $NaAE_3S$ (28% AI) was not irritating to the vaginal mucosal tissue of three dogs examined 24 hr after treatment. Applied undiluted, this product produced a slight redness in two of three dogs and a deeper, more diffuse irritation of the tissue in the third test animal (22a).

Thus, sodium lauryl sulfate and isotonic sodium chloride solution were tested together with a household detergent (Woolite) for their irritant activity on gastric and esophageal mucosa of cats by Berenson and Temple (26) (see Table 14). (See Sec. II.B.3 for description of the method, test concentration of sodium lauryl sulfate 10 and 20%). The histological sections were scored according to the following classification:

0 = no damage
\+ = surface cell desquamation
\+ + = surface cell desquamation associated with edema, congestion, and inflammation
\+ + + = extensive mucosal disruption with ulceration and necrosis

The results are summarized in Table 14. Wetting with physiological sodium chloride solution local loss of the most superficial layer of mucosa of the esophagus but had no effect on the gastric mucosa. The 10% SLS solution produced moderate injury to the esophagus, consisting of intramucosal edema and congestion and loss of the superficial cell epithelial layers. In the stomach somewhat more marked reactions were observed (loss of surface mucosal cells), vascular congestion, submucosal edema, and occasional focal ulceration). Treatment with 20% SLS solution resulted in more extensive damage, particularly extensive submucosal edema and disruption and erosion of the superficial mucosa of both the esophagus and stomach.

Ciuchta and Dodd (27a) examined the relative irritancy of sodium, ammonium, and triethanolamine salts of C_{12} AS and sodium lauryl polyehter (12) sulfate in a mouse writing test. Mice were injected intraperitoneally (i.p.) with 0.2 ml of various concentrations of aqueous solutions of these surfactants and observed until a positive response was elicited for a period of 5 min. The calculated concentrations to produce writing in 50% of the animals were 0.086% (SLS, ALS), and 0.1% (TEALS), and 0.23% (SLES).

TABLE 13 Summary of Results Regarding Compatibility of Alkyl Sulfates and Alkyl Ether Sulfates on Mucous Membranes of Rabbit Eyes (without Sodium Lauryl Sulfate and Own Investigations)

Anionic surfactants	Conc. % tested	Time of exposure, irrigation after	Conjunctival and corneal reactions: Conjunctiva	Conjunctival and corneal reactions: Cornea	Assessment by means of Scheme	See Table[a]	Ref.
Fatty alcohol sulfate (35% AS)	5	4 sec	+ + +	—	4	5	45
	10	4 sec	+ + +	—	4	5	45
	20	4 sec	+ + +	+	4	5	45
Sodium alkyl sulfate, coconut alcohol-based	1	No rinse	+ to 48 hr[b]		—[c]	11	37
Dobanol 23-based	1	No rinse	+ to 48 hr[b]		—	11	37
Coconut alcohol-based	25	No rinse	+ + + to 72 hr	—	—	11	37
Dobanol 23-based	25	No rinse	+ + + to 72 hr	—	—	11	37
Sodium alkyl sulfate (Duponol ME dry)	0.1	No rinse	Blepharospasm		—	—	40
	0.25	No rinse	(+)	—	—	—	40
	0.5	No rinse	(+)	(+)	—	—	40
	1	No rinse	+	+	—	—	40
	100	No rinse	+ + +	+ +	—	—	40
Sodium alkyl sulfate (2-ethylhexyl, Tergitol 08, 40% AS)	0.1	No rinse	Blepharospasm		—	—	40
	0.25	No rinse	(+)	—	—	—	40
	0.5	No rinse	+	(+)	—	—	40
	1	No rinse	+ +	+	—	—	40
	100	No rinse	+ + +	+ +	—	—	40
Sodium alkyl sulfate (branched C_{14}, Tergitol 4, 28% AS)	0.1	No rinse	Blepharospasm		—	—	40
	0.25	No rinse	(+)	—	—	—	40
	0.5	No rinse	+	(+)	—	—	40
	1	No rinse	+ +	+	—	—	40
	100	No rinse	+ + +	+ +	—	—	40

Sodium alkyl sulfate (branched C_{17}, Tergitol 7, 27% AS)	0.1	No rinse	Blepharospasm		—	—	40
	0.25	No rinse	(+)	—	—	—	40
	0.5	No rinse	+	(+)	—	—	40
	1	No rinse	+ +	+	—	—	40
	100	No rinse	+ + +	+ +	—	—	40
Sulfate of synthetic alcohol (syn. AS C_{10}-C_{18})	0.01-0.1	No rinse	0/0[d]	—	1	9/10	33
	0.5	No rinse	8/0	—	1	9/10	33
	1	No rinse	10/2	—	1	9/10	33
	5	No rinse	46/23	—	1	9/10	33
Sodium *n*-octyl sulfate (C_8)	2.5	No rinse	5.9	—	1	8	32
Sodium *n*-decyl sulfate (C_{10})	2.5	No rinse	16.5	—	1	8	32
Sodium *n*-dodecyl sulfate (C_{12})	2.5	No rinse	25.7	—	1	8	32
Sodium *n*-tetradecyl sulfate (C_{14})	2.5	No rinse	23.1	—	1	8	32
Sodium *n*-hexadecyl sulfate (C_{16})	2.5	No rinse	21.4	—	1	8	32
Sodium *n*-octadecyl sulfate (C_{18})	2.5	No rinse	14.9	—	1	8	32
Sodium decyl sulfate (C_{10})	0.01-0.1	No rinse	0/0[d]	—	1	9	33
	1	No rinse	4/0	—	1	9	33
	5	No rinse	30/9	—	1	9	33
Sodium dodecyl sulfate (C_{12})	0.5	No rinse	2/0	—	1	9	33
	1	No rinse	4/0	—	1	9	33
	5	No rinse	30/12	—	1	9	33
Sodium tetradecyl sulfate (C_{14})	0.5	No rinse	2/0	—	1	9	33
	1	No rinse	4/0	—	1	9	33
	5	No rinse	21/12	—	1	9	33
Sodium hexadecyl sulfate (C_{16})	0.5	No rinse	2/0	—	1	9	33
	1	No rinse	4/0	—	1	9	33
	5	No rinse	30/16	—	1	9	33
Sodium octadecyl sulfate (C_{18})	0.5	No rinse	2/0	—	1	9	33
	1	No rinse	4/0	—	1	9	33
	5	No rinse	30/16	—	1	9	33

TABLE 13 (Continued)

Anionic surfactants	Conc. % tested	Time of exposure, irrigation after	Conjunctival and corneal reactions: Conjunctiva	Conjunctival and corneal reactions: Cornea	Assessment by means of Scheme	See Table[a]	Ref.
Ammonium alkyl sulfate, coconut alcohol-based	1	No rinse	+ to 48 hr[b]	—	—[c]	11	37
Dobanol 23-based	1	No rinse	+ to 48 hr[b]	—	—	11	37
Coconut alcohol-based	25	No rinse	+ + + to 72 hr	—	—	11	37
Dobanol 23-based	25	No rinse	+ + + to 72 hr	—	—	11	37
Alkanolamine alkyl sulfate (MEA),[e] coconut alcohol-based	1	No rinse	(+) to 48 hr	—	—	11	37
Dobanol 23-based	1	No rinse	(+) to 48 hr	—	—	11	37
Coconut alcohol-based	25	No rinse	+ + + >72 hr	+	—	11	37
Dobanol 23-based	25	No rinse	+ + + >72 hr	+	—[c]	11	37
Alkanolamine alkyl sulfate (TEA),[e] coconut alcohol-based	1	No rinse	(+) to 48 hr[b]	—	—	11	37
Dobanol 23-based	1	No rinse	(+) to 48 hr	—	—	11	37
Coconut alcohol-based	25	No rinse	+ + + >72 hr	—	—	11	37
Dobanol 23-based	25	No rinse	+ + + >72 hr	—	—	11	37
Alkanolamine alkyl sulfate (TEA)[e]	46	No rinse	31 (24 hr) 26 (72 hr)	—	1	—	11

Alkanolamine alkyl sulfate (TEA)[e]	1	No rinse	1.5	—	7	—	36
	5	No rinse	2.0	—	7	—	36
	25	No rinse	4.0	—	7	—	36
	5	30 sec	1.5	—	7	—	36
	25	30 sec	3.0	—	7	—	36
Sodium alkyl ether sulfate (2EO), coconut alcohol-based	1	No rinse	+ to 72 hr	—	—	11	37
Dobanol 23-based	1	No rinse	+ to 72 hr	—	—	11	37
Sodium alkyl ether sulfate (C_{12}, 2EO)	0.5	No rinse	—	—	—	—	9
	1	No rinse	(+)	—	—	—	9
	2	No rinse	+	—	—	—	9
Sodium alkyl ether sulfate (C_{12}, 2EO)	2	No rinse	8-9	—	1	—	53
Sodium alkyl ether sulfate (C_{12}, 2EO)	8	No rinse	+ +	—	—	—	9
Sodium alkyl ether sulfate (2EO), coconut alcohol-based	25	No rinse	+ + + + 7 days	+	—	11	37
Dobanol 23-based	25	No rinse	+ + + + 7 days	+	—	11	37

[a]Further details of results see named table.
[b]The reactions had disappeared after 48 hr.
[c]The authors have a special scheme; see named table or reference.
[d]The first number gives the 24-hr value, the second one gives the 48-hr value.
[e]MEA = monoethanol amine salt; DEA = diethanol amine salt; TEA = triethanol amine salt.
Grade of reactions: (+) = very slight reactions; + = slight reactions; + + = moderate reactions; + + + = strong reactions; + + + + = very strong reactions.

TABLE 14 Histological Injury of Feline Esophagus and Stomach Following Treatment with Ingredients of Household Detergents

Test substances	Esophagus 0[a]	+	+ +	+ + +	Stomach 0	+	+ +	+ + +
0.9% NaCl	6[b]	4			5			
Woolite		3	1		1			3
10% LS[c]			4				3	1
20% LS			2	2				4

[a]Histologic gradation, see text.
[b]Combined samples: pledget and irrigation techniques.
[c]Sodium lauryl sulfate.
Source: Ref. 26.

Examination of the irritancy of different surfactants on the upper respiratory tract of guinea pigs, mice, and rabbits was described by Ciuchta (27). The method is briefly discussed in Sec. II.B.3. Aerosols of 12-25% aqueous surfactant solutions were tested on groups of animals ranging from four to eight per group by 2-5 min continued inhalation. The substances and results of these tests are summarized in Table 15 for rabbits and mice. It is evident that the surfactants in specific concentrations possess irritant activity, with the sole exception of sodium lauryl ether sulfate on rabbits. At such high concentrations the findings described are not surprising. The guinea pig model yielded inconsistent results at the dose level tested.

To estimate the acute irritation in the upper respiratory tract, Ciuchta and Dodd (27a) exposed Swiss albino mice (heads only), in body plethysmo-

TABLE 15 Inhibition of Respiration after Inhalation of Aerosolized Aqueous Solutions of Anionic Surfactants

	Inhibition of respiration, %	
Test animal:	Rabbits	Mice
Conc. of aerosolized solution:	15 + 25%,	15%,
Conc. of anionic in air:	130 + 175 μg/liter^{-1}	73 μg/liter^{-1} air
Ammonium lauryl sulfate	50-60	35-40
Sodium lauryl sulfate	50-60	35-40
Tiethanolamine lauryl sulfate	50-60	20
Sodium methyl cocoyl taurate	27-35	10-12
Sodium lauryl ether sulfate	0	10-12

Source: Ref. 27.

graphs to various concentrations of aerosolized solutions of the sodium, ammonium, or triethanolamine salts of C_{12} AS for 2 min. The percent change from the average control respiratory rate was determined for each concentration at peak inhibition (stimulation of the upper respiratory tract by irritants produces reflex inhibition of the respiratory rate). A 50% reduction in respiratory rate occurred at chamber concentrations of 88 (SLS), 114 (ALS), 135 (TEALS) μg/liter, and 1794 μg/liter (SLES).

The irritative effects of topically applied sodium lauryl sulfate and other chemicals in the buccal mucosa were investigated in a rat model by Ahlfors and Larsson (22b). Rats exposed to one cheek-painting with a 0.2% solution of SLS did not show any response. At the higher concentration (2% solution of SLS), half of the animals showed an increased cellularity, mainly of a mononuclear cell type, but caused no ulcer.

3. Alkyl sulfonates and alkylaryl sulfonates

The alkyl sulfonates (ASF) have not attained the same practical degree of importance as have the alkylaryl sulfonates (AAS). Among the AAS are the alkyl benzene and alkyl naphthalene sulfonates, substances belonging to the former being more frequently used for technical purposes. Details of the substances examined, as found in the literature, often only refer to "sulfonate or alkylaryl sulfonate," without giving a more exact chemical definition. Aerosol OT (a sulfosuccinate), dodecyl benzene sulfonate (DBS), and α-olefin sulfonate (AOS) have all been examined several times. Most of the others which follow are only mentioned once or twice.

Olson et al. (36) investigated the effect of various anionic surfactants, including several sulfonates on rabbit eyes in 1, 5, and 25% solutions by instillation of two drops into each eye. The eyes of some of the animals were rinsed after 30 sec. The evaluation of the irritant action was carried out using Scheme 7 and the results are listed in Table 16. As can be seen from Table 16, the irritant activity was considerably reduced by rinsing. Further, the most varied chemical substances lead to reactions of differing intensity. At present an explanation for this fact cannot be offered because surface tension lowering is certainly not the only contributing parameter.

Table 5 presents, among other aspects, results of Meinicke (45), which include AAS. In these results this product (substance A) did not differ markedly in 5 and 10% concentrations from the commercial fatty alcohol sulfate (substance B). At higher concentrations, however, the AAS caused irreversible damage.

Kästner and Frosch (15a) compared 16 anionic surfactants regarding their eye irritation potential. Two sulfonates (secundary alkane sulfonate = sASF and alkyl benzene sulfonate = ABS) were among these surfactants. Each substance was tested in concentrations of 5, 10, and 20% active ingredient

SCHEME 7 Scale for Scoring Ocular Lesions

Index	Reaction (objective observations)	Pain evaluation (reactions by animal)
1	No effect	No response
2	Very slight effect, disappearing within 24 hr. May consist of appreciable pain initially and some conjunctival irritation. No corneal injury	No more than a few blinks. Normal in a minute or so
3	Slight effect, disappearing within a week. May consist of appreciable conjunctival irritation and pain but no corneal or internal injury	Blinks and tries to open eye. Reflexes close eye
4	Moderate effect, consisting of superficial corneal injury which clears within a week. Probably no internal effects on eye. No loss of vision expected. Conjunctival irritation may be severe	Holds eye shut and puts pressure on lids. May rub eye with paw
5	Severe effect, consisting of serious corneal injury from which recovery will be prolonged. Some impairment of vision may be expected. Internal injury may be observed. Conjunctival irritation may or may not be severe	Holds eye shut vigorously. May squeal
6	Very severe effect, consisting of total loss of vision due to serious injury to the cornea or internal structure of the eye. Conjunctival irritation may or may not be severe	Holds eye shut vigorously. May squeal, claw at eye, jump, and try to escape

Source: Ref. 36.

with and without irrigation after application and in concentrations of 10 and 20% with irrigation 10 sec after treatment. The results of all five groups were summarized and a ranking was evaluated. ABS caused the strongest reactions, sASF caused similar strong reactions as sodium lauryl sulfate (SLS).

Modde et al. (52) and Tronnier et al. (53) examined the effects of *n*-dodecyl benzene sulfonate on rabbit eyes in 2% and 0.05 M concentrations (based on WAS). It was found that triethanolamine salts of *n*-dodecyl benzene sulfonic acid possesses somewhat better compatibility than does the sodium salt of *n*-alkyl sulfonic acid.

TABLE 16 Effects of Surfactants on Rabbit Eye[a]

	Conjunctivitis or corneal injury				
	Unwashed			Washed	
Surfactant conc.	1%	5%	25%	5%	25%
Dodecylated diphenylether disulfonate magnesium salt	1.5	3.5	5	3	4
Dodecyl benzene sulfonate triethanolamine salt	1.5	3	5	2.5	3.5
Sodium dioctyl sulfosuccinate	1.5	3	5	3	4
Sodium dodecylated diphenylether monochlorosulfonate	2	2.5	4.5	2	4
Sodium dodecylated diphenylether disulfonate	2	3.3	4	3.5	3.7
Dodecylated diphenylether disulfonate triethanolamine salt	1.5	2	4	2	2
Sodium dodecylated ether disulfonate: Sodium lauryl sulfate, 1:1 ratio	2	2	3.5	2	3.5

[a]Using scheme 7 for evaluation.
Source: Ref. 36.

The mucous membrane irritation of nAOS (C_{15}-C_{18}) was compared by Oba et al. (31) to linear alkyl sulfonate (lASF) and tetrapropylene benzene sulfonate (TPBS) (0.01-4% concentrations; applied volume 0.1 ml; three animals per group). The evaluation of the reactions was carried out according to Scheme 8. The results of this test are listed in Table 17. A distinct concentration dependence of irritating activity on rabbit eyes was observed. Of

SCHEME 8 Standard of Evaluation of Investigated Results (Rating) [according to Ref. 31]

Evaluation	Investigated results
A	No aspect
B	Slight congestion in any tissues of eyelid mucous membrane, sclera, or nictitating membrane within 2 hr
C	Considerable congestion or edema in any tissues described above within 2 hr
D	Serious congestion and edema, unusual increase of secretion, prolonged photophobia, turbidity of cornea, disappearance of corneal reflex, abnormal size of pupil and yet after 24 hr any aspect above-mentioned is observed

TABLE 17 Evaluation of Rabbit Eye Irritation by Eye Irritation Threshold Test

Conc. of sample (%) (0.1 ml instilled)	Sample and type of evaluation					
	AOS		TPBS		lASF	
	Individual	In gross	Individual	In gross	Individual	In gross
4	D	D	—	—	—	—
2	DCD	D	D	D	—	—
1	DD	D	DDD	D	D	D
0.5	CC	C	CC	C	DD	D
0.1	ABB	B	BB	B	BCC	C
0.05	AA	A	BAA	A	BB	B
0.01	—	—	—	—	AA	A

Source: Ref. 31.

the three substances tested lASF caused the most intensive mucous membrane irritation followed by TPBS. The substance with the mildest irritations was AOS.

Higher concentrations (10-40%) of AOS are moderately to severely irritating to rabbits' eyes. In a typical example, 0.1 ml of 10% $C_{14\text{-}16}$ AOS instilled in the conjunctival sac of rabbits produced slight to moderate erythema of the conjunctiva, moderate chemosis of the lids, slight discharge, and discrete corneal opacity within 24 hr. Although some improvement was seen by 72 hr, corneal opacity remained (22a).

AOS concentrations of 1% are no ocular irritants in rabbits, but concentrations greater than 5% are capable of producing reversible, mild to severe ocular irritation (51d).

A series of AOS defined carbon chain length was examined and compared to a mixture of α-olefin sulfonates (AOS mixture) and to a lASF (33). Scoring was carried out according to the scheme of Draize as well as to Scheme 8 of Oba et al. (31). The results are summarized in Tables 18a, 18b, and 19. These experiments also showed that AOS was more compatible than the lASF. No dependence of the degree of irritancy on carbon chain length was observed.

Draize et al. (2) investigated the local compatibility of dioctyl sodium sulfosuccinate (DSS) with mucous membranes. Two testing methods were used:

1. For testing of the substance on the penile mucosa of rabbits 0.2 ml of the substance in aqueous solution was applied. Erythema and edema were evaluated on the basis of the scores 0-2 (maximum point score per animal

TABLE 18a Evaluation of Rabbit Eye Irritation by Draize Method[a]

Sample	Time after instillation, hr	Conc., %					
		0.01	0.05	0.1	0.5	1	5
IASF	24	0	0	0	1	7	48
	48	0	0	0	0	2	25
AOS mix	24	0	0	0	0	2	14
	48	0	0	0	0	0	5
C_{10} AOS	24	0	0	0	0	2	17
	48	0	0	0	0	0	8
C_{12} AOS	24	0	0	0	0	2	17
	48	0	0	0	0	0	8
C_{14} AOS	24	0	0	0	0	2	17
	48	0	0	0	0	0	12
C_{16} AOS	24	0	0	0	0	2	17
	48	0	0	0	0	0	12
C_{18} AOS	24	0	0	0	0	2	17
	48	0	0	0	0	0	12

[a]Theoretical total maximum = 110. From Ref. 33.

is 4). The reactions were assessed 1, 2, and 24 hr after application. DSS in 10% concentration caused reactions, scored with 2 points on the average for the erythema and edema; 1, 2, and 24 hr after application in 2% concentration with 2 points on the average at the first and with 1 point on the average at the second evaluation.

2. For testing on the eye mucosa of rabbits 0.1 ml of 0.5, 2, and 10% solutions was applied into the conjunctival sac. Scoring was carried out according to Draize. The results are listed in Table 20.

TABLE 18b Evaluation of Rabbit Eye Irritation by Draize Method[a]

Sample	Time after instillation, hr				
	24	48	72	96	120
IASF	48	25	10	4	0
AOS mix	14	5	2	0	0
C_{10} AOS	17	8	2	0	0
C_{12} AOS	17	8	2	0	0
C_{14} AOS	17	12	4	0	0
C_{16} AOS	17	12	4	0	0
C_{18} AOS	17	12	4	0	0

[a]Theoretical total maximum = 110. Sample concentration 5%.
Source: Ref. 33.

TABLE 19 Evaluation of Rabbit Eye Irritation According to Scheme 8

	Conc., %					
Sample	0.01	0.05	0.1	0.5	1	5
lASF	A	B	C	D	D	D
AOS mix	A	A	B	C	D	D
C_{10} AOS	A	A	B	C	D	D
C_{12} AOS	A	A	B	C	D	D
C_{14} AOS	A	A	B	C	D	D
C_{16} AOS	A	A	B	C	D	D
C_{18} AOS	A	A	B	C	D	D

Source: Ref. 33.

In addition to the cited sulfates listed on page 164, Leopold (40) investigated also the sulfonates Aerosol OT and Aerosol OS according to the method described for sulfates and observed symptoms on rabbit eyes similar to those described for the sulfates (on pages 167 + 168). The results found for these sulfonates lie within the limits established for the products Tergitol 4 and Tergitol 7 as well as Duponol ME dry.

According to Bellows and Gutmann (44), Aerosol OS when tested in concentrations of 0.5-1% did not cause any corneal or conjunctival reactions in the rabbit eye. Concentrations above 2% irritated the eye mucosa; concentrations from 5 to 10% caused corneal opacity which decayed within 24 hr.

Grant (15) reported a severe corneal edema after 2-4 min exposure with a soap consisting of 80% Entsufon (sodium octyl phenoxyethoxyethyl ether sulfonate).

Swan (54) states that the application of the undiluted Aerosol OT into the rabbit eye only causes superficial inflammatory reactions.

The effects on rabbit eyes of two alkylaryl sulfonates with an average molecular weight of 352 (designated as D-40) and 375 (designated as OWA =

TABLE 20 Eye Irritation in Rabbits[a]

		Average eye injury	
Substance	Conc., %	1 hr	24 hr
Dioctyl sodium sulfosuccinate	0.5	4	2
	2.0	9	2
	10.0	26	24

[a]Draize method. Theoretical total maximum = 110 points.
Source: Ref. 2.

Oronite Wetting Agent) were investigated by Hine et al. (55). They used the method of Draize et al. (2) and applied 0.1 ml of a 5% aqueous solution on the cornea of the eyes of healthy rabbits. The product OWA proved to be somewhat more irritating than the product D-40.

Kambe (22) investigated sodium alkyl benzene sulfonate (SABS) in concentrations up to 30% on rabbit eyes (applied volume 0.1 ml). The results found 24 and 48 hr after application are presented in Table 21. As can be seen, under these conditions mild corneal and conjunctival reactions were noticeable first at an SABS concentration of 10%. A 20% SABS solution caused reactions which were clearly more severe if the eyes were not rinsed out after application. This effect was magnified with the 30% SABS solution.

Hopper et al. (43) investigated the mucous membrane tolerance on rabbit eyes with different anionic surfactants; 1% solutions (based on WAS content) were applied. The substances tested and the results are given in Table 22. Table 23 presents a summary of all the cited alkyl sulfonates and the test results.

In a vaginal mucosal irritation study, 1 ml of a 1% aqueous solution of AOS (derived from $C_{14\text{-}16}$ α-olefins) did not cause vaginal mucosal irritation in dogs examined at 24 hr (22a).

TABLE 21 Slit-Lamp Microscopic Findings after Instillation of Aqueous ABS Solution in Rabbit Eyes

Group	Cornea	Conjunctiva
1% ABS 24 hr	No change	No change
4% ABS 24 hr	No change	No change
4% ABS 24 hr washing	No change	No change
4% Soap 24 hr	No change	No change
10% ABS 24 hr	No change	No change
10% ABS 24 hr washing	No change	No change
10% ABS 48 hr	One or more pinpoint lesions 1	Chemosis 1 Redness 1
20% ABS 24 hr	Grade several lesions larger than pinpoints 2	Chemosis 2 Redness 2
20% ABS 24 hr washing	No change	No change
30% ABS 24 hr	Opacity 3 Edema 3	Hemorrhage 3 Edema 3
30% ABS 48 hr	Opacity 4 Edema 4	Hemorrhage 4 Edema 4

Note: Grades 1-4: 1 = minimal sign; 2 = mild; 3 = moderate; 4 = severe.
Source: Ref. 22.

TABLE 22 Irritation Produced by Anionic Surfactants (1% Solution) on Rabbit Cornea

Agent[a]	pH of solution	Photophobia	Discharge	Inflammation	Edema	Cloudy cornea	Total
Aerosol OT	6.0	0	0	6	0	0	6
Areskap 100	8.3	0	0	2	0	0	2
Aresket 300	6.5	0	1	6	0	0	7
Aresklene 400	8.8	0	2	4	0	0	6
Igepon T	7.2	0	0	2	0	0	2
Santomerse 3	6.1	2	0	8	0	0	10
Santomerse D	7.2	3	4	8	0	0	15
Sulfatex	6.4	0	0	0	0	0	0

[a]Chemical structure of these substances (see Table 2).

Scale for evaluating the irritation:

Photophobia	1	
Discharge	1	*Observations at*:
Mild inflammation	1	5 min
Severe inflammation	2	10 min
Edema	3	1 hr
Cloudy cornea	4	24 hr

Number of positive observations were multiplied by the rating in the scale for evaluating irritation to give the number appearing in the table.
Source: Ref. 43.

TABLE 23 Summary of Results Regarding Compatibility of Alkyl and Alkylaryl Sulfonates on Mucous Membranes of Rabbit Eyes

Anionic surfactants	Conc. % tested	Time of exposure, irrigation follows	Conjunctival and corneal reactions: Conjunctiva	Conjunctival and corneal reactions: Cornea	Assessment by means of Scheme	See Table[b]	Ref.
Sodium alkyl benzene sulfonate (SABS)	1	No rinse	0	0	—[a]	21	22
	4	No rinse	0	0	—	21	22
	10	No rinse	0/1[c]	0/(+)[c]	—	21	22
	20	No rinse	2	1	—	21	22
	20	Washing	0	0	—	21	22
	30	No rinse	3/4	3/4	—	21	22
Linear alkylate sulfonate (lASF)	0.01	No rinse	1		8	17	31
	0.05	No rinse	2		8	17	31
	0.01	No rinse	3		8	17	31
	0.5	No rinse	4		8	17	31
	1	No rinse	4		8	17	31
Linear alkylate sulfonate (lASF)	0.1	No rinse	0/0		1	18	33
	0.5	No rinse	1/0		1	18	33
	1	No rinse	7/2		1	18	33
	5	No rinse	48/25		1	18	33
n-Alkyl sulfonate sodium salt	2	No rinse	12-14 (1.3)[d]		2	—	53
	0.05 M	No rinse	11.5 (1.7)		2	—	53
Decyl benzene sodium sulfonate (Santomerse D)	1	No rinse	15	0	—	22	43
Dedecyl benzene sodium sulfonate (Santomerse 3)	1	No rinse	10	0	—	22	43
n-Dodecyl benzene sulfonate triethanolamine salt	2	No rinse	8-9 (0.85)[d]		2	—	53
	0.05 M	No rinse	10.2 (1.5)		2	—	53

TABLE 23 (Continued)

Anionic surfactants	Conc. % tested	Time of exposure, irrigation follows	Conjunctival and corneal reactions: Conjunctiva	Conjunctival and corneal reactions: Cornea	Assessment by means of Scheme	See Table[b]	Ref.
Dodecyl benzene sulfonate triethanolamine salt	1	No rinse	1.5		7	16	36
	5	No rinse	3.0		7	16	36
	25	No rinse	5.0		7	16	36
	5	30 sec	2.5		7	16	36
	25	30 sec	3.5		7	16	36
Tetrapropylene benzene sulfonate (TPBS)	0.01	No rinse	0		8	17	31
	0.05	No rinse	1		8	17	31
	0.1	No rinse	2		8	17	31
	0.5	No rinse	3		8	17	31
	1.0	No rinse	4		8	17	31
	2.0	No rinse	4		8	17	31
Alkylaryl sulfonate (35% AS)	5	4 sec	+ +	—	4	5	45
	10	4 sec	+ + +	—	4	5	45
	20	4 sec	+ + + +	+ + +	4	5	45
Alkylaryl sulfonate (Mol wt 352, (D-40)	5	No rinse	8/2 (1/24 hr)		1	—	55
Poly alkylaryl sulfonate (Mol wt 375, OWA = Oronite Wetting Agent)	5	No rinse	5/6 (1/24 hr)		1	—	55
Monobutylphenylphenol sodium monosulfonate (Areskap 100)	1	No rinse	2	0	—[a]	22	43
Monobutyldiphenyl sodium monosulfonate (Aresket 300)	1	No rinse	7	0	—	22	43

Dibutylphenylphenol sodium disulfonate (Aresklene 400)	1	No rinse	6	0	—	22	43
Sodium dodecylated diphenyl ether disulfonate	1	No rinse	2	—	7	16	36
	5	No rinse	3.3	—	7	16	36
	25	No rinse	4	—	7	16	36
	5	30 sec	3.5	—	7	16	36
	25	30 sec	3.7	—	7	16	36
Dodecylated diphenyl ether disulfonate magnesium salt	1	No rinse	1.5	—	7	16	36
	5	No rinse	3.5	—	7	16	36
	25	No rinse	5	—	7	16	36
	5	30 sec	3	—	7	16	36
	25	30 sec	4	—	7	16	36
Dodecylated diphenyl ether disulfonate triethanolamine salt	1	No rinse	1.5	—	7	16	36
	5	No rinse	2	—	7	16	36
	25	No rinse	4	—	7	16	36
	5	30 sec	2	—	7	16	36
	25	30 sec	3	—	7	16	36
Sodium dodecylated diphenyl ether monochloro sulfonate	1	No rinse	2	—	7	16	36
	5	No rinse	2.5	—	7	16	36
	25	No rinse	4.5	—	7	16	36
	5	30 sec	2	—	7	16	36
	25	30 sec	4	—	7	16	36
Sodium dioctyl sulfosuccinate disulfonate	1	No rinse	1.5	—	7	16	36
	5	No rinse	3	—	7	16	36
	25	No rinse	5	—	7	16	36
	5	30 sec	3	—	7	16	36
	25	30 sec	4	—	7	16	36
Dioctyl ester of sodium sulfosuccinic acid (Aerosol OT)	1	No rinse	6	0	—[a]	22	43

TABLE 23 (Continued)

Anionic surfactants	Conc. % tested	Time of exposure, irrigation follows	Conjunctival and corneal reactions		Assessment by means of Scheme	See Table[b]	Ref.
			Conjunctiva	Cornea			
Dioctyl sodium sulfosuccinate	0.5	No rinse	4/2[c]	—	1	20	2
	2	No rinse	9/2	—	1	20	2
	10	No rinse	26/24	—	1	20	2
n-α-C_{15}-C_{18} olefin sulfonates (AOS)	0.01	No rinse	0	—	8	17	31
	0.05	No rinse	1 (A = 1)[e]	—	8	17	31
	0.1	No rinse	2 (B = 2)	—	8	17	31
	0.5	No rinse	3 (C = 3)	—	8	17	31
	1	No rinse	4 (D = 4)	—	8	17	31
	2	No rinse	4	—	8	17	31
	4	No rinse	4	—	8	17	31
α-Olefin sulfonates (AOS)	0.1	No rinse	0/0[c]		1	18	33
	0.5	No rinse	0/0		1	18	33
	1	No rinse	2/0		1	18	33
	5	No rinse	14/5		1	18	33
α-Olefin sulfonates with differing chain length (C_{10}, C_{12}, C_{14}, C_{16}, C_{18})	0.1	No rinse	0/0		1	18	33
	0.5	No rinse	0/0		1	18	33
	1	No rinse	2/0		1	18	33
	5	No rinse	17/5-12		1	18	33

[a]The authors have a special scheme; see named table or reference.
[b]Further details of results, see named table.
[c]The first number gives the 24-hr value, the second one gives 48-hr value.
[d]Reactions expressed as NaLS index, in ( ) reactions corresponding to sodium lauryl sulfate (NaLS) = 1.
[e]Transformation of the grades of reactions (A-D) given by Oba et al. (31) into number values.

4. Anionic surfactants excluded from Secs. 1-3

The anionic surfactants now described belong to the groups of fatty acid protein condensates (PFC), fatty acid methyl taurate, fatty acid sarcosides, and fatty acid isethionates. The esters of sulfosuccinic acid are also included. However, as they contain a sulfonate group, they are described under sulfonates. The mucous membrane irritant properties of this group are only occasionally described in the literature. Data of some new anionic surfactants published in the last ten years shall be added to this chapter.

Meinicke (45) investigated in addition to the already cited fatty alcohol sulfate and an alkylaryl sulfonate also a commercial PFC (product C) containing the same WAS content as the sulfate and the sulfonate. All these products were tested comparatively for their mucous membrane irritant properties, and it was determined that this product, referred to as product C, did not possess any such irritant properties at a 20% concentration. Further, it was shown that the mucous membrane irritations of the sulfates and the sulfonates can be strongly reduced by the presence of this product C (see Table 5). Lechnitz (30) also points out the good mucous membrane tolerance of fatty acid protein condensates. In a row of 16 anionic surfactants tested for their eye irritancy, the PFC proved as the substance with the best mucous membrane compatibility (15a).

The investigations of Modde et al. (52) and Tronnier et al. (53) encompassed two protein fatty acid condensates (potassium salts, molecular weights 750 and 900). They behaved better in the tests than the other anionic surfactants tested at the same time.

A sodium salt of fatty acid methyl taurate was among the substances tested by Tronnier et al. (53). It was somewhat better regarding its mucous membranes tolerance in the tests than sodium lauryl sulfate (SLS) but not better than lauryl glycol ether sulfate. Sodium coco methyl tauride has a better mucous membrane compatibility than SLS, ammonium lauryl sulfate, and triethanolamine lauryl sulfate, but was more irritating than sodium lauryl polyether(12) sulfate tested in the upper respiratory tract irritancy test and the mouse writing test (27a).

According to Martin et al. (38), Igepon TN74 (sodium *N*-methyl-*N*-acyl laurate) does not cause any irritation to the mucous membranes of the rabbit eyes when tested in a concentration of 1.1% active substance. However, in this concentration level it did cause an increase in the threshold of pain.

Modde and Schuster (56) concerned themselves with the dependence of mucous membrane compatibility on molecular weight of fatty acid protein condensates. They found that the irritant activities of 2% solutions is very small. Ten percent solutions caused moderate reactions which decreased with increasing molecular weight. The better compatibility of products of higher molecular weight was noticeable, in so far as the milder effects caused by such products decayed faster than those of low molecular weight products.

Scholz (9) in a series of experiments detailed on page 168 reported the results of experiments with the sodium salt of coconut fatty acid sarcoside (technical) (product B) and the sodium salt of coconut fatty acid methyl taurate (technical) (product C). The test concentrations used were 0.5-8% and the applied volume 0.05 ml per eye. The results showed that there are differences between the products B and C, as far as strength and duration of the mucous membrane irritation are concerned. The findings at various concentrations do not always run parallel to each other. At a concentration of 8%, redness and swelling do not proceed parallel to each other, the redness lasting significantly longer than the swelling. Likewise the concentration dependence of both products differ. With decreasing concentration the swelling disappears more rapidly than the redness.

According to Hopper et al. (43), Sulfatex (sodium and potassium salts of a sulfated fatty alcohol amide) does not possess mucous membrane irritant properties for the rabbit eye at a concentration of 1%. Oleic acid methyl taurate (Igepon T) caused mild inflammatory symptoms when applied in 1% solution into the rabbit eye (see Table 22).

Guillot et al. (56a) tested the ocular irritancy of 56 chemicals in the rabbit eye, with and without rinsing using three procedures (proposed by AFNOR, and OECD and French authorities). Ammonium isethionate was tested undiluted without rinsing and was classified as "slightly irritant."

Serrano et al. (27b) tested several surfactants, among others dioctyl sulfosuccinate, in four concentrations (2.5-20%) on rabbit eyes and could not find a significant better compatibility of the succinate in comparison to the other tested surfactants. In contrast to these findings, other authors (15a) who also compared a row of anionic surfactants found that the sulfosuccinic esters have a low eye-irritating potential.

Nine surfactants were tested by Matsuura et al. (29a) for irritancy on eye mucosa. On direct instillation of 10% surfactant solutions into rabbit and guinea pig eyes the irritancy was highest in cationic surfactants. Sodium lauroyl sarcosinate had a similar irritancy as sodium lauryl ethoxysulfate, but sodium polyoxyethylene monoalkyl sulfosuccinate caused only mild irritations, which had disappeared within 2 days.

Ciuchta (27) observed that respiration by inhaling of aerosolized solutions of sodium methyl cocoyl taurate is obstructed at 27-35% solutions with rabbits and 10-12% solutions with mice (see page 184 and Table 15). Table 24 summarizes the results of mucous membrane irritant properties of these groups of anionic surfactants.

A new group of anionic surfactants (acyl lactylates) was tested for acute eye irritation in accordance with techniques specified in the regulations for

TABLE 24 Summary of Results Regarding Compatibility of Different Anionic Surfactants on Mucous Membranes of Rabbit Eyes

Anionic surfactants	Conc. % tested	Time of exposure, rinse follows	Conjunctival reactions	Assessment by means of Scheme	Axiliary table[a]	Ref.
Oleic acid methyl taurate (Igepon T)	1	No rinse	2	—[b]	22	43
Coco fatty acid methyl taurate sodium salt (technical)	0.5	No rinse	0.5	—	—	9
	1	No rinse	1.5	—	—	9
	2	No rinse	2	—	—	9
	8	No rinse	2.5	—	—	9
Fatty acid methyl taurate sodium salt	2	No rinse	7-8 (0.75)[c]	2	—	53
	0.05 M	No rinse	7.9 (1.16)	2	—	53
Coco fatty acid sarcoside sodium salt (technical)	0.5	No rinse	1	—	—	9
	1	No rinse	1	—	—	9
	2	No rinse	2	—	—	9
	8	No rinse	2.5	—	—	9
Protein fatty acid condensate (Mol wt 750)	2	No rinse	0.4-1.1	2	—	35
Protein fatty acid condensate potassium salt (Mol wt 750)	2	No rinse	0.6 (0.06)	2	—	53
	0.05 M	No rinse	2.7 (0.40)	2	—	53
Protein fatty acid condensate potassium salt (Mol wt. 900)	2	No rinse	0.5 (0.05)	2	—	53
	0.05 M	No rinse	1.3 (0.19)	2	—	53
Protein fatty acid condensate (Mol wt 500-1200)	2	No rinse	1-3	2	—	56
	10	No rinse	3-13	2	—	56
Protein fatty acid condensate (35% AS)	5	No rinse	0	4	5	45
	10	No rinse	0	4	5	45
	20	No rinse	0	4	5	45
Sodium and potassium salts of a sulfated fatty alcohol amide (Sulfatex)	1	No rinse	2	—[b]	22	43

[a]Further details of results, see named table.
[b]The authors have a special scheme; see named table or reference.
[c]Reactions expressed as NaLS index, in ( ) reactions corresponding to sodium lauryl sulfate (NaLS) = 1.

the enforcement of the FHSA. Sodium and calcium stearoyl-2-lactylate were nonirritating, even in a 100% concentration, sodium capryl and lauroyl lactylate were nonirritating in a concentration of 10%, and sodium isostearoyl-2-lactylate caused slight irritation in a concentration of 15% (30a).

5. Mixtures containing anionic surfactants as major constituents

Some authors investigated trademarked products which contained as major constituents anionic surfactants for their mucous membrane tolerance. Burton (7) examined also a shampoo which contained 15% monoethanolamine lauryl sulfate in undiluated form on the rabbit eyes, using a method he had developed (see Sec. II.B.2). After application of the shampoo swelling of the cornea developed rapidly and persisted for 2-3 days, followed by a deswelling to the original corneal thickness.

Opdyke et al. (50) tested two household synthetic detergents for mucous membrane compatibility on rabbit eyes according to the method of Draize.

TABLE 25 Responses of Rabbit Eyes to Aqueous Solutions and Powders

Detergent type	Conc. %	Number of rabbits	Mean maximal score	Return to normal: Number of rabbits	Return to normal: Number of days
1	10	18	8.5	10	2
				5	3
				2[a]	5
	100[b]	9	12	2	1
				6	2
				1	3
2	10	18	12	6	1
				8	2
				3	3
				1	4
	20	6	14	3	3
				1	4
				2	5
	25	3	21	3	4
	100[b]	6	12	1	1
				4	2
				1	4

[a]A slight transient corneal haziness.
[b]Three milligrams dry powder.
Source: Ref. 50.

The detergents comprised almost identical composition and, among other features, 20% anionic surfactants (either sodium alkyl benzene sulfonate = type 1, or equal parts of sodium alkyl benzene sulfonate and sodium alkyl sulfate = type 2); 0.1 ml or 3 mg powder was instilled into the conjunctival sac. Scoring was carried out after 1 and 24 hr, and consequently in 24-hr intervals until the reactions had subsided. The results of the examination are listed in Table 25. Significant differences between the two detergents concerning their irritant properties for mucous membranes were not seen.

Beckley (18) examined a liquid detergent consisting of 12% ethyl alcohol, 38% alkyl benzene sulfonate, and 50% water for its irritancy to the eye mucosa of rabbits, dogs, monkeys, and humans. The product was applied (0.1 ml) undiluted into the eyes. The eyes were either rinsed after 4 sec or not rinsed. The results of the animals experiments are given in Table 26. The

TABLE 26 Scores of Ocular Lesions Produced by Instillation of Liquid Detergent into Eyes of Several Species of Animals

Time of reading	Tissue	Eyes unflushed[a] Rabbit	Eyes unflushed[a] Dog	Eyes unflushed[a] Monkey	Eyes flushed[b] Rabbit	Eyes flushed[b] Dog	Eyes flushed[b] Monkey
1 hr	Cornea	33.3	40.0	20.0	15.0	40.0	0
	Iris	10.0	5.0	2.5	10.0	0	0
	Conjunctiva	12.0	4.0	1.0	10.0	0	0
24 hr	Cornea	33.3	45.0	10.0	13.3	20.0	0
	Iris	10.0	5.0	0	10.0	0	0
	Conjunctiva	12.0	4.0	0	10.0	0	0
48 hr	Cornea	26.7	30.0	5.0	6.7	20.0	0
	Iris	10.0	5.0	0	10.0	0	0
	Conjunctiva	10.0	4.0	0	8.7	0	0
72 hr	Cornea	21.7	30.0	0	5.0	20.0	0
	Iris	10.0	5.0	0	8.3	0	0
	Conjunctiva	9.3	0	0	7.3	0	0
4 days	Cornea	10.0	10.0	0	1.7	10.0	0
	Iris	10.0	5.0	0	6.7	0	0
	Conjunctiva	10.0	0	0	4.7	0	0
7 days	Cornea	6.7	0	0	1.7	0	0
	Iris	8.3	0	0	3.3	0	0
	Conjunctiva	6.7	0	0	2.7	0	0

[a]The numbers in these columns are the mean Draize (3) scores of six rabbits, six dogs and four monkeys.
[b]The numbers in these columns are the mean Draize (3) scores of three animals of each species. The eyes were flushed 4 seconds after instillation of the sample in the case of rabbits and dogs, 2 seconds in the case of monkeys.
Source: Ref. 18.

severest reactions were obtained with the rabbits whereas the mildest irritations were observed with monkeys. The corneal irritations with monkeys disappeared after 2 days, whereas in the case of dogs 7 days elapsed before the effects had decayed. The rabbit eyes still showed damage after 7 days. In no case inflammatory symptoms or damage were established on the cornea or iris of the human eyes, although the undiluted product had been instilled. In contrast to the animal eyes all treated human eyes were rinsed 4 sec after application, yet not only the unrinsed eyes but also the rinsed eyes of rabbits and dogs showed corneal changes. Only the monkey eyes reacted similarly to the human eyes.

A bath additive containing 35% sodium lauryl sulfoacetate was tested for eye irritation. Six rabbits were treated in one eye with 0.1 ml of a 1% solution (0.35% surfactant). Slight conjunctival redness was observed 1 hr after treatment and had dissipated within 48 hr. A single dose of 0.1 ml of a 10% aqueous solution of a milk bath containing 30% sodium lauryl sulfoacetate was tested for ocular irritation potential. All rabbits had minimal conjunctival irritations at 1 and 24 hr, and no irritations at 48 hr (45a).

The investigations described on page 179 concerning the esophageal and gastric mucosa compatibility, as carried out by Berenson and Temple (26), included also a household detergent (Woolite). The results indicated that this detergent was less irritating than 10% lauryl sulfate on the esophageal mucosa, but more irritating on the gastric mucosa (see Table 14).

III. TESTING OF SKIN COMPATIBILITY OF ANIONIC SURFACTANTS

A. Basis for the Test

As already mentioned in Sec. I, anionic surfactants are used for the most varied purposes. When a completely mechanical cleaning process of objects is not possible, human hands come into contact with the cleaning agents, and during cleansing of the body direct contact occurs.

In order to avoid possible injury to the potential user of surfactants caused by skin contact when using the cleaning agent according to the instructions, it is necessary to test the local compatibility of anionics before they come into general use. Before these substances come into contact with the human skin, they must be tested using animal subjects. Even though all the results obtained in animal experiments cannot be directly extrapolated to humans, an estimation of the compatibility of a new substance can be made if experimental results using various models on different animal species are available. In this context, it is useful to compare the results found to those for substances which are chemically similar or which find the same use, and which have been well-proven in practice. In the following sections various experimental models used to judge skin reactions are described in detail.

B. Testing Methods, Judgment Criteria, and Judgment Schematics

When laying down the methods for determining skin compatibility, the conditions of contemplated product use must form the starting point of such considerations. It is thereby decisive as to whether the substance, which is only one component of a final product, is in a product that only comes into contact with the human skin for short periods of time or very occasionally, or is in a product that comes into contact with the skin for either longer periods of time or frequently for short periods of time. As animal experiments where contact over long periods of time can be carried out only at great trouble, it is common practice to test products at a higher concentration for short periods of time, as localized effects are determined by the application time and concentration. It is often necessary in animal experiments to choose the concentration of the comparison product (or the test conditions) so that moderate reactions occur. The product under test using comparable conditions can then be seen to be more, less, or equally compatible.

The surface epidermis layers of the skin are an effective barrier against external agents. However, it is often damaged skin that comes into contact with such substances. Thus, animal experiments are frequently carried out using a scarified skin surface. The protective barrier function of the epidermis can also be lowered or canceled by applying the substances to epidermis samples which are covered by a nonporous strip, thus resulting in high humidity, so that the cells of stratum corneum swell up. This method is known as the occlusive epicutaneous test as described by Draize et al. (2,13) and is recommended as an official testing method (3,5). A variation (4-hr exposure time, only intact skin, semi-occlusive test conditions) of the Draize skin test is described in the OECD Test Guideline No. 404 (56b) and is used to test the primary skin irritation of chemicals for purpose of classification and labeling. A further method for predamaging the skin uses formaldehyde (57). The response of the skin, i.e., the formation of an inflamed area, can be read off directly or can be marked distinctively visible by application of a dyestuff (58). Substances can be tested not only for their irritant properties but also for their acanthogenic, comedogenic and sensitizing properties.

The albino rabbit is the classical experimental animal; it is easy to handle and facilitates the side-by-side testing of several substances. A major disadvantage is one common to all furred animals, namely, the skin area must be clipped prior to testing. One of the first descriptions of testing substances for skin compatibility was by Draize et al. (2).

The more detailed method, as described in 1944 (2), is an epicutaneous test under air exclusion on the intact and abraded skin of the albino rabbit clipped free of hair. This method is also recommended in the Code of Federal Regulations, Title 21, Chapter 1, § 191.11 (5). For details of the procedure, the reader is referred to the original publication. Since the appearance of the original publication, numerous modifications have been described, all of which are more or less useful if used for product comparison. Of special importance is the evaluation of the skin reactions as shown in Scheme 9. Rostenberg (59) criticized this evaluation scheme because of the difficulty of extrapolation to the human and the single application of the test product, which allows one to detect only primary irritants. The value of the Draize test when applied to mild irritants has been questioned by various authors (60-62) and it has been said that such a test gives only very rough information (63). The use of a covering material which is nonporous to air creates conditions which only exceptionally correspond to those found in practice (64). In contrast, Levenstein (65) noted good agreement between the results found for various cosmetics, including a shampoo, when tested on rabbit and human skin with the patch test method of Draize (2).

The OECD Test Guideline No. 404 (56b) is a modification of the Draize test. There are some important differences from the original Draize test. First, the method of patch application was enlarged from an occlusive to a semi-occlusive patch and second, the application time was reduced from 24 hr to 4 hr. Third, the application is carried out only on the intact skin. This

SCHEME 9 Evaluation of Skin Reactions [according to (2)]

Erythema and eschar formation	
No erythema	0
Very slight erythema (barely perceptible)	1
Well-defined erythema	2
Moderate to severe erythema	3
Severe erythema (beet redness) to slight eschar formation (injuries in depth)	*4*
Total possible erythema score	*4*
Edema formation	
No edema	0
Very slight edema (barely perceptible)	1
Slight edema (edges of area well defined by definite raising)	2
Moderate edema (edges raised approximately 1 mm)	3
Severe edema (edges raised more than 1 mm, and extending beyond area of exposure)	*4*
Total possible edema score	*4*
Total possible score for primary irritation	8

OECD Test Guideline is used for determining skin irritation potential, and for classification of substances and preparations. The substance (0.5 ml of liquids or 0.5 g of solids or semisolids) is applied in a single dose to the test area of 6 cm^2 of several experimental animals. Each animal serves as its own control. The degree of irritation is read and graded after specific intervals. The duration of observations should be sufficient to evaluate fully the reversibility of the effects observed.

In case of testing solids the test substance should be moistened with water. Liquid test substances are generally used undiluted. At least three healthy adult animals (the albino rabbit is the preferred species) are used. The signs of erythema and oedema on the treated skin area are graded and recorded to the Draize scheme (Scheme 9) at 30 and 60 min, and then at 24, 48, and 72 hr after patch removal. Further observations may be needed, as necessary, to establish reversibility. These test results are used for the classification of irritant and corrosive substances to the skin within a scope of the EEC Dangerous Substances Directive (13b) and leads to the labeling with the risk phrases R 38 (Irritant to the skin) or R 34 (Causes corrosion). Definitions of irritant and corrosive are given in the Article 2.2.1 of the Sixth Amendment (13b). Most of the undiluted anionic surfactants (LABS, AS, AOS) have to be classified as "irritant" (22a,51d). AES with a higher number of EO molecules are less irritant (51d).

Testing for relative primary irritation, which in the case of repeated application on the human skin leads to a so-called skin fatigue, can, according to Finkelstein (57), also be undertaken. Using this method, even products showing a very mild irritant action can be recognized and, applying the Trypan blue test of Hoppe et al. (58), the erythema formation can be followed by means of injected Trypan-blue solution. The results of both test methods are apparently in good agreement with those results obtained in tests using the upper arm of human subjects (patch test) (57).

Another method for testing skin compatibility using a solitary application is described by Barail (66).

The animal is injected intradermally, the method being recommended as a preliminary test for absolute and relative primary irritation. A single dose of the test substance, suitably diluted, is injected intradermally and the irritation at the site of injection 24 hr later is judged. The intradermal injection method, according to Barail, has proved to be particularly useful in those cases where a substance has to be tested and compared to others as regards its primary irritation properties. Using a point scheme, the skin irritation may be relatively easily and reproducibly judged. The white mouse is a suitable animal for this test. But a rabbit intradermal test is described by Pittz (66a). The advantage of using rabbits consists in testing a concentration row on each animal.

The repeated application in the animal experiment often corresponds closely to practical conditions of use. The results of such experiments thus allow one to know better what to expect regarding skin compatibility in the case of humans than results based upon single application.

The repeated application can be carried out such that the product is applied to the intact or abraded skin at a definite concentration once or several times per day. The treated skin area may remain uncovered (open patch test) or may be covered with a bandage or plaster (closed patch test). The covering material can either be porous to air (unocclusive patch test) or nonporous (occlusive patch test) (67). As a consequence of the covering a so-called microclimate results, which varies according to the nature of the covering, thus influencing the results of the test. The severity of the test method increases from the open to the closed test and further from the nonocclusive to the occlusive test (59,67).

Gilman et al. (67a) studied the influence of concentration, exposure duration, and patch occlusivity (occluded versus semi-occluded patch) upon rabbit primary dermal irritation indices for four detergent granular products. They found a clear correlation between skin irritation and concentration. The paste (moistened with water = 83% a.i.) not the undiluted granular products was most irritating, the 10% aqueous dilution the least irritating. The severity of the effects was correlatable with the increase in exposure time between 4 and 24 hr. The occluded patch sites had higher irritation scores than the semi-occluded sites at all time intervals.

For repeated application, rabbits, guinea pigs, rats, and mice are used. However, hairless animals, especially hairless mice, which do not have to be clipped or depilated, are being used in increasing numbers.

For products which come into contact with large areas of human skin, e.g., bath additives, the immersion test is suitable whereby guinea pigs (66a, 68) or other animals are immersed into an aqueous preparation of the test substance for a fixed period of time daily. Hairless mice are also suitable subjects for this test (69). When presenting the results of such examinations concerning skin compatibility of surfactants, it is essential to describe briefly the methods used.

In the last 15 years, alternative methods have been developed. The principal aim of such tests is to reduce the number of, or to replace, experimental animals for the assessment of the skin irritation potential of chemicals and preparations. The published techniques can be classified into those that use established cell lines of animal (mouse embryo fibroblast cell lines) or human (epidermal keratocytes) origin (13w,69a-e), those that use skin tissue (13w, 69b,69f,69g) or special parts of skin (69c,69h-l), or those approaches that are based on physicochemical models (69m,n).

In general, techniques using cells do not retain the selectivity of in vivo skin tissue as a barrier to the penetration of many chemicals. The cell culture methods are based on the cytotoxicity component, using mouse fibroblasts, designated to measure the effect of test compounds on cell proliferation, on cellular metabolism by measurement of alterations in uptake of a specific substrate (^{3}H-uridine, neutral red), and on cell-membrane integrity by measurements of the levels of leakage of enzymes (hexosaminidase, lactate dehydrogenase, acid phosphatase) or chromium-51. Another in vitro model is based on the keratinocyte differentiation for examining the ability of test compounds to impair or induce the stratification and keratinization characteristics of an epithelial cell line.

Techniques using skin tissue retain the natural species-specific permeability characteristics of in vivo skin in their primary response to irritant substances. As donor animals, white mice (69f,69g) and hairless mice (13w, 69b) are used. Skin preparations can be kept viable for short periods of time in appropriate culture systems. However, in contrast to permanent cell lines, they have to be prepared freshly before every series of experiments. Treatment with the test chemical takes place exclusively at the epidermal side of the skin explants, ensuring to take into account penetration mechanisms. The degree of cellular injury caused by the test substance is reflected by inhibition of uptake and probable incorporation of isotope-labeled amino acids from the culture medium into epidermal DNA and protein, respectively, and by measurement of leakage of intracellular enzymes into the culture medium in a dose and time-related manner. Bartnik et al. (13w) and other authors (69b,d,g) compared the test results of cell and skin cultures in vitro tests with in vivo data and found a more or less good correlation. Cell and skin cultures provide good results when used for comparative testing within the same class of chemicals. But both methods have limitations with respect to chemical properties of the product to be tested (69b).

A new in vitro test for the skin compatibility of surfactants was developed by investigating the swelling behavior of isolated pig epidermis. The treatment of epidermis strips with aqueous solutions of various surfactants (5 alkyl sulfates, 2 alkyl sulfonates, 8 alkyl ether sulfates, and some others) caused different swelling values depending on the structure of the surfactants. Comparison with skin compatibility found on animals and humans reveals a good correspondence between these parameters (69i).

Putterman et al. (69h) investigated the swelling of squares of guinea pig stratum corneum after contact with several surfactants and studied the detergent-skin interactions for predicting detergent penetration of skin and possible subsequent skin irritancy. Highest levels of swelling were observed with the anionic surfactants sodium laurate and sodium lauryl sulfate.

Oliver, Pemperton, and Rhodes (69j-l) describe in several publications the in vitro rat epidermal slice technique, a method that has been developed for identifying chemicals with potential to cause a corrosive lesion in animal skin in vivo. This potential has been correlated with the ability to lyse stratum corneum and has been measured as a lowering of the electrical resistance of the skin slice. An intralaboratory double-blind trial with 34 corrosive chemicals and 36 irritants showed the technique to have total sensitivity (i.e., no false negatives) and a specifity of 88%. The method has a high reproducibility for identifying skin-corrosive chemicals. The model can be used as a pre-screen for conventional animal tests with the advantage of providing objective and quantifiable information.

Prottey et al. (69o) measured the surfactant actions upon human skin under realistic conditions. When the stratum corneum of panelists was exposed to dilute solutions of various surfactants under realistic exposure conditions, the changes observed in stratum corneum acid phosphatase specific activity have been found to correlate very closely with the visual, macroscopic changes such as dryness and flakiness that are elicited in skin as a result of surfactants. This method monitors denaturation of stratum corneum proteins, which is an important feature of skin surfactant interactions, and serves as an effective, noninvasive predictive tool for skin irritancy and mildness of surfactants.

The Skintex method, published by Gordon et al. (69n), is based on a physicochemical model of dermal irritation. The absorption through a permeability barrier and the direct reactivity with a protein matrix are quantitated spectrophotometrically. Test samples were applied undiluted and at different dose levels. The Primary Dermal Irritation Indices (PDII) of known skin irritants were used to calibrate the response. No false-negative results were obtained. The results indicate the potential value of the Skintex methodology as an in vitro screen for dermal irritation.

Young et al. (69m) proposed a method by which substances or preparations may be classified as irritant or corrosive to skin, without being tested on animals, when the irritant or corrosive properties are due to acidity/alkalinity of substances or preparations. This method is not useful for classifying surfactants because the irritant potential of surfactants does not depend mainly on their acidity or alkalinity.

Apart from their compatibility properties, other properties of surfactants have also been examined. Thus, testing for acanthogenic activity using repeated application on guinea pigs and for comedogenic activity using repeated application on the inner side of the auricle of the rabbit ear have both been described.

C. Generalized Changes in the Skin Caused by Anionic Surfactants

The term "generalized changes" should be understood as including gross-visible and histological-visible skin changes, the influencing of skin swelling, the effect upon the pH value of the skin, and the effect upon its healing properties, due to the action of anionics. These effects are dependent on the type of surfactant, concentration and timefactors, and the condition of the skin.

1. Macroscopic and microscopic changes of the skin

Common acute skin irritation usually consists of a local inflammatory response characterized by erythema and/or edema. However, other reactions of an acute or chronic nature may lead to dryness, roughness, flaking, and epidermal thickening without noticeable inflammation. In addition to visual and tactile phenomena, an individual may experience transient pruritus, stinging, burning, or related sensations.

Mild, macroscopic, and visible changes that can be caused by the influence of specific anionic surfactants under suitable conditions are characterized by dryness of the skin and by roughness with somewhat increased scaling. In the case of moderate skin irritation a slight or well-defined erythema and edema formation (skin redness and epidermal induration) are seen. These processes can be accompanied by a hyperceratosis and/or acanthosis and a reduction in elasticity. More pronounced skin irritation causes skin erosion with the formation of fissures.

Rhagades and bleeding may also be observed. The consequence of such reactions is eschar or crust formation. Should a very severe irritant come into contact with the skin, then death of the tissue can result, i.e., necrosis occurs. The necrotic skin is later rejected, after the tissue defect has been isolated from within and the epithelization has occurred, leaving a scar. These macroscopic, visible changes proceed hand-in-hand with histologically demonstrable changes. The living epidermis cells initially show faint signs of an intra- and intercellular edema and somewhat enlarged blood vessels in the papillary dermis. A vacuolation in the epidermis cells is then to be seen. The next step is a more or less marked leucocyte infiltration of the epidermis which can be followed by infiltration of blood together with a visible edema of the stratum spinosum (mild spongiosis). At this stage there is a clear edema and inflammation in the papillary dermis. A more severe lesion is characterized by a severe crust formation, which consists of leucocytes, nuclear debris, and blood infiltration. At the same time a clear intra- and intercellular edema is to be seen with leucocyte infiltration (polymorphic nuclear leucocytes). A more drastic reaction results in the appearance of pyknotic nuclei and achromie of the upper layers, or necrosis which extends to the papillary dermis. Severe symptoms of inflammation and bleeding are also to be found in the papillary dermis. A single application of mouse skin with 10% sodium lauryl sulfate solution has been shown by Lansdown and Grasso (70) to suffice in causing marked macroscopic and histological changes. A mild swelling and slight erythema were found. The histological changes consisted of intracellular edema, nuclear pyknosis, and necrosis, whereby a dependence of the severity of the reaction on contact time was established. Electron microscopical examination by Tovell et al. (71) of rat epidermis which had been treated with 10% sodium lauryl sulfate solution six times on three consecutive days showed an increase in the intracellular lipids of most cells observed, spongiosis and vesiculation, partial detachment of basal cells from the basal lamina, condensation of the tonofibrils within the cells of the stratum basale and stratum spinosum, epidermal edema, as well as a thickening of the epidermis and the stratum corneum. Macroscopic observation showed coarsening, reddening, and edema of the skin.

Epidermal application of soaps with synthetic $C_{10\text{-}16}$ fatty acids using guinea pigs for 14-15 days resulted in inflammation and impaired permeability of the skin capillaries. $C_{17\text{-}20}$ soaps do not show these effects. Inanov et al. (72) also found that mixtures of detergents containing alkyl benzene sulfonates, alkyl sulfonates, and alkyl sulfates, when applied epicutaneously in 1-6% concentration to guinea pig skin accelerate the absorption of intracutaneously injected, physiological NaCl solution. Histological examination showed a focal proliferation of the epidermis.

Although anionics do not cause any specific irritant symptoms when applied to the skin of experimental animals, reactions caused by soaps differ from those caused by alkyl sulfates and alkyl sulfonates. Some surfactants cause more exsudative, others more infiltrative reactions. The various experimental animals used also show varying reactions. For these reasons the discussion of the individual surfactants must also describe the type of reactions found.

2. Degreasing and swelling of the skin by anionic surfactants

Synthetic surfactants have a degreasing effect upon the skin, which can cause dryness, roughness, and increased desquamation. On the other hand, alkali soaps in particular are responsible for a certain swelling of the skin.

According to Hoffmann (73), washing the skin has various consequences, dependent upon the degree of degreasing. Predominant removal of the emulsifiable components results in an increased water loss from the skin surface. More vigorous degreasing causes a temporary swelling, finally resulting in an excessive drying of the skin. Skin which is morbidly changed exhibits an increased tendency to swell after washing. The permeability of the skin in both the swollen and the dry, rough state is such that penetration of inanimate and animate injurious substances is facilitated. These processes are involved in the triggering off of many skin complaints arising from working conditions.

According to Schneider (74), water alone is capable of exerting an unfavorable influence on the physicochemical state of the skin. This effect becomes more noticeable as the more active detergents and soaps are brought together with water into contact with the skin.

A degreasing of the skin is achieved to a much higher degree when using sulfonation and fatty acid condensation products than soap (75). It was also found that the sulfonation and fatty acid protein condensation products formed from fats of low molecular weight cause more severe degreasing and irritant symptoms than wetting agents based upon high molecular weight substances.

The influence of wetting agents upon the penetration of fatty substances into the skin was studied by Duemling (76), who found that the loss of lipids resulted in a loss of the water barrier and that its spontaneous regeneration occurred parallel to the reappearance of the extractable lipids. This is a clear demonstration of the function of the lipids: protection of the function of the water barrier. However, according to Götte (77), the passage of water through the stratum corneum is not hindered by the skin sebum. The body is normally resistant to the penetration of water by way of the barrier epidermis, the protective action of which is, however, removed by degreasing (78). Schulz and Szakall (79) have studied the penetrating ability of aqueous solutions of pure alkyl sulfates and saturated soaps. The C_{12} derivative penetrated the skin most readily, whereas the permeation of solutions of the C_8 and the C_{18} substances was scarcely different from that of pure water.

According to Geller (80), soap solutions form an emulsion with the skin fats, which then gets between the dirt particles and skin, thus loosening them from the skin surface. Cleansing the skin by means of surfactants was found by Götte and Herzberg (81) to remove, in addition to the pigmented soiling, fat-containing soiling. It is thereby unavoidable that natural skin lipids are also removed. Such a removal of fats from the uppermost skin layer can result in a change in the pH of the skin surface, thus disturbing the protective buffer effect. An even more severe initial defatting and a longer contact time of detergent with skin has been shown by other authors to cause a deep-seated swelling of the epidermis, which can be accompanied by an adsorptive or salt-type binding of the surfactant ions to the skin.

As regards swelling of the skin, there are marked differences between the effect found for soap and surfactants. Geller (80) attributes to soaps the property of causing skin swelling, so that soaps retard the rapid drying of the skin. A softening of the skin, especially at the skin folds, occurs, which also causes a blockage of the sebaceous and sweat glands, so that favorable conditions for harmful pyogenic organisms results. Soaps mainly have a swelling and alkalizing effect upon the skin, syndets also resulting in defatting. The degree of defatting is dependent on concentration. The syndets scarcely result in swelling of the skin, thus favoring a drying out of the skin folds, which is advantageous in the case of fungus infections.

Stüpel and Szakall (82) have confirmed that soap solutions result in a well-defined swelling of the skin. Their pH value of 9-11 and higher temperatures cause severe, lasting damage to the skin. According to Schulz (83), the skin swelling is an integral factor in the transport of substances to the deeper layers of the cutis, where inflammatory reactions can occur.

Implantation experiments using the sodium salt of caprylic acid (C_7H_{15}-COONa) showed that this substance causes a swelling of the subcutis and the stratum corneum (84). Concomitant with this swelling, the penetration relationships in the skin blood vessels are changed, with all the resultant consequences. Götte (85) showed that above the critical micelle concentration the swelling of the basal layers of the epidermis is more marked than that of the stratum corneum. Choman (86) was able to show that active detergents only had a drying action when below the critical micelle concentration. If this value is exceeded, then a swelling maximum is found for C_{12} compounds, for both soap and alkyl sulfates, when using calves' skin as substrate. The C_{16} and C_{18} compounds did not result in any damage to the collagen structure (see Fig. 2).

Putterman et al. (86a) investigated the effect of detergents on swelling of guinea pig stratum corneum. Squares of stratum corneum 20 $\times$ 20 mm were soaked in water for 1 hr. After lifting out of water, their dimensions were determined. The squares were then immersed in the surfactant solutions for 16 hr. After this time the dimensions were again determined. Swelling is expressed as the percent change in area after exposure to the second solutions as compared to the first solvent, water. Several anionic surfactants tested at concentrations of 0.05 M (this concentration is above the critical micelle concentration) showed ability to produce in-plane swelling (increase surface area) of guinea pig stratum corneum squares. Sodium laurate, sodium lauryl sulfate, and sodium oleate produced marked effects (16.7, 13.1, 15.1%). Sodium myristyl ether sulfate, sodium hexanoate, and sodium lauryl triether sulfate had no effect or little effect. The cationic and nonionic surfactants examined caused little or no swelling as well. To gain insight into the mechanism of swelling, the authors evaluated the effects of protein denaturants and concluded that protein denaturants per se do not cause stratum corneum swelling but that swelling is due to a reversible conformation change resulting from cooperative binding of the detergent.

Some investigators used the swelling behavior of isolated stratum corneum after contact with surfactants in order to predict the skin irritancy of the test substances. They proposed to use the swelling effect as an in vitro test to replace skin-rritancy tests on animals (13h,i).

3. Influence on pH value of the skin by anionic surfactants

The influence on the pH value of the skin by anionic detergents has been almost exclusively tested on human skin, and in this connection the alkaline-

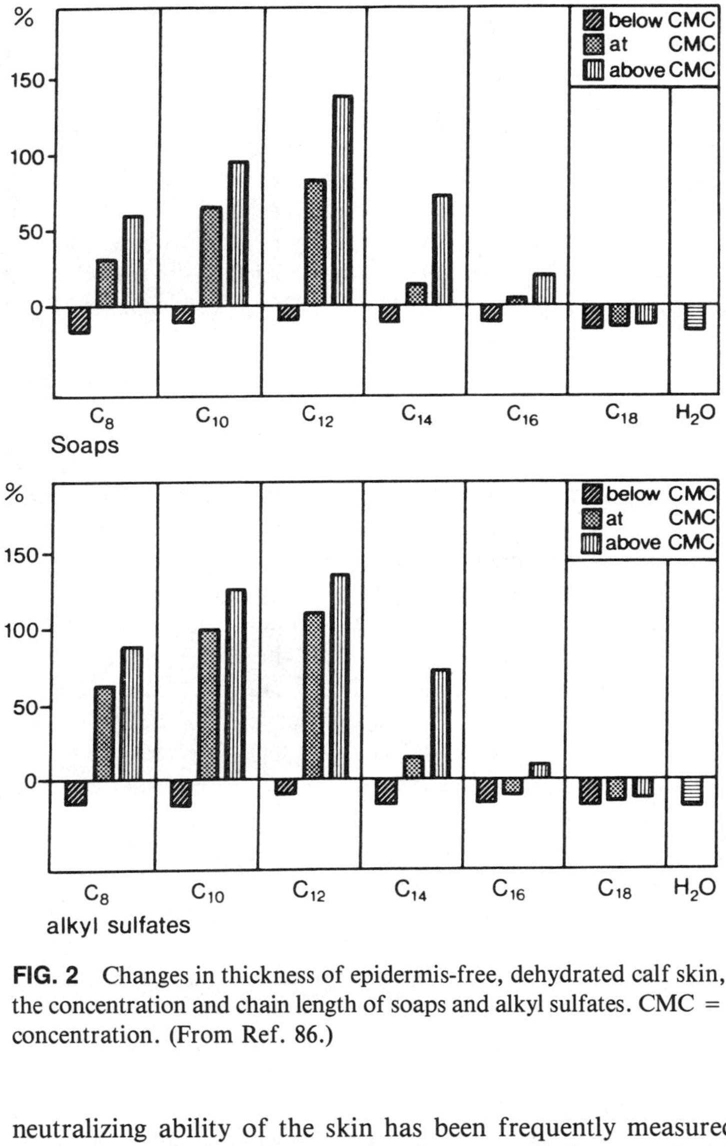

FIG. 2 Changes in thickness of epidermis-free, dehydrated calf skin, dependent on the concentration and chain length of soaps and alkyl sulfates. CMC = critical micelle concentration. (From Ref. 86.)

neutralizing ability of the skin has been frequently measured (52,73,82). After extended periods of contact between alkaline or other irritants with skin, the ability of the skin to neutralize alkaline substances is reduced or can disappear. Therefore the time necessary for the skin to neutralize a drop of alkaline solution is measured. If the time is lengthened this is an indication of a lowered buffer capacity of the skin. Either the skin has been rendered alkaline or the buffer capacity is reduced due to a dissolving out of substances necessary for the neutralization.

Neuhaus (87) has observed that after washing the skin with soap, a lengthening of the alkali neutralization time results. Washing results in the removal of acidic surface products contained in sweat and sebum, thus lowering the buffer capacity for alkalies. These effects are also observed for syndets, which reduce the buffer capacity because of swelling of the skin and the washing away of the fatty acids. The changing of the actual pH value of the skin after washing is not a neutralization in the chemical sense of the word but is simply the consequence of the physical removal of the acidic skin components by washing. The cleansing operations on the skin also remove part of the natural lipids. If such a removal occurs, then a pH change of the skin surface and a resultant temporary disturbance of the buffer protection is possible (81).

4. Reactions between anionic surfactants and skin proteins

Apart from certain physicochemical properties (solubilization, micelle formation), the anionic surfactants have the ability to form compounds with proteins. They can therefore solubilize the less soluble proteins such as zein, or can themselves remain on the skin due to formation of chemical compounds with the skin keratin. In addition to the ability to bind the proteins, the anionic surfactants promote the dissolution of the proteins out of the skin, cause a release of sulfhydryl groups from the sclero proteins, react upon the various enzymes of the skin, and finally can denaturize proteins. Most studies that concern themselves with such properties of surfactants were carried out either in vitro or on human skin. The following therefore concerns itself with those experiments that are either of a general nature or based upon results obtained from animal experiments.

According to Götte (77), the epidermis reacts with anionic surfactants, whereby their anions react with the basic groups of the proteins to form salts. This reaction has practical importance because in unfavorable cases, at high concentrations, the outer epidermal layers are partly dissoluted, so that the skin becomes mildly sticky, finally becomes brittle, and the upper cutaneous scales are prematurely lost. A mild alkaline milieu reduces salt formation, and the incorporation of ether, ester, corbonamide, and amino groups in the surfactant molecules, or the addition of nonionic surfactants, reduces the tendency to salt formation, thus leading to substances or combinations which have better skin compatibility properties. The tendency to form salts can be subjected to preliminary testing using zein, a water-insoluble protein from the outer covering of maize, despite the difference between this protein and skin proteins (88). The binding of alkyl sulfonates to skin proteins can be demonstrated with the help of the dyestuff indigotetrasulfonate (89,90). The dyeing of the skin with this dye does not occur if the basic groups are blocked by anionic surfactants.

The binding ability of surfactants to skin proteins is dependent upon the chain length of the surfactants. The quantities of surfactant absorbed on the skin and hair can be experimentally measured (91). The surfactants bound to the skin form a molecular film on the skin and can only be removed with difficulty by washing (92). Their effect upon the skin therefore extends beyond mere contact time.

Sven-Gösta Blohm (89) showed that sodium dodecyl sulfate (SDS) reacts vigorously with proteins. The reaction occurs stepwise. Initially, the basic groups in the proteins are occupied by the negatively charged dodecyl sulfate ions to form a stable salt. This complex attracts additional SDS ions, resulting in the formation of a higher complex which causes a destruction of the protein structure thus leading to denaturization. SDS is also strongly bound to the protein in that it cannot be removed by rinsing. SDS is considered to be the strongest protein-denaturizing substance from all the anionic surfactants.

Imokawa et al. (93) assume that skin roughness is a consequence of the binding of the anionics to the skin proteins, which leads to a denaturization of the skin proteins. In order to test the binding and thus the denaturizing power of the various anionics, they measured the optical rotation of beef serum in the presence of these substances. The specific rotation of the beef serum albumin was taken to be a measure of the denaturization. A remarkable correlation between the ability of the anionics to cause skin roughness and their influence upon the specific rotation of the beef serum was found, thus leading to the conclusion that the adsorption of the surfactants onto the keratin is a necessary initial step in the production of skin roughness. This beef serum albumin test is regarded as a pretest for skin-roughening potency in the absence of tests on human skin. The alkyl sulfates (AS) C_8-C_{14}, the alkyl benzene sulfonates (ABS) C_8-C_{16}, and the α-olefin sulfonates (AOS) C_{12}-C_{16} were all tested.

The flocculation of water-soluble proteins such as albumins and globulins proceeds parallel to the protein-binding ability of anionic surfactants and their protein denaturization properties (94). On the other hand, Foster (95) showed that the water-insoluble zein is solubilized by the addition of such surfactants (see Ref. 91). It is therefore to be expected that contact of anionic surfactants with the skin will result not only in binding to the skin keratin but also in a dissolution of proteins from the skin by these surfactants.

Carruthers et al. (96) examined the extraction of proteins from cow snout epidermis, which is about 1.5-mm thick, is practically hairless, and is low in lipids by sodium lauryl sulfate (0.02, 0.05, and 0.1 M after 6, 7, and 8 days at 28°C), and found that concentrations above 0.02 M at the isoelectric point did not dissolve more proteins than at a 0.02 M concentration. In contrast, at pH 7 the amount of protein dissolved from the epidermis was dependent upon concentration.

Prottey and Ferguson (97) studied the extraction of proteins and amino acids from guinea pig skin by washing one side of the flank of the animal with surfactant solution, the other side being washed with water under the same conditions, as a control. The wash solutions were then analyzed for proteins and amino acids. The examinations showed that undetectably small quantities of soluble proteins were dissolved from the skin by pure water. The four sulfated surfactants, on the other hand, dissolved considerable quantities, especially SLS. SLS was also the most effective substance in dissolving amino acids from the skin. The other anionic surfactants did not dissolve more amino acids than did water.

Earlier examinations by the same authors showed that significant quantities of amino acids and proteins are only extracted from the stratum corneum when the surfactants are present above the critical micelle formation concentration. The effect of the chain length upon the elution of amino acids and proteins from the skin was checked using a homologous series of sodium alkyl sulfates and sodium soaps. It was shown that AS with a chain length of C_{12}-C_{13} dissolved more proteins and amino acids than other chain lengths (both shorter and longer). SDS was the most effective substance. The sodium soaps exhibited similar behavior. Among the homologous series, sodium myristate dissolved the largest quantity of proteins from the skin, whereas the largest quantity of amino acids was dissolved by sodium laurate. The ability to dissolve proteins and amino acids is a function of the nature of the polar groups as well as the length of the lipophilic chains, the nature of the polar groups in the surfactant appearing to be of more importance than the length of the lipophilic chain as regards the ability of the surfactant to extract substances from the stratum corneum.

In addition to the solubilization of proteins by surfactants, the liberation of free sulfhydryl (SH) groups by anionic surfactants in contact with the skin has been examined. Tronnier (98) has reported an increase in the number of SH groups after contact of the skin with anionic detergents, whereby syndets apparently are more effective than soaps in this context. The dependence of the liberation of SH groups upon the chain length of the surfactants was the subject of a study by Blohm (89). The maximum number of SH groups was found in the case of C_{12} alkyl sulfate. Scott and Lyon (99) assume that the compact keratin structure is loosened by the surfactant without the disulfide bridges being split. A higher pH value is not responsible for the number of SH groups liberated. Various authors have determined the amino-nitrogen content of the surfactant solutions after having treated the skin, the isolated epidermis, or zein with them. The action of anionic surfactants upon the proteins have been examined under this method (15a,63,89,91,97, 99-101).

The action of anionic detergents upon cod skin collagen, which, in contrast to collagen from mammals, is more easily solubilized by surfactants, has also been studied (102).

According to Harrold (103), soaps result in only a minimal increase in the number of free sulfhydryl groups. ABS and AS exhibit larger effects which are concentration-dependent, whereby the latter have a greater action than the former.

The denaturizing effect of sodium lauryl sulfate and dodecyl benzene sulfonate upon the epidermal protein enables metallic sensitizers to penetrate the epidermis more readily. The work of Harrold (103) supports these findings.

5. Influence on skin metabolism of anionic surfactants

It has been established by various authors that single or repeated applications of anionics to the skin resulted in changes that could be attributed to alterations in the activities of various enzymes. The activities of certain enzymes were therefore measured after they had been brought into contact with anionics or after the skin had been treated with them. Enzymes from the gastrointestinal tract have also been examined in a similar manner. Such experiments were carried out by Freeman et al. (105) using sodium alkylaryl sulfonate (AAS); they found that almost 100% of the amylase, lipase, trypsin, and pepsin activities were inhibited, and phosphatase activity by almost 25%, when using 0.05-0.2% AAS. However, feeding experiments using rats and dogs showed that the inactivation in vivo is not significant enough to disturb food digestion (dogs of approximately 15 kg body weight received 1 g AAS daily for a 6-month period; rats 5 g/kg^{-1} body weight daily for 65 days in the food). No difference in body weight between the test and control groups was found.

The measurement of the inhibition of saccharase activity by anionic surfactants was recommended as a screening test for skin compatibility by Wilmsmann (106) because a parallelism between the results of epicutaneous patch tests and the rabbit eye test had been demonstrated. The author distinguishes between strong and weak enzyme-active surfactants, whereby the protein fatty acid condensates belong to the weakly active surfactants. Using the method of Wilsmann (106), Modde and Schuster (56) measured the activity of various protein fatty acid condensates and established a dependence upon the molecular weight. Substances with a molecular weight of >700 do not exhibit any saccharase inhibition activity. A parallelism to the rabbit eye test was evident where it was likewise found that substances with a molecular weight of >700 exhibit only minimal irritant action.

Modde et al. (52) also examined the influence of various surfactants on fermentation activity, especially saccharase inhibition. The effects of 2%

sodium lauryl sulfate (SLS) and 0.05 M SLS solutions were used as standards and set as 1. The surfactants *n*-dodecyl benzene sulfonate triethanolamine salt, *n*-alkyl sulfate, fatty acid methyl tauride, lauryl glycolether sulfate as Na salt all had effects similar to that of SLS saccharase activity. Only fatty acid protein condensate (potassium salt MG750) did not have any influence upon the saccharase activity. Tovell et al. (71) suggest that SLS causes a proliferation of the basal cells, which is closely associated with an increase in DNA synthesis. The intracellular fat storage and thickening of the stratum basal and stratum spinosum occurring as consequences of the treatment of rat skin with SLS were described as resulting from cellular hypoxia. This itself was caused by direct cell damage combined with damage of the mitochondrial oxidation.

Brown (107) observed that when sodium alkyl benzene sulfonate (branched C_{12} chain) was applied in a 2% active detergent solution to mouse skin for 28 consecutive days, an increase in the activity of glucose-6-phosphate dehydrogenase, succinic dehydrogenase, and DNA occurred. In contrast, phosphogluconate dehydrogenase, alkaline and acid phosphatase and monoamine oxidase retained unchanged activity. Prottey and Hartop (108) have reported a study of changes in the glyceride and lipid metabolisms of rats when their intact skin was repeatedly treated with soap solutions. The germinative cellular layer of the stratum basale reacts to a soap treatment with an increased triglyceride synthesis and a decreased phospholipid synthesis. These changes could be detected after treatment with soap only twice daily, the effects being more readily observable after treatment four times daily. The changes in glyceride metabolism are connected with morphological skin changes. However, the sebaceous glands showed neither morphological changes (irritation) nor changes in lipid synthesis. The increase in the triglyceride synthesis proceeded parallel to a thickening of the stratum granulosum and the severity of the skin irritation.

6. Studies concerning permeability of anionics in and through the skin

The skin damage or irritation caused by surfactants is closely connected with the ability of surfactants to penetrate the living epidermis. Should the surfactants penetrate even deeper, then the connective tissues and blood vessel cells of the corium are also irritated. As not all anionics cause the same degree of skin irritation, it may be assumed that those substances of mild irritant action do not penetrate as rapidly and deeply as those that have more severe irritant properties. Various authors have studied the question of penetration, either by means of in vitro examinations (92,109) or using the intact animal (92,97), and have found a far-reaching parallelism between penetration and irritant action. The penetration also seems to be strongly dependent upon the chain length of the alkyl chain of the anionics (92,110).

Furthermore, the animal species used in such tests also appears to play a major role. Bettley (23) showed that the penetration of anionics is more rapid in small laboratory test animals than in humans. It was suggested that the explanation lies in the fact that the haired laboratory animals possess a less well-formed and thinner epidermis than humans. Kröper (75) studied model experiments involving the migration of potassium soaps of definite chain length through a cellophane membrane and found that diluted solutions of the potassium soaps of saturated, higher molecular weight fatty acids can migrate through such membranes. Laurate migrated completely, myristate to only 50%, and palmitate and stearate even less. The presence of a double bond favored ease of dialysis. In the case of fatty alcohol sulfates and fatty acid condensation products, it was found that ease of dialysis increased with decreasing molecular size of the fatty acids. Ease of dialysis parallels the empirically observed irritant action in the case of potassium soaps and other anionic surfactants.

Bettley (110) also sees a connection between its penetrative power through the epidermis and irritation potency in the case of soaps. Blank (111) has already demonstrated differences in the irritant action of soaps and attributed this to differing fatty acid compositions. The soaps can penetrate the epidermal barrier layer and some penetrate more readily than others.

Bettley (110) carried out experiments on human skin both in vivo and in vitro and observed that C_{12} soap under the experimental conditions used penetrated the epidermis to the extent of 24.4% of the original quantity and lead to irritant reactions in 18 cases of total 28 (64%) whereas octanoate and oleate penetrated much less (5.3 and 2.1%) and palimitate hardly at all (0.15%). Laurate soaps had a higher epidermal permeability than sodium lauryl sulfate (SLS) and a greater effect upon the waterbinding capacity and skin barrier than SLS. An explanation for the greater effect of the C_{12} chain is to be found in the molecular length of 17 Å, approximately that of cholesterol, a component of many lipoprotein membranes. Moreover its length also corresponds to the phospholipid sphingomyelin (112).

Prottey and Ferguson (97) examined the penetration of ^{14}C-labeled surfactants with a chain length of C_{12} in in vivo experiments with guinea pigs. It was found that sodium laurate penetrated the skin most of all and that sodium lauryl isethionate and sodium lauryl triethoxy sulfate penetrated much less. Sodium lauryl sulfonate had the lowest degree of penetration. Another publication (112) claims that various adsorption rates of surfactants of skin keratin parallels skin irritancy. Sodium cetyl sulfate, which is adsorbed only slightly, has less irritant action, which is in contrast to the rapidly adsorbed SLS. A similar behavior is observed in the case of sodium oleate in comparison with sodium laurate, which is a more powerful irritant. The irritant action of anionics can be moderated by the addition of nonionics, which occupy positions in the keratin. The permeability of n-alkyl sulfates and soaps of definite chain length through the epidermis was examined by Dugard et al. (113), who found that the permeability of the soaps and the alkyl sulfates (AS) reached a maximum at C_{12}. In the case of soaps, C_{14} was less effective than C_{12}. With the sulfates, the difference between C_{12} and C_{14} was larger, C_8 and C_{10} showing only very low or zero penetration. The degree of penetration was found to decrease as follows: $C_{12} > C_{14} \gg C_{16} > C_{10} > C_8$.

Howes (92) interpreted the irritancy of surfactants in terms of their ability to penetrate the stratum corneum. His in vitro results obtained using rat skin treated with SLS, sodium lauryl isothionate, dodecyl benzene sulfonate (DBS), and a C_{18} soap showed that these did not penetrate at all when measured 24 hr after application. However, a C_{16} soap showed a low degree of penetration, C_{14} and C_{12} marked penetration, and the greatest degree of penetration was measured using a C_{10} soap. Only 30% of the soaps and sodium lauryl isothionate were adsorbed on the skin of the rat. SLS and DBS were adsorbed to approximately 70%. Using human skin, the results obtained for soaps were similar to those found using rat skin (maximum at C_{12} soaps), but it was also shown that SLS and sodium lauryl isothionate could penetrate the human epidermis. The quantities adsorbed onto the epidermis were at a maximum with SLS (75%) and were of the order of 30-50% for the other anionics. The following degrees of penetration were established: $C_{12} > C_{10} = C_{14} > C_{16} = C_{18}$.

Szakall and Schulz (79) also confirmed by their studies using AS and soaps that the C_{12} substances are most effective at penetrating the skin, and that a parallelism exists between irritant action and permeation. The relationship between adsorption and skin roughness caused by various substances using the callus was studied by Imokowa (114). The quantities adsorbed on the callus decreased in the order alkyl sulfate (AS) > alkyl benzene sulfonate > α-olefin sulfonate > alkyl polyoxyethylene sulfate. Callus was regarded being a representative model for epidermal keratin. The adsorption is both concentration and chain-length-dependent (increasing adsorption, C_8 AS < C_{10} AS < C_{14} AS < C_{12} AS).

Nilzen and Wilkström (115) showed that some anionics have the ability to activate the eczematous properties of simple chemical substances. Corresponding experiments on guinea pigs have shown that sodium dodecyl sulfate (SDS) is capable of increasing the penetration of chromate and nickel ions through the skin. The ability to increase the degree of penetration was regarded as the cause of the activation of the eczematous properties of some substances. Similar results were obtained by Vinson and Choman (104), using SDS and sodium dodecyl benzene sulfonate as anionics and nickel sulfate. All such experiments, described in greater detail in the section dealing with sensitization of anionics, have, however, shown that the observed effect always parallels the irritant effect of the anionic and is thus clearly concentration-dependent.

7. Influencing of wound healing by anionic surfactants

Scheinberg et al. (116) studied the effect of various surfactants upon wound healing using guinea pigs. The animals were anesthetized with sodium pentobarbital and standardized, circular wounds made in their right and left flanks. Such wounds when left untreated were 80% healed after 10-11 days, the healing process being complete after 11.6-11.7 days. The application of 1% aqueous solution of SLS to the wound had little effect whereas the application of a 1% SLS solution in a 2% methyl cellulose (MC) preparation caused a noticeable quickening of the healing process (after 8 days 80%, after 9.2 days 100% healed). Even clearer was the effect of applying a 2% SLS solution in 2% MC suspension (80% after 7 days, 100% healing after 8.5 days). In contrast to SLS, the cationics and nonionics did not show any positive effects upon the rate of wound healing. The shortening of the healing period was statistically significant in the case of SLS solutions in MC suspensions when compared with the controls. The positive effect of a 2% SLS solution in an MC suspension could possibly attain therapeutic use in the treatment of peptic ulcers.

D. Special Changes in Skin Caused by Anionic Surfactants

The following section deals in some detail with special changes of the skin, such as acanthotic and comedonic effects and effects upon the hair cycle.

1. Acanthotic effects

The appearance of epidermis proliferation as a result of increased cell division is known as acanthosis. It may be caused by various factors and can be quantified by measuring the thickness of the epidermis. As cell divisional processes are involved, a counting of the mitoses is suitable as well.

The acanthotic effect can be demonstrated using either guinea pigs or rats. The substance to be tested is repeatedly applied to the skin of the flank of guinea pigs or rats, or to the teats of male guinea pigs. There are no significant differences in the results obtained with substances not possessing hormone character when using either guinea pigs or rats. After repeated application, skin samples are removed from the test animals and from the corresponding area of the control group animals, from which histological slides are prepared and the epidermis thickness for both groups of animals is measured.

The quotient from both values is known as the acanthosis factor. The results of such examinations with various anionics are shown in schedule 24.

The carboxylic acids of chain length C_6-C_{18} show acanthotic activity, according to Schaaf and Gross (117). The strongest effect was shown by myristic acid, the weakest by stearic acid. A clear acanthotic effect is also shown by heptyl alcohol and decyl alcohol, as well as by unsaturated fatty acids, especially ricinoleic and oleic acids (117). These results are in agreement with those found by Brun et al. (118) who tested the pure saturated and unsaturated fatty acids on the flank of guinea pigs. A maximum acanthotic factor was found for lauric acid, followed by capric acid in second place and myristic acid in third place (see Fig. 3). Landsdown and Grasso (70) repeatedly applied sodium lauryl sulfate (SLS) in 1 or 2% concentrations to mice.

In addition to acanthosis and hyperkeratinization, nuclear debris was found in the keratin layer. Analogous findings were reported by Tovell et al. (71), who washed rats repeatedly with 1% SLS solution. A thickening of the epidermis and stratum corneum were established.

Marzulli and Maibach (119) carried out a 16-day accumulative irritation test with a 1% sodium lauryl sulfate solution using rabbits. An increase in skin fold thickness without skin reddening was observed, which indicates an epidermal rather than a dermal reaction.

Vinson and Cohman (104) applied 0.1, 0.2, 0.5, and 1% SLS or sodium dodecyl benzene sulfonate solutiuons 40 times to the dorsal areas of female

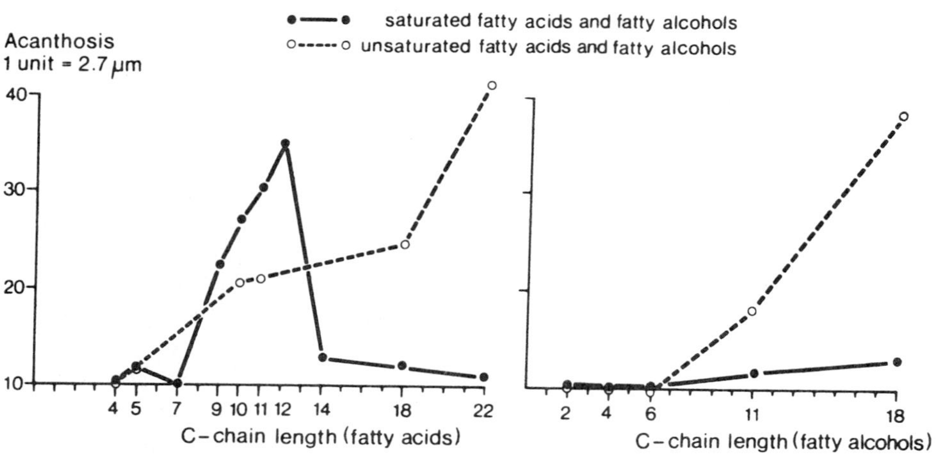

FIG. 3 Acanthotic effects of saturated and unsaturated fatty acids and saturated and unsaturated fatty alcohols (5% solution in absolute alcohol) dependent on chain length. (From Ref. 118.)

guinea pigs and established an acanthotic effect by histological examination. This is regarded as a necessary precondition for the penetration of nickel into the skin. This publication also describes experiments carried out with soap (8%), which was applied over a 10-day period. Apart from a moderate acanthosis, no other effects were seen.

Bear et al. (120) are of the opinion that skin acanthosis is responsible for rendering the skin more sensitive to allergic and primary irritating, eczematous substances. The repeated application of soap results in a skin thickening but does not, however, facilitate the allergic nickel reactions, as insoluble nickel soaps are formed.

Experiments using New Zealand rabbits, described by Ingram and Grasso (121), whereby repeated application (daily 5 hr; 5 days; patch test; concentration up to 2.4%) using SLS in 0.15-0.5% solutions resulted in clear acanthosis and hyperkeratosis. At higher concentrations, primary irritant effects were of major importance under the experimental conditions used.

Prottey and Ferguson (97) examined the irritant effect of surfactants on rat skin and applied 0.25 M solutions (5-10%) twice daily. After 3 days of such treatment, a marked thickening of the epidermis in the case of SLS was seen, whereas only a slight thickening with sodium laurate was observed. In contrast, sodium lauryl triethoxysulfate, sodium lauryl isothionate, and sodium lauryl triethoxylate caused no epidermal thickening (acanthosis).

Butcher (122) treated the dorsum area of young rats once per day for 2 weeks with potassium salts of oleic, linoleic, ricinoleic, and stearic acids; sodium oleate was also tested. The soap concentrations were all 10%. The mitotic rate of the animals was measured. Sodium oleate showed a clear effect in addition severe hyperplasia of the stratum spinosum and proliferation of the stratum basale being found. The stratum granulosum showed seven to eight cell layers, only one to two such layers being normal. Potassium soaps of oleic and linoleic acids were almost as effective as oleic acids. In contrast, the potassium soap of ricinoleic acid showed a much less severe effect.

Ivanov et al. (72) applied 1-6% solutions of alkyl sulfates and alkyl sulfonates to guinea pig skin for 14-15 days. These applications lead to a local proliferation of the epidermis. The sodium salt of sulfosuccinic acid monoether did not cause such reactions. (See Table 27.)

2. Comedogenic and comedolytic effects

Testing for comedogenic and comedolytic effects is carried out in the external ear canal of the rabbit. In addition to the classic comedone-causing substances such as chlorinated naphthalenes and tetrachlorobenzodioxin, there are numerous other chemicals which also possess this effect.

An examination of cosmetics (soaps, shampoos, and face creams) and cosmetic raw materials was undertaken by Mills and Kligman (123) and Kligman

TABLE 27 Acanthosis Factors (AF) and Irritant Activity (IA) of Several Anionic-Active Surfactants Measured on Flank Skin of Guinea Pigs or Dorsum Area of Rats

Product	Conc. %	Applications	AF	IA	Species	Ref.
Na oleate	Undiluted	5-15	+ +	+	Rat	122
K oleate	Undiluted	5-14	+ + +	+	Rat	122
K ricinoleate	Undiluted	5-14	±	—	Rat	122
K linoleate	Undiluted	5-14	+ +	+	Rat	122
K stearate	Undiluted	5-14	+ +	+	Rat	122
Na sulforicinoleate	Undiluted	10	5	+ + + +	GP[a]	1
	15	10	2.8	±	GP	1
	4.7	10	1.3	—	GP	1
Na lauryl sulfate	20	10	1.9	+	GP	1
	25	7-13	3.5	—	GP	122
Aerosol OT	20	10	3.3	+ + + +	GP	1
	10	10	2.5	±	GP	1
	2	10	1.8	—	GP	1
Ultravon CO	25	10	2.1	±	GP	1
	10	10	1.4	—	GP	1
	2	10	1.0	—	GP	1
Diisobutylnaphthalene	0.1	1	+	—	Mouse	84
sulfonic acid	1.0	1	+	—	Mouse	84
sodium salt	5.0	1	+	—	Mouse	84
(as implantation in connective tissue)						

[a]GP = guinea pig.
Irritant Activity: — = no activity, + = mild skin irritation; ± = very mild activity; + + + + = very severe skin irritation.
Source: Ref. 1.

and Mills (124). The soaps and shampoos were applied as 25% solutions 10 times. From the 21 soaps examined, 6 showed more or less distinct comedogenic properties. The group of rabbits treated with 25% shampoo showed no evidence for comedogenic activity in 8 cases, the rest showing weak to strong comedogenic activity. Most of the soaps and shampoos examined as 25% solutions possessed skin irritant properties under the used test conditions. They caused gross edema and scaling, and microscopic infiltration of the dermis with neutrophiles was prominent. The irritation itself did not lead to comedo formation; on the other hand, comedogenic soaps were always irritating as well. A shampoo lead to comedo formation without symptoms of irritation. Most soaps which possessed comedogenic properties contained bacteriostatic compounds, e.g., hexachlorophene. Sulfur-containing soaps were also comedogenic. The author attributed to soaps the property of being the most frequent cause of acne detergicans. The more unsaturated fatty acids they contain, the stronger the comedogenic activity of the soaps. Comedogenicity is also dependent upon chain length. The salt of lauric acid has the strongest comedogenic properties of all of the fatty acids tested. However, an important role in the comedogenic activity of soaps is played by the bacteriostatic components such as hexachlorophene and chlorinated salicylanilides. A comparison of soaps with and without hexachlorophene demonstrated this clearly. Sodium lauryl sulfate does not show comedogenic properties (124).

Mill and Kligman (125) examined various substances for cosmedolytic activity as well, among which were 5% aqueous sodium laurate and 5% aqueous sodium lauryl sulfate, using rabbits which had comedones induced by crude coal tar. A positive reaction was not found.

3. Effect upon hair cycle

Because shampoos primarily contain anionics as surfactant basis, it is a justifiable question as to what effect such substances have upon the hair cycle. The hair cycle is basically divided into two phases, the anagen (growth) phase and the telogen (rest) phase. The transition from the one to the other is known as the katogen phase. The most detailed study concerning the effect of certain anionics upon the hair cycle of rats has been published by Rubisz-Brzazinska et al. (126). They carried out their experiments using the rat with sodium lauryl ether sulfate. The product was applied and rubbed in once daily for 65 consecutive days in various concentrations to the back of the animals. This treatment when carried out with the undiluted and 50% surfactant (the content of the active substance in the sodium lauryl ether sulfate was 60%) caused macroscopic visible features of toxic inflammation in the form of redness, scaling, and keratinization of the hair follicles. Histological examination revealed hyper- and parakeratosis, and a thinning of the epidermis and erythrocytes outside the vessels. The animals in the group which were treated with 15% surfactant did not show any macroscopic changes of the treated skin areas but did exhibit histological evidence of mild irritation symptoms. Application of 1.5% surfactant preparations showed neither macroscopic nor histological changes. The examinations simultaneously carried out concerning the hair cycle were begun in the telogen phase. The results are shown in Table 28.

As can be seen from the table, the 60% group exhibited a shortening of the anagen phase both in induced and spontaneous hair cycles, which was clinically evident by diffuse alopecia being found not only on the applied areas but also around these areas. In the case of the 30% group, the induced cycle proceeded regularly, whereas the spontaneous cycle showed changes analogous to those found in the 60% group. The 9 and 0.9% concentration groups did not exhibit any effect upon the hair cycle of the rats. It may be assumed that the observed hair cycle effects possess a connection to the degree of irritancy.

Kuhn (84) describes the effect of diisobutylnaphthalene sulfonic acid (sodium salt) in 0.1, 1, and 5% concentrations after being implanted subcutaneously with the help of aga. In addition to a thickening of the epidermis, an increasing cornification occurred at all stages. The hair follicles were also affected in this way. The lumen filled up with cornified cells but resulting in a new formation of the rejected hairs.

TABLE 28 Course of Induced and Spontaneous Hair Cycles in Rats under Influence of Various Concentrations of Sodium Lauryl Ether Sulfate[a]

Animal group (treatment with —% AS)[b]	Induced cycle (artificial)	Spontaneous cycle (physiological)
60	Shortening of the anagen phase, premature telogen synchronization.	Shortening of the anagen phase, premature telogen synchronization
30	Regular	Shortening of the anagen phase, premature telogen synchronization
9	Regular	Regular
0.9	Regular	Regular

[a]Treated once daily for 65 consecutive days.
[b]The surfactant used contained 60% active substance.
Source: Ref. 126.

According to Schmid (127), the repeated cutaneous application of a mixture consisting of surfactants and solvents (1 and 10% concentration; 3 times weekly; 24 times in 8 weeks) resulted in disturbance of hair growth in mice, whereas the application of surfactants alone did not result in any macroscopic findings being reported.

Muto and Tatsuma (128) applied repeatedly various surfactants, among which were soaps and alkyl benzene sulfonates, to the skin or in the subcutaneous tissue and found by electron microscopy of skin samples, among other changes, that hair formation changes.

E. Influence of Structure, Chain Length, and Degree of Ethoxylation and Saturation of Alkyl Rest Upon Skin Compatibility

As already pointed out several times in other sections, the number of C atoms in the alkyl rest has a large influence on the properties of this class of substances. In this context, studies carried out with homologous series are of special interest in enabling such relationships to be more clearly illuminated.

Various concentrations of commercially available alkyl ether sulfates with alkyl chains consisting of even or odd numbers of carbon atoms have been tested for their primary irritancy to rabbit skin in vivo by Pemperton and Rhodes (128a). Six ether sulfates (2 with even—C_{12}-C_{14}—and 4 with odd—C_{13}-C_{15} and C_9-C_{15}—carbon numbers, 3 with 2, and 3 with 3 mol ethoxylates) were tested in concentrations of 4, 8, 12, 16, 20, and 24% according to the standard OECD procedure. The level of irritation showed dose-dependent increases in the concentration range 4-12% with a plateauing of response of moderate to severe irritation occurring between 12 and 24%. The 3 EO even (carbon numbered) surfactant was the least irritant whereas the 2 EO even surfactant was the most irritant, however, the differences were small. The data support the conclusion that the intensity and duration of the skin irritancy response to even or odd (carbon numbered) alkyl ether sulfate surfactants are statistically indistinguishable. Although there were no differences between even and odd carbon number surfactants, there was a marginal trend for 3 EO surfactants to produce less erythema than the 2 EO surfactants, with the differences being most marked with the comparison of even carbon numbered surfactants.

In a 24 hr closed-patch test on guinea pig skin, Imokawa (128b) demonstrated the less skin irritancy of C_{12} alkyl polyoxyethylene sulfate in comparison to the C_{12} alkyl sulfate (AS).

According to Schott (31a), the greatest effects among the sodium alkyl sulfates (NaAS) ranging from octyl to octadecyl were shown by the dodecyl sulfate. In the series of soaps ranging from sodium octanoate to stearate, sodium laurate displayed the highest activity. These results were gained by different test methods (irritant action on human skin, swelling effect of human skin or dermal collagen of calf).

Brun et al. (118) examined saturated and unsaturated carboxylic acids of chain length C_4-C_{22} and saturated and unsaturated fatty alcohols with a chain length of C_2 and C_4-C_{18} for acanthotic activity using guinea pigs. The thickness of the epidermis was used as a measure of such activity. It was found that in the case of the saturated, aliphatic carboxylic acids, acanthotic activity first commenced at C_9. The maximum activity was attained with lauric acid (C_{12}), a clear drop in activity being seen at C_{14}; the higher homologues did not possess any such activity. The acanthotic activity of the unsaturated fatty acids (one double bond) is weak at C_5 and increases continuously up to C_{22}. The saturated fatty alcohols show hardly any acanthotic activity. The simple unsaturated fatty alcohols are acanthotic from C_{11}, a maximum being found at C_{18} (see Fig. 3). Similar results were obtained by Schaaf and Gross (117). They report that all carboxylic acids from C_6-C_{18} are acanthotic, the strongest effect being shown by myristic acid, the weakest by stearic acid. This differs from Brun et al. (118) insofar as these authors reported a maximum with lauric acid and that capric acid had a decidedly higher acanthotic activity than myristic acid. According to Schaaf and Gross (117), heptyl and decyl alcohol also exhibit acanthotic activity. The unsaturated fatty acids, especially ricinoleic and oleic acid, also possess unmistakable acanthotic activity (117). Mills and Kligman (123) report that of the saturated fatty acids, the potassium and sodium salts of lauric acid possess the most severe acanthotic activities. A higher percentage of unsaturated fatty acids also play a role.

Soaps with a high coconut oil fatty acid content are characterized by an intensive cleansing activity. However, according to Hoffmann (73), they are responsible for causing the most skin irritations. The skin-irritating properties of those soaps is attributed to the lauric acid part (129).

The activity of various soap solutions on rat skin was studied by Prottey et al. (130). The products used, together with the results obtained after the application had been carried out five times, are shown graphically in Fig. 4. As can be seen from this figure, sodium laurate evidently possesses the strongest irritancy properties, the result being a complete necrosis of the epidermis. The soap of chain length C_{10} also causes an above-average irritant reaction.

In addition to the irritant action of sodium soaps, Schulz and Rose (131) have also examined NaAS of chain lengths C_{12}-C_{18} in 0.1 and 0.25% aqueous solutions using human test subjects by means of the closed cup-test (contact time 22-24 hr). The most frequent positive results were found with sodium laurate and sodium dodecyl sulfate (SDS). The following order of irritant activity was found:

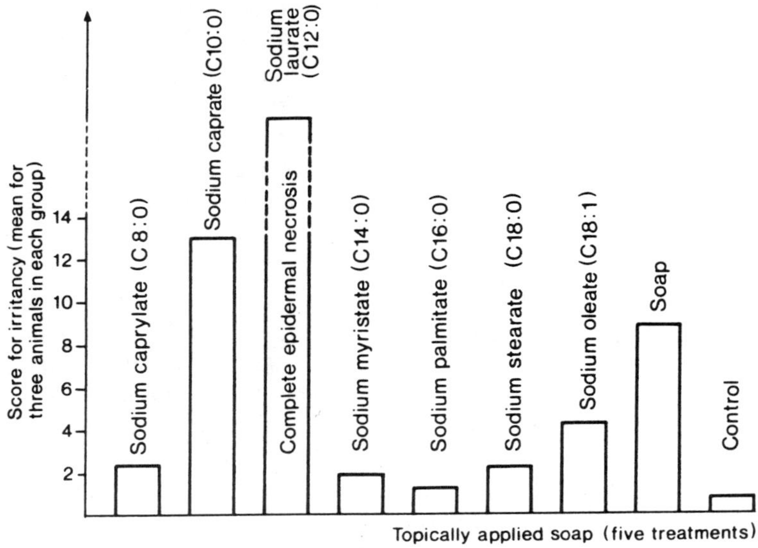

FIG. 4 Irritancy of sodium soaps of fatty acids. All individual fatty acid sodium soaps were 0.25 M, pH 9.8. The soap mixture was 8%, pH 9.8. (From Ref. 130.)

$$\text{Sodium soaps: } C_{12} > C_{14} > C_{10} \underbrace{> C_8, C_{16}, C_{18}}_{\text{no reactions}}$$

$$\text{Sodium alkyl sulfates: } C_{12} > C_{14} > C_{10} > C_8 \underbrace{> C_{16}, C_{18}}_{\text{no reactions}}$$

In general the alkyl sulfates were more aggressive to the skin than were the soaps.

As a measure of skin damage, one can use the extraction of amino acids and proteins from the skin. According to Prottey and Ferguson (97), SDS has the greatest elution effect upon skin proteins, followed by sodium tridecyl sulfate and sodium tetradecyl sulfate. The weakest eluting activity was shown by the sulfates having C_9 and C_{15} chain length. Among the soaps, sodium laurate possessed the greatest eluting power for amino acids and sodium myristate for proteins. The weakest activity was shown by sodium caprylate. Using the intracutaneous test of Barail (66) as an aid, Gloxhuber and Schulz (69) studied the dependence of skin reactions on chain length in a homologous series of AS. The individual substances were injected in increasing concentrations into the shaved skin of abdomens of white mice (0.05 ml per application). The results were read off after 24 hr. They are shown graphically in Fig. 5. As can be seen, the irritant activity is primarily dependent upon the concentration, and only in a secondary manner upon the chain length of the sulfate used. In this study the maximum irritant activity was not at C_{12} but at C_{10}, a much less severe irritant action being shown by C_8, whereas the longer chain lengths show a gradual decrease in irritant activity.

Gloxhuber (132) also examined the dependency of skin-irritating properties on the chain length of *n*-alkyl sulfonates using the method of Barail (66). The most irritant activity was again with the compound with a C_{10} chain (see Fig. 6). Imokawa et al. (93) assume that skin roughness is a consequence of the binding of the anionics to the skin proteins, thus resulting in their denaturation. In order to test the binding and thus denaturing capacity of the various anionics, they combined them with bovine serum albumin and determined the optical rotation. The alkyl sulfates (AS) C_8-C_{14}, the alkyl benzene sulfonates (ABS) C_8-C_{16}, and the α-olefin sulfonates (AOS) C_{12}-C_{16} were tested. The following order of denaturizing power and strength of skin roughness were found.

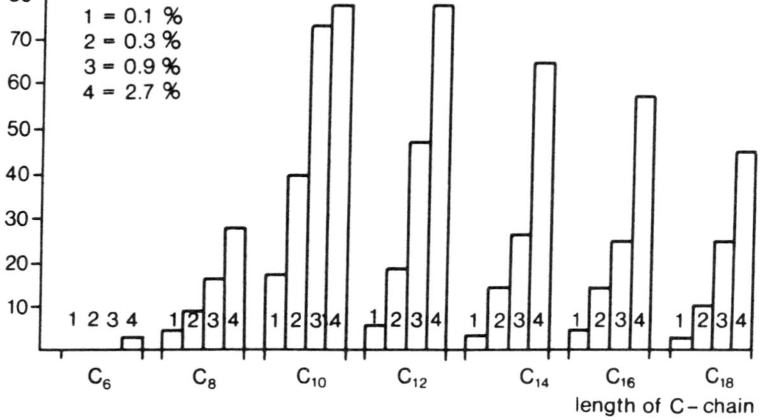

FIG. 5 Results of tissue compatibility testing of alkyl sulfates of different chain lengths (C_6-C_{18}) by intracutaneous injection. Judgment: bleeding 0-8 points; skin defect 0-8 points. The represented result per preparation and concentration is calculated from the sum of single-point evaluations of the animal groups. (From Ref. 69.)

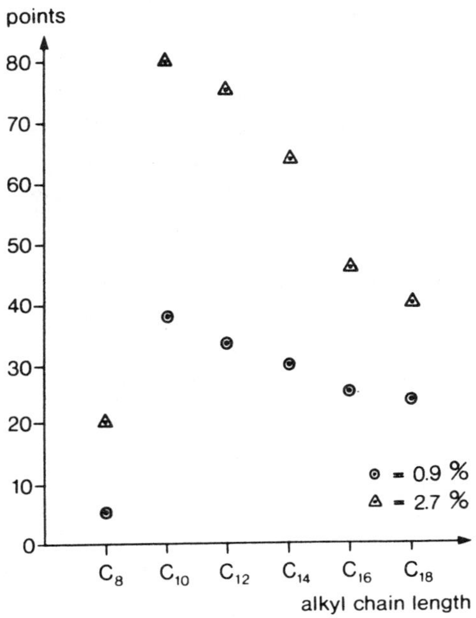

FIG. 6 Degree of tissue reaction following intercutaneous injection: two concentrations (0.9 and 2.7%) of *n*-alkyl sulfonate with different chain lengths (C_8-C_{18}) (intercutaneous injection volume 0.05 ml). Maximum reaction: 80 points. (From Ref. 132.)

for AS $C_{12} > C_{10} > C_{14} > C_8$

for ABS $C_{12} > C_8 > C_{16} > C_{14}$ and

for AOS $C_{12} > C_{14} > C_{16}$

for the skin roughness, and for the bovine serum albumin denaturing potency,

for AS $C_{10} > C_{12} > C_{14} > C_8$

for ABS $C_{12} > C_8 > C_{14} > C_{16}$ and

for AOS $C_{12} > C_{14} > C_{16}$

In summary, it can be said that compounds possessing saturated C chains (C_{10}-C_{14}) show a maximum in their activity on the skin, whereas anionics of shorter and longer chain lengths possess less skin-damaging activity. Anionics with unsaturated alkyl rests appear to show stronger effects than those of the same chain length which are saturated.

F. Skin Compatibility of Special Anionic Surfactants

Studies which have been carried out with specific products are described in the following paragraphs. In many cases the trade names are used because the products studied cannot be chemically defined in an unambiguous way and actually represent mixtures of substances chemically closely related to each other. Their properties are dependent on the starting material used and on the production process. The methods used often differ markedly from each other, so that a short description in any case is necessary, as occurred in the preceding section on the activity of surfactants upon the mucous membranes. Were this not done, then differing results obtained with the same substances and concentrations could not be understood. In addition, one must not forget the variation in results which is in any case a characteristic of biological experiments.

1. Soaps

The skin compatibility of a substance is related to the ability of the substance to penetrate the skin. As a model for such a process, Kröper (75) used a diffusion chamber equiped with a cellophane membrane. He established a dependence of the dialysis speed upon the C chain length of the soaps, fatty alcohol sulfonates, and fatty acid condensation products studied. The smaller the molecules, the more rapidly they are dialyzed. Substances with double bonds exhibit a more rapid dialysis rate than those with saturated C chains.

Dugard and Scheuplein (113) studied the permeability of soaps and alkyl sulfates of defined chain length through isolated human epidermal membranes and found that it is a maximum at C_{12}. As for soaps, those possessing a C_{14} chain were only slightly less effective than those with a C_{12} chain. Howes (92) also found in his studies with radioactive labeled soaps using rats and isolated epidermis from rats and humans that the soaps pass through the skin to a degree dependent upon their chain length. A C_{18} soap showed no penetration after 24 hr, a C_{16} soap penetrated only slightly, and a C_{14} soap penetrated to a slightly greater extent. The greatest penetration of the human epidermis was found with C_{10} and C_{12} soaps. With intact animals, the following order of rank of penetration after single application was found: $C_{12} > C_{10} = C_{14} > C_{16} = C_{18}$. However, the penetration of the intact skin was very limited. The action of repeated soap applications to the skin was examined by Prottey and Hartop (108) using rats (8% soap solution applied daily two to four times). This study not only reported macroscopic skin changes; glyceride and lipid metabolism was also studied. The soaps used were prepared from fatty acids derived from coconut oil and tallow. After 7 days of application, erythema, edema, dryness, and scaling were observed. Histological examination revealed a thickening of the stratum corneum. Parakeratosis and hyperplasia of the malpighian layer with evidence of pyknotic and vacuolated cells were also seen. Further it was shown that the germinative cells of the stratum basale responded to soap treatment by a selective synthesis of more triglycerides.

Brown (107) studied, in addition to other surfactants, a soap (pure fatty acid soap, pH 10.2) using various animal models (occlusive patch test on albino rabbits, 1% active material aqueous solution, 6-hr contact time; on albino rats, 5% active material aqueous solution, 16-hr contact time; open patch test on rabbits and guinea pigs, 1% active material; open patch test on hairless mice, 2% active material) and human skin.

The soap tested did not cause skin irritation on the uncovered skin even after repeated application. However, when the site of application was covered, reactions were seen both in the human and animal subjects.

As part of a teratogenic study, soaps (0.5 ml, 0.3, 3, and 30%) were applied repeatedly to mice, rats, and rabbits (133). No local reactions were observed on the animals treated with 0.3 and 3% soap solutions. However, the rats and mice treated with 3% soap solution showed an increased irritability. Those rats and mice treated with 30% soap solution developed necroses. This irritant action caused cracking and bleeding of the skin when rabbits were used.

According to Hoffmann (73), soaps that contain coconut oil often cause skin irritations because coconut oil contains 47-48% lauric acid and 17% myristic acid. The salts of lauric acid are apparently especially responsible for causing skin irritation.

The C_{12} soap (sodium carboxylate) was tested for skin irritancy in comparison to alkyl sulfates, alkyl sulfonates, and alkyl polyoxyethylene sulfates in a 24-hr closed patch test on the inner surface of the forearm (132a). C_{12} soap caused less irritation score and roughness score than C_{12} alkyl sulfate and C_{12} α-olefin sulfonate, but stronger reactions than C_{12} 3-alkyl polyoxyethylene sulfate and C_{12} alkyl benzene sulfonate.

In a wide-reaching study, Prottey and Ferguson (97) tested sodium laurate in 0.25% M concentration (application twice daily for 3 days) using weanling rats and the open patch test method. The treatment resulted in erythema, edema, a mild degree of epidermis thickening, and a drying out of the skin. These findings were evident on the first day and strangely became partly less evident as time progressed. Opdyke et al. (50) reported mild skin reactions after cutaneous application of soap (5%) over a period of weeks and months using guinea pigs.

Ivanov et al. (72) reported the local effects of soaps (16% formulation) that contained C_{10}-C_{16} fatty acids on guinea pigs after 14-15 days of application. Inflammation in the skin capillaries at the site of application and to a lesser extent at the neighboring sites was observed. The C_{17}-C_{20} soaps had no effect upon the skin capillaries. The local compatibility is associated with various properties of the surfactant. A connection between the irritant effect of soap and the epidermal permeability has been assumed by Bettley (110). The degree of alkalinity also has an effect. Blank (111), who had already demonstrated differences in irritant activity in 1939, attributed these differences to varying fatty acid composition of the soaps. A substance can only have an irritant effect when it penetrates the epidermis in sufficient quantities. This penetration can occur with differing degrees of rapidity, so that the penetration speed is a measure of the irritant action. Bettley (110) has carried out experiments on human skin, both in vivo and in vitro, and found that under the experimental conditions used, 24.4% of the original quantity of laurate penetrated the epidermis and lead to irritant reactions in 18 cases of a total of 28 (64%). The other soaps, octanoate, oleate, and palmitate, which were markedly less irritating, penetrate the skin to a much lesser extent. Using the bath test [a method described by Gloxhuber and Schulz (69)], sodium laurate was tested for its skin compatibility in comparison to a standard soap, which contained 83% tallow fatty acid and 17% coconut oil fatty acid. Hairless mice were used as test animals (34). The skin reactions (erythema, edema, skin necrosis) were evaluated using a point system (scale 0-5), the average values calculated and expressed as percentages of the maximum possible reaction (5 points = 100%). The percent values are given in Table 29.

Kaminsky et al. (132b) tested two different commercially obtained bar soaps (Soap A and B) as 20, 10, 5, 2.5, and 1.25% aqueous solutions on the intact and abraded skin of rabbits under occluded conditions and 24-hr exposure time. They used two patch methods (the Hill Top Chamber = HTC with 0.4-ml test solution and the Gaze Patch with 0.5-ml test solution). For scoring the skin reactions, the Draize scale was used (Scheme 9). The primary irritation index for each test solution was determined in the following manner. The mean erythema values for both the abraded and intact test sites at both 24 and 72 hr (4 values) were calculated. The mean edema values were

TABLE 28a Primary Irritation Indexes (PII) for Each Solution of Soap A and B

	Soap A		Soap B	
Concentration (%)	HTC	Gauze Patch	HTC	Gauze Patch
20	5.46	6.21	5.50	5.09
10	4.51	5.38	4.12	4.08
5	2.13	2.09	3.25	2.55
2.5	2.54	1.54	2.54	1.91
1.25	1.04	0.63	1.21	0.96

Source: Ref. 132b.

calculated in the same manner. These 8 values were then summed and divided by 4 to give the PII (Primary Irritation Index). The 20% solution of both soaps produced irritation that was concerned severe, 10% solution produced moderate or severe irritations, 5% caused moderate, 2.5% mild to moderate, and 1.25% mild irritations (Table 28a).

As can be seen in Table 29, the 0.35% sodium laurate solution causes somewhat more severe reactions on the hairless mice than does a 1% soap solution. However, the 0.3% sodium laurate solution is somewhat less effective than the 1% soap solution. It may be concluded from these results that the C_{16} and C_{18} fatty acids, which were the main components of the soap, are noticeably less irritant than the C_{12} fatty acid. The most important findings obtained from the skin of animals after application of soaps, already discussed in this and earlier sections, are summarized in Table 30.

2. Skin compatibility of alkyl sulfates and alkyl ether sulfates

The alkyl sulfates (AS) are the most intensively studied substances among the anionics, especially sodium dodecyl sulfate (SDS) or sodium lauryl sulfate (SLS). In addition to the AS, the alkyl ether sulfates (AES) are also important as active washing ingredients.

(*a*) *Skin compatibility of alkyl sulfates.* Finkelstein et al. (134) using the methods described in Sec. III.B tested SLS in various concentrations with a 16-hr exposure time, using the occlusive patch test method. A pretreatment involving the triple application of 20% formaldehyde solution and the application of 0.5% trypan blue solution after treatment to the axilla of the test animals was carried out. The evaluation scale given by these authors (nondyed treated skin 0%, maximum dyed skin 100%) is used to present the results in Table 31 for the dependence of the irritation strength upon surfactant concentration. With these results in hand, a standard had been achieved against which other products can be compared. According to this method, substances with low irritancy can be recognized as well. Table 31 shows a linear correlation between the concentration of SLS and the irritating potency.

TABLE 29 Average Point Score of Five Mice per Experiment, Expressed as Percent of Maximum Possible Reaction (Bath Test)

		After baths				
Product	Conc., %	1st	2nd	3rd	4th	5th
Na laurate	0.30	0	4	16	24	36
	0.35	0	12	24	36	52
	0.35	0	8	16	36	45
Soap	1.0	0	4	16	32	40

TABLE 30 Summary of Experiments Carried Out with Soaps to Establish Their Skin Compatibility Together with Results

Soap specification	Conc., %	Species	Application (patch test)	Frequency and duration of applications	Test results	Ref.
Pure fatty acid soap (pH = 10.2)	1	Rabbit	Occlusive	Once daily for 6 hr 3 days	Skin irritation (mild)	
	5	Rat	Occlusive	1 × 16 hr	Skin irritation (very mild)	
	1	Rabbit	Open	5 times weekly 23 appl.	No reaction	107
	1	GP[a]	Open	5 times weekly 23 appl.	No reaction	
	2	hl. mouse	Open	7 times weekly 28 appl.	No reaction	
Standard soap	0.3	Rat	Open		No reaction	
	3.0	Rat	Open		No reaction	
	30	Rat	Open	1 time daily	Severe local irritation	
	0.3	Mouse	Open	0.5 ml for 14 days	No reaction	133
	3.0	Mouse	Open		No reaction	
	30	Mouse	Open		Severe local irritation	
Conventional soap (coconut oil and tallow)	8	Rat	Open	2-4 times daily: 30 min total 14 appl.	Erythema, edema, scaling, alteration of fat metabolism	108
Various soaps	25	Rabbit ear	Open	1 time daily 10 days	Most of the 21 tested soaps caused skin irritation; 6 of them were comedonic	123
Commercial toilet soap	8	GP	Occlusive	1 time daily 10 days	Moderate thickening of the skin (acanthotic effects)	104

Sodium soap ($C_7H_{15}COONa$)	5	Mouse	Implantation	1 time with the help of agar implant	Swelling of subcutis and corium, hyperkeratosis	84
Soap	5	GP	Open	1 time daily several weeks	Moderate skin reactions	50
	8	Rat	Open	2-4 times daily several days	Erythema and edema, middle intensity	130
Sodium salts of defined fatty acids	0.25 M	Rat	Open	2-4 times daily several days	C_{12} soap most irritant, followed by C_{10}, C_8, C_{14}, C_{16}, and C_{18}, the last ones midly irritant	130
Na oleate	0.25 M	Rat	Open	2-4 times daily several days	More severe irritation than the C_{18} soap	130
K soaps of oleic, linoleic, ricinoleic acids, Na oleate	10	Rat	Open	1 time daily 10 days	Clear irritant action of all soaps, acanthotic effect	122
Na laurate	0.25 M	Rat	Open	Twice daily 3 days	Severe edema and erythema after only 2 applications	97
Na laurate	0.35	hl. mouse	Bath test	Daily 1 hr 5 days	Moderate erythema and edema	34
Soap (coconut oil tallow fatty acids)	1	hl. mouse	Bath test	Daily 1 hr 5 days	Moderate erythema and edema	34

[a]GP = guinea pig.

TABLE 31 Evaluation of Irritancy of a Series of Concentrations of SLS

Solution	Irritancy score, %
H_2O	0-5
1% SLS	20
2% SLS	40
4% SLS	60
6% SLS	80
8% SLS	100

Source: Ref. 134.

The effect of SLS on mouse skin after single and repeated applications was studied by Landsdown and Grasso (70). After a single application (0.5 ml 2 cm^{-2}; 10%), the reaction with 10% SLS solution dependent upon contact time was found to result in numerous foci of cellular degeneration in the epidermis. In addition, necrosis was evident, and dermal changes consisted of edema, leucocyte infiltration, and some engorgement of superficial blood vessels. After repeated application (once daily for 2.5 or 9 days; 1 and 2.5% SLS solutions), acanthosis and hyperkeratosis were observed; nuclear debris occurred in the keratin layer. In general, the histological changes preceded the visible ones by 1-2 days. The authors consider on the basis of their findings that SLS is a potent, primary irritant for both human and animal skin. Similar findings were reported by Gisslen and Magnusson (135) for the guinea pig.

SLS was examined by Nixon et al. (136), who applied a 50% aqueous solution to intact and abraded skin of rabbits, guinea pigs, and test persons. The solution was kept in contact with the skin for 4 hr under a plaster. Using a scheme from Edwards (137), the authors determined a primary irritancy score as mean of the reactions on abraded and intact skin of 2.1 for rabbits, 0.4 for guinea pigs, and 0.9 for test persons. SLS in 50% concentration was described as a moderate irritant for rabbits, as mildly irritating for test persons and as hardly irritating for guinea pigs. This finding does not agree with the results of Landsdown and Grasso (70), which is certainly due to the different test methods used.

The application of linear and 30% branched-chain AS ($>$90% C_{12}) to skin of adult rats as 20 and 30% solution produced scabs in 2-3 days. At the 20% concentration, repeated painting for 16 days resulted in skin ulcers that persisted for the entire test period. At the 30% level, the ulcers worsened with time, and 5 of 6 animals died between 8 and 15 days. The rats also exhibited tissue damage on the tongue and oral mucosa as a consequence of licking surfactant from their own backs (137a).

Occluded, 24-hr exposure of the rabbits to 2, 10, or 20% aqueous solutions of the sodium, ammonium or triethanolamine salts of C_{12ave}, AS produced moderate to severe skin irritation. The sodium salt was the most irritating, particularly at the lowest concentration (27a).

Cumulative, open patch test, application of 0.1 ml of a 2% aqueous solution of C_{12} AS (100% pure) to the shaved backs of guinea pigs twice daily for a total of 9 treatments resulted in moderate to severe cumulative skin irritation. A skin irritation score of 3 of a possible 4 points was recorded (128b).

Brown and Muir (37) examined and compared the skin compatibility of five various AS derived either from coconut alcohol (even C number, mainly lauryl and myristyl alcohol) or from the synthetic alcohol Dobanol 23 (a similar alcohol, technically synthesized containing odd in addition to even C number). The following methods were used.

1. *The occlusive patch test.* Application of 0.1, 1, and 25% surfactant solutions to shaved back of rabbits on 3 consecutive days with a daily contact time of 6 hr. The erythema and edema were judged according to Hunter et al. (138) after i.v. application of sulphan blue. The final visual evaluation was carried out 7 days after the first application and a skin sample taken for histological examination.
2. *The open patch test.* Albino rabbits and albino guinea pigs were treated daily (5 times per week) with 0.1, 1, or 25% surfactant solution for 4.5 weeks.

The skin reactions observed are summarized in Table 32. The 0.1% solutions did not result in any irritant action, the 1% solutions behaving similarly in causing no or very mild skin irritation. The 25% preparations caused marked skin irritation in all tests in both animal species. The ethanolamine salts had somewhat more favorable properties than the other compounds, possibly because of their higher molecular weights.

Electron microscope examinations in the epidermis of clipped rat skin following six washings over 3 days (6 times on 3 days) with 1% SLS solution were carried out by Tovel et al. (71). After application, an increased deposition of lipid was observed in most cells; a marked spongiosis and vesiculation, and partial detachment of basal cells from the basal lamina with rupture of the basal membrane were observed. Further, a condensation of tonofibrils within the cells of the stratum basale and stratum spinosum, protein deposition between the deeper epidermal cells (epidermal edema, blister formation), thickening of the epidermis and stratum corneum, and reduction in the number of keratohyalin granules were observed. The basal cell nuclei contained dispersed chromatin. There was also an apparent increase in the number of ribosomes. No lymphocytes, macrophages, or neutrophiles were

TABLE 32 Skin Irritant Effects in Rabbits and Guinea Pigs[a]

Concentration active matter in water (% w/v)	Product	Dobanol 23-based			Coconut alcohol-based		
		Rabbit covered	Rabbit uncovered	Guinea pig uncovered	Rabbit covered	Rabbit uncovered	Guinea pig uncovered
0.1	Monoethanolamine	0	0	0	0	0	0
1	alcohol sulfate	0	0	0	0	0	0
25		3	2	2	3	2	2
0.1	Triethanolamine	0	0	0	0	0	0
1	alcohol sulfate	0	0	0	0	0	0
25		3	2	2	3	2	2
0.1	Ammonium	0	0	0	0	0	0
1	alcohol sulfate	2	0	0	2	0	0
25		3	3	3	3	3	3
0.1	Sodium alcohol	0	0	0	0	0	0
1	sulfate	2-3	0	0	2-3	0	0
25		3	3	3	3	3	3
0.1	Sodium alcohol (2EO)	0	0	0	0	0	0
1	ethoxy sulfate	1	0	0	2	0	0
25		4	4	4	4	4	4

[a]Findings based on gross and microscopic examination of the skin.
Irritancy rating: 0 = nonirritant; 1 = very slightly irritant; 2 = slightly irritant; 3 = irritant; 4 = very irritant.
Source: Ref. 37.

observed. The changes seen in the deeper layers of the epidermis such as intracellular spongiotic, vesiculations and edema, intracellular lipid deposition, thickening of the stratum basale and stratum spinosum could have their cause in the cellular hypoxia as a result of direct cell damage or of dermal capillary damage, or as a result of direct damage to the mitochondral oxidative processes. SLS activates the basal cells to proliferate, i.e., an increase in DNA synthesis results.

The effect of detergents on swelling of the stratum corneum was investigated by Putterman et al. (69h), and Zeidler and Reese (69i). Sodium laurat and sodium oleat caused the highest swelling, SLS some less. Lower swelling was measured after contact with triethanolamine lauryl sulfate and sodium dodecyl benzene sulfonate (SDBS), but sodium lauryl ether and triether sulfate caused only a little swelling, and sodium myristyl ether sulfate none (69h). Among 16 anionic surfactants (alkyl and alkyl ether sulfates, and some others), SLS caused the highest swelling, followed by monoethanolamine lauryl sulfate, sodium $C_{12\text{-}18}$ AS, ammonium lauryl sulfate, triethanolamine lauryl sulfate, and sodium alkyl benzene sulfonate (SABS). A lower swelling was observed with secondary alkyl sulfonate, sodium salt, sodium lauryl 2EO-sulfate, sodium lauryl myristyl 3EO sulfate and sulfosuccinate sodium salt. A protein fatty acid condensation product had only minimal effect on the swelling, and sodium/magnesium alkyl 6EO sulfate had nearly no effect (69i).

Within a study of toxicity determination, Carson and Oser (139) reported their observations concerning the skin compatibility of SLS on rabbits and guinea pigs. The acute dermal toxicity was determined with a 33% SLS solution and varying quantities of this concentration were applied for a 24-hr period to the intact and abraded skin of rabbits. The post observation time was 14 days. The skin compatibility was evaluated according to Draize (2). The quantities applied to the skin, which lay between 150 and 2000 mg kg^{-1}, scarcely influenced the skin irritation because skin reactions are dependent on concentration, not dose.

The primary irritation score for all doses lay between 3 and 4.2 (that means 37.5-52.5% of the maximum possible reactions). The changes established at the end of the observation period were, however, not all the same. They were characterized by slight scaling, scaling, wrinkling and folding of the skin, up to leathery changes.

In a study of Imokawa (128b) on skin irritating of monoalkyl phosphate anionic surfactants 4 alkyl sulfates (C_8, C_{10}, C_{12}, C_{14}) were tested for comparison. In the 24-hr closed patch test on intact back skin of guinea pig skin C_{12} AS produced stronger reactions than alkyl benzene sulfonate (ABS), alfa-olefin sulfonate, alkyl polyoxyethylene sulfate and monoalkyl phosphate. Also Kästner and Frosch (15a) demonstrated that the alkyl sulfates are more irritating than the alkyl ether sulfates, applied once intracutaneously or repeated epicutaneously.

In a 90-day test, 6, 60, and 150 mg kg^{-1} of SLS (open patch test, once daily, 5 times weekly for 13 weeks; after application the animals were placed in restrictive cages for 6 hr) were applied to the intact skin of rabbits (four animals) and scarified skin (two rabbits) (139). The dosages stated were in 0.5 ml water so that concentrations of 1.2, 12, and 30% resulted. The primary irritation score was determined during the maximum of the skin reactions at the end of the applications; furthermore, the day was noted on which the maximum skin reactions were observed. As can be seen in Table 33, the irritation score was dose- and therefore concentration-dependent. The time of appearance of the maximum irritation effect varied between individual animals within any one group, the variation being particularly large at low doses. In connection with the question concerning the influence of surface-active substances upon contact sensitization, Gisslen and Magnusson (135) reported experiments using SLS and ABS with guinea pigs. The concentrations of 0.1, 1, and 5% (occlusive patch test) for a 24-hr contact time are used for the applications. After removal of the patches, skin biopsies were taken and examined histologically. Necrotic epidermal changes and migration of neutrophilic granulocytes, slight spongiosis, and hydropic degeneration of the basal cell layer were found. Total necrolysis of the epidermis occurred at 5% concentration. The corium also suffered changes consisting of cell migration of neutrophiles, transformation of the collagen fibrils, etc. From these results it may be assumed that the sensitization-promoting effect of the anionics first occurs at those concentrations which cause a toxic dermatitis. Marzulli and Maibach (119) have studied the skin-irritating properties of various substances, among which was sodium lauryl sulfate (rabbits, hairless rats, and hamsters) using various methods and species. The results obtained were compared with the results of the human skin.

TABLE 33 Primary Irritation Score According to Draize During and After Repeated Application of SLS Solutions in Rabbits

				Primary irritation score				
		Number of rabbits		Maximum		On day	Final score	
Doses, mg kg^{-1}	Conc., %	Intact	Abraded	V	$\bar{x}$	V	V	$\bar{x}$
0	0	2	1		0			0
6	1.2	4	2	0-2	1.33	4-54	0-1	0.33
60	12	4	2	4-5	4.17	4-12	1-2	1.5
150	30	4	2	5-6	5.17	11-18	2-4	3.17

V = variation; $\bar{x}$ = mean.
Source: Ref. 139.

Although, according to the authors, rabbit skin is four or more times more permeable than human skin for various substances, the occlusive patch test methods increased the permeability of skin by a factor of about 4. The authors left the skin of the rabbits uncovered during the repeated applications whereas the skin of human test subjects was covered with a nonporous plaster. Thus the authors attempted to attain a comparability of the permeability of human and rabbit skin. It was shown that the hairless animals used were poorly suited for such tests. Further, a so-called cumulative irritation test was done. On each day of treatment, erythema (score 0-4) and edema (measurement of the skin fold thickness) were evaluated. For each rabbit four different products or concentrations were applied simultaneously (contact time daily 24 hr). An analogous experiment using six human test subjects was carried out. The samples were applied to the back every day for 21 consecutive days and were covered with plaster. The application was stopped when the first reactions were seen. According to the results obtained, a 1% SLS preparation is not expected to cause reactions in humans. The treatment with SLS was remarkable inasmuch as a relatively large increase in skin fold thickness was observed. This finding requires that the possibility of SLS having an acanthogenic activity must be considered. Histological examinations of excised skin samples treated with 1% SLS solution revealed an epidermal thickening; samples treated with 4% SLS solution already showing necrotic changes.

Similarly, Philipps et al. (60) reported their results obtained by testing for skin compatibility using normal and abraded skin of rabbits and humans (contact time 24 hr). A 10% SLS solution was tested. The contact time in a continuous, closed patch test was 21 days. Under these conditions the substance possessed markedly irritant properties. Patch tests involving single applications were also undertaken.

Opdyke and Burnett (68) detail another method for testing the skin compatibility of cosmetic raw materials: the guinea pig immersion test. This test gives results of good reproducibility. Ammonium lauryl sulfate (ALS), ammonium myristyl sulfate, and various ethoxylated derivatives thereof were subjected to this test in 0.25 and 1% concentrations. The results are presented in Table 34. As can be seen, ALS when tested under the conditions of the immersion test resulted in several fatalities. This has certainly less to do with the resorption of the material than with the large area of skin damage.

In the course of a broad study of sensitization testing Nilzen and Wilkström (115) observed the action of a 1% SLS solution on guinea pig skin in an 8-day open-skin test. Only slight macroscopic effects were seen, but microscopic examination revealed thickening of the epidermis as well as intra- and extracellular edema.

An interesting comparative study of Davies et al. (140) addressed itself to the question of which animal species is most similar to the human under

TABLE 34 Results of Evaluation of Mildness to Skin of Some Alkyl Sulfates and Their Ethoxylated Derivatives Tested by Guinea Pig Immersion Test

Material	Conc., %	Gross skin grades[a]
Ammonium lauryl sulfate	0.25	All 3 died
Ammonium lauryl sulfate 1EO[b]	0.25	4, 4, 4
Ammonium lauryl sulfate 3EO	0.25	9, 9, 9
Ammonium lauryl sulfate 7EO	0.25	10, 10, 10 (normal)
Ammonium myristyl sulfate 4EO	0.25	8, 8, 9
Ammonium myristyl sulfate 6EO	0.25	10, 10, 10
Ammonium lauryl sulfate 3EO	1	6, 7, 8
Ammonium lauryl sulfate 7EO	1	9, 9, 10

[a]Judgment of skin reactions with help of a 10% scale (10 points = normal skin, no reactions; less points = skin reactions, various grades; 1 point = strongest skin damage).
[b]Number of ethylene oxide molecules.
Source: Ref. 68.

path test conditions. SLS was used as test substance (evaluation according to Draize, contact time 24 hr). It was established that the mouse, rabbit, and piglet were the most sensitive to SLS, dogs and guinea pigs showing milder reactions. The human and minipig both showed only very mild reactions. In the case of the baboon, SLS in 1 + 5% concentrations did not cause any skin irritation at all. Table 35 shows which species exhibit reactions similar to those of the human and which animals exhibit more severe or weaker reactions than humans do after treatment with SLS. This table clearly shows that there are marked differences in the reactions exhibited by various species. According to these results, rabbits, which are often used for studying the skin-irritating properties of substances, are not the best animals for such tests.

TABLE 35 Relative Irritancy, Humans vs Seven Other Species[a]

Species	Similar to humans	More than humans	Less than humans
Mouse	—	1 and 5%	—
Guinea pig	1 and 5%	—	—
Rabbit	1%	5%	—
Minipig	5%	—	1%
Piglet	—	1 and 5%	—
Dog	5%	—	1%
Baboon	—	—	1 and 5%

[a]Tested substance was sodium lauryl sulfate 1 and 5%.
Source: Ref. 140.

Ingram and Grasso (121) have also studied the problem of extrapolating results obtained with animals to humans, also using SLS as the test substance. The test used was a 5-hr semiocclusive patch test on rabbits, with daily application for 5 days. At a concentration of 0.0096%, no visual reactions were seen.

At 0.038-2.4% concentrations, the first signs of an erythema usually appeared after the first 5 hr. The maximum erythema and edema symptoms were reached on the second or third day, the edema then receding and the erythema becoming darker and less obvious. A linear dependence of irritancy (cumulative score) upon the logarithm of concentration was found. Most of the skin areas which had been treated with 0.038-0.1% SLS solution exhibited mild acanthosis and hyperkeratosis together with mild infiltration; but no destruction of the epidermis was observed. Between concentrations of 0.15-0.5%, a marked acanthosis and hyperkeratosis together with signs of epidermal damage (spongiosis and isolated cell damage) appeared. In most of the patch areas, slight epidermal destruction and slight to moderate dermal infiltration were seen. Concentrations of SLS above 0.6% caused various degrees of epidermal destruction with moderate to severe dermal infiltration. Parakeratosis and loss of the granular layer were not usually found after 5 days exposure to SLS. In other tests a hyalinization of dermal collagen was seen after application of concentration above 0.5% SLS on 1-3 days.

In a similar study, Prottey and Ferguson (97) applied sodium lauryl sulfate in 0.25 M concentration (twice daily; 3 consecutive days) to rat skin and left it there uncovered. The twice-daily application did not result in any skin reactions, whereas the sixfold treatment caused moderate reactions (erythema, edema, scaling and cracking of the stratum corneum, as well as severe thickening of the epidermis).

Fischer and Maibach (141) examined the action of various substances upon the mitotic activity of the human epidermis. There was apparently no direct relationship between gross inflammation and mitotic response. A 1% SLS solution produced a mild erythema but caused a marked increase in mitotic activity, which peaked at about 48 hr. A 0.5% SLS solution and a 2% SLS solution both caused less mitotic activity than the 1% SLS solution (see Fig. 7). At present, there is no explanation for these findings.

A comparative testing of skin compatibility was published by Olson et al. (36). The experiments were carried out on rabbits using a 14-day period of contact on the intact skin or three applications on the abraded abdominal skin (1, 5, and 25% solutions). The results are given in Table 36, which shows that SLS was one of the strongest irritants tested.

Poor agreement was found in the results obtained by Brown (107), who tested a number of various surface-active substances, among which were

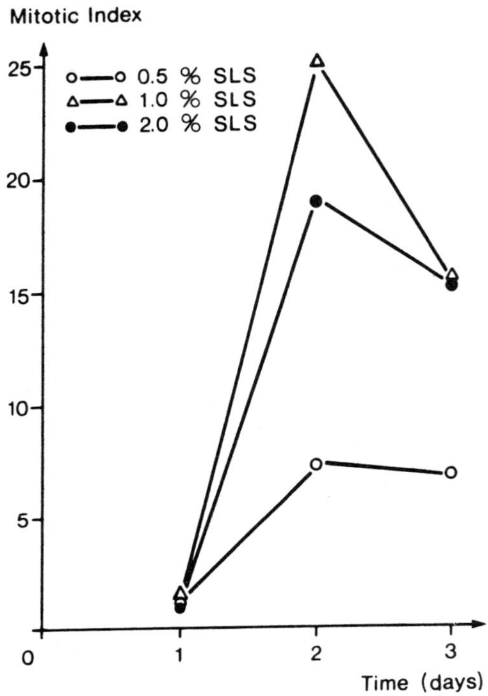

FIG. 7 Mitotic response to sodium lauryl sulfate (SLS). (From Ref. 141.)

alkyl and alkyl benzene sulfonates. The substances were compared in various skin tests using animal experiments. With a sodium salt of a broad-cut coconut alcohol sulfate (C_{12} and C_{14}), he found that a 7-day occluded exposure to a 1% solution was highly irritating to rabbits, although a 0.1% solution was not. In rats, a 5% solution of the AS did not elicit a response after occlusion for 16 hr. In uncovered exposures on rabbits and guinea pigs, a 1% solution of surfactant was not irritating. Similarly, repeated percutaneous exposure to hairless mice for 28 days showed no histochemical skin changes.

A comparative testing of lauryl sulfate, lauryl ether sulfates, and alkyl benzene sulfonates using the guinea pig immersion test demonstrated that alkyl benzene sulfonates are less irritating than sodium alkyl sulfates in 0.25% concentration. In a consumer test involving a housewives panel, a good correlation was found between the reactions observed using guinea pigs and those observed on the housewives' hands (142). Vinson and Choman (104) addressed themselves to the question of whether SLS influences the penetration

TABLE 36 Average Skin Irritation after Repeated Application on Rabbit Ear and Intact and Abraded Abdominal Skin

	Effects of surfactants					
	Rabbit ear (index)			Rabbit abdominal skin (index)		
Conc., %	1	5	25	1	5	25
Sodium dioctyl sulfosuccinate	1	3.5	6	4	5	6
Sodium lauryl sulfate	1.5	3.5	6	3.5	5	6
Triethanolamine lauryl sulfate	1.5	4	4.5	3	5	5.5
Sodium dodecylated diphenyl ether disulfonate: sodium lauryl sulfate 1:1 ratio	1	3	5.5	1.5	4	5.5
Dodecyl benzene sulfonate triethanolamine salt	1	3	5	1	3	4.5
Sodium dodecylated diphenyl ether monochloro sulfonate	1	3	4	1.5	3	4.5
Sodium dodecylated diphenyl ether disulfonate	1	1	2	1.5	2.5	3
Dodecylated diphenyl ether disulfonate magnesium salt	1	1	1.5	1	1.5	3
Dodecylated diphenyl ether disulfonate triethanolamine salt	1	1	1.5	1.5	2	2

Index reactions:
1 Essentially no irritation to intact skin
2 Very slight hyperemia on intact skin, likely more irritating to abraded skin
3 Slight hyperemia on intact skin, may or may not burn abraded skin
4 Moderate hyperemia on intact skin, may or may not burn abraded skin
5 Burn from two to four 24-hr applications to intact skin
6 Burn from one 24-hr application to intact skin

Source: Ref. 36.

of other substances through the skin. They used guinea pigs to whose backs the test solutions were applied (4 times within 1 hr, afterwards closed patch test, repeated for a 10-day period). The following surfactants in the given concentrations were examined.

Sodium lauryl sulfate: 0.1, 0.2, 0.5, and 1.0%
Sodium dodecyl (branched) benzene sulfonate: 0.2, 0.5, and 0.1%
Sodium acyl (coco) isothionate: 1.0%
Sodium acyl (coco) n-methyl taurate: 1.0%
Commercial toilet soap: 8.0%

To some of the animals, nickel sulfate ($NiSO_4$) was simultaneously applied.

After a 10- to 14-day interval of no treatment, three daily challenge applications were applied. The animals which had been simultaneously treated with nickel sulfate showed allergic responses, which indicated an increase in resorption caused by the surfactants. The 1% SLS solution alone caused reactions after the fourth application which reached a maximum effect at the 10th application. The 0.5% SLS solution caused a skin reaction only after the 6th treatment. Whereas the 0.2% solution caused mild reaction after the 8th application, the 0.1% solution caused no reactions at all. All the concentrations tested showed an acanthotic effect. Meinicke (143) compared the skin compatibility of protein fatty acid condensation products (PFC) with that of the alkylaryl sulfonates (AAS) and fatty alcohol sulfates (FAS), which are usually used trade products, using rabbits and an epicutaneous test method (24-hr occlusive patch test). The reactions noted were 24, 48, 72 hr after application of the plaster.

The comparison of the skin reactions after application of 5, 10, and 20% solutions showed that, next to AAS, the most severe reactions were caused by FAS (papular reactions). Iimori et al. (144) published experiments with linear alkyl sulfonates (lASF), α-olefin sulfonates (AOS), sulfates of synthetic alcohols (AS), and a formulated detergent for kitchen use. The anionics were applied to the clipped back skin of guinea pigs and left for 10 or 20 min. For 24 hr at 1 week after the application, the histopathological reactions of the skin were observed. At the 30% concentration level, AOS caused markedly less severe reactions than lASF and AS, whereby the reactions caused by lASF and AS were comparable. The 1% solutions did not cause any changes under the experimental conditions used.

Various alkyl sulfates were examined by us (34) for skin-irritating properties using various experimental models. The experiments were carried out on rabbits (occlusive patch test), hairless mice (single open application or repeated application without removal of the test substance), or according to the method of W. Burckhardt (145-147), using repeated "wetting" (rubbed in with a glass rod at 30-sec intervals, 60 times within 30 min, to a circular skin section on the rabbit back) and various concentrations. The products thus tested were sodium dodecyl sulfate (Texapon K-12 and Texapon L-100), sodium alkyl sulfate (C_{12}-C_{18}; Texapon Z highly conc., Texapon V highly conc.), ammonium alkyl sulfate (C_8-C_{14}; Texapon A-400), monoethanolamine lauryl myristyl sulfate (Texapon MLS), and triethanolamine alkyl sulfate (C_8-C_{14}; Texapon TH).

As shown in Tables 37 and 38, the 1 and 2% and sometimes the 5% concentrations* of these anionics did not cause any skin reactions on the test animals under various experimental conditions. Texapon Z highly conc. was not without reactions in a 24-hr closed patch test at 5% concentration; however, the same concentration tested in a Burckhardt test did not cause any reactions on the rabbit skin. In contrast, Texapon V highly conc. did not cause any skin irritation with rabbits using the same conditions and concentration. Texapon A-400 when tested in an undiluted form on hairless mice in single application without subsequent removal of the product from the skin caused severe skin reactions with initial necrosis. Texapon TH in undiluted form also caused skin reactions put to the Burckhardt test, the same being found for Texapon MLS in 50% aqueous solution. Tables 37 and 38 summarize the results obtained with sodium lauryl sulfate relevant to skin compatibility and those carried out with other alkyl sulfates mentioned.

(b) Skin compatibility of alkyl ether sulfates. Alkyl ether sulfates (AES) are classified as slight to moderate irritants (53c). They are generally regarded as possessing better skin compatibility than the alkyl sulfates (AS). For this reason they are mainly used where they come into intimate contact with the human skin (dishwashing liquid for washing up by hand, shampoos, bubble baths, etc.).

AE_3S (28% AI) was classified as a primary skin irritant according to the Draize procedure when applied undiluted to the intact and abraded skin of rabbits. This undiluted material produced moderate to severe erythema with eschar formation but no dermal injury in the depth. Edema ranged from barely perceptible to slight for intact skin and from barely perceptible to moderate with abraded skin (22a).

AES with a higher degree of ethoxylation are less irritant. Ciuchta and Dodd (27a) found that occluded, 24-hr exposure of rabbits to 2, 10 or 20% aqueous solutions of sodium lauryl polyether(12) sulfate (C_{12} AES) produced minimal to no skin irritation. No graded response was seen; primary skin irritation scores were about 1 (of a possible 8 points) for all three test concentrations.

Brown (107) included in his tests for skin compatibility of various anionics the sodium salt of sulfated coconut ethoxy (3EO) alcohol. This product was

*Concentration based upon WAS content.

TABLE 37 Summary of Experiments Carried Out with Sodium Lauryl Sulfate Relating to Skin Compatibility and Their Results[a]

Ref.	Conc., %	Species	Application (patch test)	Frequency and duration of application	Results of test
134	1-8	Rabbit	Occlusive	Once 16 hr	Erythema, 1% = 20%, 8% = 100% of irritancy score
70	10	Mouse	Open	Once 2 hr	Mild swelling and erythema, histological changes
70	2.5	Mouse	Open	Once daily; 6 days	Edema, erythema, and histological changes
70	1.0	Mouse	Open	Once daily; 7 days	Edema, erythema, and histological changes
71	1.0	Rat	Open	Twice daily; 3 days	Coarsening, redding, and edema
139	33	Rabbit	Open	Once 24 hr (intact and abraded skin)	Severe erythema and edema
135	0.1	GP[b]	Occlusive	Once 24 hr	No reaction
135	1.0	GP	Occlusive	Once 24 hr	Clear histological changes, necrosis of the epidermis
135	5.0	GP	Occlusive	Once 24 hr	Clear histological changes, necrosis of the epidermis
119	1.0	Rabbit	Open	Once daily; 14 days	10% of the max. irritancy score
119	4.0	Rabbit	Open	Once daily; 14 days	37% of the max. irritancy score
60	10	Rabbit	Occlusive	Once 24 hr	Mild, irritant action
115	1.0	GP	Open	Once daily; 8 days	Minor skin changes, irritation score according to Draize <2
140	1.0	Mouse, GP, among others	Occlusive	Once 24 hr	0-1 primary irritation index according to Draize
140	5.0	Mouse, GP, among others	Occlusive	Once 24 hr	0-2 primary irritation index according to Draize

121	0.0096-2.4	Rabbit	Occlusive	Five times 5 hr; 5 days	No reactions at low concentrations. At higher concentrations: erythema, edema, acanthosis, hyperkeratosis
36	1.0	Rabbit	Open	14 days contact, once daily; 14 days	Mild to moderate skin reactions
36	5.0	Rabbit	Occlusive	14 days contact, once daily; 14 days	Moderate skin reactions
36	25	Rabbit	Occlusive	14 days contact, once daily; 14 days	Severe skin reactions
142	0.25	GP	Immersion test		Severe skin irritation (cracking of the skin)
104	0.1	GP[b]	Occlusive	Once daily; 24 hr contact time for 10 days	Score 0 (maximal score = 4)
104	0.2	GP	Occlusive		Score 1
104	0.5	GP	Occlusive		Score 3
104	1.0	GP	Occlusive		Score 4
97	0.25 M	Rat	Open	Twice daily; 3 days	After 2 applications no reactions, after 6 applications distinct erythema and edema, severe thickening of the epidermis, clear scaling and cracking of corneum
34[c]	1.0	Rabbit	Occlusive	Once 1 hr	No reaction
34	1.0	Rat	Open	Twice daily; 5 days	No reaction
34[d]	1.0	Rabbit	Occlusive	Once 1 hr	No reaction
34	1.0	Rat	Open	Twice daily; 5 days	No reaction
136	50	Rabbit	Occlusive	Once 4 hr	Primary irritation index: 2.1
136	50	GP	Occlusive	Once 4 hr	Primary irritation index: 0.4

[a]The term sodium lauryl sulfate is to be understood as meaning those alkyl sulfates which are named as sodium lauryl sulfate in the literature cited.
[b]GP = guinea pig.
[c]Texapon K-12 (C_{12} pure; 90% active substance).
[d]Texapon L-100 (C_{12} pure; 99% active substance).

TABLE 38 Summary of Experiments and Results Concerning Skin Compatibility with Alkyl Sulfates (Excluding Sodium Lauryl Sulfate)

Substance	Conc., %	Species	Application (patch test)	Frequency and duration of applications	Results of test	Ref.
Sodium alcohol sulfate (1)	0.1 1.0 25	Rabbit GP[a]	Occlusive	Once daily, 6 hr for 3 days	0.1%: Rabbit and GP: (1)-(4) score: 0	37
Ammonium alcohol sulfate (2)	0.1 1.0 25	Rabbit GP	Open	5 times weekly for 4-5 weeks	1.0%: Rabbits: (1)+(2) score: 2-3 GP: (1)-(4) score: 0	37
Monoethanolamine alcohol sulfate (3)	0.1 1.0 25	Rabbit GP	Open	5 times weekly for 4-5 weeks	25%: Rabbit and GP: (1)-(4) score: 2-3 (maximal score = 4)	37
Triethanolamine alcohol sulfate (4)	0.1 1.0 25	Rabbit	Open	5 times weekly for 4-5 weeks		37
Ammonium lauryl sulfate	0.25	GP	Immersion test	Once daily; 4 hr for 3 days	Severe dermal toxicity	68
Alkylsulfate of defined chain length (C_6-C_{18})	0.1	Mouse	Intracut.	Once 0.05 ml/mouse	Various severe skin reactions around injection spot, C_{10} most severe, C_6 weakest reaction. See under Sec. III.E.	69
	0.3	Mouse	Intracut.	Once 0.05 ml/mouse	Various severe skin reactions around injection spot, C_{10} most severe, C_6 weakest reaction. See under Sec. III.E.	69

	0.9	Mouse	Intracut.	Once 0.05 ml/mouse	Various severe skin reactions around injection spot, C_{10} most severe, C_6 weakest reaction. See Sec. III.E.	69
	2.7	Mouse	Intracut.	Once 0.05 ml/mouse	Various severe skin reactions around injection spot, C_{10} most severe, C_6 weakest reaction. See Sec. III.E.	69
Lauryl sulfate triethanolamine salt	1	Rabbit	Occlusive	14 days contact time	Mild to moderate	36
	5	Rabbit	Open	Once daily for 14 days	Moderate	36
	25	Rabbit	Open	Once daily for 14 days	Moderate	36
Sodium secondary sulfate (C_8-C_{18}) (T_5)	1	Rabbit	Occlusive	Once daily; 6 hr; 3 days	Occlusive: T_5: Mild skin reactions T_9: Rabbit strong, rat slight	107
	5	Rat	Occlusive	Once 16 hr	Occlusive: T_5: Mild skin reactions T_9: Rabbit strong, rat slight	107
Na salt of sulfated broad-cut coconut alcohol (T_9)	1	Rabbit	Open	5 × weekly 23 applications	Open: Rabbit and GP: mild skin reactions	107
	1	GP	Open	5 × weekly 23 applications		107
	2	hl. mouse	Open	7 × weekly for 4 weeks	hl. mouse: T_9: Mild skin reactions T_5: Obvious skin reactions	107

TABLE 38 (Continued)

Substance	Conc., %	Species	Application (patch test)	Frequency and duration of applications	Results of test	Ref.
Fatty alcohol sulfate (35% AS)	5	Rabbit	Occlusive	Once 24 hr	Irritant activity score 2-3	143
	100	Rabbit	Occlusive	Once 24 hr	Irritant activity score 4 (maximal score = 4)	143
Sulfate of synthetic alcohols	10	GP	Open	10 and 120 min contact time	No reactions	144
	30	GP	Open	10 and 120 min contact time	Mild skin irritation	144
	100	GP	Open	10 and 120 min contact time	Mild skin irritation	144
Sodium alkyl sulfate (Texapon Z highly conc.)	5	hl. mouse	Open	Twice daily for 1 week	No reaction	34
	5	Rabbit	Occlusive	24 hr contact time	At the point of removing plaster, mild skin reddening: 24 hr later, no reaction	34
	5	Rabbit	Open	Burckhardt test[b]	No reaction	34
Sodium alkyl sulfate (Texapon V highly conc. need.)	5	Rabbit	Occlusive	24 hr contact time	No reaction	34
	5	Rabbit	Open	Burckhardt test[b]	No reaction	34

Ammonium alkyl sulfate	2	Rabbit	Open	Burckhardt test[b]	No reaction	34
(Texapon A-400)	100	hl. mouse	Open	One application	Skin reddening and commencement of necrosis	34
	5	hl. mouse	Open	Once daily for 5 days	No reaction	34
	5	Rabbit	Occlusive	24 hr contact time	No reaction	34
Monoethanolamine alkyl sulfate	10	hl. mouse	Open	Twice daily for 5 days	After tenth application creased skin, otherwise no reaction	34
(Texapon MLS)	100	Rabbit	Occlusive	6 hr contact time	No reaction	34
	50	Rabbit	Open	Burckhardt test[b]	24 hr after application, mild reddening; 48 hr after application, no reaction	34
Triethanolamine lauryl sulfate	10	hl. mouse	Open	Twice daily for 5 days	No reaction	34
	5	Rabbit	Occlusive	24 hr contact time	No reaction	34
(Texapon TH)	100	Rabbit	Open	Burckhardt test	At end of treatment mild; 24 hr later clear reddening; 48 hr later reversion of the reactions	34
	5	Rabbit	Open	Burckhardt test[b]	No reaction	34

[a]GP = guinea pig.
[b]For 30 min in time intervals of 30 sec rubbed into the skin 60× with a glass rod.

tested using both the closed and open patch test on various animals (rabbits, rats, guinea pigs, hairless mice). The results obtained in the closed patch test showed that this product was one of the least irritating substances, whereas in the open patch test it was one of the strongest irritants.

Brown and Muir (37) examined the skin compatibility of various AES which were derived from coconut alcohol or from the synthetic alcohol Dobanol 23, which were otherwise completely identical in comparison with sodium alcohol (2EO) ethoxysulfate (0.1, 1, and 25% concentrations). They used the close patch test on rabbits and the open patch test on both rabbits and guinea pigs, thus determining that the alkyl ether sulfate in 0.1 and 1% concentration was equally good or better than the alkyl sulfates as far as the skin compatibility of the animals used was concerned. It was, however, a strong irritant at 25% concentration and was the worst of five substances tested at this concentration.

Using the immersion test Opdyke and Burnett (68) tested the skin compatibility of various ethoxylated products among other surfactants. The results obtained from testing various ammonium lauryl ether sulfates and ammonium myristyl ether sulfates are summarized in Table 34. As can be seen, a clear dependence of skin compatibility upon the degree of ethoxylation exists. At concentrations of 0.25 and 1%, the products with 6 and 7 ethoxys were clearly better tolerated on the skin of guinea pigs than those products with 3, 4, or 1 ethoxy.

The effect of sodium lauryl ether sulfate (60% active substance) upon the skin and the hair cycle of rats was studied by Rubisz-Brzezinska et al. (126). The product was applied to the shaved back skin once daily for 65 consecutive days undiluted, and in 50, 15, and 1.5% dilutions. The undiluted product caused skin reddening, scaling, and keratinization of the hair follicles after 8 days. Histologically, hyper- and parakeratosis, thinning of the epidermis (at a later stage a thickening of the epidermis occurred), atrophia of the granulosa cell layer, and inflammatory reactions in the dermis were identified. The 50% preparation resulted in erythema, scaling, and keratinization after 2 weeks. No gross changes of the skin were found after application of the 15% solution. In *Schimmel Briefs* (142), sodium lauryl ether (5EO) sulfate was not only described as possessing good skin compatibility properties in the immersion test using guinea pigs, but also the property of alleviating the irritation in the case of AS and alkyl benzene sulfonates.

In various tests, published in the last 10 years (15a,27a,69h,69i,128b), the better compatibility of AES in comparison to AS or/and alkyl sulfonates is demonstrated.

The open application of sodium lauryl triethoxysulfate (0.25 M solution, twice daily for 3 days) to rats caused only a slight erythema and edema, according to Prottey and Fergusson (97).

Cumulative, open patch test application of 0.1 ml of a 2% aqueous solution of $C_{12\text{-}13}$ AE_3S (99.8% pure) to the shaved backs of guinea pigs twice daily for a total of 9 treatments resulted in slight to moderate skin irritation. A skin irritation score of 1.25 of a possible 4 points was recorded (128b).

McMillan et al. (148) discuss the results of experiments they carried out using various surfactants and surfactant mixtures in the patch test of Draize, the 5-day dermal test with rabbits, the immersion test with guinea pigs, and the 21-day human repeat irritation test. Among the anionics tested was Sipon ES (3.5 M ethoxylated sodium lauryl sulfate), which lead to relatively severe reactions when applied in 5% concentrations in the closed patch test. However, in the open test (5-day dermal test), only a mild erythema and edema were caused; likewise, in the guinea pig immersion test, the compound had only moderate irritant properties (see Table 39). Some AES were tested by us (34) for their skin compatibility using hairless mice and white rats, the substances being applied once, or repeatedly, using the open patch test method. The sodium lauryl myristyl di- and triglycol ether sulfates in undiluted form do not cause skin reactions after a contact time of 1 or 2 hr.

In contrast, the repeated application of Texapon N-25 (twice daily, applied to the backs of hairless mice without removal) caused severe reactions after the third application. Treatment with a 25% aqueous solution occurred only after the 6th application. A twice-daily application of the 10% product over a 5-day period did not result in any skin reactions. Treatment with Texapon K-14S special in 10% concentrations twice daily for 5 days also did not cause any reactions on the skin of hairless mice. The higher ethoxylated compound Texapon ASV was applied once daily to the backs of hairless mice and not removed. The undiluted product caused reddening of the skin after the second application, after the following applications the reactions became stronger. The 50% concentration did not lead to any reaction after repeated application during two consecutive weeks. The higher ethoxylated products seem to have a better skin compatibility than the lower or not ethoxylated compounds. This result corresponds with the observations have been seen on the mucous membranes of rabbit eyes (see Sec. II.G). Further results are given in Table 39. In the table, the other experiments described in this section together with their results are collected.

3. Skin compatibility of alkyl and alkylaryl sulfonates

The irritating potential of alkyl sulfonates (ASF) is nearly the same as that one of alkyl sulfates (AS). In some studies the ASF have a lower (128b,132a), in others nearly the same (15a) potential. Quack and Reng (148a) found that a 1% solution of secondary alkane sulfonate is tolerated without any reactions on the rabbit and human skin under the condition of a closed patch test. Tests for irritation using the Draize procedure have shown that undiluted

TABLE 39 Summary of Experiments Carried Out With Alkyl Ether Sulfates to Establish Skin Compatibility Together with Results

Substance		Conc., %	Species	Application (patch test)	Frequency and duration of applications	Results of test	Ref.
Natrium salt of sulfated broad-cut		1	Rabbit	Occlusive	Once daily; 6 hr for 3 days	Middling skin irritation	107
coconut ethoxy		5	Rat	Occlusive	Once 16 hr	Mild skin irritation	107
(3EO) alcohol		1	Rabbit	Open	5 × weekly, 23 applications	Middling skin irritation	107
		1	GP[a]	Open	5 × weekly, 23 application	Middling skin irritation	107
		2	hl. mouse	Open	7 × weekly, 28 application	Middling skin irritation	107
Sodium alcohol (2EO) ethoxy-		0.1	Rabbit	Occlusive	Once daily; 6 hr for 3 days	0.1%: Rabbit and GP: score 0	37
sulfate		1.0	GP	Open	5 × weekly, 4-5 weeks	1.0%: Rabbit, occlusive: score 1-2	37
		25	Rabbit	Open	5 × weekly, 4-5 weeks	25%: Rabbit and GP: score 4 (maximal score = 4)	37
Ammonium	1ETO	0.25	GP	Immersion test	3 × 4 hr	Score 4[b]	68
lauryl	3ETO	0.25	GP	Immersion test	3 × 4 hr	Score 9	68
sulfate		1.0	GP	Immersion test	3 × 4 hr	Score 7	68
	7ETO	0.25	GP	Immersion test	3 ×4 hr	Score 10	68
		1.0	GP	Immersion test	3 × 4 hr	Score 9.3	68
Ammonium	4ETO	0.25	GP	Immersion test	3 × 4 hr	Score 8.3	68
myristyl sulfate	6ETO	0.25	GP	Immersion test	3 × 4 hr	Score 10	68

Sodium lauryl ether sulfate (60% active substance)	100	Rat	Open	Once daily for 65 days	On 12th day, severe reaction	126
	50	Rat	Open	Once daily for 65 days	On 14th day, clear reaction	126
	15	Rat	Open	Once daily for 65 days	On 65th day, mild reaction	126
	1.5	Rat	Open	Once daily for 65 days	Without any effect	126
Sodium lauryl triethoxysulfate	0.25 M	Rat	Open	Twice daily for 3 days	After 6 applications, mild erythema and edema	97
3.5 M ethoxylated sodium lauryl sulfate (Sipon ES)	5	Rabbit	Open	Once daily; 4 hr for 5 days	Erythema 1.7; edema 0.7 (maximum = 4)	148
	5	Rabbit	Occlusive	Once 24 hr contact time	Primary irritation index: 4.8 (maximum = 8)	148
	5	GP[a]	Immersion test	3 times, 3 hr for 3 days	7.5 (scoring scheme, see[b])	148
Natrium lauryl ether sulfate	100	Rat	Open	Once 1 hr contact time	No reaction	34
	100	hl. mouse	Open	Twice daily	After third application, severe reaction	34
Texapon N-25 ($C_{12}+C_{14}$; 2EO)	25	hl. mouse	Open	Twice daily for 5 days	After sixth application, skin reddening	34
	10	Rat	Open	Twice daily for 5 days	No reaction	34
	10	hl. mouse	Open	Once daily for 5 days	No reaction	34
Texapon N-40 ($C_{12}+C_{14}$; 2EO)	100	hl. mouse	Open	Once 1 hr contact time	No reaction	34
	0.1	hl. mouse	Open	Once daily for 10 days	No reaction	34
Texapon K-14S special ($C_{12}+C_{14}$; 3EO)	10	hl. mouse	Open	Twice daily for 5 days	No reaction	34
	100	hl. mouse	Open	Once 2 hr contact time	No reaction	34
Texapon ASV (C_{12}-C_{18}; 7EO)	100	hl. mouse	Open	Once daily for 5 days	After second application, skin reddening	34
	50	hl. mouse	Open	Once daily for 10 days	No reaction	34

[a]GP = guinea pig.
[b]Scoring scheme: score 10 = no reaction; score 1 = strongest reaction.

commercial samples of linear alkylbenzene sulfonate (lABS) applied as a moistened powder are primary skin irritants. In contrast, at a concentration of 1%, which is above normal domestic use levels, lABS is nonirritating for rabbit skin (22a).

The studies of Olson et al. (36), already mentioned on page 245, also include sulfonates. As can be seen in Table 36, the sulfonates showed an irritant activity to a lesser degree than sodium lauryl sulfate and sodium dioctyl sulfosuccinate, which were also tested at the same time. Furthermore, it can be seen from the table that a far-reaching agreement between the results obtained using the rabbit ear and those obtained on the abdominal skin of the rabbits exists.

Two sodium salts of alkyl benzene sulfonate (C_{12}, highly branched carbon chain; and C_{10}-C_{13}, unbranched carbon chain) were examined in comparable skin compatibility tests by Brown (107). The latter product was somewhat less skin-compatible than the former.

In the previously mentioned teratogenic tests, Palmer et al. (133). tested lABS on mice, rats, and rabbits by repeated percutaneous administration during pregnancy. Whereas a 0.3% solution was without effect, a 3% solution caused marked local skin reactions. The reactions began with an erythema and an edema with peak response on the sixth day for mice, from the fourth to fifth day for rats, and from the sixth to seventh day for rabbits. In the case of rabbits, this peak response was characterized by cracking and bleeding of the skin; the mice exhibited necrosis. In a similar study, Masuda et al. (149) treated mice by cutaneous administration with lABS. The 3.4% solution applied caused marked skin reactions within 12 days.

Gisslen and Magnusson (135) examined various surfactants, among which were alkyl benzene sulfonates (ABS), for their activity upon guinea pig skin, especially with regard to their sensitization-promoting effect. They applied 0.1, 1, and 5% concentrations using the occlusive patch test on the skin of the test animal over a period of 24 hr. From the results it was concluded that surfactants only exhibit a sensitization-promoting effect if they initiate toxic reactions.

In agreement with the results of Gisslen and Magnusson (135), Bühler (150) found that 0.2% tetrapropylene benzene sulfonate (TPBS) is not an irritant when repeatedly applied to the skin of guinea pigs under conditions of the closed patch test (6 hr contact time per application).

Gloxhuber (132) examined the dependence of the skin irritating properties of nASF upon their chain length using the method of intracutaneous injection with mice, as described by Barail (66). Figure 6 is a graphic representation of the results obtained. From this it may be seen that the severest skin-irritating properties are shown by C_{10} rather than C_{12} alkyl sulfonates.

Various anionics were tested for their skin compatibility using the immersion test method with guinea pigs. Among others, ABS were tested and were judged to be better than the AS (142).

Vinson and Choman (104) examined the question as to whether sodium dodecyl (branched) benzene sulfonate (DDBS) influences the permeation of other substances through the skin. After repeated patch test application, it was found that the 0.2% DDBS did not cause any reaction until after the tenth application, whereas the 0.5 and 1% DDBS solutions caused only a slight erythema. All three concentrations tested gave an indication of acanthogenic activity.

Kuhn (84) studied the compatibility of surfactants upon mouse skin with an agar implantation method. When testing diisobutylnaphthalene sulfonic acid sodium salt (5% in agar), thickening of the epidermis, keratinization, and other changes in the lower layers of the skin together with severe tissue reactions occurred.

Undiluted alkyl aryl sulfonate caused bullons and necrotic changes when applied once for a 24-hr period, as reported by Meinicke (143). A 5% solution caused papular reactions.

Application of an occluded patch containing 0.1 ml of a 2% aqueous solution of 97.9% pure C_{12} lABS to the shaved backs of guinea pigs for 24 hr resulted in moderate to severe skin irritation; i.e., skin irritation score of 3 out of a possible 4 (128b).

Kuchinka (151) tested several types of anionic surfactants and compared their skin-irritating properties. He found that the product with the least irritating activity was a soap, followed by a fatty alcohol sulfate (Gardinol) and an alkyl sulfonate (Mersolat D). An alkylaryl sulfonate (AAS) with short chain length (Nekal BX) caused the most severe skin irritations.

Ivanov et al. (72) reported that 1-6% solutions of ABS and ASF accelerate the absorption of physiological saline solution injected into the treated skin areas, causing focal proliferation of the epidermis. Furthermore, moderate stagnation of circulation in the upper dermal layer associated with perivascular proliferation of histiocytes, fibroblasts, and lymphoid cells was observed.

Iimori et al. (144) carried out experiments with surfactants including linear alkyl sulfonates (lASF) and a-olefin sulfonates (AOS). These sulfonates were applied in 1, 10, and 30% concentrations to the shaved backs of guinea pigs and were then left for 10 or 120 min. Twenty-four hours or 1 week after application, the hitological skin changes were examined. The 1% solutions did not cause any changes in the skin area treated. In the case of the 10 and 30% solutions, AOS caused markedly less severe reactions than did lASF.

AOS are slightly to severely irritating to rabbit skin but data are inconsistent, probably due to such factors as purity, method of production, etc. (51d).

TABLE 40 Summary of Results Obtained with Alkyl and Alkyl Aryl Sulfonates, Regarding Their Skin-Compatibility

Substance	Conc., %	Species	Application (patch test)	Frequency and duration of applications	Results of test	Ref.
Dodecyl benzene sulfonate triethanolamine salt (1)	1 5 25	Rabbit	Open (rabbit ear)	Once daily for 14 days (rabbit ear)	1%: mild skin irritation (1)-(5)	36
Sodium dodecylated diphenyl ether monochlorosulfonate (2)	1 5 25	Rabbit	Occlusive (rabbit belly)			
Sodium dodecylated diphenyl ether disulfonate (3)	1 5 25	Rabbit	Occlusive (rabbit belly)	Once daily per 24 hr contact time; 14 days (rabbit belly)	5%: mild (3)-(5) to moderate (1) + (2) skin reactions	36
Dodecylated diphenyl ether disulfonate magnesium salt (4)	1 5 25		Occlusive (rabbit belly)			
Dodecylated diphenyl ether disulfonate triethanolamine salt (5)	1 5 25	Rabbit	Occlusive (rabbit belly)		25%: mild (4) + (5), moderate (2) + (3) to strong skin irritation (1)	36
Natrium salt of alkyl benzene sulfonate (C_{12} branched)	1	Rabbit	Occlusive	3 times 6 hr for 3 days	Middling skin irritations	107
	5	Rat	Occlusive	Once 16 hr	Middling skin irritations	107
Natrium salt of alkyl benzene sulfonate (C_{10}-C_{13} unbranched)	1	Rabbit	Open	5 times weekly; 23 appl.	Slight skin irritations	107
	1	GP[a]	Open	5 times weekly; 23 appl.	Slight skin irritations	107
	2	hl. mouse	Open	7 times weekly; 28 appl.	Mild skin irritations	107

Linear alkyl benzene sulfonate	0.03	Mouse	Open	Once daily for 12 days	0.03 and 0.3%: without reactions; 3% causes severe skin reactions in all three species	133
	0.03	Rat	Open	Once daily for 14 days	0.03 and 0.3%: without reactions; 3% causes severe skin reactions in all three species	133
	3.0	Rabbit	Open	Once daily for 16 days	0.03 and 0.3%: without reactions; 3% causes severe skin reactions in all three species	133
	0.85	Mouse	Open	Once daily for 0.5 ml for 13 days	No reactions	149
	3.4	Mouse	Open	Once daily per 0.5 ml for 13 days	Severe skin reactions	149
Alkyl benzene sulfonate	0.1	GP[a]	Occlusive	Once 24 hr	No reaction	135
	1.0	GP	Occlusive	Once 24 hr	Clear histological changes and necrosis of the epidermis	135
	5.0	GP	Occlusive	Once 24 hr	Clear histological changes and necrosis of the epidermis	135
Tetrapropylene benzene sulfonate	0.2	GP	Occlusive	Repeated applications per 6 hr contact time	No reaction	150
n-Alkyl sulfonates of defined chain length (C_8-C_{18})	0.9	Mouse	Intracut.	Once 0.05 ml per mouse	Order of rank of irritating activity (first the strongest, last the weakest irritant) $C_{10} > C_{12} > C_{14} > C_{16} > C_{18} > C_8$	132

TABLE 40 (Continued)

Substance	Conc., %	Species	Application (patch test)	Frequency and duration of applications	Results of test	Ref.
n-Alkyl sulfonates of defined chain length (C_8-C_{18}) (continued)	2.7	Mouse	Intracut.	Once 0.05 ml per mouse	Order of rank of irritating activity (first the strongest, last the weakest irritant) $C_{10} > C_{12} > C_{14} > C_{16} > C_{18} > C_8$	132
Sodium dodecyl (branched) benzene sulfonate	0.2	GP	Occlusive	Once daily 24 hr contact time for 10 days	Score 0	104
	0.5	GP	Occlusive	Once daily 24 hr contact time for 10 days	Score 1 (maximal score = 4)	104
	1.0	GP	Occlusive	Once daily 24 hr contact time for 10 days	Score 1	104
Diisobutylnaphthalene sulfonic acid sodium salt	0.1	Mouse	Implantation	Once with the help of agar implantation	Marked thickening of epidermis, marked keratinization, destruction of connective tissue and musculature, ulcerations	84
	1.0	Mouse	Implantation	Once with the help of agar implantation	Marked thickening of epidermis, marked keratinization, destruction of connective tissue and musculature, ulcerations	84
	5.0	Mouse	Implantation	Once with the help of agar implantation	Marked thickening of epidermis, marked keratinization, destruction of connective tissue and musculature, ulcerations	84

Sodium salt succinic acid half-ester sulfonate (E-4161)	20	Rabbit	Open	Once daily, 4 hr; 5 days	Erythema: 0.8; edema: 0 (maximum = 4)	148
	20	Rabbit	Occlusive	Once 24 hr contact time	Primary irritation index: 0.2 (maximum = 8)	148
	20	GP[a]	Immersion test	3 Times, 5 hr	Score 8.7 (scoring scale see Table 34)	148
Alkylaryl sulfonate (35% AS)	100	Rabbit	Occlusive	Once 24 hr contact time	Bullons and superficial necrotic changes	143
	5	Rabbit	Occlusive	Once 24 hr contact time	Papular reactions (score: 2 of maximal 4)	143
Linear alkyl sulfonate	1	GP	Open	10 and 120 min contact time	1%: No reaction	144
	10	GP	Open	10 and 120 min contact time	10%: slight to moderate reactions	
	30	GP	Open	10 and 120 min contact time	30%: moderate reactions	
α-Olefin sulfonate	1	GP	Open	10 and 120 min contact time	1%: no reaction	144
	10	GP	Open	10 and 120 min contact time	10%: no reaction	144
	30	GP	Open	10 and 120 min contact time	30%: mild reaction	144
Alkylaryl sulfonate D-40 (mol wt 352)	100	Rabbit	Occlusive	Once 24 hr contact time	D-40: No reaction	55
OWA (mol wt 375)	100	Rabbit	Occlusive	3 times 24 hr	OWA: mild reactions (intact skin), slight and moderate erythema and edema (abraded skin and repeated application)	55
	5	Rat	Open	Once daily for 30 days	Rat: No reaction	
	5	Rabbit	Open	Once daily for 30 days	One rabbit slight erythema, others no reactions	55

[a]GP = guinea pig.

The majority of data concerned with the dermal irritation of AOS show it to be slightly to severely irritating to rabbit skin. Three of the AOS products tested were classified as primary irritants (22a).

Cumulative, open-patch test application of 0.1 ml of 2% aqueous solution of either $C_{16\text{-}18}$ AOS or C_{12} to the shaved backs of guinea pigs twice daily for a total of nine treatments resulted in nil-to-slight and slight-to-moderate cumulative skin irritation, respectively. Skin irritation scores of 0.42 and 1.67 of a possible 4 points were recorded for these two samples (128b).

Hine et al. (55) examined 2 alkylaryl sulfonates with the trade names D-40 (average molecular weight 352) and OWA (average molecular weight 375) using a closed patch test for their skin compatibility on rabbits (undiluted, 24-hr contact time). Moderate erythema, edema, and scabbing of the skin resulted. However, the skin of all the animals rapidly returned to normal and the damage was not severe. In another experiment, the same AAS were applied repeatedly to the skin of rats and rabbits; 5% solutions were applied over 30 days. A mild erythema was found only in individual cases. The undiluted products did not cause severe irritation when applied in a patch test to either intact or abraded skin. Table 40 summarizes the results of the experiments described in this section on skin compatibility with alkyl sulfonates.

4. Skin compatibility of other anionic surfactants

In this section all anionics are grouped which do not belong to the classic three groups: soaps; alkyl sulfates (AS), including alkyl ether sulfates (AES); and alkyl sulfonates (ASF). These are the sulfosuccinic acid esters, methyl taurides, fatty acid protein condensation products, and some other types.

Draize et al. (2) showed that dioctyl sodium sulfosuccinate (DSS) was a moderate irritant when tested as a 2% aqueous solution on rabbits under conditions of the occlusive patch test for a contact time of 24 hr. Olson et al. (36) tested DSS for skin compatibility on rabbits in a closed (abdominal skin) and open (rabbit ear) patch test. The concentrations tested were 1, 5, and 25%. Whereas the 1% concentration caused mild reactions on the rabbit ear, it caused moderate skin irritation of the abdominal skin of the rabbit. The 5% concentration caused middling to severe and the 25% concentration maximum skin irritation in both animal tests.

In a row of 16 anionic surfactants tested with different test methods (intracutaneous and epicutaneous application), a sulfosuccinic ester and a protein fatty acid condensation product had low irritating potential and proved to be better skin compatible than AS, ASF, and AES (15a). Sulfosuccinates were also tested for their primary skin irritancy by Hein (151a) in a concentration of 10% on the intact skin (Draize method) and proved as mild irritants (primary irritation indices: 0.35-0.87).

Vinson and Choman (104) examined the question as to whether anionics influence the permeation of other substances through the skin. Among the surfactants tested were sodium acyl (coco) isethionate and sodium acyl (coco) *n*-methyl taurate, both applied in 1% concentration to the shorn skin of guinea pigs (daily, 1-hr applications, for 10 days). The guinea pig skin did not show any signs of irritant action between the several applications and after the tenth application.

According to Modde and Schuster (56), protein fatty acid condensates of molecular weights 600, 750, 900, and 1200 were tested in 2% concentration on rabbit skin where they did not cause any reaction under conditions of a 24-hr occlusive patch test. At 0.05 M concentration, the protein fatty acid condensate with a molecular weight of 600 gave a simple positive reaction which had disappeared after 24 hr. Protein fatty acid condensates with a molecular weight greater than 700 showed very weak reactions, under the irritation stage 1. Skin compatibility increased with increasing molecular weight.

Prottey and Fergusson (97) studied, among others, sodium lauryl isethionate. After 2 and 6 applications, no skin reactions were seen, excepting a moderate surface skin dryness.

Meinicke (143) compared the effect of protein fatty acid condensation products in the epicutaneous test with rabbits with the usual market fatty alcohol sulfates and alkylaryl sulfonates (AAS) (24-hr patch test). The tested protein fatty acid condensation product (35% active substance) did not cause any skin reactions when tested in 5, 10, and 20% concentrations. The product tested in undiluted form under the same conditions caused papular to maculopapular-vesicular reactions. The protein fatty acid condensation products were on the average better than the AS and the AAS. These results are in agreement with the test results published by Kästner and Frosch (15a). Table 41 contains a summary of the experiments discussed in this section together with their results.

In the last 15 years, some new anionic surfactants have been developed and their irritating potential was compared with AS, ASF, and AES.

Acyl lactylates (sodium and calcium salts of stearoyl-, lauroyl- and capryl lactylates) had primary skin irritation indices of 0-1.13 when tested in concentrations of 10, 15, or 100% (30a).

Undiluted sodium lauryl sulfoacetate moistened with 0.9% saline was applied to the shaved backs of New Zealand Rabbits (0.5 g test substance, semiocclusive patches, 24-hr exposure time). Test sites were scored for erythema and edema 30 min and 24 hr after patch removal. The mean primary irritation score was 2.7 (Draize scheme = max. 8.0). Undiluted sodium lauryl sulfoacetate was a moderate skin irritant (45a).

Monoalkyl phosphate sodium salts with chain length of C_8 to C_{18} caused less skin roughness and skin irritation than alkyl sulfates, soaps, alkyl ben-

TABLE 41 Summary of Experiments and Results Concerning Skin Compatibility with Various Anionics

Substance	Conc., %	Species	Application (patch test)	Frequency and duration of applications	Results of test	Ref.
Sodium dioctyl sulfosuccinate	2	Rabbit	Occlusive	Once 24 hr	Primary irritation index: 2.7	2
	1	Rabbit	Open (ear)	Once daily for 14 days	Ear: 1%: 1, 5%: 3.5, 25%: 6	36
	5	Rabbit	Occlusive (belly)	Once daily for 14 days	Belly: 1%: 4, 5%, 5, 25%: 6 at maximal score = 6	36
	25	Rabbit	Occlusive (belly)	Once daily for 14 days	Belly: 1%: 4, 5%: 5, 25%: 6 at maximal score = 6	36
Sodium acyl (coco) isethionate	1	GP[a]	Occlusive	Once daily per 24 hr contact time for 10 days	No reaction	104
Sodium acyl (coco) *n*-methyl taurate	1	GP	Occlusive	Once daily per 24 hr contact time for 10 days	No reaction	104
Protein fatty acid condensates (mol wt 600-1200)	2	Rabbit	Occlusive	24 hr contact time	No reaction	56
	0.05 M	Rabbit	Occlusive	24 hr contact time	Mol wt: 600 mild, higher mol wt: very weak reaction	56
Sodium lauryl isethionate	0.25 M	Rat	Open	Twice daily for 3 days	No reaction; superficial drying	97
Protein fatty acid condensates (35% AS)	5	Rabbit	Occlusive	Once 24 hr contact time	No reaction	143
	10	Rabbit	Occlusive	Once 24 hr contact time	No reaction	143
	20	Rabbit	Occlusive	Once 24 hr contact time	No reaction	143
	100	Rabbit	Occlusive	Once 24 hr contact time	Score 2-3 (maximal score = 4)	143

[a]GP = guinea pig.

zene sulfonates, and α-olefin sulfonate tested with closed patch test on human skin (132a). The same results were found with 24-hr closed patch test and with the cumulative open patch test on guinea pig skin (128b,151b).

5. Skin compatibility of mixtures containing anionic surfactants as major components

This section describes the skin-irritating properties of mixtures that consist completely of anionics or contain anionics as major components.

Meinicke (143) tested various mixtures of fatty alcohol sulfates (FAS), alkylaryl sulfonates (AAS), and protein fatty acid condensation products (PFC) and protein hydrolysates (PH) using the epicutaneous test in rabbits. He applied the mixtures in an occlusive patch test (24-hr contact time). The mixtures were applied either undiluted or as 5, 10, and 20% concentrations. Mixtures of FAS and PFC, as well as AAS and PFC, FAS and PH, and AAS and PH in equal parts (35% active substance), did not show any skin irritant properties up to 20% concentration, with the exception of FAS and PFC mixtures, whereas FAS and AAS alone at these concentrations caused marked skin irritation. The concentrated mixtures lead, however, to severe skin irritations in all cases. The experiments showed that the severe skin irritation of FAS and AAS can be considerably alleviated by the addition of PFC and PH. The protective action is explained by assuming that the keratin molecules in the skin are occupied by the less irritating PFC and PH molecules, this largely obviating the bonding to the strong irritant surfactants.

Mills and Kligman (123) examined, in addition to soaps, 15 shampoos for comedogenic activity. Comedogenicity was assessed by applying 25% aqueous suspensions of the agents once daily for 5 days a week for 2 weeks to one external ear canal of each of two rabbits. Of the 15 shampoos examined, eight were histologically noncomedogenic, the others having a more or less marked activity. Most of the shampoos examined were skin irritants under the test conditions used. These provoked gross edema and scaling, microscopic infiltration of the dermis with neutrophils being prominent. The irritation itself did not lead to comedone formation. One shampoo caused comedone formation, which occurred in the absence of skin irritation.

MacMillan et al. (148) tested Sipon ES (3.5 M ethoxylated sodium lauryl sulfate), E-4161 (sodium salt succinic acid half-ester sulfonate), and a mixture of both substances with and without certain additives, using a previously described 5-day skin test, a patch test according to Draize (2) using rabbits, an immersion test as described by Opdyke and Burnett (68) on guinea pigs, and a 21-day continuous test using human test subjects. It was shown that Sipon ES causes relatively severe skin reactions whereas E-4161 causes very mild skin reactions. The mixture of both anionics caused a middling erythematous reaction in the 5-day dermal test, but an edema was not formed. In contrast, the primary irritation index of the mixture lay higher than that for Sipon ES. The result of the guinea pig immersion test was also closer to that of the Sipon ES, which had poorer skin compatibility properties. An addition of 2% modified lauric alkanolamide or lauric myristic diethanolamide scarcely influenced the skin-irritating properties of the anionic mixture. The addition of amine oxide MO to Sipon ES clearly increased its irritant properties most especially in the guinea pig immersion test.

Five shampoos were tested by Jacobi and Ortmann (152) that contained sodium lauryl ether sulfate as surfactant in concentrations of 4-30% for their skin compatibility according to Draize (3). Rabbits with intact and abraded skin were used (time of exposure 2 or 24 hr). The five shampoos were tested in undiluted form for 2 and 24 hr contact time, and finally in 10% aqueous solution with a 24-hr contact time. Two shampoos in undiluted form showed only slight reactions in a 24-hr closed patch test, whereas three others showed moderate reactions. All five shampoos were mild irritants when tested in indiluted form with 2 hr contact time or as 10% aqueous solutions with a 24 hr contact time.

A liquid shampoo, consisting of 20% triethanolamine alkyl sulfate and other components, was tested on rabbits, guinea pigs, and human subjects for irritant activity on both intact and abraded skin by Nixon et al. (136). The undiluted compound was tested under patch-testing conditions (4 hr-time of exposure). The authors characterized the shampoo as moderately irritating with rabbits and guinea pigs and as mildly irritating in humans.

Although sodium lauryl sulfoacetate proved as moderately irritating, a bath product in powdered form containing 35% of this surfactant was applied undiluted (500 mg of the product with 150 mg surfactant) daily for four days on the backs of rabbits and did not produce any irritation. On the contralateral site a 1% solution of the bath additive (0.35% sodium lauryl sulfoacetate) was applied in the same manner. No irritation was observed at the sites treated with the powdered bath product. All sites treated with the 1% solution had slight erythema on day 2 but were normal at day 7 (45a).

Opdyke et al. (50) examined the skin compatibility of two detergents on people and guinea pigs using local application (patch test) and intradermal injection. Both detergents contained not only anionics (type 1, 20% sodium alkyl benzene sulfonate (SABS) and type 2, 20% of a mixture of equal parts of SABS and sodium alkyl sulfate), but also equal quantities of phosphate (50%), sodium sulfate (13%), and sodium silicate (6%). The guinea pigs were treated 5 times per week for 4 weeks with 5% aqueous solution, and finally for 6 months with 7.5% aqueous solution, on both the intact and mildly abraded abdominal skin, in the same way. The skin of the guinea pigs remained fairly normal during the repeated treatment. There was no indication of sensitization, cumulative irritation, or enhanced penetration as a result of repeated exposure to either product under the variety of conditions used (methods and species).

G. Comparative Studies Concerning Skin Compatibility of Various Anionic Surfactants and Anionic Groups

Section III.F makes it clear that various authors have been primarily interested not in the skin compatibility of individual substances but rather in the comparison between individual substances. The goal of numerous studies was to establish which surfactants possessed the best skin compatibility characteristics or which surfactants could have these properties improved by the addition of other substances. The following description of previously mentioned publications is given in terms of this point of view.

Olson et al. (36) compared the skin compatibility of alkyl sulfates (AS), alkyl sulfonates (ASF), and a succinic acid ester on rabbits using both open and closed test methods and found that diphenyl ether disulfonate possessed the best skin compatibility and that sodium lauryl sulfate (SLS) and sodium dioctyl sulfosuccinate (DSS) the worst (see Table 36).

Brown (107) studied soaps, AS, ether sulfates, and ASF using various models and could not find a uniform tendency in the compatibility properties. The sulfonates turned out to be the least irritating in some models, whereas they were judged to be the most irritating in other models.

Brown and Muir (37) compared various AS and an alkyl ether sulfate (AES) which were derived either from coconut fatty alcohol or from the synthetic alcohol Dobanol 23. Skin compatibility was tested, and in contrast to various other authors Brown and Muir found the ethoxylated compound to be the most irritating (see Table 32).

Ciuchta and Dodd (27a) tested the skin irritancy according to the standard Draize skin irritancy procedure (intact and abraded skin, occlusive conditions, 24-hr exposure time) of various surfactants in three concentrations each (2, 10, and 20%). Sodium lauryl polyether(12)sulfate (SLES) was the least irritating at all three concentrations, SLS, and ammonium lauryl sulfate (ALS) were the most irritating surfactants, while the other surfactants (sodium coco methyl tauride, triethanolamine lauryl sulfate) were intermediate, although they did not necessarily stay within the same irritancy rations for all concentrations.

Ammonium lauryl and ammonium myristyl ether sulfates were tested by Opdyke and Burnett (68) using guinea pigs in the immersion test. Both showed better skin compatibility properties than ALS, which was also tested. A clear correlation was found between the degree of ethoxylation and skin compatibility, inasmuch as with an increasing degree of ethoxylation a decrease in the skin irritation occurred. This was valid for both the lauryl and the myristyl ether sulfates (see Table 34).

TABLE 42 Order of Precedence of Skin Irritating Properties of Various Anionic Surfactants[a]

Substances and precedence	Conc., %	Species	Application (patch test)	Frequency and duration of application	Ref.
Dodecylated diphenyl ether disulfonate triethanolamine salt = dodecylated diphenyl ether disulfonate magnesium salt < sodium dodecylated diphenyl ether disulfonate < sodium dodecylated diphenyl ether monochlorosulfate < dodecyl benzene sulfonate triethanolamine salt < lauryl sulfate triethanolamine salt < sodium lauryl sulfate = sodium dioctyl sulfosuccinate	1	Rabbit ear	Open	1× daily; 14 days	36
	5	Rabbit (belly)	Occlusive	1× daily; 14 days	36
	25	Rabbit (belly)	Occlusive	1× daily; 14 days	36
Na salt of alkyl benzene sulfonate (C_{12} highly branched) (T_1), Na salt of alkyl benzene sulfonate (C_{10}-C_{13}) (T_2), sodium secondary alkyl sulfate (C_8-C_{18}) (T_5), soap (T_6), Na salt of sulfated broad-cut coconut ethoxy (3EO) alcohol (T_7), Na salt of sulfated broad-cut coconut alcohol (T_9)					107
$T_5 = T_6 < T_1 = T_7 < T_2 < T_9$	1	Rabbit	Occlusive	1× daily 6 hr; 3 days	107
$T_6 = T_7 = T_9 < T_5 < T_1 < T_2$	5	Rat	Occlusive	1× 16 hr	107
$T_1 = T_2 = T_5 = T_6 = T_9 < T_7$	1	Rabbit	Open	5×/week, 23 applications	107
$T_1 = T_2 = T_5 = T_6 = T_9 < T_7$	1	GP[b]	Open	5×/week, 23 applications	107
$T_2 = T_6 = T_9 < T_1 = T_5 = T_7$	2	hl. mouse	Open	7×/week, 28 applications	107

Monoethanolamine alcohol sulfate (1), triethanolamine alcohol sulfate (2), ammonium alcohol sulfate (3), sodium alcohol sulfate (4), sodium alcohol (2EO) ethoxy sulfate (5)					37
1 = 2 < 5 < 3 < 4	1	Rabbit	Occlusive	1× daily, 6 hr; 3 days	37
1 = 2 = 3 = 4 < 5	25	Rabbit	Occlusive	1× daily, 6 hr; 3 days	37
1 = 2 < 3 = 4 < 5	25	Rabbit	Open	5×/week; 4.5 weeks	37
1 = 2 < 3 = 4 < 5	25	GP	Open	5×/week; 4.5 weeks	37
Igepon A = Igepon T < dodecyl benzene sulfonate < sodium lauryl sulfate	1	GP	Occlusive	1× daily 24 hr contact time 10 days	104
Alkyl benzene sulfonate < alkyl sulfonate		GP	Immersion test	3× 4 hr contact time	142
α-Olefin sulfonate (AOS), linear alkyl sulfonates (lASF), sulfates of synthetic alcohols (AS)	1	GP	Open	10 and 120 min time of exposure	144
AOS < lASF < AS	10	GP	Open	10 and 120 min time of exposure	144
	30	GP[b]	Open	10 and 120 min time of exposure	144
Ammonium lauryl sulfate 7EO (0.25%) = ammonium myristyl sulfate 6EO (0.25%) < ammonium lauryl sulfate 7EO (1%) < ammonium lauryl sulfate 3EO (0.25%) < ammonium myristyl sulfate 4EO (0.25%) < ammonium lauryl sulfate 3EO (1%) < ammonium lauryl sulfate 1EO (0.25%) < ammonium lauryl sulfate (0.25%)	0.25 1.0	GP GP	Immersion test Immersion test	1× daily 4 hr; for days 1× daily 4 hr; for days	68 68
Sodium lauryl sulfate (1), sodium laurate (2), sodium lauryl triethoxy sulfate (3), sodium lauryl isethionate (4)					97
1 = 3 = 4 < 2	0.25 M	Rat	Open	Twice on 1 day	97
4 < 3 < 2 < 1	0.25 M	Rat	Open	2× daily; 3 days	97

TABLE 42 (Continued)

Substances and precedence	Conc., %	Species	Application (patch test)	Frequency and duration of application	Ref.
Sodium palmitate < sodium myristate < sodium stearate ⩽ sodium caprylate < sodium oleate	0.25 M	GP	Open	2-4× daily for some days	130
< soap < sodium coprate < sodium laurate	0.8%	GP	Open	2-4× daily for some days	130
Protein fatty acid condensates < alkylaryl sulfonate < fatty alcohol sulfate	5	Rabbit	Occlusive	1× 24 hr	143
Alkylaryl sulfonate D-40 (mol wt 352) < alkylaryl sulfonate OWA (mol wt 375)	100	Rabbit	Occlusive	1× 24 hr or 3× 24 hr with intervals	55
n-Alkyl sulfonates					
$C_8 < C_{18} < C_{16} < C_{14} < C_{12} < C_{10}$	0.9 + 2.7	Mouse	Intracut.	1× 0.05 ml/mouse	132
Alkyl sulfates					
$C_6 < C_8 < C_{16} = C_{18} < C_{14} < C_{12} < C_{10}$	0.9	Mouse	Intracut.	1× 0.05 ml/mouse	69
$C_6 < C_8 < C_{18} < C_{16} < C_{14} < C_{12} = C_{10}$	2.7	Mouse	Intracut.	1× 0.05 ml/mouse	69

[a]Arranged such that first-mentioned substance causes the least reactions, the following substances react stronger and the last substance reacts strongest.

[b]GP =*f* guinea pig.

Pemberton and Rhodes (128a) examined whether there is a difference between those alkyl sulfate surfactants with alkyl chains consisting of even or odd numbers of carbon atoms and could not find such a difference regarding skin irritancy.

Kästner and Frosch (15a) compared a series of 16 anionic surfactants regarding their skin irritating potential with different test systems (human repeated patch test, single intracutaneous and repeated cutaneous application on white or hairless mice). The authors found that the AS, ASF and alkyl benzene sulfonate (ABS) are more irritating than the AES, a sulfosuccinic acid ester, and a protein fatty acid condensation product (PFC).

ABS were described as possessing better skin-compatible properties than the ASF (142). This statement is based on results obtained in the guinea pig immersion test. A skin compatibility experiment between sodium laurate, sodium lauryl isothionate, SLS, and sodium lauryl triethoxysulfate was described by Prottey and Fergusson (97). Prottey et al. (130) compared various soaps with each other. Whereas sodium palmitate caused the mildest reactions, the most severe reactions were caused by sodium laurate. This was followed by sodium caprate and a formulated soap (see Fig. 4).

Meinicke (143) compared fatty alcohol sulfate (FAS), alkylaryl sulfonate (AAS), and PFC in a 24-hr closed patch test using rabbits and ascertained that FAS was a more severe irritant than AAS, whereas PFC possessed only mild skin irritation properties.

Iimore et al. (144) tested synthetic alcohol sulfates (AS), linear alkyl sulfonates (lABS), and α-olefin sulfonates (AOS) and found that AOS had the best skin compatibility properties and that lABS and AS were clearly more irritating. AS was judged as being somewhat inferior to lABS.

Table 42 summarizes the results described in this section, or experiments already described elsewhere, which relate to skin compatibility of various anionic surfactants.

IV. TESTING FOR SENSITIZING PROPERTIES

A. Results Concerning Sensitizing Properties of Anionic Surfactants

The question as to whether anionics possess sensitizing properties has not been discussed very frequently, but there has been some experimental work on the subject some times. A number of experiments have been carried out on formulated products which contained anionics as major components. These experiments will also be described in this section.

According to Götte (77), Jambor and Suskind (153) did not find any indications that soaps or synthetic cleansers were sensitizers. Sodium lauryl sulfate (SLS) was investigated by Ahlfors and Larsson (22b) in a Sprague-Dawley rat model and no sensitizing capacity was noted.

Hoffmann (73) examined the question as to whether soaps and syndets (alkylaryl and fatty acid sulfonates) were sensitizers using the guinea pig. No effects were found. Similarly, Schneider (74) did not find any evidence for sensitizing properties of synthetic detergents which contained large quantities of anionics from experiments carried out over months under practical washing conditions. An incompatibility for active washing substances does not lie in a hypersensitivity but rather, according to Schneider, in a slow physical and chemical "wearing away" (fatigue) of the skin.

Okamoto et al. (153a) discussed the promoting effect of surfactants on skin toxicity caused by chemical agents with respect to an increasing phenomenon of challenge response. Based on the experimental findings, the authors suggested that surfactants could promote sensitization, not by their surface activation abilities but by their primary irritating action. Consequently, it is meaningful to employ these surfactants experimentally as promoting agents, since surfactants such as SLS have the capacity to emulsify and solubilize haptens and also to irritate the skin.

Opdyke et al. (50) tested two detergents which consisted of 20% anionics, 50% phosphate, etc., and a soap for sensitizing properties. Neither the detergents nor the soap caused reactions which could be interpreted as allergic. Although alkyl ether sulfates (AES) are valid as nonsensitizer (53c) and they have been in worldwide use for many years without sensitizing problems being known, allergic reactions attributed to these products were described in the Scandinavian countries in the 1970s. It was assumed that the substance responsible was not lauryl ether sulfate (LES) itself but rather a byproduct, and an intensive search was made. Walker et al. (154) initially examined lauryl ether sulfates of various origins according to the method of Magnusson and Kligman (155) and found that a product with the code number LES 13-035 gave positive results but no other samples.

Sensitized guinea pigs gave negative cross-challenge results when treated with other alkyl ethoxysulfates. Petrol ether extracts from the charge LES 13-2035 gave positive allergic reactions in sensitized guinea pigs. Byproducts formed during the production of LES were held to be responsible for the sensitizing properties.

Magnusson and Gilje (156) also examined in this context lauryl ether sulfate which had been used in liquid dishwashing agents in Scandinavia and had caused contact hypersensitivity in housewives. The experiments on sensitized persons and guinea pigs lead to the result that the allergen is neither chemically bound to the lauryl ether sulfate molecule nor present in the actual raw material, the fatty alcohol. LES contains small quantitites of unsulfated material which can be extracted with petrol ether. As purified LES does not contain the sensitizing substance, it is presumed that it is formed during the manufacturing process.

Gloxhuber et al. (157) could also show that the reactions were caused by a component in the above-specified charge of LES, which could be extracted with petrol ether. In agreement with Walker et al. (154), they came to the conclusion that pure LES was not a sensitizer. Extracts from various LES samples, manufactured elsewhere, were free of such sensitizing agents. It appears that such a sensitizing component is formed during the production of LES under conditions that deviate greatly from the normal production practice.

The observations described above were confirmed by Sylvest et al. (158).

The Lauryl Ether Sulphate Allergy Committee (LESAC), an organization of European experts which had been created with the goal of clearing up the allergic reactions caused by LES, in 1974 came to the conclusion that the sensitizing agent was an impurity of specific batches of LES. The impurity in AES (LES 13-2035) was identified by several investigators (158a-c) as unsaturated sultones such as 1-dodecene-1,3-sultone and 1-tetradecene-1,3-sultone, and as chlorsultones as 2-chloro-1,3-dodecanesultone and 2-chloro-1,3-tetradecanesultone. The authors speculated that perhaps small quantities of dodecene and tetradecene were present during the sulfanation reaction and were carried throughout the manufacturing process. Ritz et al. (158d) tested these unsaturated and halogenated sultones in guinea pigs and found them to be very potent sensitizers. The pure LES had no sensitizing property.

Morikawa et al. (158e) tested the skin-sensitizing potential of a series of alcohol ethoxy sulfates (AES) in guinea pigs. All animals were induced by intradermal (i.d.) injections (8% aqueous solution) and topical application (20% aqueous solution, 48 hr, occluded) and challenged 2 weeks later by topical application (8% aqueous solution). None of the 11 tested ethoxy sulfates induced skin sensitization in guinea pigs. In addition, two samples of a C_{12} AE_3S sulfated either with sulfur trioxide or chlorosulfonic acid were tested. The chlorosulfonic acid preparation was not a skin sensitizer (0/10) in guinea pigs while 15/15 animals gave positive reactions with the sulfur trioxide preparation. The allergenic material was isolated and identified as 1-dodecene-1,3-sultone, formed as a by-product during industrial synthesis.

Five ethanolamine or sodium sulfates derived from coconut alcohol or synthetic alcohol Dobanol 23 were tested for sensitizing properties in a series of comparative tests by Brown and Muir (37). The topical epicutaneous applications did not lead to a positive reaction in a single case. Intradermal treatment with mono- or triethanolamine salts of the surfactants did sensitize, whereas the ammonium salt did not. Thus, the authors suggest that sensitization may be due to the presence of free amines rather than the surfactant.

In two other studies (22a,155) using several different procedures, confirmation was obtained that AS are not sensitizing materials.

Hine et al. (55) tested the two alkylaryl sulfonates D-40 (mol wt 352) and OWA (mol et 375) for sensitizing properties using guinea pigs according to the method described by Landsteiner and Jacobs (159). The experiments did not show any signs of sensitizing properties. Skin sensitization studies in guinea pigs indicate α-olefin sulfonate (AOS) are generally nonsensitizers; a few positive responses were either unexplainable, attributed to incomplete hydrolysis of the product, or irreproducible (51d).

In our own laboratories we have tested for sensitizing properties a series of anionics which were to be used in cosmetics preparations (shampoos and bubble baths) (34). The testing was done according to the method of Magnusson and Kligman (155). The following products were tested:

Texapon ALS	=	ammonium lauryl (C_{12}-C_{16}) sulfate (30% AI)
Texapon MGS	=	lauryl (C_{12}-C_{16}) sulfate magnesium salt (30% AI)
Texapon T-42	=	triethanolamine dodecylmyristyl sulfate (42% AI)
Sulfopon T-55	=	tallow alcohol sulfate sodium salt (55% AI)
Texapon N-40	=	dodecylmyristyl ether (2EO) sulfate sodium salt (28% AI)
Texapon K 14 S special	=	laurylmyristyl ether (3,6EO) sulfate sodium salt (28% AI)
Texapon ASV	=	mixture of sodium and magnesium salts of alkyl (C_{12}-C_{18}) ether (7EO) sulfate (29% AI)

The challenge application did not cause topical or general reactions. In summary, it may be said that the seven Texapon types tested did not possess any sensitizing properties.

Bay and Danneman (159a) carried out a program to evaluate the contact skin-sensitization potential of commercial AOS to quantify this ingredient as safe for use in a high-duty liquid dishwashing detergent. The program consisted in guinea pig and human sensitization studies, diagnostic patch tests and a 9-month product use test with a prototype AOS-containing dishwashing detergent. The degree of response appeared to be dose-related to the amount of unsaturated sultone present in the material being tested.

Allergic contact dermatitis from synthetic detergents is rare, but a case report about a 46-year-old man with hand dermatitis was published. The man had a positive skin reaction after patch testing with 0.5 and 5.0 sulfosuccinate derivate, but 20 control subjects had negative results. A guinea pig sensitization test was carried out with this surfactant. No evidence of sensitization was found in the Magnusson and Kligman test at 50% (159b).

A cream shampoo containing 2.1% sodium lauryl sulfoacetate was tested for sensitization in guinea pigs by a modification of the method of Buehler. A 10% aqueous dilution of the cream shampoo was topically administered underr occlusive patches in 0.5-ml doses to the shaved backs of guinea pigs. Three inductive patches were applied at weekly intervals, and challenge patches were applied 2 and 3 weeks after the last inductive patch. No erythema was observed after any inductive or challenge patches. It was concluded that this product did not cause sensitization under these conditions (45a).

B. Sensitizing-Promoting Effects of Anionic Surfactants

As shown in Sec. IV.A, anionic surfactants do not possess any sensitizing properties. Nevertheless, it is said that they promote sensitization by other substances.

According to Gisslen and Magnusson (135), some surfactants aid the penetration of other substances through the skin because of the surfactants' primary irritant activity. This has the effect of promoting contact sensitization.

Vinson and Choman (104) tested the influence of the following anionics upon the penetration of 0.5 and 1% nickel sulfate in guinea pig skin: sodium lauryl sulfate (SLS), 0.1, 0.2, 0.5, and 1%; sodium dodecyl (branched) benzene sulfonate (DDBS), 0.2, 0.5, and 1%; sodium acyl (coco) isethionate, 1.0%; sodium acyl (coco) n-methyl taurate, 1.0%; commercial toilet soap, 8.0%. They established that the eczemateous properties of nickel sulfate could be increased by the addition of anionics. Further, the experiments showed that soaps, SLS, and sodium dodecyl benzene sulfonate are irritating under the experimental conditions used. The irritating property was to be seen both macroscopically and histologically. Nickel sulfate alone in 5% concentration was not irritant, but after combination with 0.1% NaLS (which alone also did not cause any reactions) reactions were caused on guinea pig skin that were not caused in combination with soap.

Okamoto et al. (153a) demonstrated with different anionic surfactants that their primary irritating potential promote sensitization. They suggest that the surfactants present in ointments, creams and lotions, with their ingredients being carefully established for safety, promote contact sensitization by any haptens that may happen to contaminate cosmetic products.

Bear et al. (120) and Skog (160) maintain that an acanthosis renders the skin more sensitive for allergens. According to Vinson and Choman (104), soaps also show an acanthotic effect but do not promote the nickel reaction. It is assumed that insoluble nickel soaps are formed that do not penetrate the dermis. The authors were further able to show that the sensitization by nickel is dependent on the concentration, from both the available nickel ions and the surfactant concentration.

When the nickel ion concentration was reduced to 0.1% and the DDBS concentration to 0.5%, no allergic reactions were seen. The other anionics tested (sodium acyl isethionate and sodium acyl *n*-methyl taurate) did not show such effects. In order to explain such penetration-promoting effects, it was assumed that the surfactants have the effect of denaturing epidermal protein, which may result in the easier passage of the metal sensitizer through the epidermis. The results of the experiments carried out by Harrold (103) support this suggestion.

Wahlberg (161) did experiments with the goal of determining how penetration of sodium chromate (Na_2CrO_4) was influenced by SLS. He used two approaches. He either applied radioactive chromate in 0.017 or 0.239 M concentrations together with 0.1, 1, or 10% SLS upon the flank of guinea pigs or he initially treated the flank with a 1% SLS solution and then applied radioactive chromate in 0.017 or 0.239 M concentrations. He then ascertained which penetration of chromate ions had occurred, either in mixture with SLS or after pretreatment, in comparison with the application of chromate solution alone. It could be shown that SLS either in a pretreatment or as carrier resulted in an increase in the penetration of chromate ions. The concentration of the carrier as well as the chromate solutions are important to the penetration rate. In the case of the 0.017 M solutions SLS did not have any influence upon the penetration but had a marked influence at a concentration of 0.239 M. The 10% SLS solution accelerated the penetration more than the lower SLS concentrations did. The same effects were observed with mercury(II) chloride ($HgCl_2$) dissolved in soap and alkylaryl sulfonate (162). An increase in adsorption was only seen in the case of the highest of the five mercury concentrations tested. Wahlberg (161) is of opinion that a precondition for the formation of a sensitization is that the allergen or surfactant or both cause an irritation. However, because the irritant effect is concentration-dependent, the penetration and thus the sensitization-promoting effect of surfactants must also be dependent on concentration.

Magnusson (163) confirms the concentration dependence of the effect of anionics in the promotion of contact sensitization. A 0.1% concentration did not have an influence upon the degree of sensitization. However, a 1% concentration increased the sensitization of a number of substances, and a 5% concentration had an ever clearer stimulating effect upon sensitization.

Nilzen and Wilkström (115) also studied the effect of SLS on the sensitization of guinea pigs by chromium and nickel. Only with difficulty chromium and nickel salts cause sensitization reactions in guinea pigs. If, however, surfactants, e.g., SLS, in higher concentrations act on the skin at the same time, then the sensitization reactions can be induced. The authors confirmed that the SLS alone does not sensitize, but guinea pigs can be sensitized when treated with potassium chromate and nickel sulfate and SLS simultaneously.

It is assumed that this effect is caused by an increase in skin permeability or that the metabolic processes are so influenced by SLS that the combination of allergen with protein can occur more rapidly. The histological effects, such as spongiosis, vascularization, and leucocyte invasion, were similar in appearance of allergic response to those found in human subjects.

Sodium dodecyl benzene sulfonate has an adjuvant effect upon aerosol allergic sensitization with subtilopeptidase A, a proteolytic enzyme of Bacillus subtilis. Both local and systemic antibody responses were accelerated and prolonged by use of detergent. This adjuvant effect was only seen when an inactive form of the enzyme was used. The detergent, however, was able to potentiate early clinical responses upon initial exposure to the active enzyme. Female Hartley-strain guinea pigs were used in the experiments (163a).

The findings of Nilzen and Wilkström (115) oppose those of Bühler (150), who did not find any evidence that tetrapropylene benzene sulfonate (TPBS) promoted sensitization by certain metal salts when applied in 0.2% concentration. Alkyl isethiocyanate (0.75%), mercury chloride (5%), nickel sulfate (50%), nickel dinitrate (50%), cobalt chloride (50%), each dissolved in 0.2% TPBS solution, did not cause a positive response in a single case as a result of the challenge application. An explanation cannot be offered.

The concentrations at which the above experiments lead to positive results lay in regions which scarcely occur under practical conditions. It may therefore be concluded that the surfactants in the concentrations which are found in washing, dishwashing, and cleansing agents do not have an influence in the development of chromate, nickel, or cobalt allergies in housewives.

LIST OF ABBREVIATIONS

The following abbreviations have been used throughout the text:

AAS	Alkylaryl sulfonate(s)
ABS	Alkylbenzene sulfonate(s)
lABS	linear ABS
SABS	Sodium alkyl benzene sulfonate(s), alkyl benzoic sulfonic acid Na salt
AES	Alkyl or alcohol ethoxy sulfate(s), alkyl ether sulfate(s)
AE_3S	Alkyl ethoxy (3EO) sulfate(s)
NaAES	Sodium alkyl ethoxy sulfate(s)
AI	active ingredient
ALS	Ammonium lauryl sulfate
AOS	alpha-olefin sulfonate(s), α-olefin sulfonate(s)
nAOS	*n*-α-olefin sulfonate(s)
AS	Alkyl sulfate(s)
AS%, % AS	active substances in %, % washing active substances
nAS	*n*-alkyl sulfate(s)
NaAS	Sodium alkyl sulfate(s)

ASF	Alkyl sulfonate(s)
lASF	linear alkyl sulfonate(s)
nASF	n-alkyl sulfonate(s)
ave	average; used to designate broad-cut-derived alkyl sulfates
DBS	Dodecyl benzene sulfonate
nDBS	n-dodecyl benzene sulfonate
SDBS	Sodium dodecyl benzene sulfonate
DDBS	sodium dodecyl (branched) benzene sulfonate
DEA	Diethanolamine
DSS	Dioctyl sodium sulfosuccinate
FAS	Fatty alcohol sulfate(s)
lABS	linear ABS
lAS	linear AS
lASF	linear alkyl sulfonate(s)
LES	Lauryl ether sulfate(s)
LS	Lauryl sulfate(s)
MEA	Monoethanolamine
Na	Sodium
NaAES	Sodium alkyl ethoxy sulfate(s)
PFC	Protein fatty acid condensation products
sAFS	secundary alkane sulfonate(s)
SABS	Sodium alkyl benzene sulfonate(s), alkyl benzoic sulfonic acid Na salt
SAS	Sodium alkyl sulfate(s)
SCMT	Sodium coco methyl tauride(s)
SDBS	Sodium dodecyl benzene sulfonate
SDS	Sodium dodecyl sulfate
SLES	Sodium lauryl polyether(12) sulfate
SLS	Sodium lauryl sulfate
TEA	Triethanolamine
TEALS	Triethanolamine lauryl sulfate
TPBS	Tetrapropylene benzene sulfonate
WAS	washing active substances

REFERENCES

1. F. Schaaf, *Probleme Dermatologischer Grundlagenforschung*, A. Hüthig Verlag, Heidelberg, Chap. II.N, 1969, pp. 93-113, 398-399.
2. J.H. Draize, G. Woodward, and H. Calvery, *J. Pharmacol. Exp. Ther., 82,* 377-390(1944).
3. J.H. Draize, *Appraisal of the Safety of Chemicals in Foods, Drugs and Cosmetics,* Association of Food and Drug Officials of the United States Business Office,

Bureau of Food and Drugs, Austin, Texas, Chap. Dermal Toxicity, pp. 46-48, Texas State Dept. of Health, 1959.

4. U.S. Food and Drug Administration, Test for eye irritants, Sec. 191.12 of Title 21, Chap. I of the Code of Federal Regulations. Fed. Reg., *29*, 13009.
5. Code of Federal Regulations, Test for eye irritants, Title 21, Chap. I, Sect. 191.12, 1970.
6. K. Neumann and J. G. Aalbers, *Fette, Seifen, Anstrichm., 62,* 1053-1057 (1960).
7. A.B.G. Burton, *Food Cosmet. Toxicol., 10,* 209-217(1972).
8. J.H. Beckley, *Amer. Perfum. Cosmet., 80,* 51-54(1965).
9. J. Scholz, *Aesthet. Med., 12,* 343-350(1963).
10. K.L. Russell and S.G. Hoch, *Proc. Sci. Sect. Toilet Goods Ass., 37,* 27-32 (1962).
11. C.S. Weil and R.A. Scala, *Toxicol. Appl. Pharmacol., 19,* 276-360(1971).
12. C.P. Carpenter and H.F. Smyth, *Amer. J. Ophthalmol., 29,* 1363-1372(1946).
13. J.H. Draize, *Food Drug Cosmet. Law J., 10,* 722-732(1955).

13a. OECD (1987). Acute Eye Irritation/Corrosion. In *OECD Guidelines for Testing of Chemicals.* Section 4, No. 405. OECD, Paris.

13b. EEC (1984): Commission Directive of 25th April 1984 adopting to technical progress for the 6th time Council Directive 67/548/EEC on the Approximations of Laws, Regulations and Administrative Provisions relating to the Classification, Packaging and Labelling of Dangerous Substances. Part B. Methods for the determination of toxicity. *Off. J. Eur. Comm., L 251,* 271, 1.

13c. S.J. Williams, in *Animals and Alternatives to Toxicity Testing,* M. Balls, R.J. Riddell, and A.N. Worden (Eds.), Academic Press, London, 1983, pp. 371-377.

13d. S.J. Williams, *Fd. Chem. Toxic., 23,* 189-193(1985).

13d1. V.M. Seabaugh, S.P. Bayard, R.E. Osterberg, W.K. Porter, D.F. McCaulley, C.A. Hoheisel, R.M. Hehir, and G.W. Bierbower, *Amer. J. Public Health, 67,* 367-369(1977).

13d2. M.R. Gilman, E.M. Jackson, D.R. Cerven, and M.T. Moreno, *J. Toxicol.—Cut. & Ocular Toxicol., 2* (2 & 3), 107-117(1983).

13d3. S.C. Gad, R.D. Walsh, and B.J. Dunn, *J. Toxicol.—Cut. & Ocular Toxicol., 5*(3), 195-213(1986).

13e. J.F. Griffith, G.A. Nixon, R.D. Bruce, P.J. Reer, and E.A. Bannan, *Toxicol. Appl. Pharmacol., 55,* 501-513(1980).

13f. W.M. Grant, *Toxicology of the Eye,* 2nd ed. Charles C. Thomas, Springfield, Ill., 1974.

13g. F.E. Freeberg, J.F. Griffith, R.D. Bruce, and P.H.S. Bay, *J. Toxicol.—Cut & Ocular Toxicol., 1*(3), 53-64(1984).

13h. F.E. Freeberg, G.A. Nixon, P.J. Reer, J.E. Weaver, R.D. Bruce, J.F. Griffith, and L.W. Sanders III, *Fundam. Appl. Toxicol., 7,* 626-634(1986).

13i. F.E. Freeberg, D.T. Hooker, and J.F. Griffith, *J. Toxicol.—Cut & Ocular Toxicol., 5*(2), 115-123(1986).

13j. G.J. Oliver, A.P. Walker, W. Kästner, M. York, B.J.E. Simpson, W.J. Bontinck, *J. Toxicol.—Cut. & Ocular., 8*(1), 51-57(1989).

13k. A.B.G. Burton, M. York, and R.S. Lawrence, *Fd. Cosmet. Toxicol., 19,* 471-480(1981).
13l. W.E. Parish, *Fd. Chem. Toxic., 23,* 215-227(1985).
13m. M.C. Scaife, *ATLA, 12,* 157-162(1985).
13n. C.K. Muir, *ATLA, 11,* 129-134(1984).
13n1. M. Bell, P.M. Holmes, T.M. Nisbet, M. Uttley, and N.J. van Abbe, *Int. J. Cosmetic Science, 1,* 123-131(1979).
13o. M.C. Scaife, *Int. J. Cosmetic Science, 4,* 179-193(1982).
13p. H. North-Root, F. Yackovich, J. Demetrulias, M. Gacula Jr., and J.E. Heinze, *Toxicology Letters, 14,* 207-212(1982).
13q. K.Y. Chan, *Alternative Methods in Toxicology, Vol. 3: In Vitro Toxicology,* Mary Ann Liebert, Inc. Publishers, New York, 1985.
13r. E. Borenfreund and O. Borrero, *Cell Biol. Toxicol., 1,* 33-39(1984).
13s. M.C. Scaife, *Fd. Chem. Toxic., 23,* 253-258(1985).
13t. J.A. Shadduck, J. Everitt, and R.A. Meccoli, *Soap/Cosmetics/Chemical Specialties, 61,* 36-56(1985).
13u. R.J. Riddell, R.H. Clothier, and M. Balls, *Fd. Chem. Toxic., 24,* 269-471 (1986).
13v. K. Hockley and D. Baxter, *Fd. Chem. Toxic., 24,* 473-475(1986).
13w. F. Bartnik, W. Kästner, K. Künstler, and W. Sterzel, *Seifen, Öle, Fette, Wachse, 114,* 41-47(1988).
13x. J.B. Price, M.P. Barry, and I.J. Andrews, *Fd. Chem. Toxic., 24,* 503-505 (1986).
13y. A.P. Walker, *Fd. Chem. Toxic., 23,* 175-178(1985).
13z. H.E. Kennah, I.S. Hignet, P.E. Laux, J.D. Dorko, and C.S. Barrow, *Fundam. Appl. Toxicol., 12,* 258-268(1989).
14. H. Herrmann and F.H. Hickman, *Johns Hopkins Hosp. Bull., 82,* 182(1948).
15. W.M. Grant, *Toxicology of the Eye,* Charles C. Thomas, Springfield, Ill., 1962, pp. 510-516.
15a. W. Kästner and P.J. Frosch, *Fette, Seifen, Anstrichm., 83,* 33-46(1981).
16. R.E. Davies, S.R. Kynoch, and M.P. Liggett, *J. Soc. Cosmet. Chem., 27,* 301-306(1976).
17. E.V. Buehler and E.A. Newmann, *Toxicol. Appl. Pharmacol., 6,* 701-710 (1964).
18. J.H. Beckley, *Toxicol. Appl. Pharmacol., 7,* 93-101(1965).
19. J.H. Beckley, T.J. Russell, and L.F. Rubin, *Toxicol. Appl. Pharmacol., 15,* 1-9(1969).
19a. J.-Cl. Etter and A. Wildhaber, *Pharm. Acta Helv., 59,* 8-15(1984).
20. L.M. Carter, G. Duncan, and G.K. Rennie, *Exp. Eye Res., 17,* 409-416(1973).
21. A.S. Weltman, S.B. Sparber, and T. Jurtshuk, *Toxicol. Appl. Pharmacol., 7,* 308-319(1965).
22. J. Kambe, *Acta Soc. Ophthalm. Jap., 78*(10), 1109-1129.
22a. A.D. Little, Inc., Cambridge, MA, U.S. Department of Commerce National Technical Information Service PB-301193(1977).
22b. E.E. Ahlfors and A. Larsson, *Scand. J. Dent. Res., 96*(5), 428-434(1988).
23. F.R. Bettley, *Brit. J. Derm., 80,* 635-642(1968).

24. E. Wannenmacher, *Fétte, Seifen, Anstrichm., 74,* 177-180(1972).
25. A.K. Reng, *Parfüm. Kosmet., 57,* 307-316(1976).
26. M.M. Berenson and A.R. Temple, *Clin. Toxicol., 8*(4), 399-404(1975).
27. H.P. Ciuchta, Toxicology Research Unit, Gillette Medical Evaluation Laboratories, Rockville, Maryland.
27a. H.P. Ciuchta and K.T. Dodd, *Drug Chem. Toxicol., 1*(3), 305-324(1978).
27b. J.J. Serrano, H. Delonca, L.T. Viet-Nga, M. Luitkus, *Parfums, Cosmetiques, Aromes, 13,* 55-60(1977).
27c. B.H. Arthur, S.C. Pennisi, L.C. Dipasquale, T. Re, J. Dinardo, G.L. Kennedy, H. North-Root, D.A. Penney, H.J. Sekerke, *J. Toxicol.—Cut. & Ocular Toxicol., 5,* 215-227(1986).
28. J.H. Draize and E.A. Kelley, *Drug Cosmet. Ind., 71,* 36(1952).
29. L.W. Hazleton, *Proc. Sci. Sect. Toilet Goods Ass., 17,* 5-10(1952).
29a. T. Matsuura, T. Tsuyosi, and Y. Masamoto, *J. Soc. Cosmet. Chem. Japan., 17,* 153-163(1983).
30. V. Lechnitz, Kosmet. *Monatsschr., 20,* 8-9(1972).
30a. L.J. Murphy and F. Baiocchi, *Drug Cosmet. Ind., 5,* 35-45(1978).
31. K. Oba, A. Mori, and S. Tomiyama, *Rinsho Derma Tokyo, 11,* 30-36(1968); Yukagaku, *17*(11), 629-634(1968).
32. H. Daweke, *Arch. Klin. Exp. Derm., 209,* 520-524(1959).
32a. G.W. Panzer and L.J. Nehmsmann, *Cosmetics & Toiletries, 95,* 21-28(1980).
33. M. Iimori, T. Ogata, and K. Kudo, *Yukagaku, 21*(6), 334-337(1972).
34. W. Kästner, unpublished.
35. H. Tronnier and W. Schneider, *Arch. Klin. Exp. Derm., 229,* 40-53(1967).
35a. M. York, R.S. Lawrence, and G.B. Gibson, *Int. J. Cosmetic Science, 4,* 223-234(1982).
36. K.J. Olson, R.W. Dupree, E.T. Plomer, and V.K. Rowe, *J. Soc. Cosmet. Chem., 13,* 469-476(1962).
37. V.K.H. Brown and C.M.C. Muir, *Tenside, 7,* 137-139(1970).
38. G. Martin, J.H. Draize, and E.A. Kelley, *Proc. Sci. Sec. Toilet Goods Ass., 37,* 2-3(1962); Drug Cosmet. Inc., *91,* 30-31(1962).
38a. R.S. Lawrence, M.H. Groom, D.M. Ackroyd, and W.E. Parish, *Fd. Chem. Toxic., 24,* 497-502(1986).
39. M. Ginsburg and J.M. Robson, *Brit. J. Ophthal., 33,* 574-579(1949).
40. I.H. Leopold, *Arch. Ophthal., 34,* 99-102(1945).
41. A.J. Lehman and J.H. Draize, *Ass. Food Drug Officials U.S. Quart. Bull., 23,* 150-152(1959).
42. J.B. Feldman, P. deLong, and C.P. Brown, *Arch. Ophthal., 40,* 668-679(1948).
43. S.S. Hopper, H.R. Hulpieu, and V.V. Cole, *J. Amer. Pharmaceut. Ass., 38,* 428-432(1949).
44. J.G. Bellows and M. Gutmann, *Arch. Ophthal., 30,* 352-357(1943).
45. K. Meinicke, Fette, Seifen, *Anstrichm., 62,* 107-112(1960).
45a. Anonym. Final Report on the Safety Assessment of Sodium Lauryl Sulfoacetate, *J. Amer. Coll. Toxicol., 6,* 261-277(1987).
46. J. Regnier, M.D. Thesis, Paris, 1929, as cited by I.C. Geddes, Brit. J. Anesthesia, *27,* 609(1955).

47. L.S. Harris, Y. Kahanowicz, and M. Shimmyo, *Opthalmologica, 170,* 320-325(1975).
48. H. Bloching, *Haushaltwaschmittel,* Ciba-Geigy AG, Basel, 17-24(1975).
49. D.B. Feldman, D.P. Rall, and J.A. Moore, *JAMA, 221,* 1055(1972).
50. D.L. Opdyke, F.H. Snyder, and H.L. Rubenkoenig, *Toxicol. Appl. Pharmacol., 6,* 141-146(1964).
51. F.M. Marzulli, Personal communication, as cited by L.S. Harris, Y. Kahanowicz, and M. Shimmyo, *Opthalmologica, 170,* 320-325(1975).
51a. J.A. Shadduck, J. Everitt, and R. A. Meccoli, *HAPPI, 22*(11), 57-64(1985).
51b. T.F.M. Ferguson and C. Prottey, *Fd. Cosmet. Toxicol., 14,* 431-434(1976).
51c. G.K. Dossou and B. Vinsonneau, *Ann. Fals. Expert. Chim., 75,* 417-424(1982).
51d. A.D. Little, Inc., Supplement, A report by M.M. Goyer, J.H. Perwak, A. Sivak, and P.S. Thayer, 1981.
52. H. Modde, G. Schuster, and H. Tronnier, *Tenside, 2,* 368-373(1965).
53. H. Tronnier, G. Schuster, and H. Modde, *Arch. Klin. Exp. Derm., 221,* 232-249(1965).
53a. A.W. Finkstein, cited in *Soap/Cosmetics/Chemical Specialties,* 70-76(1977).
53b. C. Prottey and T.F.M. Ferguson, *Fd. Cosmet. Toxicol., 14,* 425-430(1976).
53c. W.E. Adam and K. Neumann, *Fette, Seifen, Anstrichm., 80,* 392-397(1978).
53d. W.E. Adam and K. Neumann, *Fetta, Seifen, Anstrichm., 82,* 367-370(1980).
54. K.C. Swan, *Amer. J. Ophthalmol., 27,* 1118-1122(1944).
55. C.H. Hine, H.H. Anderson, I. Finely, and J.G. Moore, *J. Amer. Pharmaceut. Ass., 42,* 489-495(1953).
56. H. Modde and G. Schuster, *Parfüm. Kosmet., 46,* 1-3(1965).
56a. J.P. Guillot, J.F. Gonnet, and C. Clement, *Fd. Chem. Toxic., 20,* 573-582 (1982).
56b. OECD (1981). Acute dermal irritation/corrosion. In *OECD Guidelines for Testing of Chemicals.* Sec. 4, No. 404, OECD, Paris 2.
57. P. Finkelstein, K. Laden, and W. Miechowski, *J. Invest. Derm., 40,* 11-14 (1963).
58. J.O. Hoppe, E.B. Alexander, and L.C. Miller, *J. Amer. Pharmac. Ass., Sci.* Edit., *39,* 147(1950).
59. A. Rostenberg, *Drug. Cosmet. Ind., 88,* 592-594(1961).
60. L. Phillips, II, M. Steinberg, H.I. Maibach, and W.A. Akers, *Toxicol. Appl. Pharmacol., 21,* 369-382(1972).
61. M.M. Rieger and G.W. Battista, *J. Soc. Cosmet. Chem., 15,* 161-172(1964).
62. R.O. Carter and J.F. Griffith, *Toxicol. Appl. Pharmacol., 7,* Suppl. 2, 60 (1965).
63. M.K. Polano, *J. Soc. Cosmet. Chem., 19,* 3-20(1968).
64. A.B.G. Lansdown, *J. Soc. Cosmet. Chem., 23,* 739-772(1972).
65. I. Levenstein, *J. Soc. Cosmet. Chem., 15,* 377-384(1964).
66. L.C. Barail, *J. Soc. Cosmet. Chem., 11,* 241-249(1960).
66a. E.P. Pittz, *Cosmetics & Toiletries, 98,* 51-65(1983).
67. J.V. Klauder, *J. Soc. Cosmet. Chem., 11,* 249(1960).
67a. M.R. Gilman, R.A. Evans, and S.J. De Salva, *Drug Chem. Toxicol. 1*(4), 391-400(1978).

68. D.L. Opdyke and C.M. Burnett, *Proc. Sci. Sect. Toilet Goods Ass., 44,* 3-4 (1965).
69. C. Gloxhuber and K.H. Schulz, *Detergency,* in *Theory and Test Methods (Part 2)*, W.G. Cutler and R.C. Davis (Eds.), Marcel Dekker, New York, 1975, pp. 695-706.
69a. V.A. De Leo, L.C. Harber, B. Kong, B, and S. De Salva, *Alternative Methods in Toxicology Vol. 3: In Vitro Toxicology,* Mary Annn Liebert Inc., Publishers, New York, 1985, pp. 469-481.
69b. K. Künstler, F. Bartnik, N.-P. Lüpke, and S. Wallat, *Reprints of the XIVth I.F.S.C.C. Congress, Barcelona, 1,* 11-23(1986).
69c. W.E. Parish, *Fd. Chem. Toxic., 24,* 481-494(1986).
69d. P.A. Duffy, O.P. Flint, T.C. Orton, and M.J. Fursey, *Fd. Chem. Toxic., 24,* 517-518(1986).
69e. M.A.E. Mol, J. van Genderen, and O.L. Wolthuis, *Fd. Chem. Toxic., 24,* 519-520(1986).
69f. J. Kao, J. Hall, and M. Holland, *Toxicol. Appl. Pharmacol., 68,* 206-217 (1983).
69g. R.G. Helman, J.W. Hall, and J.Y. Kao, *Fundam. Appl. Toxicol., 7,* 94-100 (1986).
69h. G.J. Putterman, N.F. Wolejsza, M.A. Wolfram, and K. Laden, *J. Soc. Cosmet. Chem., 28,* 521-532(1977).
69i. U. Zeidler and G. Reese, *Ärztliche Kosmetologie, 13,* 39-45(1983).
69j. G.J.A. Oliver, M.A. Pemberton, C. Rhodes, *Fd. Chem. Toxic., 24,* 507-512 (1986).
69k. G.J.A. Oliver and M.A. Pemberton, *Fd. Chem. Toxic., 24,* 513-515(1986).
69l. G.J.A. Oliver, M.A. Pemberton, and C. Rhodes, *Toxic. in Vitro, 2,* 7-17 (1988).
69m. J.R. Young, M.J. How, A.P. Walker, and W.M.H. Worth, *Toxic. in Vitro, 2,* 19-26(1988).
69n. V.C. Gordon, C.P. Kelly, and H.C. Bergman, Manuscript presented at the Society of Toxicology 28th Annual Meeting, Atlanta, Georgia, Feb. 27th to March 3rd(1989).
69o. C. Prottey, D. Oliver, and A.C. Coxon, *Int. J. Cosmetic Science, 6,* 263-273 (1984).
70. A.B.G. Lansdown and P. Grasso, *Brit. J. Dermatol., 86,* 361-373(1972).
71. P.W.A. Tovell, A.C. Weaver, J. Hope, and W.E. Sprott, *Brit. J. Dermatol., 90,* 501-506(1974).
72. V.V. Ivanov, E.A. Ievleva, and L.I. Sych., *Vestn. Dermatol. Venerol., 49,* 62-66(1975).
73. H. Hoffmann, *Fette, Seifen, Anstrichm., 60,* 367-369(1958).
74. W. Schneider, *Fette, Seifen, Anstrichm., 59,* 38-41(1957).
75. H. Kröper, *Fetta Seifen, 44,* 298-301(1937).
76. W.W. Duemling, *Arch. Dermatol. Syphilol., 43,* 264(1941).
77. E. Götte, *Tenside, 4,* 209-217(1967).
78. A. Szakall, *Berufsdermatosen, 6,* 171-192(1958).
79. A. Szakall and K.H. Schulz, *Fette, Seifen, Anstrichm., 62,* 170-175(1960).

80. L. Geller, *Kosmet. Monatsschr., 19,* 7-8(1970).
81. E. Götte and J.J. Herzberg, *Fette, Seifen, Anstrichm., 59,* 747-750(1957).
82. H. Stübel and A. Szakall, *Die Wirkung von Waschmitteln auf die Haut,* A. Hüthig Verlag, Hiedelberg, 1957.
83. J. Scholz, *Arch. Exper. Pathol. Pharmakol., 232,* 241-244(1967).
84. O. Kuhn, *Arch. Klin. Exper. Dermatol., 203,* 311-322(1956).
85. E. Götte, *Kolloid-Z., 117,* 42-47(1950).
86. B.R. Choman, *J. Invest. Dermatol., 40,* 177(1963).
86a. G.J. Putterman, N.F. Wolejsza, M.A. Wolfram, and K. Laden, *J. Soc. Cosmet. Chem., 28,* 521-532(1977).
87. H. Neuhaus, Fette, Seifen, *Anstrichm., 53,* 522(1951).
88. E. Götte, *Aesthet. Med., 15,* 313-319(1966?.
89. S.G. Blohm, *Acta Dermato-Venereologica* (*Stockholm*), *37,* 269-275(1957).
90. B.R. Choman, *Brit. J. Invest. Dermatol., 37,* 263-271(1961).
91. E. Götte, *Aesthet. Med., 12,* 146-154(1963).
92. D. Howes, *J. Soc. Cosmet. Chem., 26,* 47-63(1975).
93. G. Imokawa, K. Sumura, and M. Katsumi, *J. Amer. Oil Chem. Soc., 52,* 484-489(1975).
94. F.W. Putnam, *Advances in Protein Chemistry,* Vol. 4, Academic Press, New York, 1948, pp. 80-118.
95. J.F. Foster, *J. Phys. Colloid Chem., 53,* 75(1949).
96. C. Carruthers, D.L. Woernley, A. Baumler, and B. Kress, *J. Invest. Dermatol., 25,* 89-101(1955).
97. C. Prottey and T. Ferguson, *J. Soc. Cosmet. Chem., 26,* 29-46(1975).
98. H. Tronnier, *Parfüm. Kosmet., 58,* 60-65(1977).
99. E.J. van Scott and J.B. Lyon, *Brit. J. Invest. Dermatol., 21,* 199-203(1953).
100. J. Malaszkiewicz and Ch. Gloxhuber, *Arch. Klin. Exp. Dermatol., 237,* 652-661(1970).
101. E. Götte, Vortrag IV. *Int. Kongr. Granzflächenaktive Stoffe, Brüssel,* 1964, C/1-10.
102. K.H. Gustavson, *Acta Chem. Scand., 4,* 1171-1184(1950).
103. S.P. Harrold, *Brit. J. Invest. Dermatol., 32,* 581-588(1959).
104. L.J. Vinson and B.R. Choman, *J. Soc. Cosmet. Chem., 11,* 127-137(1961).
105. S. Freeman, M.W. Burril, Tsan-Wen Li, and A.C. Ivy, *Gastroenterology, 4,* 332-343(1945).
106. H. Wilmsmann, *Fette, Seifen, Anstrichm., 65,* 958-964(1963).
107. V.K.H. Brown, *J. Soc. Cosmet. Chem., 22,* 411-420(1971).
108. C. Prottey and P.J. Hartop, *J. Invest. Dermatol., 61,* 168-179(1973).
109. J. Scala, Don E. McOsker, and H.H. Reller, *J. Invest. Dermatol., 50,* 371-379(1968).
110. F.R. Bettley, *Brit. J. Derm., 75,* 113-116(1963).
111. I.H. Blank, *Arch. Dermatol., 39,* 811-824(1939).
112. W.R. Markland, *Norda Briefs,* 466(1975).
113. P.H. Dugard and R.J. Scheuplein, *J. Invest. Dermatol., 60,* 263-269(1973).
114. G. Imokawa, K. Sumura, and M. Katsumi, *Yukagaku, 23*(11), 719-725(1974).
115. Å. Nilzen and K. Wikström, *Acta Dermato-Venerologica, 35,* 292-299(1955).

116. S. Scheinberg, S.P. Bralow, and H. Necheles, *Surgery, 24,* 972-979(1948).
117. F. Schaaf and F. Gross, *Dermatologica, 106,* 170-175 and 357-378(1953).
118. R. Brun, E. Bujard, and W. Jadassohn, *Dermatologica, 115,* 35-39(1957).
119. F.N. Marzulli and H.I. Maibach, *Food Cosmet. Toxicol., 13,* 533-540(1975).
120. R.L. Baer, S.A. Rosenthal, and C. Sims, *J. Invest. Dermatol., 27,* 249(1956).
121. A.J. Ingram and P. Grasso, *Brit. J. Dermatol., 92,* 131-142(1975).
122. E.O. Butcher, *J. Invest. Dermatol., 16,* 85-90(1951).
123. O.H. Mills and A.M. Kligman, *Arch. Dermatol., 111,* 65-68(1975).
124. A.M. Kligman and O.H. Mills, *Arch. Derrmatol., 106,* 843-850(1972).
125. O.H. Mills and A.M. Kligman, *Animal Models in Dermatology,* Churchill Livingstone, London and New York, 1975, pp. 176-183.
126. J. Rubisz-Brzezinska, D. Musialowicz, S. Szezurek, and K. Ludyga, *Przeg. Dermatol.,* LXIII, *1,* 25-30(1976).
127. O. Schmid, *J. Soc. Cosmet. Chem., 21,* 835-843(1970).
128. H. Muto and Y. Tatsuma, *Aichi Ika Daigaku Igakukai Zasshi, 2*(1), 13-17 (1974); ref. *Chem. Abstr., 80,* 69(1974).
128a. M.A. Pemberton and C. Rhodes, *Toxicology Letters, 44,* 85-90(1988).
128b. G. Imokawa, *J. Am. Oil Chem. Soc., 56*(5), 604-609(1979).
128c. H. Schott, *J. Pharmac. Sciences, 62*(2), 341-343(1973).
129. B.E. Emery and L.D. Edwards, *J. Amer. Pharm. Ass., 29,* 251-254(1940).
130. C. Prottey, P.J. Hartop, and T.F.M. Ferguson, *J. Soc. Cosmet. Chem., 24,* 473-492(1972).
131. K.H. Schulz and G. Rose, *Arch. Klin. Exper. Dermatol., 205,* 254-260(1957).
132. Ch. Gloxhuber, *Arch. Toxicol., 32,* 245-270(1974).
132a. G. Imokawa, H. Tsutsumi, and T. Kurosaki, *J. Am. Oil Chem. Soc., 55,* 839-843(1978).
132b. M. Kaminsky, M.M. Szivos, and K.R. Brown, *J. Toxicol.—Cut. & Ocular Toxicol., 5*(2), 81-87(1986).
133. A.K. Palmer, M.A. Readshaw, and A.M. Neuff, *Toxicology, 4,* 171-181(1975).
134. P. Finkelstein, K. Laden, and W. Miechowski, *Toxicol. Appl. Pharmacol., 7,* 74-78(1965).
135. H. Gisslen and B. Magnusson, *Acta Derm-Venereol., 46,* 269-274(1966).
136. G.A. Nixon, C.A. Tyson, and W.C. Wertz, *Toxicol. Appl. Pharmacol., 31,* 481-490(1975).
137. C.C. Edwards, *Fed. Reg., 37,* 27.635-27.636(1972).
137a. M. Sadai and N. Nisuno, *Nitsuhi Kaishi, 82*(4), 207-221(1972).
138. C.G. Hunter, V.K. Brown, and L.W. Ferrigan, *Brit. J. Industr. Med., 23,* 137-141(1966).
139. S. Carson and B.L. Oser, *J. Soc. Cosmet. Chem., 15,* 137-147(1964).
140. R.E. Davies, K.H. Harper, and S.R. Kynoch, *J. Soc. Cosmet. Chem., 23,* 371-381(1972).
141. L.B. Fischer and H.I. Maibach, *Contact Dermatitis, 1,* 273-276(1975).
142. Schimmel Briefs No. 299 (1960).
143. K. Meinicke, *Arch. Klin. Exp. Dermatol., 211,* 370-378(1960).
144. M. Iimori, E. Iwata, and Y. Sato, *Yukagaku, 20*(9), 584-586(1971); *J. Jap. Oil Chem. Soc., 20,* 584-586(1971).

145. W. Burckhardt, *Dermatologica, 129,* 37-46(1964).
146. W. Burckhardt, R. Schmid, and P. Schmid XIII. *Int. Congr. Dermatol. Bd., 1,* 224-226(1967).
147. W. Burckhardt, *Berufsdermatosen, 18,* 179-188(1970).
148. F.S.K. MacMillan, R.R. Rafft, and W.B. Elvers, *Animal Models in Dermatology,* Churchill Livingstone, London and New York, 12-22(1975).
148a. J.M. Quack and A.K. Reng, Fette, *Seifen, Anstrichmittel, 78*(5), 200-206 (1976).
149. F. Masuda, K. Okamoto, and K. Inoue, *J. Food Hyg. Soc. Jap., 15,* 349-355(1974).
150. E.V. Buehler, *Arch. Dermat., 91,* 171-177(1965).
151. R. Kuchinka, Seifen, *Öle, Fette, Wachse., 76,* 9-11(1950).
151a. H. Hein, *Parfümerie und Kosmetik, 59,* 73-82(1978).
151b. G. Imokawa, *J. Soc. Cosmet. Chem., 31,* 45-66(1980).
152. O. Jacobi and E. Ortmann, *Berufsdermatosen, 19,* 75-83(1971).
153. J.J. Jambor and R.E. Suskind, *Brit. J. Invest. Dermatol., 24,* 379-385(1955).
153a. K. Okamoto, S. Tsuchiya, and K. Nakamura, *Dev. Toxicol. Environ. Sci., 12,* 159-168(1986).
154. A.P. Walker, G.K. Ashforth, R.E. Davies, E.A. Newman, and H.L. Ritz, *Acta Dermatovener* (*Stockholm*), *53*, 141-144(1973).
155. B. Magnusson and A.M. Kligman, Thomas, Springfield, Ill., 103-123(1970), *J. Invest. Dermatol., 52,* 268(1969).
156. B. Magnusson and O. Gilje, *Acta Dermatovener* (*Stockholm*), *53*, 136-140 (1973).
157. C. Gloxhuber, M. Potokar, S. Braig, H.G. van Raay, and G. Schwarz, Fette, Seifen, *Anstrichm., 76,* 126-129(1974).
158. B. Sylvest, N. Hjorth, and B. Magnusson, *Contact Dermatitis, 1,* 359-362 (1975).
158a. W.E. Lindup and P.T. Nowell, *Food Cosmet. Toxicol., 16,* 59-62(1978).
158b. D.S. Connor, H.L. Ritz, R.S. Ampulski, H.G. Kowollik, P. Lim, D.W. Thomas, and R. Parkhurst, *Fette, Seifen, Anstrichmittel, 77*(1), 25-29(1975).
158c. D.S. Connor, H. Kowollik, and D.W. Thomas, *J. Am. Oil Chem. Society, 53,* 182-185(1976).
158d. H.L. Ritz, D.S. Connor, and E.D. Sauter, *Contact Dermatitis, 1,* 349-358 (1975).
158e. F. Morikawa, M. Tejima, S. Fujii, and S. Hyakutake, *Arerugi, 27*(7), 684-661(1978) quoted from 51d. A.D. Little, Inc., Supplement, A report by M.M. Goyer, J.H. Perwak, A. Sivak, and P.S. Thayer(1981).
159. K. Landsteiner and J. Jacobs, *J. Expl. Med., 61,* 643-656(1935).
159a. P.H.S. Bay and P.J. Danneman, *JAOCS, 62*(4), 646(1985).
159b. N.J. Reynolds and R.D.G. Peachey, *Contact Dermatitis, 22,* 59(1990).
160. E. Skog, *Acta Dermato-Venereol.* (*Stockholm*), *38*(1), 682-685(1958).
161. J. E. Wahlberg, *Acta Derm.-Venereol.* (Stockholm), *48*, 549-555(1968).
162. J.E. Wahlberg, *Acta Derm.-Venereol.* (Stockholm), *45*, 335(1965).
163. B. Magnusson, *Acta Derm.-Venereol.* (Stockholm), *45*, 325(1965).
163a. R.J.F. Markham and B.N. Wilkie, *Int. Archs Allergy Appl. Immun., 51,* 529-543(1976).

6

Dermatological Observations (Humans)

W. MATTHIES Henkel KGaA, Düsseldorf, Germany

I. INTRODUCTION

The oldest anionic surfactant traditionally in use for cleaning human skin is soap. Soap has existed for at least 5000 years. Particular hints on the uses of soap were found on Sumeric clay tables from about 2500 B.C., at which time soap was made by blending plant oils and potash. By the beginning of the 19th century, the expensive potash was substituted by soda, what made soap easily producible and commonly available (82). During the following decades, various attempts were made to alter natural raw materials in order to get improved washing properties. This resulted in a series of new derivatives, the most prominent of which were the fat alcohol sulfates and the alkylbenzene sulfonates (Böhme Chemie, 1928).

By the 1930s and 1940s different alkyl sulfates were systematically investigated for skin compatibility (40) and found to be more compatible than traditional types of soap. Today, these types of tensides are still used frequently in cleaning products for skin, fibers, and surfaces. Table 1 gives groups of derivatives presently in use for various purposes and, therefore, possible sources of contact with human skin.

In reviewing the former dermatological literature for evidence of the effects of anionic surfactants on the skin and mucous membranes of humans and for any evidence of allergic reaction to them, one finds very little information dealing with specific types of organic detergent or tenside (1). Even the name tenside is uncommon in dermatological tests, being more familiar to the specialist chemist, who on occasion has introduced it to dermatologists as a more convenient term to describe the general class of surface-active compounds of which synthetic detergents form a very large part (2).

That there is little information is hardly surprising when one considers that the prime responsibility of dermatologists and general practitioners is the clinical care of patients, and single, named tensides are not available in the home. Patients will have been using products containing many additional constituents as well as the tenside, and the complete product is referred to simply as detergent. However, useful information can come from newer reports discussing special effects of selected anionics to the skin and from epidermiological surveys of the last years.

As this review centers around anionic detergents, prominence will be given to them and, indeed, this could also be said to reflect their predominance in the market place. As soap was the first anionic detergent, it will also be useful to consider its use where this is helpful in understanding other cases said to be related solely to the use of tensides or "synthetic" detergents. Such consideration will be quite relevant to the overall purpose of looking at anionic detergents because, in succession, over the years various anionics have been available for washing clothes, household fabrics, and surfaces. First soap, then in succession alkyl sulfates such as sodium lauryl sulfate, secondary alcohol sulfate, alkylaryl sulfonate, linear alkyl benzene sulfonate, and finally the above-mentioned derivatives, which have been designed to overcome the nuisance of foam in the environment and to improve skin compatibility. During this time, particularly in Europe, soap has remained predominant as a personal washing aid. It is probable, therefore, that everyone who is exposed to a fabric-washing anionic will also be exposed to daily use of soap.

TABLE 1 Types of Anionic Surfactants Currently in Use

Sulfates	
R—CH$_2$—OSO$_3$ Na	Fatty alcohol sulfate (FAS)
R—CH$_2$—O—(CH$_2$—CH$_2$—O)—SO$_3$ Na	Fatty alcohol ether sulfate (FAES)
CH$_2$—COOR | CH-OH | CH$_2$—OSO$_3$ Na	Monoglyceride sulfate
Sulfonates	
R—CH$_2$—CH=CH—CH$_2$—SO$_3$ Na	Oleine sulfate
R-CH—COOCH$_3$ | SO$_3$—Na	alpha-sulfo fatty acid ester
R \\ CH—SO$_3$ Na / R	sec. Alkansulfonate
SO$_3$ Na / CH—COONa | CH$_2$—COOR	Sulfosuccinate
R—CO-N—CH$_2$—CH$_2$—SO$_3$ Na | CH$_3$	
R-CO—O—CH$_2$—CH$_2$—SO$_3$ Na	Fatty acid isothionate
Carboxylates	
R—COONa	Soap
RO—(CH$_2$—CH$_2$—O)$_n$—CH$_2$—COONa	Ethercarboxylate
CH$_3$ | R—CO—N—CH$_2$—COONa	Sarcosinate
COONa | (CH$_2$)$_2$ | CH—COONa | NH—COR	Acylglutamate

II. EFFECTS OF ANIONIC SURFACTANTS ON THE SKIN

Since the action of anionic surfactants on the skin includes different objectives, it may be helpful for the understanding of interactions between surfactant and stratum corneum as its most important protective component to summarize some critical aspects of skin function.

A. The Barrier

The most important function of the skin is the maintenance of a permeability barrier between inner and outer milieu. To survive as an aqueous body in a terrestrial arid environment, mammals, but also plants, arthropods, reptiles, and birds have been forced to develop cellular mechanisms to regulate integumental water loss (83,84). Although every group has developed its own regulatory system the use of lipids for this purpose is common.

Human epidermis consists mainly of two components: protein-rich keratinocytes in different stages of differentiation and a lipid-rich intercellular substance. The organization of these compounds is comparable to a brick wall with several layers of cells (bricks) attached by a lipidic (mortar) (see Fig. 1) (84).

By the 1950s Stüpel and Szakall (86) had already defined the "barrier" as a continuous membrane positioned in the stratum lucidum, which was removable by tape stripping. According to Kligman the complete stratum corneum works as a barrier with an increasing degree of efficacy from upper to deeper layers (1964).

These older results focused primarily on the role of the horny material and its barrier function. More recent results, focusing on the localization of different lipids as parts of the intercellular substance within the intercellular space of several cell layers, help to understand the complex mechanism of blockade of the transcellular and paracellular pathway, as well.

According to Elias and co-workers (87-90) multilamillar-directed lipids play the major role in the permeability barrier function. These lipids derive from a specific organel of the keratinocyte, the "lamellar body" or "Odland body." They are synthesized in the stratum spinosum and excreted by exocytosis in the stratum granulosum (85,91). While the intercellular substance of the deeper layer also contains phospholipids, cholesterolsulfate, and sugars, the composition changes to minimal portion of ionic lipids and high content of neutral lipids in the upper layers of the stratum corneum (for details see Table 2). This variation in the composition of the intercellular substance explains the phenomenon of selective permeability and shows the prominent role of the stratum corneum as a hydrophobic barrier.

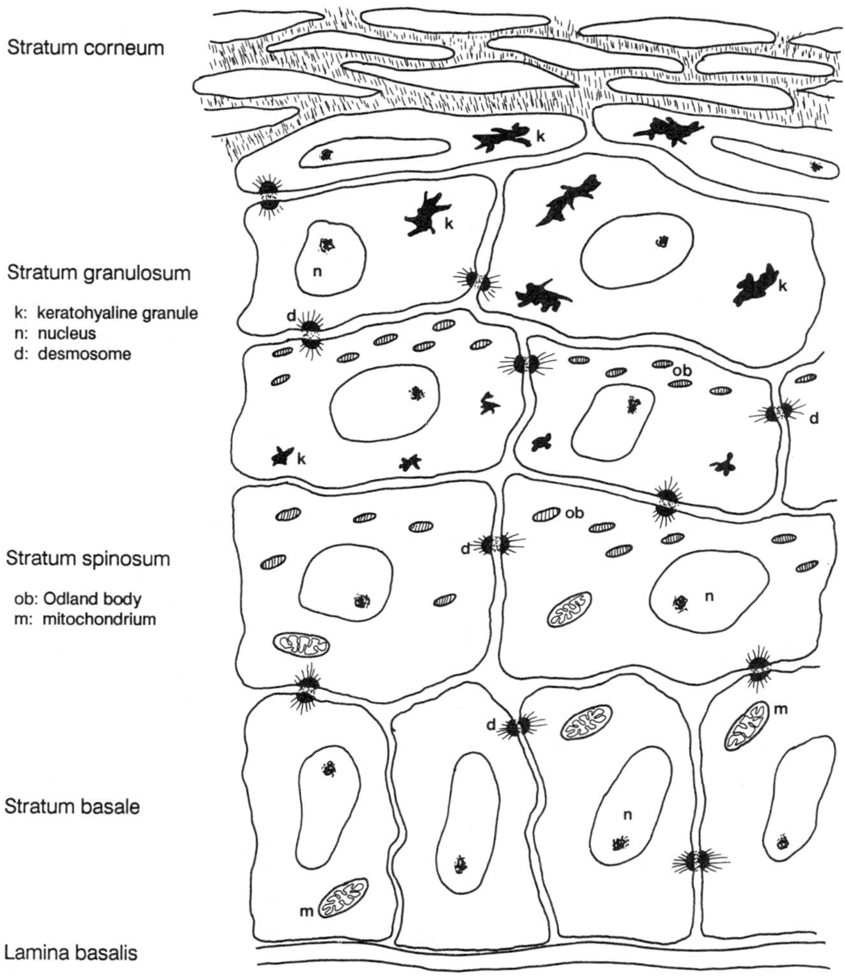

FIG. 1 Scheme of the structure of the epidermis.

As the absolute content of lipids in the stratum corneum of different parts of the body varies (e.g., face 7-10%, abdomen 6-8%, legs 2-4%, palms and soles 1-2%, (85,90) as well as the thickness of the stratum corneum, the barrier effect against hydrophilic substances varies also (87,91). This has to be considered when performing measurements of specific functional skin parameters in order to assess irritation potential of surfactants.

TABLE 2 Variation of Lipid Composition During Epidermal Differentiation

	Stratum basale/ spinosum	Stratum granulare	Stratum corneum
Polar lipids	45	25	5
Cholesterol sulfate	3	6	2
Neutral lipids	51	57	78
Free steroles	11	12	14
Free fatty acids	7	9	19
Triglycerid	12	25	25
Sterole/waxester	5	5	5
Squalene	5	5	5
n-Alkane	4	4	6
Spingolipids	7	12	18
Glucosylceramides	4	6	traces
Ceramides	4	9	18

Source: Ref. 85.

B. Water Content and Regulation

Since the water content of the stratum corneum is an important factor to maintain its viscoelastic properties (45), the corneum has to keep a certain amount of water in order to guarantee extensibility and flexibility for the movements of the skin. Three main factors are responsible for this purpose (after Ref. 92):

1. The binding state of water molecules with the keratin components
2. Epidermal lipids as protectors of hydrosoluble substances and plasticisers of the stratum corneum, mainly situated in the intercellular space
3. The "natural moisturizing factors" (NMF), water-soluble substances (e.g., lactic acid, urea, pyrollidone carbonic acid), which strongly influence the equilibrium rate of the stratum corneum with the environment.

The influence and importance of sebum as a probable fourth factor for skin function remains unclear; indeed sebum consists of different fractions of lipids that are excreted by sebaceous glands covering most of the skin surface (227) (see Table 3). This led to former interpretation as "water-lipid-coat," or "acid coat" due to its slightly acidic pH, and is still present in todays practitioner's knowledge (216,224). Nevertheless, the role of sebum as a contributor to stratum corneum plasticity seems no longer acceptable, since

TABLE 3 Skin Lipid Pattern in Adults

	Sebum (%)	Epidermis (%)	Surface lipids (%)
Squalene	12	<0.5	10
Sterol esters	<1	10	2.5
Sterols, nonesterified	0	20	1.5
Wax esters	23	0	22
Triglycerid	60	10	25
Di- and monoglycerides	0	10	10
Fatty acids, nonesterified	0	10	25
Glycolipids and phospholipids	0	30	0
Unidentified	5	10	4

Source: Ref. 227.

Sebum does not build a closed film on the skin (93,94)
Children before puberty possess a fully intact function of the skin barrier without a significant production of sebum (95)
Areas on the body surface with higher sebum production do not have a better barrier function (85).

Kligman et al. suppose that sebum production may be a reminiscence from the human evolution (96,97) without any directed function. However, this sebum "coat" may contribute in certain cases to generate the surface milieu, the surface pH and may have emulsifying or, to a lesser extent, protective properties after contact with chemicals.

In summary, the prominent factors of skin function, which anionic surfactants may act on, are surface lipids from sebaceous glands, horny material from keratinocytes, intercellular lipids of different fractions and types, "natural moisturizing factors," and cellular membranes of living cells.

Affecting diverse items of this system anionic surfactants may provoke changes of the barrier function, of the water content, of the surface milieu, etc. The clinically predominant observation of "irritation" after intense use or experimentally induced prolonged contact, which was formerly the nearly unique approach to characterize skin compatibility of and reaction against surfactants, consequently subsumes different aspects of skin-surfactant interaction.

Since new equipment and methods have been made available during the last decade, a great number of publications on dermal function and investigation of the influence of application of anionics has appeared. Besides the patch test technique with its visual scoring of reactions, these new approaches are helpful to a major understanding of reactions and lead to suggestions why derivatives of soaplike materials act so differently and show various degrees of compatibility to the skin.

C. Experimental Results: Adsorption, Absorption, Permeation

The interaction between anionic surfactants and human skin with respect to adsorption and penetration phenomena has already been intensively studied together with investigations about the irritation potential of various anionic derivatives (32,100,121,122). Observations that different fatty acids varied considerably in their irritant effect on the intact skin in patch testing experiments led to various studies indicating that alkyl sulfates, alkyl benzene sulfonates, and related derivatives irritated human skin depending on their carbon chain length. Since, generally speaking, a substance applied to the skin can only irritate if it is absorbed in sufficient quantity to reach the Malpighian cells in a toxic concentration (123), these findings indicate a different capacity of permeability. In fact, C_{12} components revealed a maximum permeability as well as a maximum water influx catalyzing effect (127) compared to shorter (C_8/C_{10}) or longer (C_{14}, C_{16}, C_{18}) chain lengths (123,125,128). This corresponds with patch test results which indicated that the C_{12} compound was a stronger irritant than shorter or longer analogues (40,100,121, 128).

Schulz and Rose showed that the comparable fatty acids and sodium soaps were more compatible than the corresponding alkyl sulfates (125,128). Studies with labeled radioactive potassium octanoate, potassium laurate, potassium oleate, and potassium palmitate on isolated human epidermis showed a relatively free penetration of laurate (24%) as compared with octanoate (5%) and oleate (2%), while palmitate penetrated very much less (0.15%) (123). Also, dodecylsulfate was shown to penetrate even after application of only 0.005 M a solution (129), while a similar study with laurate and stearate indicated capable penetration of laurate but nearly no penetration of stearate (126). Further comparative studies with potassium caprylate, potassium laurate, potassium oleate, sodium laurylsulfate, sodium dodecyl benzene sulfonate, sodium lauroyl sarcosinate, cationics, and surfactants of other types investigating the permeability through the human epidermis in vitro after treatment with 0.04 M solution again showed the striking effect of laurate soap, while caprylate soap lacked any effect; sodium lauryl sulfate and sodium dodecyl benzene sulfate had an about equal activity, sarcosinate was almost inactive. In this study it was also stated that there was no direct relation between pH and increase of permeability (131).

Due to its penetration properties sodium lauryl sulfate has been chosen for experimental purpose as an enhancer of penetration (67,166). Sodium lauryl sulfate was the most effective detergent for increasing absorption compared to benzalkonium chloride (cationic), or dodecanol and ethylen oxide. Sodium lauryl sulfate was also shown to enhance the absorption of solvents in different systems, as it was investigated by Boman et al. (209). Binding of anionic surfactants on keratin causes denaturation and liberation of SH-groups following the unfolding of the protein (98,130,226). Head group polarity determines the capability to denature protein, to extract compounds like amino acids from the stratum corneum, and to penetrate into deeper layers to the living cells. Within the homologous series of sodium alkyl sulfates, especially those with chain length 11, 12, and 13 were found to remove amino acids and soluble proteins from the skin (98). Again, C_{12} showed the strongest effect. However, in another study, a significant dependency with respect to irritation on even or odd numbers of carbon atoms could not be found (116).

Protein extraction from the skin of guinea pigs after washing with various surfactants of C_{12} carbon chain length showed considerable activity for sodium lauryl sulfate, sodium lauryl monoethoxy sulfate and sodium lauroyl isethionate while sodium lauryl triethoxy sulfate and sodium laurate revealed little and laurylmonoethoxylate and lauryltriethoxylate no activity (98).

Further studies, using in vitro and in vivo methodologies for investigation of percutaneous absorption of various surfactants after use related contact times of 15 min, showed no measurable penetration of different anionics up to 24 hr post application but increasing levels of measurable material after 24 hr and more. This indicates a certain association of surfactant with the skin and a penetration after a probable breakdown of barrier function as a consequence of later penetration of compound formerly deposited in the stratum corneum. The penetration rates in this experiment ranked $C_{12} > C_{10} > C_{14} > C_{16} > C_{18}$ (99).

Associated with the effect of adsorption and deposition of anionic surfactants on the skin, even after rinsing, may be the induction of tightness. The clinical knowledge that facial skin frequently answers with tightness after washing with an anionic surfactants, caused a study in which different types of anionic surfactants were investigated for their lipid-removal capacity, adsorption ability, and ability to release amino acids and urocanic acid from the horny layer. Among these factors, the degree of tightness was shown to be related on both the lipid removal and the adsorption ability. The conclusion was that residual surfactant present after removal of the surface lipid was primarily responsible for inducing the sensation. The degree of induced skin tightness ranked sodium lauryl sulfate > sodium laureth-2 sulfate > potassium myristate = TEA-cocoyl glutamate > TEA-lauryl phosphate (105). An additional aspect of this study was the measurement of the specific activity of acid phosphatase in the stratum corneum, which was stated to be very closely related to dryness and roughness.

A further study (110) also used acid phosphatase measurement in order to have a histochemical parameter for the action of surfactant on the stratum corneum. The conditions employed were a mild exposure (hand dipping, 0.2% surfactant solution, 10 min, volunteers) followed by tape stripping of the treated skin and a measurement of the specific activity of enzyme as a marker of skin-surfactant interaction. Compared to the effect of water, various surfactants led to greater reduction of acid phosphatase activity. Furthermore, a relation between the degree of reduction of enzyme activity and the clinically induced skin dryness for different surfactants was obtained (110).

Pathophysiological consequence of penetration of anionic surfactants into the epidermis is the induction of an inflammatory process accompanied by epidermal hyperplasia and increased stratum corneum turnover (102,119). Comparative investigation with different skin irritants [neat dimethylacetamide (DMAC)], 5% hydrochloric acid, 1 and 5% benzalkoniumchloride, and 0.5, 1, and 2% sodium lauryl sulfate) revealed no effect after 24 hr application time for DMAC, HCI, or 1% benzalkoniumchloride, while the 5% benzalkoniumchloride solution produced a 10-fold increase in mitosis at day two, accompanied by intense erythema and blistering. One percent sodium lauryl sulfate produced much less erythema but a greater increase in mitotic activity (25-fold). This shows sodium lauryl sulfate being a compound that interacts and damages cell membranes and, therefore, induces cell division. This effect was clearly concentration-dependent, as 0.5% sodium lauryl sulfate was insufficient to produce a maximal mitotic response while 2% was sufficiently toxic to prevent cells dividing and perhaps destroying them (102).

Histologically, the influence of treatment with alkyl benzene sulfonate and soap using 2% solutions to human skin has been studied by Skog (27). He applied both solutions for 4 min twice daily to a total of 40 applications. This repeated exposure did not give any gross naked-eye effect, but biopsy showed a histological reaction. A marked thickening of the prickle-cell and granular-cell layers was seen and was similar to a greater response noted in the biopsy material from guinea pigs similarly treated. The reaction was related to the concentration applied, and the human skin tolerated higher concentrations than did the rabbit or guinea pig (28).

Another aspect of action of surfactant investigated by the example of sodium lauryl sulfate and alkyldimethylbenzammoniumchloride has been studied by Frosch and Czarnetzki (113). Neutrophil migration, basophil histamine release, and lysosomal enzyme release after treatment with these agents in different dilutions was studied in human neutrophils from different donors. The striking finding was a dose-related chemotactic and chemokinetic response at dilutions from 10^{-3} to 10^{-8} g/ml. This indicates a direct influence of the surfactants used on peripheral leucocytes and elucidates the mechanism of inflammatory response with erythema and infiltration after epicutanous application (113).

Another effect of anionic surfactant interaction with the epidermis may be a lysosome labilizing action, which was described by Imokawa and Mishima (104). According to these authors, epidermal lysosomes, which are responsible for inducing desquamation of the horny layer cells, might be affected by anionic surfactants in the sense of a labilization that was assumed to be related to skin roughness. As a result, this study showed that the labilizing ability of surfactants decreased in the order of cationic, anionic, and nonionic properties, and varied with its alkyl chain length. Furthermore, the study demonstrated that even very little concentration of surfactant (e.g., 10 min treatment with 1% surfactant solution once a day during 5 days) corresponded with in vitro effects seen at 5×10^{-6} or 5×10^{-5}g/ml) (104).

D. Barrier Function, Water Household

Many different attempts have been made to describe the barrier function of the stratum corneum and the influence of anionic surfactants onto it. Already Burckhardt (101) stated that the use of detergents led to removal of a so far hypothetic "regulation factor for the hydration of the stratum corneum," with the consequence that, after prolonged and intense contact, a dry chapped skin and at least an eczemalike dermatitis would arise (101).

While earlier works focused on hygroscopic water-soluble substances within the stratum corneum as an important factor of water-binding ability, assuming these substances as being protected by a semipermeable lipid membrane within the stratum corneum (52,53,54), this hypothetic factor meanwhile has been identified as the above-described variety of lipids in the stratum corneum. From clinical observations in atopic patients it is known that typically the dry and chapped skin is correlated with a disturbed water permeability barrier, a diminished stratum corneum hydration, and an increased susceptibility to irritants (162,163,164). The typical difference between normal skin and this atopic skin seems to be an impaired biosynthesis of ceramides and acylceramides, showing the great importance of these fractions for a normal barrier function (147). But other conditions of the skin that also give rise to imperfect formation of the stratum corneum will show an increased permeability to water, as has been demonstrated for ichthyosis, psoriasis, and fatty acid deficiency (48,49).

Application of surfactants at concentrations above the critical micelle concentration (CMC) may lead to lipid solubilization. By the example of sodium lauryl sulfate and linear alkyl benzene sulfonate at several dilutions above and below the CMC (0.24 and 0.04%) the amount and kind of removed lipids were investigated in dependency of the time. Lipid removal increased as a function of time only between 30 sec and 10 min; no additional lipid was removed by increasing incubation time. No lipids were removed at concentrations below the CMC; above the CMC, cholesterol and free fatty acids were removed by both surfactants but no ceramides or triglycerides. In the second part of the same study, both surfactants were applied in the soap chamber test after Frosch and Kligman (165) to volunteers using concentrations below and above the CMC. At both concentrations sodium lauryl sulfate produced moderate/intense erythema while linear alkyl benzene sulfonate produced moderate/intense erythema at 1%, and mild/moderate erythema at 0.1%. Whenever these two surfactants are similar in structure sodium lauryl sulfate removed 4% of the total stratum corneum lipids and linear alkyl benzene sulfonate removed 7%. As well, the irritation potential of both substances differed slightly. Interestingly, sodium lauryl sulfate as the compound with the more irritating potential, was the surfactant with the higher CMC. As the concentration and penetration of surfactant monomers appears to be the primary factor for irritation and lipid removal only occurs above the CMC, the authors suggested that surfactant-protein interactions rather than surfactant-lipid interactions are involved in the induction of inflammation (146), whereas a minor degree of denaturation and surface lipid disruption might result in roughness and chapping.

Another study also showed that even extensive treatment of the skin with sodium lauryl sulfate led to a small amount of solubilization of stratum corneum-derived lipid, only. However, the lipid composition of the skin after such a treatment was altered—specifically ceramides and cholesterol, which are responsible for the barrier function. These changes correlated with surface roughness and a reduction of hydration (168).

Various chemical methods have been used to demonstrate the presence of amino acids and nitrogen in detergent solutions contacting skin but without naked eye changes being apparent (26,37,38). In one such experiment the claim was made that there was a relationship between the irritant effects of surfactant on skin and the elution of nitrogenous constituents from the skin. Differences similar to those demonstrated by more conventional tests of irritancy were obtained, e.g., the C_{12} sulfate was shown to have a worse effect than other members of a homologous series (39).

For the measurement of hydration of different skin layers, a numerous methods have been used: electrical skin impedance (107,108), electron spectroscopy (111), infra red spectroscopy (134,157), optothermal spectroscopy (167), gas chromatography with mass spectroscopy (GC/MS) (138), and other techniques (145). As a result of surfactant-lipid interaction as well as removal of water-soluble substances of the upper layers (free amino acids, pyrrolidone carbonic acid, urocanic acid, urea, anorganic electrolytes, etc.), a decrease of horny layer bound water can be found. Using high concentrations of different surfactants (5 g/liter) to provoke degreasing and dehydration of the skin changes of the electric skin impedance were measured as an expression of the change of the stratum corneum water content. Sodium dodecyl sulfate, alkyl sulfonate, and a sulfosuccinate produced an 2.5-fold increase in impedance compared to nontreated skin. In contrast, fatty alcohol ether sulfate and an acid amide betaine had a lower activity. This effect was interpreted as a loss of stratum corneum water due to a removal of natural moisturizing factors (108). Interestingly, the effect of pure water as a comparative treatment was nearly equal with an 2-fold increase compared to nontreated skin. This shows the remarkably strong influence of pure water—which may also produce primary hydration of the skin, remove soluble substances and, after penetration into deeper layers, accompanied by a breakdown of the barrier function, may irritate the cells in the deeper layers (108, 144).

Hydration of the corneum has also been shown to increase the rate of penetration of water-soluble chemicals. The more water-soluble esters of salicylic acid in one series studied were found to penetrate hydrated skin more rapidly than dry skin (50). Other workers have shown that ethyl nicotinate absorption is increased after one exposure to water for 30-45 min. The average factor difference was 12 in the areas exposed to hot soaks and 6 in the case of cold soaks (51). In cases of heavy occupational exposure to water it will be expected that the corneum may become hydrated and be increasingly vulnerable to penetration by chemicals used in daily tasks.

Solvents can be shown by in vitro experiments to affect the corneum and make it less likely to hold water, which is necessary to maintain flexibility (52). Temperature has also been shown to be a factor in making the stratum corneum less extensible (53). Surfactants also, when acting for several hours under in vitro conditions, have been shown to have an effect on the stratum corneum (54). Recently performed measurements of frictional properties of the skin related to age, sex, anatomical region, stratum corneum hydration, and transepidermal water loss indicate that frictional properties of the skin are dependent on more than the water content, suggesting sebum secretion as one more possible factor (155).

A further study with sodium lauryl sulfate and sodium laureth sulfate (6.5 EO) confirms the higher skin lipid removal activity of sodium lauryl sulfate compared to an ethoxylated compound and indicated different influences on the lipid composition of the stratum corneum depending on the layer concerned (111). The stratum corneum obviously contains more "true" intercellular lipid within the layers nearer to the surface (which is thought to be highly ordered) while in the deeper layers the content of intercellular lipid substance decreases instead of an increasing number and shape of corneocytes (157). Impairment of surface lipids, only, or "natural moisturizing factors" may lead to minor effects such as roughness and chapping. After more intense or prolonged contact as it is produced by 24-hr patch testing, repeat-insult patch tests, or similar techniques (109,112,115,117,118,120, 139,159,165), a loss of barrier function will happen expressed by increase of transepidermal water loss (TEWL) or increased surface vapor loss (SVL).

With the help of improved technical equipment and standardized measurements many data have been collected that show the influence of anionic surfactant, especially sodium lauryl sulfate, on the barrier function (25,46,47, 132,133,136,139,143,145).

Demonstrating the effectiveness of the technique in vivo, two subjects were examined who soaked their arms for 1 hr in 5% sodium lauryl sulfate daily for 87 days. No gross morphological changes were apparent at this high concentration but the water loss increased some three times to 0.71 and 0.74 mg hr^{-1}/cm^{-2}, respectively, at the end of the 3-month period. This technique can therefore inform the clinician and the researcher of an abnormal stratum corneum in an otherwise apparently normal skin (47).

To find the lowest surfactant concentration inducing barrier damage sodium lauryl sulfate as an anionic surfacant, benzalkonium chloride as a cationic surfactant, sorbitan monolaurate as a nonionic surfactant and cocoamphoglycinate as an amphoteric surfactant were compared at concentrations of 0.05-0.25-0.5%, 0.005-0.025-0.05%, 3-15-30% and 3-15-30%, respectively, in a 24 hr skin chamber test on the forearms of ten male patients. Increased values of TEWL were observed for all compounds tested compared with distilled water. Statistically significant increases in TEWL were observed at: 0.25% sodium lauryl sulfate, 0.05% benzalkonium chloride, 30% sorbitan monolaurate, and 15% cocoamphoglycinate. At these concentrations, only benzalkomium chloride showed visible skin irritation suggesting different events in inducing skin irritation and barrier damage (139).

Repeated low-dose exposures with sodium lauryl sulfate, dodecyl benzene sulfonate, sodium dodecyl sulfate, sodium lauryl ether sulfate, sodium lauryl ether sulfosuccinate, sodium laurate, and some nonionics and amphoterics at concentrations of 12.5mM, applied twice daily for 45 min for 3 weeks caused increases of the TEWL value in time, which was interpreted as an cumulative irritative action on the epidermal barrier. This increase was different in degree for the various solutions. Thus, it was possible to rank these compounds according to their irritational potential: sodium lauryl sulfate > sodium dodecyl sulfate > cocoamphobetaine > dodecyl benzene sulfonate > sodium lauryl ether sulfate > sodium lauryl ether sulfosuccinate > sodium laurate > other nonanionics (117).

E. Irritant Potential, Surface Effects, pH Changes

An irritant may be defined as a substance that, in most people, is capable of producing cell damage if applied for a sufficient time and in sufficient concentration. In a small proportion of individuals there may be a reaction to concentrations too low to affect the majority so exposed. As used by dermatologists in the context of contact dermatitis, the term "irritant dermatitis" has been said to describe the effect caused by a strong irritant on the skin (21). There seems little doubt, from the review of dermatological opinion and observations, that detergents and those containing anionic surfactants are seldom if ever the cause of such irritant dermatitis or of an acute irritant dermatitis (21). On the occasions when they have been linked with dermatitis, it has been of the type classified as a cumulative insult dermatitis which usually develops "after repeated insult by weak irritants over a long period" (22).

Individual tensides do possess the potential to irritate the skin and there is a wealth of literature demonstrating how, by exaggerating the concentration and method of application, the potential can be demonstrated to enable comparisons between individual tensides to be made. Although seldom related to use conditions in the home or industry, this work does illuminate some of the factors surrounding the interaction of detergents with human skin (23-26).

One of the most effective methods to compare the potential of tensides is the patch test technique with all its variations. Used in a standardized way and with an adequate number of subjects to overcome the problems of interindividual variation (35,232), inregional variation (158), different skin type (229,230,232) or other factors (232,235,237), it offers the opportunity to compare a number of different compounds within the same subject at the same moment.

In a series with 16 different anionic surfactants (five alkyl sulfates, seven ether sulfates, two sulfonates, a succinate ester and a protein-fatty acid condensate), Kästner and Frosch (238) treated 20 healthy subjects with a 0.5% aqueous solution. Using the chamber assay techniques according to Frosch and Kligman (239,165), they applied simultaneously all compounds for 24 hr the first day and 6 hr each the following 4 days. The strongest reactions were found for the alkyl sulfates and secondary alkane sulfonate; moderate reactions were observed for alkyl benzene sulfonate, ammonium lauryl ether sulfate and sodium lauryl ether sulfate (2 EO), slight reactions were seen in the other types of ether sulfates, while the sulfosuccinate and a special alkylether sulfate (6 EO) evoked no reactions. These rankings have been repeatedly confirmed by our group (Matthies, 1989, unpublished data), even after one single 24 hr application, only.

A similar test with monoalkyl phosphate analogues of different chain length and several other anionics was performed by Imokawa and Tsutsumi (122). They compared the irritation score with a roughness score and found high irritation and roughness for alkyl sulfate, α-olefin sulfonate, lower irritation but high roughness for alkyl benzene sulfonate and low roughness and low irritation values for C_{12} monoalkyl phosphate and C_{12} monoalkylphosphate TEA. This illustrates the role of the type of derivative and terminal group, respectively, for their potential to irritate the skin. The role of ethoxylation and use of different salts is shown for different in vivo and in vitro methods by Sterzel et al. (254). The influence of the vehicle was demonstrated by Bergstrasser and Eaglestein (161). Occlusive patches containing a hydrophilic ointment with 1% sodium lauryl sulfate (SLS) and a cream formula with the same content of SLS for 16 hr produced irritation in all subjects within 3 to 5 days of the treatment. Once stimulated, the irritation progressed despite cessation of the treatment and evolved into painful fissures and lamellar exfoliation of scale, which is not typical for treatment with aqueous SLS alone.

Skog (27) applied alkyl benzene sulfonate and soap to human skin using 2% solutions applied for 4 min twice daily to a total of 40 applications. This repeated exposure did not give any gross naked-eye effect but biopsy showed a histological reaction. A marked thickening of the prickle-cell and granular-cell layers was seen and was similar to a greater response noted in the biopsy material from guinea pigs similarly treated. The reaction was related to the concentration applied and the human skin tolerated higher concentrations than did the rabbit or guinea pig (28).

An important series of observations that came closer to use conditions and that also employed individual tensides was reported by Smeenk (32), who obtained appropriate anionic detergents from industry and was able to comment directly on the effect of the surfactant. By using a closed patch test to exaggerate exposure and a 1% solution of the surfactant, he and his colleagues were able to rank the detergent actives in terms of irritant response of skin. When using an arm immersion test similar to the Suskind one, they also used exaggerated levels of detergent and by so doing produced evidence of irritant response. By reducing the concentration they also demonstrated loss of this effect and also a residual dry or scaling skin which could be produced by sodium lauryl sulfate. The effects they achieved at the low-level arm immersion concentration were such that they could not classify them on one scale of response; therefore, the effects were very minimal but certainly demonstrated that there were differences between the various actives examined in terms of a direct irritant effect on skin. Smeenk also concluded that usage tests or immersion tests are to be preferred for testing the irritancy of detergents on human skin. This is in agreement with industrial practice (23,33).

The influence of pH value on surfactant-induced skin irritation was investigated by Antoine et al. (240) with SLS applied at three different pH values (pH 5, 7, 9). No significant change of functional property of the skin could be objectified by measurement of the sebaceous secretion, electrical conductance, transepidermal water loss, and cutaneous blood flow due to the increasing pH of the detergent (240). This corresponds with a finding of Murahata et al. (241) who studied the relationship between alteration of skin-surface pH and the development of skin irritation within a range of pH 4.0-10.5. His results indicated that pH is not a primary factor in determining the irritation potential for surfactant solutions.

More use-related tests with alkaline soaps and sour syndets, which were performed as arm immersion test or use tests, also revealed no or only little effect of the pH value (140,145,242) and on the skin compatibility. In infants, comparative tests with acid syndets or alkaline soap showed no influence on the pH value of the skin of older infants. In younger infants the syndet displaced the pH for a short time towards acid pH values. In the long run, washing with syndet kept the skin pH for a longer time within the physiological range than washing with ordinary soap (243). According to Hartmann (247), in adults the skin flora represents a very stable system that reconstitutes quickly even after daily rinsing or bathing procedures. Thus, the pH of a surfactant plays only a minor role in eliciting an irritative reaction and the normal use of anionic baths does not impair the surface properties significantly, even in neonates.

A further study in babies that were bathed in an anionic and amphoteric bath additive or with an ordinary baby soap focused on changes of the skin flora of the neonates. The yield of *Staph. aureus, Staph. epidermidis, Pseudomonas spp., Candida spp.*, and coliforms from skin swabs was compared in two groups of 22 babies each. No differences were detected in the isolation rate of these organisms between the two groups (246).

Though one may demonstrate differences in certain effects of soap and synthetic detergents on skin, they are overwhelmingly similar in the instance where the same alkyl radical is involved, e.g., comparing the soap, sodium laurate, or the synthetic detergent, sodium lauryl sulfate, with other soaps or synthetics, both can be shown to be significantly more active. It is not surprising, therefore, that a 5% solution of sodium lauryl sulfate, which is equivalent to over 100 times the use concentration, has been used regularly to inflame human skin during sensitization-testing regimes (41). It is also reasonable to suppose that continued exposure to 100% detergent could give rise to skin conditions. This is exemplified by a report that two housewives suffered dried and fissured pulpitis of the second and third phalanges of the finger of the right hand as the result of excessive exposure to powdered detergent. They used the fingers to remove excess washing powder collected on the washing machine either from the container holding the powder or from the ledge which secured the washing machine lid (42).

It would appear from all the published work using many different methods for examining effects of detergents on skin that there is a potential for a variety of effects, many of which may have a direct bearing on the probability of irritation being produced, but all are in direct relation to concentration and the time of exposure.

Cracking and scaling of the corneum occurring during the winter months independent of any applied chemicals has been well recognized in man and is commonly referred to as skin chapping. One of the pioneering studies of human corneum and its response to climate was that by Gaul and Underwood, who found that a low dewpoint led to a greater incidence of chapping. The low dewpoint in their study was accompanied by a low air temperature (44). In effect, a very cold dry air alone could be expected to cause chapping.

III. IRRITANT CONTACT DERMATITIS AND ANIONICS: ETHIOLOGY, CLINICAL INVESTIGATIONS

Eczematous dermatitis of the hands ascribed to soaps and synthetic detergents used has long been studied. Since most soaps and detergents were used in the household and the people concerned were mainly housewives, this type of disease was named housewives' eczema. When we read in a treatise on diseases of the skin in 1921 (3) that in housewives' dermatitis there is exposure to wet, heat, soap, soap powders, polishing and scouring materials, and handling and cleaning of vegetables and meats, this can be taken as some evidence of circumstances surrounding the onset of the condition. The same author indicates that the condition may vary in degree and extent from erythematous, vesicopapular, and vesicular character to a markedly inflammatory oozing and crusted eczema rubrum. More recently, in 1973 (4), housewives' eczema was described as occuring principally on the fingers, webs, and dorsa of the hands, beginning with mild dryness, redness, and scaling and proceeding to fissuring, crusting, and eventually chronic eczema. "Such eczema may be intensified by common household irritants such as bleaches, waxes, polishes and turpentine (4)." Excessive exposure to soap and detergents is part of the background quoted. It will be appreciated, therefore, that the condition has probably not changed much over years but synthetic detergents were not invented at the time of its earlier description. However, the anionic detergent soap was used at that time and no doubt might well have been considered a principal agent.

It is also important to appreciate that ethiological factors in hand dermatitis are complex and it is often difficult to evaluate their relevant importance. When the skin is subjected over a long period to several different causes of injury, there is a greater possibility that the total effect of the injuries will be sufficient to provoke irritant dermatitis (214). Such dermatitis frequently starts under finger rings, where soap, detergents, chemicals, dirt, and bacteria easily congregate in the damp environment (5). Thus, we can see that there has been a hand condition present over the years, during which time a similar variety of materials have been used by the housewife, but dermatologists have appreciated the difficulties in diagnosis.

One of the first dermatologists to examine, in a very precise and careful way, the circumstances surrounding the use of detergents was Suskind (6). His main purpose was to examine the effect of detergents but, by virtue of the method employed, he separated very clearly the action of the detergents from the possible effects of the environment in which they were used. This work has, of course, been referred to frequently in subsequent dermatological literature but it is worth considering in detail, first for the reasons the test is being carried out, and second, for the results and their significance for a general consideration of tensides under the terms of this review.

Suskind drew attention to the fact that, prior to the "detergent era" and more recently as well, so-called soap dermatitis was a common "inspection" diagnosis for a variety of eczematous dermatoses of the hands. In the tests he reported, the patients, who were predominantly those with housewives' dermatitis, were removed from their work environment and then exposed exclusively and repeatedly to solutions of washing products, both soap, and an all-purpose synthetic detergent with a pH of 10. The exposure was not entirely passive as the subject squeezed a small sponge rubber ball during exposure. Of some significance was the fact that at the end of each immersion period the hands were not rinsed but were patted dry with a clean cotton towel. The concentration of the product was 0.5% and the duration of exposure 30 min on two occasions daily or 10 min on six occasions daily, using a temperature of immersion of 105° F. The condition of their hands was observed both by the naked eye and by using a stereoscopic microscope and numerical scoring.

The authors had expected that a significant proportion of persons with hand dermatitis would be made worse and were therefore surprised to observe the contrary. They concluded from these observations that soap and detergent immersion alone might not be a significant primary provoking stimulus or aggravating factor in housewives' dermatitis. The conclusions from this moderately large sample of patients taken into the hospital confirmed the earlier results on a much smaller sample reported by Jambor (7). The relevance of the levels of exposure used in the hospital situation were examined in the light of a survey to establish the conditions under which 267 Cincinnati housewives used cleaning agents. It was reported that the total duration of exposure of their hands each week was 131 min for dishwashing,

29 min for laundry, and 32 min for miscellaneous washings, with the collective figure of 192 min each week. These were less than the exposures used in the hospital experiment (8).

Thus, if detergents alone are not the sole agents, there must be other common factors in the conditions surrounding the use of detergents. Certainly, all uses must involve water, and water alone has been shown to have some effect on the corneum. There will also be the conditions of ambient temperature and humidity, these being factors that have also been shown to affect the corneum. The housewife or industrial worker will not always use the recommended concentration and this could be critical. Occasions when these various factors all combine and interact adversely toward the skin could explain some of the few cases reported by dermatologists and others as being due to detergents. These have usually been related to the use of a new product but have not increased in numbers throughout the general use of the product even when no subsequent change in formulation has taken place (9).

When housewives' dermatitis in the Stockholm area was investigated, with particular attention being paid to the part that might be played by detergents, results somewhat similar to those seen by Suskind were obtained. Blohm found in his series of some 200 patients that the dosage instructions given by the manufacturer were followed by only 25% of respondents, whereas 75% of the patients and 80% of the controls used guesswork. Analysis of the wash liquor received from the homes of the participants showed concentrations up to 10 times greater than recommended. The authors strongly advocated an educational program stressing that people should not exceed the recommended dose when using soap and detergent preparations, particularly as the recommended level was for optimum value and no improvement would be achieved by simply increasing the concentration. They also considered that the findings reported by Suskind were supported by the practice at Karolinska Sjukhuset, where therapeutic baths containing liquor or carbonis detergents and salts were given for periods of 15 min to control a variety of skin conditions (10). This is also in agreement with further work on the pH of household detergents which concluded that repeated immersions in buffered alkaline phosphate solutions of concentrations and pH equivalent to alkaline or "built" household detergents in most instances did not appear to adversely affect persons with dermatoses of the hands. It was further concluded that the alkalinity of common laundering and dishwashing solutions is not a provoking factor in any role played by tensides or soap products in housewives' dermatitis (11).

Observations similar to those of Blohm were made by Lyon, who found indications of overdosage in a study of dermatitis and detergents. His figures suggested a possible excessive use of some 10-20 times recommended levels of use (12).

Another comment on contact dermatitis in women in Sweden comes from Fregert, who considered that in women the occupations that dominate are those that cause irritant contact dermatitis rather than allergic. The onset of hand exzema occurs at an earlier age in women than it does in men. It is also probable that working at home with water may contribute to earlier occurence. In a series of 1752 patients they found that 50% of the women were engaged in cleaning work or nursing, with frequent contact with water. He also comments that detergents and washing-up liquids in concentrations prescribed by the manufacturer have a fairly mild effect on normal skin. Most often, however, such substances are used in too high a concentration and can then cause considerable damage (5,19).

In 1961, an examination of the extent of housewives' dermatitis in Reading and the nearby area of Berkshire described its effects and suggested methods for its prevention (14). The conclusion of this study of 1438 case records was that the incidence of synthetic detergent dermatitis was not nearly as high as the public thought. It did appear, however, as a definite entity although its existence or the existence of a similar disease long preceded the introduction of modern synthetic detergents. This author came to the similar conclusion as mentioned earlier, viz. greater attention to instructions of dosage, careful rinsing of hands, and avoidance of too-frequent or too-long exposure to detergent solutions. He had found that the average concentration of detergent used was three times as high as is necessary for washing dishes and six times as high for washing clothes. Another author reported that, contrary to popular belief, the incidence of dermatitis had not increased since detergents had replaced soap in the washtub and remained at about 1.5% of all new cases seen in hospital outpatient departments. With other authors he emphasized that a careful history and examination of the whole skin were essential to exclude constitutional eczemas and to eliminate a host of other irritants to which the housewife is exposed. "She will often blame the detergent because she is unaware that abrasives, disinfectants, fruit and vegetable uses may also irritate her skin" (15).

In Western Europe during the last decades, the market of washing machines was fully settled in that nearly every household possessed its own machine, and one important potential factor for hand dermatitis emerged. The remaining contact with detergents as residue on the clothes does not represent a significant risk for irritation or sensitization. Volunteers of the seborrhoic and sebostatic type as well as babies and infants were exposed to textiles that contained measurable amounts of anionic surfactants and soap after washing in normal household machines. No skin reaction was observed after wearing the clothes nor after 48-hr patch testing (253).

When dealing with the question of irritants on human skin in America, Kligman has also pointed out that hand exzema in housewives was once attributed to soap but in the previous 30 years synthetic detergents had been blamed. Despite this, however, the author and his collaborator did not believe that the increase in the use of various washing machines had greatly reduced the amount of hand exzema (16). It is frequently stated that hand dermatitis is an occupational dermatitis of the housewife and that soaps, detergents, and the macerating effect of water are the direct cause. However, Hollander believed that these were by no means the only underlying factors. He considered that any misunderstand of this and any underestimation of the indirect or contributary causes could unduly prolong the duration of the dermatitis. Some of the contributory causes he lists included excessive or lengthy maceration in water; exposure to cold and wind; degreasing with fat solvents such as turpentine, paint, and stain removers; and harsh scrubbing with hand brushes (17).

Similarly, after a scholarly appraisal of the possibilities of soap or synthetic detergents affecting the skin, Blank (18) concluded that skin damage from the use of cleansers had probably been overemphasized. He suggested a careful double check of history of occurrence and elimination the possibility of allergy or photosensitization as being essential for assessing implication. Where there was implication, adequate therapy would probably be to reduce the extent of the contact with the cleanser rather than seeking a replacement. A year later, in England, Rowell (19) when writing on contact dermatitis from various causes considered that despite their widespread use dermatitis of primary irritant type directly resulting from soap and detergents was relatively rare. He took the view that, when the skin is eczematous from any cause, it will be irritated by soap and water cleansing. This was in contradistinction to the view held by Stoughton that the use of soap for routine handwashing and bathing did not influence the course of exzematous disease while patients were on a standard therapeutic regimen (20).

The same obervation of a beneficial use of syndets containing anionic surfactants was reported by Raab (213), who observed 20 healthy volunteers with sensitive skin and 30 dermatological patients who were given these products as skin cleansers. The group of dermatological patients consisted of the following diagnoses: acne vulgaris (7x), balanitis (4x), various eczemata (19x). All lesions could be cleaned with the products without any sign of irritation or impairment by the treatment. All patients showed a high degree of compliance and were content. Even in balanitis, an excellent compatibility was found.

Meanwhile interest has increased in the differential diagnosis of contact dermatitis, and data have been collected from different authors dealing with the occupational aspect of dermatitis and have general aspects about the number and distribution of dermatitic reactions due to the use of household products, cosmetics, and toiletries (175-179,184,185,206,215-220). In the above-mentioned 10-year study with 1752 patients considered to have occupational dermatoses, Fregert (215) divided them into allergic and nonallergic-type dermatitis. Housewives' dermatitis was excluded is this study. Nevertheless, contact dermatitis was localized on the hands in 94% of women and 84% of men, three quarters of an allergic type in men, and 50% in women. The most common allergens were chromium, rubber and plastic in men, and nickel, rubber, and chromium in women. Among women, 55% of the dermatitis cases were attributed to hospital and cleaning work, while among men, metal, concrete, and building work accounted for 52% of contact dermatitis cases.

In an investigation among 541 members of a hospital cleaning staff a prevalence rate of 15.3% of skin diseases was found. Seventy-five percent of the occupational diseases were irritant dermatitis, 21% allergic contact dermatitis, and 4% monilia of the finger webs. The most common causes were found to be alkaline substances, acids, and sodium perborate as well as hypochlorite and hypobromide besides surfactants (206). A recent study with 1,385 subjects, selected out of 20,000 individuals who had received a questionaire and considered themselves having hand eczema, revealed in total a prevalence of hand eczema of 11.8%. The most frequent exposures were to water, detergent, dust, and dry dirt. Other relevant factors were: coins, foodstuffs, oils, chemicals, solvents, plants, soil, paints, glues, and cement (218). Combined with the preceding study the subjects were also investigated for the consequences of having hand eczema and for possible predictive factors to develop hand dermatitis. Comparing the different types of hand eczema allergic dermatitis seemed throughout to cause more serious consequences. The most predictive factor for developing hand eczema was a history of childhood eczema, second was female sex, followed by occupational exposure (219,220).

Summarizing these data concerning the use of anionic surfactants as part of household products or as occupational material, the dependency between exposure factors such as concentration, time, humidity, temperature, occupation, host factors like anatomical region, age, sex, genetic background, and, on the other hand, the expression of dermatitis of exposed skin becomes obvious. It is clear that females are more often concerned with housewives' dermatitis, wet labor in cleaning professions, or contacting detergents and disinfectants in the hospital cleaning staff. On the other hand, the high incidence of cement as an irritant agent in males also shows the coincidence of wetness, mechanical factors and the chemical itself.

Since the absolute number of cases clearly and exclusively related to contact with anionic surfactant is minimal, it can be summarized that anionic surfactants are safe if properly used. In fact, it has been pointed out that virutally any substance can be an irritant under certain conditions (223), but as the above mentioned example shows, even present dermatoses can be treated with a surfactant without any harm to the skin if used correctly (213).

Furthermore, many attempts have been made to optimize the skin compatibility of anionic surfactant-containing products. Besides a careful selection of a suitable derivative primarily blending of anionics with anionics, nonionics, amphoterics, and protein hydrolysates has led to a variety of options in order to reduce the irritation potential of the most frequently used standard anionics without loss of performance (248-252,254).

IV. ALLERGIC CONTACT DERMATITIS AND ANIONICS

One of the earliest suggestions that an anionic detergent could cause allergic contact dermatitis came from Sweden in 1954, with the report of a serious outbreak of dermatitis that was considered to be due to a particular detergent powder (66). The product contained 6% lauryl sulfate and 20% alkylaryl sulfonate and the author reported that the lauryl sulfate used in the preparation had allergic and primary toxic properties. With continued widespread use of sodium lauryl sulfate, it is clear that the anionic detergent itself, if manufactured correctly, is not a sensitizing chemical.

Some 15 years later another outbreak of dermatitis in Scandinavia was related first to use of a particular detergent (69) and then also was subsequently shown to have been caused by impurities produced during incorrect manufacture of the chemical (70). During the two years required to investigate the reason for the outbreak, the same chemical produced by other manufacturers was satisfactorily marketed in Scandinavia, demonstrating very clearly that the problem had been one of manufacturing impurity.

Additionally, performed tests in 1973-1975 with the identical batches of Laurylethersulfates (known as LES 13-2035, which had been in storage for about 6 years) in comparison with 17 different batches of LES and LES-containing products showed only one positive reaction with LES 13-2035, while with the other 17 samples tested no positive reactions were observed (169,170). Further investigation (171) revealed two components, characterized as 1-dodecene-1,3-sultone and 1-tetradodecene-1,3-sultone, as two of the sensitizers. Another two components, 2-chloro-1,3-dodecene sultone and 2-chloro-1,3-tetradodecene sultone were tentatively identified (70,173).

Reviewing the data of this case, Lindup and Nowell (174) summarized that the evidence was fairly convincing, that unsaturated sultone and chlorosultone contaminants were the cause of the allergic dermatitis provoked by LES 13-2035.

A similar case was reported by Sylvest et al. who examined a group of patients who reacted against an sodium lauryl ether sulfate-containing product. In contrast to these positive responses, no reaction was obtained using purified sodium lauryl ether sulfate as single component. Thus, this case was clearly attributable to impurities of the sodium lauryl ether sulfate compound (170). Since then no case with clear evidence of a causative role of an alkylethoxysulfate as such as an allergen has been reported.

For sodium lauryl sulfate (SLS) the situation is very similar. In testing for allergy, great emphasis is laid on the patch test and, it has to be appreciated, that patch testing has gone through quite considerable changes during the last two decades. Much of the early confusion was caused by irritant reactions giving false positives. Fisher reported that even a concentration of 1% of sodium lauryl sulfate acted as a mild primary irritant and may, like most soap solutions, produce erythema under a closed patch test. He regarded numerous positive reactions in his studies to be nonallergic irritant reactions (72). When detergents have been examined directly, negative results are usually obtained, but patch testing can give odd responses.

Pluss in Switzerland examined 93 patients and concluded that negative results with the synthetic detergents might be due to their neutral or almost neutral reaction and that they did not appear to be sensitizing in nature (73). It was somewhat surprising in this study to note that, although alkyl sulfates, alkyl sulfonates, alkylaryl sulfonate, and a fatty alcohol sulfonate were examined at concentrations up to 30%, positive reactions were only obtained in 7 of the 93 patients. Perhaps it was not surprising that 3 of 30 control subjects gave positive reactions at the 30% concentration. The observation was also made that when soap tests were carried out, 44 of the 93 patients gave positive results. It has, of course, long been known that under an occlusive path test soap of almost any kind will give a strong reaction, particularly at concentrations above 8%. Indeed, the term "soap effect" has been given to the wrinkled, slightly swollen skin produced after patch testing with quite low levels of soap (74).

Our own experience with LES and SLS as positive controls in standard patch tests for irritancy in thousands of single applications is that even a 0.1% solution of both substances may act as a weak irritant in very sensitive but healthy volunteers. Solutions of 1% SLS and 2% LES, respectively, provoke nearly in all cases slight to moderate erythema and/or edema. In this light a recent report of a single case of allergy against an cocobetaine and sodium lauryl ether sulfate (both tested in 2% aq. sol.) stays doubtful even with respect to the authors' tentative link with impurities (175).

Besides an inadequate concentration used for the diagnostic patch test, another cause for wrong positive patch test results may be the fact that many of the surfactants used in consumer products are already preserved as raw material so that test probes may also contain undeclared traces of common preservatives or unexpected cleavage products (255). But other origin of false results also has to be taken into consideration as contributor to misinterpretation in some cases (187-192,225,256).

Since the pioneering work of the International Contact Dermatitis Group in the use of the patch test in a standardized manner and with appropriate concentrations of materials, contact dermatitis in the housewife has frequently been investigated using patch technique as a means of determining presence or absence of allergy. Over wide-ranging investigations in various countries, the principal anionic detergents such as the alkyl benzene sulfonate, linear alkyl benzene sulfonate, and sodium lauryl sulfate have not been implicated as allergens. On two occasions, one the previously mentioned Scandinavian investigation (69) and one in Belgium (75), patch testing was able to determine impurities as causative agents in detergents. It is by the experience of such extensive testing that members of the group are aware that synthetic detergents as such have not given rise to an allergy, but are rather to be considered as cumulative irritants (76).

Indeed, it has been shown in experiments in man that pretreatment of the skin with 5% sodium lauryl sulfate for 24 hr or combined exposure of sodium lauryl sulfate and allergen is one of the most effective ways of inflaming skin in order to promote sensitization. During the conduct of this type of investigation, there has been no indication that the sodium lauryl sulfate itself is an allergen (67). This particular procedure has also been used extensively by Magnusson in his series of maximization tests and, again, has never given a positive result (68).

During this predictive testing of synthetic detergents prior to use in the market place there have been many tests carried out in a standard fashion, none of which have given evidence of sensitization from either alkyl benzene sulfonate or from sodium lauryl sulfate (33,79,80).

In early work on household soaps and synthetic detergents, American authors examined the possibility of their acting as sensitizers. One study comprised of 57 cases of dermatitis of the hands that had been attributed by the patients to household washing products. When examined critically as to the ethiology, the authors were able to determine the principal cause of the dermatitis in 30% of the cases. These included allergy to chromate, nickel, lathe-cutting oils, FD&C dyes, and lanolin. In no instance did any of the cases react significantly by patch or usage exposure to commercially available soaps and synthetic detergents (71).

Further comments on synthetic detergents or soaps and their possible role in allergy show that in general the frequency of adverse reactions after application of detergents is between 0 and 12.6% (176-182). Soaps and shampoos play a prominent role in eliciting cosmetic intolerance reactions (183-185). Nevertheless, all authors agree that dermatitis after contact with detergents is clearly due to irritation, but allergy related to synthetic detergents is an extremely rare event which is only to expect, when other components like preservatives or impurities are contained (186). The influence of other components in products and the influence of environmental factors (e.g., wet work, temperature) is illustrated as followed.

For example, when dermatitis that is associated with the use of detergent products occurs, it is now modern practice to subject ingredients to test against the affected individual. One such instance comes from Denmark and showed that sodium lauryl sulfate and sodium dodecyl benzene sulfate were not involved in a contact dermatitis affecting a particular hospital population. In this case the problem was caused by a perfume ingredient which gave only positive responses when heated to a temperature comparable to that used for cleaning. Under the same conditions of test, 20% solutions of the detergents were negative (78).

The value of participating in the diagnostic task of the dermatologist can be inferred from the results of national surveys. In a series of 4000 patients patch-tested in five European clinics, 58% had a dermatitis involving hands, 281 women with contact dermatitis of the hands were engaged in domestic work only, but it was important to note that half of these women had positive patch tests to allergens contacted in their home environment. In this series 42% cases considered as allergic reacted to allergens such as balsams, nickel, medicaments, and rubber. An important additional factor was considered to be the wet nature of the work, which would lead to hydration of the corneum (13,81). Without the use of such tests it would not have been possible to say more than that these women had a hand dermatitis of doubtful origin.

Therefore, it is the task of the future to intensify the diagnostic efforts toward dermatitis in order to get a more secure assessment of the potential of a substance to induce sensitization in man and to check old data for so far "known" allergens. From our view today, anionic surfactants belong to the group of substances that do not hold a significant allergenic potential.

V. EFFECTS OF ANIONICS ON MUCOUS MEMBRANES

When one considers the possibility of irritation on mucous membranes, one might expect to find considerable evidence of harmful effect, as there is no protective series of layers of corneal squames on top of mucous membrane nor are they protected, except by situation, from excessive loss of moisture.

Mainly three anatomically different organs possess mucous membranes that are likely to come in contact with anionic surfactants: the eye, the mucous membranes of mouth and throat, and the urogenital mucous membranes including the special epithelium of vulva and vagina.

The vast majority of detergents will be used for household tasks such as fabric washing or dishwashing, where the contact with the skin is incidental to the task being performed. Detergents used in shampoos and bubble baths, however, are expected to contact skin and may contact mucous membranes.

Therefore, many information about irritant action on mucous membranes of the eye have been collected over the years by the use of the Draize test in rabbits (193,212). The knowledge that different chain length and different number of double bonds influence the compatibility and that addition of nonionic surfactants or other components like sugars or proteins (29,45, 196) may help to improve the compatibility to eye membranes has already led to fairly well-tolerable formulations in general.

Some former cases have been reported suggesting effects relating to the use of bubble bath products. Before looking for evidence of the extent of the problem as a feature of the action of anionic detergents, it will be useful first of all to consider the medical background to this particular type of effect.

In discussing the subject of genital hygiene, an American woman doctor has drawn attention to the historical point that Louis XIV was credited with some responsibility in this matter, having desired to reduce odors among the women of his court. His court physicians recommended the vaginal douche and then the bidet was invented. From this time, washing outside the genital area was considered a good practice. The recommendation of the doctor was that a thorough cleansing of the genital area in a tub or under a shower, or sponging, was a cleansing method much more healthful than regular vaginal douching (56). This recommendation has to be balanced against an earlier report (57) in 1955 that, in a series of 280 consecutive cases of acute vulval vaginitis, 41.4% were associated with a history of having used soap products in the form of cake or liquid or in douches, bubble baths, or bath salts prior to the onset of symptoms. The author pointed out that the relatively non-specific clinical signs could also be caused by trichomonas, candida, meno-pausal and surgical estrogen deprivation, but that cases so caused had been excluded from the 41.4% cases studied. In more recent years (58), attention has been drawn to "obsessive washing with soap and water" related, along with many other causes, to inflammatory reactions in the genital area. Adults are quoted who have been taking bubble baths several times a week or pro-longed and highly perfumed bubble baths several times a week. In an earlier publication the author refers to five cases in which people had several baths per week and this he regarded as being causally related to cystitis present at the time (59). Another author recommended that, in cases presenting with symptoms of irritation of the lower urinary tract, physicians should ascer-tain if soap or detergents had been added to the bath water. He pointed out that boys might also be affected and quoted cases where bathing took place every 1 or 2 days (60). Four cases among children between 4 and 10 years of age, all with varying degrees of genital problems, have also been related to the use of detergents or perfumed soap (61).

Particularly, interest may be applied to the so-called fun products, but even here the extent of the problem does not appear to be great and appropriate warnings have been suggested as a possible way of avoiding misuse. Some perspective may be gained by a recent analysis made of bubble bath product complaints received by the Division of Cosmetics Technology, Bureau of Foods, United States (65). This indicated that only 165 complaints were received in the 7-year period 1967-1974, and that, of those relatively few complaints, 56% were against one particular product. Of 20 other products on the market, none had more than 3% of the total complaints. The majority of complaints were directed against children's bubble baths and principally affected the female child. This would suggest that there is no great problem from this use of detergents as a class, but that attention to formulation is important. Of the 165 cases, only 107 were seen by a doctor and those seen by the doctor were reviewed by an investigating dermatologist. Sixty-two cases seen by a doctor were said to be caused by the product. The absolute numbers involved were 89 female children and 31 female adults and, among the female children, only 16 out of the 89 had symptoms that could be related to genital disorders.

In a recently performed experimental study, the irritant response of vulvar skin to 3% solution of sodium lauryl sulfate was quantified by measuring the transepidermal water loss (TEWL), capitance, pH, blood flow, and color reflectance. The aqueous solution of sodium lauryl sulfate was applied to filter paper disks in Large Finn Chambers for a minimum of 24 hr. Patches with distilled water and an additional application on the forearm skin served as controls.

Surprisingly, only on the forearm skin, significant changes of all parameters were observed. In contrast, the vulvar skin showed little effect. In particular, neither the TEWL as an indicator of barrier function, the capitance as indicator of moisture nor the pH changed. The slightly elevated blood flow decreased again, rapidly. Thus, an irritating effect on the vulvar skin could not be assured (194). The authors summarized that the techniques were less suitable to quantify irritant dermatitis in the vulva than in the forearm; it seems necessary to add that obviously the reaction to 3% sodium lauryl sulfate on vulvar skin was less impressive or, as the visual scoring showed a much quicker decrease of the observed effects, the vulvar skin was able to a much larger extent to compensate the above mentioned insult.

Another special field is the application of toothpastes and mouth rinses containing anionic surfactants. Sodium lauryl sulfate is one of the most widely used synthetic detergents in dentifrices. Due to its high affinity for protein molecules and its foaming activity, sodium lauryl sulfate was thought to be useful for a proper removal of mucoproteids from the tooth surface. The benefit of sodium lauryl sulfate is clearly demonstrated as an additive, cleansing even interdental spaces that stay out of reach in normal tooth brushing. This was proved without any effect of impairing the mucous membranes of the mouth (195-197).

Other authors described the influence of anionic surfactants on the membrane surfaces as slightly irritating (198), expressed by an increase of the number of stipples of the gingiva. As stippling of the gingiva is an interindividually differing quality that may be due to a functional adaptation to mechanical stress (199,200), this finding is not necessarily being interpreted as an critical side effect.

More recent studies with toothpaste formulations with a low (<2%) or a high (about 8%) content of anionic surfactant revealed divergent results. Albers et al. (201) found that the use of toothpaste with a high content of anionic surfactant resulted in a significant increase of the sulcus bleeding index after 6 week use.

Morphological changes in the stratum spinosum were identified as an increase of filament bundles in the cytoplasm, which could be interpreted as stabilization of the cell structure by forming structural elements.

The contradiction between the clinical finding of bleeding and the electron microscopic findings were explained by the assumption "that the great increase of filament bundles leads to a lowering of the adaptability of the epithelium and consequently to increased tendency to bleed, especially in response to mechanical irritation (201).

In contrast, Krekeler and Klaiber (202) found in a three-group parallel study in volunteers, in which they used two toothpastes with a low content of anionic surfactant (1.8% and 1.9%, resp.) and one toothpaste with a high content of anionic surfactant (about 8%) that a slight increase of bleeding occured in all three groups and, to the same extent, there was a parallel increase of the plaque index. They discussed that correct tooth cleansing led to reduction of gingival inflammation without any sign of irritation as long as the actual concentration of surfactant in the saliva fluid was lower than 1% (202). This is usually the case even if a toothpaste with a high content of anionic surfactant is used.

Examples of indices of a more indirect action of surfactants on mucous membranes were given by Siegel and Gordon (203) and earlier workers [e.g., (204,205)]. In their studies, surfactants were shown to increase the permeability of the mucosa to oil-soluble compounds and small and also larger water-soluble compounds. This effect was concentration dependent. Histologically, widening of the stratum corneum was observed due to separation of layers and loss of surface layers. This finding corresponds to similar effects described for action of anionic surfactants on horny layers of normal skin (see Sec. III). Nevertheless, the relevance of those effects is not quite clear as long as clinical tests do not prove any adverse effect under conditions of intended use (202).

Very particular action of anionic surfactants as being the reason for maintainance of a food allergy, as reported recently (208), is still at the level of hypothesis, for this suggestion was made after observance of five patients with unspecific symptoms, only. Thus far, the use of anionic surfactants in toothpastes and mouth rinses stays useful and generally accepted without indicating any relevant toxicological risk for the daily user.

In summary the contact with anionic surfactants and products containing synthetic detergents shows very little problems as long as the intended use conditions are respected. Even if contact occurs involuntarily by use of shampoos or bubble bath products the incidence of serious adverse effects is very low (177-179,210,211). Ingestions and/or misuse of products resulting in closer contact with concentrated detergents to mucous membranes will lead to irritation of different degrees depending on the time of contact, the actually active concentration of surfactant, and the first aid step that was taken.

Nearly always, the first impetus is rinsing with water, which almost guarantees a quick and unproblematic recurrence of the effects. Thus, serious adverse reactions to the use of anionic surfactants are very uncommon. The only serious side effects that have been reported resulted from ingestions by children or babies who developed a bronchitis and pneumonitis due to aspiration of the product (210).

REFERENCES

1. I. Treuhoff, *Abh. Deut. Akad. Wiss. Boling Kl. Chem., 6,* 367-374(1966).
2. H.E. Garrett, *Trans. St. Johns Hosp. Dermatol. Soc., 51,* 36(1965).
3. H.W. Stelwagon and H.K. Gaskill, in *A Treatise on Diseases of the Skin for Advanced Students and Practitioners*, 9th, W.B. Saunders, 1921, p. 453.
4. A.A. Fisher, in *Contact Dermatitis*, 2nd, Lea & Febiger, 1973, p. 73.
5. S. Fregert, in *Manual of Contact Dermatitis*, Munksgaard, 1974, pp. 29-38.
6. R.R. Suskind, M.M. Meister, S.R. Scheen, and D.G.A. Rebello, *Arch. Dermatol., 88*(2), 117(1963).
7. J.J. Jambor, *J. Invest. Dermatol., 24,* 387-396(1955).
8. R.R. Suskind and H.S. Whitehouse, *Arch. Dermatol., 88* (2), 53-60(1963).
9. Leading article, *Brit. Med. J., 518* (Feb. 1970).
10. S.G. Blohm and A. Lodin, *Acta Dermatol. Venereol., 48,* 7(1968).
11. R.R. Suskind and D.J.A. Rebello, *Arch. Dermatol., 88* (2), 125-129(1963).
12. J.B. Lyon, *Practitioner, 186,* 485-492(1961).
13. S. Fregert, *Contact Dermatitis, 1*(2), 104(1975).
14. H.S. Samuel, *Practitioner, 186,* 485(1961).
15. E. Waddington, *Practitioner, 184,* 532(1960).
16. A.M. Kligman and W.M. Wooding, *J. Invest. Dermatol., 49,* 927(1957).
17. E. Götte, *Aesthetische Medizin 12* (5), 146-154(1963).
18. I.H. Blank, *Practitioner, 202,* 147(1969).
19. N.R. Rowell, *Practitioner, 204,* 678(1970).
20. R.B. Stoughton, L.W. Potts, W. Glendenning, S. Fisher, and M. Kress, *J. Amer. Med. Ass.,* 1196(1960).

21. D.S. Wilkinson, S. Fregert, B. Magnusson, H.J. Bandmann, C.D. Calnan, E. Cronin, N. Hjorth, H.J. Maibach, K.E. Malten, C.L. Meneghini, and V. Pirila, *Acta Dermatov. (Stockholm)*, *50,* 287-292(1970).
22. N. Hjorth and S. Fregert, in *Textbook of Dermatology,* A.J. Rook, D.S. Wilkinson, and F.J.G. Ebling (Eds.), 1st, Blackwell, Oxford, 1968.
23. J.D. Justice, J.J. Travers, and L.J. Vinson, *Soap Chem. Spec., 37,* 53(1961).
24. J. McL. Philp, *Labo.-Pharma.-Problems et Téchniques, 255,* 585(1976).
25. P.A. Isherwood, *J. Invest. Dermatol., 40* (3), 143-145(1963).
26. E. Götte, *Tenside 4* (7), 209-217(1967).
27. E. Skog, *Acta Dermatol., 38* (1), 1-14(1958).
28. J.G. Black and D. Howes, *J. Soc. Cos. Chem., 26,* 205(1975).
29. A. Nilzen and K. Wikstrom, *Acta Dermatol., 35,* 292(1955).
30. B. Magnusson and A.L. Kligman, in *Allergic Contact Dermatitis in the Guinea Pig,* Charles C. Thomas, Springfield, Ill., 1970.
31. J. McL. Philp, in *Comparative Physiology and Pathology of the Skin*, A.J. Rook and G.S. Walton (Eds.), Blackwell, Oxford, 1965, pp. 499-514.
32. G. Smeenk, *Arch. Klin. Exp. Derm., 235,* 180-191(1969).
33. J.F. Griffiths, J.E. Weaver, H.S. Whitehouse, R.L. Poole, E.A. Newman, and G.A. Nixon, *Food Cosmet. Toxicol., 7* (6), 58, 1(1969).
34. M. Valér, Berufsdermatosen, *2*, 83-99(1969).
35. R. Patiala, S. Kojas, and H. Kalliola, *Ann. Chir. Gyn Fenn, 43,* Suppl 5, 332-339(1954).
36. T. Ito, K. Kumihero, S. Hagashi, and T. Kariyone, *Parfum Crom. Savons, 12* (11), 550-554(1969).
37. W. Kling, *Fette, Seifen, Anstrichm., 59,* 820-822(1957).
38. E. Götte, *Proc. 4th Int. Congr. Surface Active Substances*, Brussels, 7-12 Sept. 1964, Vol. 3 (Sect. C of Congress).
39. J. Malaszkiewicz and Ch. Gloxhuber, *Arch. Klin. Exp. Derm., 237,* 642-661 (1970).
40. B.E. Emery and L.D. Edwards, *J. Amer. Pharm. Ass. Sci. Ed., 29,* 251-254 (1940).
41. A.M. Kligman, *J. Invest. Dermatol., 46,* 573(1966).
42. F. Grimalt and C. Romaguera, *Contact Dermatitis, 3,* 54(1977).
43. J.E. Kirk, *Acta Dermato.-Venereol., 46,* Suppl. 57(1966).
44. L.E. Gaul and G.B. Underwood, *J. Invest. Dermatol., 17,* 9(1951).
45. I.H. Blank, *J. Invest. Dermatol., 18,* 433(1952).
46. G.F. Burch and T. Winsor, *Arch. Int. Med., 74,* 437(194
47. H. Baker and A.M. Kligman, *Arch. Dermatol., 96,* 441-452(1967).
48. K.A. Grice and F.R. Bettley, *Brit. Med. J., 4* (5573), 195(1967).
49. M. Press, P. Hartop, and C. Prottey, *Lancet, 1,* 597-599(1974).
50. D.E. Wurster and S.F. Kramer, *J. Pharm. Sci., 50,* 288(1961).
51. E. Cronin and R.B. Stoughton, *Brit. J. Dermatol., 74,* 265-272(1962).
52. J.D. Middleton, *Brit. J. Dermatol., 80,* 437(1968).
53. J.D. Middleton and B.M. Allen, *J. Soc. Cosmet. Chem., 24,* 239-243(1973).
54. J.D. Middleton and B.M. Allen, *J. Soc. Cosmet. Chem., 20,* 399-412(1969).
55. R.R. Suskind and M. Ishihara, *Arch. Environ. Health, 11,* 529-537(1965).

56. Eva F. Dodge, *J. Amer. Med. Womens Ass., 27,* 3(1972).
57. R.J. Simmons, *Amer. J. Obs. Gynec., 6,* 447-448(1955).
58. J.J. Roberts, *J. Flor. Med. Ass., 60,* 31-35(1973).
59. H.J. Roberts, *JAMA, 201,* 125-126(1967).
60. S. Marshall, *J. Urol., 93,* 112(1965).
61. H.N. Bass, *Clin. Paed., 7,* 174(1968).
62. Editorial, *Brit. Med. J., 2,* 192(1968).
63. P.S. Porter and J.S. Lyle, *Arch. Dermatol., 93,* 402(1966).
64. R.D. Catterall, *Lancet, 2,* 830(1966).
65. Food and Drug Administration, Office of Planning and Evaluation, Consumer Safety Statistics Staff, O.P.E. Study, *17*(1975).
66. A. Nilzen, *Acta Dermato.-Venereol., 38,* 104-111(1958).
67. A.M. Kligman, *J. Invest. Dermatol., 47,* 375(1965).
68. B. Magnusson, *Acta Dermato-Venereol., 45,* 325(1965).
69. B. Magnusson and O. Gilje, *Acta Dermato.-Venereol., 53* (2), 136-140(1973).
70. D.S. Connor, H.L. Ritz, R.S. Ampulski, H.G. Kowollik, P. Lim, D.W. Thomas, and R. Parkhurst, *Fette, Seifen, Anstrichm., 77* (1), 25(1975).
71. J.J. Jambor and R.R. Suskind, *J. Invest. Dermatol., 24,* 279-386(1955).
72. A.A. Fisher, in *Contact Dermatitis,* 2nd, Henry Kimpton, London, pp. 58-59.
73. Von J. Pluss, *Dermatologica, 186*(1953).
74. A. Tzanck and E. Sidie, in *Les Dermatoses Allergique,* Maison, Paris, 1950, p. 129.
75. J.M. Lachapille and D. Tennstedt, *Contact Dermatitis, 1* (4), 260(1975).
76. N. Hjorth and D.S. Wilkinson, *Brit. J. Dermatol., 81,* 311-313(1969).
77. G. Agrup, *Acta Dermato.-Venereol., 49,* Suppl. 61(1969).
78. H.W. Rothenborg, T. Menne, and K.E. Sjolin, *Contact Dermatitis, 3,* 37-48 (1977).
79. D.L. Opdyke, F.H. Snyder, and H.L. Rubenkvenig, *Toxicol. Appl. Pharmacol., 6,* 141-146(1964).
80. J. McL. Philp, in *Fluorescent Whitening Agents*, M.V.C. Rep. 2, Miljovardscentrum, Stockholm, 1973, pp. 128-213.
81. C.D. Calnan, H.J. Bandmann, E. Cronin, S. Fregert, N. Hjorth, B. Magnusson, K. Malten, C. Meneghini, V. Pirila, and D.S. Wilkinson, *Brit. J. Dermatol., 82,* 543(1970).
82. O. Braun-Falco, H.C. Korting (Eds.), in *Hautreinigung mit Syndets*, Springer, Heidelberg, New York, pp. 3-7(1990).
83. N.F. Hadley, *American Scientist, 68,* 546-553(1980).
84. N.F. Hadley, *Prog. Lipid Res., 28,* 1-33(1989).
85. E. Proksch, *Ärztliche Kosmetologie, 19,* 424-443(1989).
86. H. Stüpel, A. Szakall, in *Die Wirkung von Waschmitteln auf die Haut,* Hüthig Verlag, Heidelberg, 1957.
87. P.M. Elias, *J. Invest. Dermatol., 80,* 44-49(1983).
88. P.M. Elias, E.R. Cooper, A. Korc, and B.E. Brown, *J. Invest. Dermatol., 76,* 297-301(1981).
89. G. Grubauer, K.R. Feingold, R.M. Harris, and P.M. Elias, *J. Lipid. Res., 30,* 89-96(1989).

90. M.A. Lampe, A.L. Burlingame, J. Whitney, M. Williams, B.E. Brown, E. Roitman, and P.M. Elias, *J. Lipid Res., 24,* 120-129(1983).
91. P.M. Elias, in *Dermatology in general medicine*, 3rd, Fitzpatrick (Ed.), McGraw Hill, New York, 342-346(1986).
92. J.L. Leveque, M. Escoubez, and L. Rasseneur, *Bioeng. Skin, 3,* 227-242(1987).
93. M. Gloor, K. Munsch, and H.C. Frierich, *Dermatol. Monatsschr., 158,* 576-581(1972).
94. M. Gloor, K. Munsch, and H.C. Frierich, *Dermatol. Monatsschr., 160,* 291-296(1974).
95. A. Yamamoto, S. Serizawa, M. Ito, and Y. Sato, *J. Invest. Dermatol., 89,* 507-512(1987).
96. A.M. Kligman and W.B. Shelley, *Brit. J. Dermatol., 30,* 99-125(1958).
97. A.M. Kligman, *Brit. J. Dermatology, 75,* 307-319(1963).
98. C. Prottey and T. Ferguson, *J. Soc. Cosmet. Chem., 26,* 29-46(1975).
99. C.D. Howes, *J. Soc. Cosmet. Chem., 26,* 47-63(1975).
100. H. Tronnier, G. Schuster, and H. Modde, *Arch. f. klin. u. experiment. Dermatol., 221,* 232-249(1965).
101. W. Burckhardt, *Dermatologica, 129,* 37-46(1964).
102. L.B. Fisher and H.I. Maibach, *Contact Dermatitis, 1,* 273-276(1975).
103. L.G. Dominguez, J.L. Parra, R. Infante, C.M. Pelejero, F. Balaguer, and T. Sastre, *J. Soc. Cosmet. Chem., 28,* 165-182(1977).
104. G. Imokawa and Y. Mishima, *Contact Dermatitis, 5,* 151-162(1979).
105. M. Kawai and G. Imokawa, *J. Soc. Cosmet. Chem., 35,* 147-156(1984).
106. E.Q. Bernstein, D.W. Coble, and E.O. Kairinen, *Scanning Electron Microscopy III,* 347-354(1979).
107. G.P. Serban, S.M. Henry, V.F. Cotty, and A.D. Marcus, *J. Soc. Cosmet. Chem., 32,* 407-419(1981).
108. G. Würbach, F. Schiller, K. Langguth, E. Weidermann, and J. Braband, *Dermatol. Monatsschar., 169,* 248-252(1983).
109. P.J. Frosch, *Ärztliche Kosmetologie, 13,* 397-406(1983).
110. C. Prottey, D. Oliver, and A.C. Coxon, *Intern. J. of Cosmet. Sci., 6,* 263-273(1984).
111. M.K. Bahl, *J. Soc. Cosmet. Chem., 36,* 287-296(1985).
112. St. Babulak, L. Rhein, D. Scala, F.A. Simion, and G. Grove, *J. Soc. Cosmet. Chem., 37,* 475-479(1986).
113. P. Frosch and B. Czarnetzki, *J. Invst. Dermatol., 88,* 52s-55s(1987).
114. E. Berardesca and H.I. Maibach, *Contact Dermatitis, 19,* 136-140(1988).
115. K. Lammintausta, H.I. Maibach, and D. Wilson, *Contact Dermatitis, 19,* 84-90(1988).
116. M.A. Pemberton and C. Rhodes, *Toxicology Letters, 44,* 85-90(1988).
117. R. Tupker, I. Pinnagoda, P.-J. Coenraads, and J.P. Nater, *Contact Dermatitis, 20,* 108-114(1989).
118. K.-P. Wilhelm, C. Surber, and H.I. Maibach, *Arch. Dermatol. Res., 281,* 273-295(1989).
119. K.-P. Wilhelm, J. Saunders, and H.I. Maibach, *Brit. J. Dermatol., 122,* 793-798(1990).

120. E. Berardesca, D. Fideli, P. Gabba, M. Cespa, G. Rabbiosi, and H.I. Maibach, *Contact Dermatitis, 23,* 1-5(1990).
121. S.-G. Blohm, *Acta Dermatol.-Venereol., 37* (4), 269-275(1957).
122. G. Imokawa, H. Tsutsumi, and T. Kurosaki, *J. Amer. Oil Chem. Soc., 55,* 839-843(1977).
123. F.R. Bettley, *Brit. J. Dermatol., 75,* 113-116(1963).
124. I.H. Blank and E. Gould, *J. Invest. Dermatol., 37,* 485-488(1961).
125. A. Szakall and K.H. Schulz, *Fette, Seifen, Anstrichm., 62* (3), 170-175(1960).
126. G. Neis, *Dermatologische Wochenschrift, 141* (18), 467-473(1960).
127. A. Szakall, *Berufsdermatosen 6* (4), 171-192(1958).
128. K.H. Schulz and G. Rose, *Archiv f. klin. exper. Dermatologie, 205,* 254-260 (1957).
129. I.H. Blank and E. Gould, *J. Invest. Dermatol., 33,* 327-336(1959).
130. K.H. Gustavson, *Acta Chem. Scand., 4,* 1171-1184(1950).
131. F.R. Bettley, *Brit. J. Dermatol., 77,* 98-100(1965).
132. H. Baker, A. Kligman, *Arch. Dermatol., 96,* 441-452(1967).
133. G.E. Nilsson, *Med. & Bio. Eng. & Comput., 15,* 209-218(1977).
134. M. Gloor, G. Hirsch, and U. Willebrandt, *Arch. Dermatol. Res., 271,* 305-313(1981).
135. M. Dahl and M.J. Roering, *J. Am. Acad. Dermatol., 11*(3), 477-479(1984).
136. R. Murahata, D. Crowe, and J. Roheim, *Int. J. Cos. Sci., 8,* 225-231(1986).
137. D. Crowe, M. Willard, R. Murahata, L. Anst, and M. Gacula, *Contact Dermatitis, 19,* 192-194(1988).
138. B. Morrison and B. Rogers, *Bioeng. Skin, 4,* 105-114(1988).
139. S. Motta, M. Bozzetti, A. Burberi, and M. Monti, *Bioeng. Skin, 4,* 271-276 (1988).
140. P. Thune, T. Nilsen, I.K. Hanstad, T. Gustavsen, and H. Lövig Dahl, *Acta Dermatol.-Venereol. (Stockholm), 68,* 277-283(1988).
141. R. Blanken, M. van Vilsteren, R. Tupker, and P. Coenraads, *Contact Dermatitis, 20,* 265-269(1989).
142. M. Gloor and T. Scheer, *Ärztliche Kosmetologie, 19,* 31-40(1989).
143. R. Tupker, P. Coenraads, J. Pinnagoda, and J.P. Nater, *Contact Dermatitis, 20,* 265-269(1989).
144. J. Wieriks, *Safety Evaluation of Water as a Cosmetic Ingredient,* Paper at Waterology Symposium, Amsterdam, April 10-12, 1989.
145. H. Zienicke, in *Hautreinigung mit Syndets*, O. Braun-Falco and H.C. Korting (Eds.), Springer, Heidelberg and New York, 1990, pp. 137-147.
146. C. Froebe, F. Simion, L. Rhein, R. Cagan, and A. Kligman, *Dermatologica, 181,* 277-283(1990).
147. B. Melnik, J. Hollmann, U. Hofmann, M.-S. Yuh, and G. Plewig, *Arch. Dermatol. Res., 282,* 549-551(1990).
148. A. Rougier, M. Rallis, P. Krien, and C. Lotte, *Arch. Dermatol. Res., 282,* 498-505(1990).
149. P. Elias, M. Williams, and S. Rehfeld, *Arch. Dermatol., 126,* 868-870(1990).
150. Ph. Wertz and D. Downing, *Science, 217,* 1261-1262(1982).
151. J. Antoine, J. Contreras, and D. Van Neste, *Dermatosen, 37,* 96-100(1989).

152. R. Murahata, R. Toton-Quinn, and M.B. Finkey, *J. Am. Acad. Dermatol., 18,* 62-66(1988).
153. E. Nannipieri, V. Carelli, G. Dicolo, I. Giorgi, and M.F. Serafini, *Int. J. of Cosmet. Sci., 12,* 21-31(1990).
154. S.E. Friberg, I. Kayali, L. Rhein, F. Simion, and R. Cagan, *Int. J. of Cosmet. Sci., 12,* 5-12(1990).
155. A. Cua, K.-P. Wilhelm, and H.I. Maibach, *Brit. J. Dermatol., 123,* 473-479 (1990).
156. A. Cua, K.-P. Wilhelm, and H.I. Maibach, *Brit. J. Dermatol., 123,* 607-613 (1990).
157. D. Bommannan, R. Potts, and R. Guy, *J. Invest. Dermatol., 95,* 403-408(1990).
158. S. Flannigan, R. Smith, and J. McGovern, *Contact Dermatitis, 10,* 123-124 (1984).
159. D. Strube, St. Koontz, R. Murahata, and R. Theiler, *J. Soc. Cosmet. Chem., 40,* 297-306(1989).
160. G. Imokawa, S. Akasaki, A. Kawamata, S. Yano, and N. Takaishi, *J. Soc. Cosmet. Chem., 40,* 273-285(1989).
161. P. Bergstresser and W. Eaglestein, *Arch. Dermatol., 108,* 218-219(1973).
162. Y. Werner, M. Lindberg, and B. Forslind, *Acta Dermatol.-Venereol. (Stockholm), 62,* 334-337(1982).
163. Y. Werner and M. Lingberg, *Acta Dermatol.-Venereol. (Stockholm), 65,* 102-105(1985).
164. Y. Werner, *Acta Dermatol.-Venereol. (Stockholm), 66,* 281-284(1986).
165. P. Frosch and A.M. Kligman, *J. Am. Acad. Dermatol., 1,* 35-41(1979).
166. J. Wahlberg, *Acta Dermato.-Venereol. (Stockholm), 48,* 549-555(1968).
167. T. Frödin, P. Helander, L. Molin, and M. Skogh, *Bioeng. Skin, 3,* 115-130 (1988).
168. A. Fulmer and G. Kramer, *J. Invest. Dermatol., 86,* 598-602(1986).
169. A. Walker, G. Ashforth, R. Davies, E. Newmann, and H. Ritz, *Acta Dermato. Venereol. (Stockholm), 53,* 141-144(1973).
170. B. Sylvest, N. Hjorth, and B. Magnusson, *Contact Dermatitis, 1,* 351-362 (1975).
171. Ch. Gloxhuber, M. Potokar, S. Braig, H.G. van Raay, and G. Schwarz, *Fette, Seifen, Anstrichmittel, 76,* 126-129(1974).
172. A. Kligman, *J. Invest. Dermatol., 46* (6), 573-583(1966).
173. D. Connor, H. Kowollik, and D. Thomas, *J. Am. Oil Chem. Soc., 53* (5), 182-185(1976).
174. W. Lindup and P. Howell, *Fd. Cosmet. Toxicol., 16,* 59-62(1978).
175. N. Van Haute and A. Dooms-Goossens, *Contact Dermatitis, 9,* 169(1983).
176. Tabulation of Cosmetic Product Experience Report (Jan. 1974-June 1975), Food & Drug Administration Division of Cosmetic Technology, Washington, D.C.
177. Cosmetic related injuries: A MODS study of NEISS, July 1st, 1977 to June 30th, 1978, National Technical Information Service US Dept. of Commerce, Springfield, Ill.

178. Consumers perception of adverse reactions to cosmetic products (PIB-242 480), Westat. Inc. prepared for Food and Drug Administration, June 1975, National Technical Information, US Dept. of Commerce, Springfield, Ill.
179. Reactions of the skin to cosmetics and toiletry products (1979), Consumers' Assoc., 14 Buckingham Street, London WC 2.
180. Z. Ngangu, M. Samsoen, and J. Fossereau, *Dermatosen Beruf Umw., 31,* 126-130(1983).
181. C. Romaguera, J. Camarasa, A. Alomar, and F. Grimalt, *Contact Dermatitis, 9,* 167-170(1983).
182. J. Ring and R. Gollhausen, in: *Hautreinigung mit Syndets,* O. Braun-Falco, H.C. Korting (Eds.), Springer, Heidelberg, New York, 1990, pp. 181-188.
183. J.P. Nater and A. de Groot, in *Unwanted effects of cosmetics and drugs used in dermatology*, J.P. Nater and A. de Groot (Eds.), 2nd ed., Elsevier, Amsterdam, 1985.
184. A. de Groot, J.P. Nater, R. Lende, and B. Rijcken, *Int. J. Cosmet. Sci., 9,* 255-259(1988).
185. W. Broekx, A. Blondeel, A. Dooms-Goossens, and G. Achten, *Contact Dermatitis, 16,* 189-194(1987).
186. O. Braun-Falco and G.P. Heilgemeir, Therapie der Gegenwart, 120, 1028-1045(1981).
187. G. Bäurle, in: *Handekzeme,* G. Bäurle (Ed.), Schattauer Vlg., Stuttgart and New York, 1986, p. 88.
188. G. Bäurle, *Zbl. Arbeitsmed., 35,* 212-216(1985).
189. H. Djawari, *Therapiewoche, 30,* 2170-2179(1980).
190. A. Fisher, Rystedt, *Contact Dermatitis, 12,* 93-98(1985).
191. A. Fisher, in: *The Safety of Patch Testing—Controversies in Dermatology,* E. Epstein (Ed.), Saunders, Philadelphia, 1984.
192. B. Lindelöf, *Contact Dermatitis, 22,* 288-308(1990).
193. J. Draize, G. Woodard, and H. Calvery, *J. Pharmacol. Exp. Ther., 82,* 377 (1944).
194. P. Elsner, D. Wilhelm, and H. Maibach, *Contact Dermatitis, 23,* 20-26(1990).
195. E. Wannenmacher, *Fette, Seifen, Anstrichm., 63* (10), 934-938(1961).
196. E. Wannenmacher, *Fette, Seifen, Anstrichm., 74* (3), 177-180(1972).
197. W. Weinert, in: *Hautreinigung mit Syndets,* O. Braun-Falco and H.C. Korting (Eds.), Springer Verlag, Berlin and Heidelberg, 39-44(1990).
198. Ch. Flechsig, Thesis (Dissertation), Universität Mainz, March 1968.
199. P. Riethe, *Dtsch. Zahnärztl. Z., 20,* 9(1965).
200. H. Greene, *J. Periodont., 33,* 176-182(1962).
201. H. Albers, G. Jung, and W. Götze, *Dtsch. Zahnärztl. Z., 33,* 559-565(1978).
202. G. Krekeler and B. Klaiber, *Die Quintessenz, 9,* 1729-1732(1982).
203. I. Siegel and H. Gordon, *Archs. Oral Biol., 30* (1), 43-47(1985).
204. N. Khalafallah, M. Gouda, and S. Khalil, *J. Pharm. Sci., 64,* 991-994(1975).
205. W.E. Sprott, *Trans. St. John's Hosp. Derm. Soc., 51,* 186-201(1965).
206. K.S. Hansen, *Contact Dermatitis, 9,* 343-351(1983).
207. M. Cowan and M. Frost, *J. Hospital Infection, 7,* 91-95(1986).
208. P. Barkvoll, G. Rolla, *J. Dent. Res., 68,* 991(1989).
209. A. Boman, I. Blute, P. Fernström, J. Carlfors, and L. Rydhag, *Contact Dermatitis, 21,* 92-104(1989).

210. J. Velvart, in: *Toxikologie der Haushaltsprodukte,* H. Huber Vlg., Bern, Stuttg., Toronto, 1989.
211. A. Temple, Soap Cosmetics, Chemical specialities, April 1978, Intermountain Poison Control Center, Salt Lake City, 44-53(1978).
212. H. Daweke, *Arch. f. klin. exp. Dermatol., 209,* 520-524(1958).
213. W. Raab, *Ärztliche Kosmetologie, 17,* 354-359(1987).
214. H. Tronnier, *Therapiewoche, 21,* 18.07(1986).
215. S. Fregert, *Contact Dermatitis, 1,* 96-107(1975).
216. M. Gloor, P. Völlm, M. Gehse, and R. Ringelmann, *Dermatosen, 33,* 86-89 (1985).
217. J. Gehse, *Seifen Öle Fette Wachse, 115,* 159-160(1989).
218. B. Meding and G. Swanbeck, *Contact Dermatitis, 22,* 13-23(1990).
219. B. Meding and G. Swanbeck, *Contact Dermatitis, 23,* 6-14(1990).
220. B. Meding and G. Swanbeck, *Contact Dermatitis, 23,* 154-161(1990).
221. A. de Groot, *Contact Dermatitis, 22,* 273-277(1990).
222. W. Gehring, *Contact Dermatitis, 22,* 292-293(1990).
223. A. Kligman, in: *Current Concepts in Cutaneous Toxicity,* V.A. Drill and P. Lazar, (Eds.), Academic Press, New York, 1980, pp. 69-94.
224. R. Neubert, D. Partyka, W. Wohlrab, B. Dettlaff, W. Fürst, and K. Taube, *Dermatol. Monatschrift, 176,* 711-716(1990).
225. A. Kligman and W. Epstein, *Contact Dermatitis, 1,* 231-239(1975).
226. S.P. Harrold, *J. Invest. Dermatol., 32* (5), 581-588(1959).
227. N. Nicolaides, *Science, 186,* 19(1974).
228. H. Maibach, K. Lammintausta, E. Berardesca, and S. Freeman, *J. Am. Acad. Dermatol., 21,* 833-835(1989).
229. Y. Linde, *Acta Dermatol.-Venereol. (Stockholm), 69,* 315-319(1989).
230. Y. Linde, A. Bengtsson, and M. Lodén, *Acta Dermatol.-Venereol. (Stockholm), 69,* 315-319(1989).
231. K. Wilhelm and H. Maibach, *J. Am. Acad. Dermatol., 23* (1), 122-124(1990).
232. K. Wilhelm and H. Maibach, *Dermatologic Clinics, 8* (1), 17-22(1990).
233. G. Piérard, *Dermatologica, 179,* 1-2(1989).
234. D. Saint-Léger, A. Francois, I. Lévêque, T. Stoudemayer, A. Kligman, and G. Grove, *Dermatologica, 178,* 151-155(1989).
235. T. Agner and J. Serup, *Brit. J. Dermatol., 121,* 323-328(1989).
236. P. Elsner, H. Oriba, and M. Maibach, *Hautarzt, 40,* 411-417(1989).
237. M. Flemming and W. Bergfeld, in: *Irritant Contact Dermatitis,* E. Jackson and R. Goldner (Eds.), Marcel Dekker, New York, 1990, pp. 41-66.
238. W. Kästner and P. Frosch, *Fette, Seifen, Anstrichmittel, 83,* 33-46(1981).
239. P. Frosch and A. Kligman, *Contact Dermatitis, 5,* 73(1979).
240. J.L. Antoine, J. Contreras, and D. van Neste, *Dermatosen, 37* (3), 96-100(1989).
241. R. Murahata, R. Toton-Quinn, and M. Finkey, *J. Am. Acad. Dermatol., 18,* 62-66(1988).
242. D. Vieluf, in *Hautreinigung mit Syndets,* O. Braun-Falco, H.C. Korting (Eds.), Springer Verlag, Berlin, Heidelberg, 1990, pp. 123-136.
243. F. Braun, D. Lachmann, and E. Zweymüller, *Hautarzt, 37,* 329-334(1986).

244. G. Graf and B. Ummenhofer, *Dermatosen, 29* (5), 139-141(1981).
245. Gy. Kiss, *Dermatosen, 29* (1), 15-18(1981).
246. M. Cowan and M. Frost, *J. Hospital Infect., 7,* 91-95(1986).
247. A. Hartmann, in: *Hautreinigung mit Syndets,* O. Braun-Falco, H.C. Korting (Eds.), Springer Verlag, Berlin, Heidelberg, 1990, 88-92.
248. W. Markland, *Norda Briefs, 466*(1975).
249. R. Goldemberg and L. Safrin, *J. Soc. Cosmet. Chem., 28,* 667-679(1977).
250. A. Domsch, H. Pospischil, G. Schuster, and H. Tronnier, *Parfümerie und Kosmetik, 61,* 325-330(1980).
251. K. Miyazawa, M. Ogawa, and T. Mitsui, *JSCCJ, 18* (2), 96-105(1984).
252. M. Cox, N. Borys, and T. Matson, *J. Am. Oil Chem. Soc., 62* (7), 1139-1143 (1985).
253. W. Matthies, A. Löhr, and H. Ippen, *Dermatosen, 38* (6), 184-189(1990).
254. W. Sterzel, F. Bartnik, W. Matthies, W. Kästner, and K. Künstler, *Toxic. in Vitro, 4* (4/5), 698-701(1990).
255. M. Rosen and A. McFarland, *J. Soc. Cosmet. Chem., 35,* 157-169(1984).
256. T. Fischer, *Contact Dermatitis, 23,* 221-223(1990).

7

Carcinogenic, Mutagenic/Genetic Toxicity, and Teratogenic Properties

KENKICHI ŌBA AND REIKO TAKEI *Lion Corporation, Tokyo, Japan*

I. CARCINOGENIC PROPERTY

Although tumors have many factors in common, a completely satisfactory definition has yet to be produced. In the majority of cases, a tumor consists of an abnormal mass of cells that may or may not be typical in their arrangement. Of greater importance is the fact that tumor cells exhibit an abnormal behavior in that they no longer respond to the mechanisms controlling cell growth and anatomical arrangement. The ensuing abnormal proliferation, completely uncoordinated with that of the normal tissue, results in a mass or tumor. An additional feature displayed by certain types of tumor is that of invasiveness, i.e., the cells of the tumor actually migrate into the surrounding tissues. In some instances, the cellular migration exceeds the abnormal proliferation; this results in a diffuse infiltrating carcinoma without the formation of a mass or tumor.

It is possible to classify tumors in various ways: macroscopically, histogenetically, histologically, etiologically, and behaviorally. A convenient and useful method divides tumors into two types, dependent upon their behavior: (1) benign, in which the cells are not invasive, and (2) malignant, in which the cells migrate into the surrounding tissues. In the latter case, the cells eventually enter lymphatics and blood vessels to be transported via the lymph or blood to other areas of the body, where they initiate secondary tumors or metastases. This local invasion and embolic spread are the two characteristics of malignant tumors.

Compounds that initiate or promote tumor growth are termed oncogenic or tumorigenic; those that lead to the production of malignant tumors are said to be carcinogenic. It is pertinent to mention that the division of tumors into benign and malignant is, in fact, an oversimplification; there are, for example, intermediate tumors that are locally invasive and, therefore, not benign, but which show little or no evidence of metastasizing. However, for the purpose of this chapter, this can be ignored.

Chemically induced carcinogenesis in humans has been substantiated as a result of epidemiological studies performed on persons who have been exposed occupationally, environmentally, or medically to a carcinogen. In most instances where a compound is known to be a carcinogen in humans, the carcinogenicity has been confirmed in laboratory animals; hence, the importance and significance of animal studies.

Since there exists the possibility that humans may directly, through contact with household detergents, or indirectly, through ingestion of contaminated drinking water, come into contact with surfactants, these compounds have been scrutinized with regard to their potential toxicity and oncogenicity. In common with other chemical products, surfactants are not pure materials; commonly found impurities include raw materials, byproducts, unreacted substances, etc. A surfactant may comprise a few or several compounds, and there may be many homologues in its hydrophobic moiety. These facts must be considered when designing studies to evaluate potential toxicity or carcinogenicity. Saffiotti et al. (1) and Hunter and Benson (2), in evaluating the potential carcinogenicity of alkyl benzene sulfonate (ABS) and α-olefin sulfonate (AOS) materials, paid particular attention to the impurities in the test samples; their data indicate that surfactants themselves present no particular hazard to animals.

The possibility that surfactants might influence solubility, dispersion, mobility, or penetration of other potential carcinogens, or that they may engage in synergistic or antagonistic reactions with carcinogens, has also been investigated (3). Although it is possible that carcinogens could be solubilized and eluted from soils by surfactants, experimental evidence does not substantiate this; solubilization of such substances to any appreciable extent will only occur at concentrations above the critical micelle formation (4,5).

A. Carcinogenicity

A chronic toxicity test does not always afford proper information on the evaluation of carcinogenic potential because of the probable insufficiency or inadequacy regarding animal numbers, dosing period, observation items, and so forth. Carcinogenic studies are usually performed in the mouse and the rat. There is no short-term test that has been successful in evaluating tumorigenic potential for chemical agents. Most of the tests involve lifetime studies or at least exposure of the animal to the chemical for a significant portion of the lifetime of the animal; tests in the rat last 2 years. In animal experiments for carcinogenicity, as with other types of toxicity, it is usually not difficult to demonstrate the action of a compound when it is strongly carcinogenic. However, many substances have been investigated which would best be classified as weak carcinogenic agents so that only when maximally tolerated doses of the compounds are administered to large numbers of animals is there significant evidence of tumorigenicity. Anionic surfactants might be suspected as being weak carcinogenic agents; it is the weak carcinogenic agent that presents the greatest challenge in the design and conduct and interpretation of toxicity tests.

Since distinct tumors are easily observed during the course of prolonged chronic toxicity tests and some researchers have examined the generation of tumors in their chronic toxicity tests on anionic surfactants, the above-mentioned reports on the carcinogenic potential of anionic surfactants are summarized in Appendix 1 together with prolonged toxicity test reports. Useful reviews on the carcinogenic potentials of anionic surfactants are available (3,6-9). Representative studies are reviewed below according to the type of anionic surfactant.

1. Sulfonate-type anionic surfactants

(a) Alkyl benzene derivatives. Saffiotti et al. (1) conducted carcinogenesis tests on tetrapropylene-derived alkyl benzene (AB) and its sulfonates (ABS). Alkyl benzenes were studied for their potential carcinogenicity by skin-painting tests using Swiss mice. In addition to the actual tumorigenicity tests on anionic surfactants, consideration was given to tests for promoting activity; these may be defined as tests for tumor induction in skin pretreated with a subcarcinogenic dose of a known carcinogen. Tests for the promoting action of alkyl benzenes were carried out by applying test material to a similar area of mouse skin previously treated with a single application of the carcinogen, 7, 12-dimethylbenz(a)anthracene (DMBA) applied as a 1% solution in mineral oil a week prior to the first application of the test sample. The results showed that seven commercial alkyl benzenes (A-G) were free from potential hazard and from carcinogenicity and five samples (A-E) revealed a promoting activity to tumor generation by DMBA. One alkyl benzene sample (0) showed both carcinogenicity and promoting activity; this was a noncommercial pilot-plant sample and contained certain impurities. Two alkyl benzenes, B and D, proved to be noncarcinogenic in a skin-painting test in New Zealand White rabbits and sample D was found to be noncarcinogenic by subcutaneous injection into C57 black mice.

Two alkyl benzene sulfonates made from samples A and D were applied twice weekly to the skin of Swiss mice at concentrations of 5 and 10% in water, respectively. ABS made from sample D was also tested for carcinogenicity on the skin of rabbits and by subcutaneous injection into C57 black mice. All these substances proved to be without carcinogenic action on the skin of the mouse or rabbit and the subcutaneous tissues of the C57 black mouse. Petroleum ether extracts of the two ABS made from sample A and D were tested by skin painting in mice to assess whether any activity carried over from the alkyl benzene to its sulfonate remained in the solvent-extractable material. These materials were essentially inactive.

Bornmann et al. (10,11) reported on prolonged carcinogenicity tests using LAS (linear alkyl benzene sulfonates) and ABS in rats; 60 male and 60 female rats received drinking water containing 0.01% of LAS or ABS for 100 weeks.

The tumor-producing rate in treated groups was nearly the same as in control groups, and no carcinogenic potential was found either in LAS or ABS.

Paynter and Weir (12) conducted a 2-year chronic toxicity test on ABS (Santomerse No. 3) in rats; the dietary levels were 0, 200, 1000, and 2000 ppm. No adverse effect was found, and gross and microscopic examination of tissues revealed no pathologic changes attributable to the ingestion of the test material.

A 2-year feeding study was carried out by Tusing et al. (13) in which 120 male and 120 female Wistar rats received 0, 0.1% (1000 ppm), and 0.5% (5000 ppm) of ABS in the diet. In addition, another 2-year study was made in which rats received ABS in their drinking water supply. The level administered (about 0.05%) represented at least a 500-fold increase in daily consumption, on a body weight basis, over the possible human figure of less than 1 ppm. In these two tests, lifetime administration of ABS to rats produced no evidence of toxicity. Microscopic evaluation of the tissues and a histopathological survey of the essential organ systems revealed no evidence of toxicity.

Tokoro et al. (14) reported 1-year studies, using rats, on ABS and a commercial light-duty liquid detergent for kitchen use. Aqueous solutions of the detergent and ABS, the concentrations being 0.15 and 0.3 ml liter^{-1} for the former (two groups) and 150 mg liter^{-1} for the latter (one group), were administered to each of the three groups, respectively. Each group was composed of sixty 78-80-day-old rats. After the 1-year experiment, the average total intake of detergent determined in each group was 1.61 ml, 3.13 ml, and 1.58 g per animal, respectively. Some surviving animals were autopsied after 5, 6, 8, and 10 months and the remainder after 1 year. Organs and tissues were examined histopathologically and in some cases electromicroscopically. However, no progressive morbid condition was found in the liver, allegedly recognized as a target organ, or in the other organs.

To summarize, these experiments gave the unequivocal result that even long-term intake of the detergent did not evoke any biologically significant change, at the level of 150 ppm in drinking water.

Buehler et al. (15) conducted a 2-year feeding study on LAS in rats in which no carcinogenic potential of LAS was demonstrated. Groups of 50 male and 50 female rats were fed the test material for 2 years at levels of 0, 0.02, 0.1, and 0.5% in the diet. Necropsies were performed after the 8th and 16th months in which 5 females and 5 males were examined grossly and major tissues were obtained for microscopic evaluation. All animals that died during the study were autopsied, when possible, and all surviving animals were sacrificed at 24 months. All tumors observed during the study were noted, examined, and identified to determine any carcinogenic effect. No gross unusual lesions or abnormalities were noted in the visceral organs. There were no microscopic differences between test and control animals. No changes or lesions considered to be indicative of a test-related response, were observed in any of the tissues. Tissues taken from animals at the final autopsy showed a high incidence of intercurrent diseases not related to the exposure of the animals to the test material. The number and distribution of neoplasms was not related to the dietary treatment. Chronic interstitial nephritis and adrenal telangiectasis were the most commonly observed degenerative conditions. The majority of the tumors were classified as fibroadenomas (subcutaneous). Histological examination of tissues from animals that died during the test did not reveal findings that could be interpreted as test-related effects.

Hiraga of the Tokyo Metropolitan Government (149) carried out a 2-year chronic toxicity and carcinogenicity study on LAS in rats receiving 0, 0.04, 0.16, and 0.60% in the diet. Autopsy and pathological examination at 2 years showed various changes in each treated group. Almost all of these findings and changes were comparable to those in control groups, and they were not considered to be indicative of LAS-related responses. Although occasional tumors were found in each group, including the control groups, they did not appear to be associated with the ingestion of LAS. The induction of preputial gland tumor, which was not related to LAS treatment was reported (150).

Hiraga and his colleague also conducted a lifetime feeding study of LAS, with and without butylated hydroxy toluene (BHT), an antioxidant, in rats (149). The dietary levels were LAS 0 ~0.2% with BHT 0~0.60%. Various types of tumors, which seemed to be spontaneous, were found in most animals in the treated and control groups. No adverse effects attributable to treatment were found. Brown cell adrenal carcinoma was induced in the BHT, LAS alone and LAS with BHT groups. The incidence of this carcinoma and of the total tumors in male rats was increased according to the dose of BHT. LAS did not effect the occurrence of tumors and LAS did not enhance the BHT toxicity.

Endo and co-workers (148) also reported a 2-year rat chronic toxicity study on LAS at a level of 0.1% in drinking water. The dosage level chosen was derived from standards for a kitchen detergent in the Japanese Food Sanitation Law. LAS did not affect gross and microscopic histopathologic parameters.

No carcinogenic activity of LAS was reported by Tiba (16), who conducted a 2-year chronic toxicity study in Wistar rats receiving 0, 0.01, 0.05, and 0.10% of LAS in drinking water.

Prolonged percutaneous exposure of rats to LAS has been reported and gave no indication of carcinogenic activity (151). A three times weekly administration of 0.005, 0.025 or 0.125 ml of LAS (active ingredient, 19.7 w/w%) per animal to the dorsal skin of SLC-Wistar rats was continued for 24 months.

LAS based detergent (0.025 ml/animal) and distilled water were applied in control groups. Each application was washed from the skin with warm water, after 24 hr. Autopsy at 6 and 12 months did not reveal any unusual changes, although various effects related to LAS and the LAS based detergent, were observed after the 24 months. The histopathologic examination of the kidney denaturation at 6, 12, and 24 months revealed no evidence of toxicity attributable to LAS and LAS-based detergent. To summarize, organ weights and a histopathological survey of the organ systems revealed no evidence of toxicity and LAS and LAS-based detergent proved noncarcinogenic in skin-painting tests in the rat.

Furthermore, two carcinogenicity tests in mice performed by Philp (17) afforded additional information on the noncarcinogenic potential of sodium dodecyl benzene sulfonate (LAS) and an anionic-type surfactant mixture (80% LAS, 18% sodium sulfates, and 2% water). The purpose of their experiments was to assess the carcinogenic potential of an optical brightener (FWA); groups treated with LAS alone were included in their experimental runs; subcutaneous injection of 10 mg FWA alone, or with 1 mg of LAS, or 1 mg of LAS alone, in 0.2 ml of water were made twice weekly to a total of 25 injections using groups of 20 males and 20 females for all treatments. The negative control group received 25 injections of 0.2 ml normal saline. After treatment, the animals were maintained to complete an 80-week experimental period. There was no evidence to suggest that FWA or LAS and the surfactant mixture had any carcinogenic potential in C57 black mice.

(b) α-Olefin derivatives. α-olefin sulfonate, or AOS, was studied for its carcinogenic potential in long-term feeding studies and skin-painting experiments with rats and mice as shown in Appendix I. No indications of increased tumor incidence were noted in these studies.

Hunter and Benson (2) conducted a study for carcinogenic potential on AOS in rats.

AOS was composed of a mixture of alkenyl sulfonate and hydroxyalkane sulfonate (60.4:39.6% w/w) obtained by hydrolysis of the sulfonation product of *n*-α-olefin (C_{14}:C_{16}:C_{18} = 25:45:30) with sulfur trioxide, the active ingredient content of which was 97.93%. It also contained 0.64% of petroleum ether soluble matter (mainly unreacted oil), 10.6% of which was mainly alkane 1,4-sultone. The material was administered to CFY rats for 104 weeks at dietary levels of 0, 1000, 2500, and 5000 ppm; each group consisted of 50 male and 50 female animals. All rats used in the study were examined for both distribution and types of tumors as reported by Hunter and Benson (2).

Pancreatic islet cell tumors in male rats were confined to the treated groups, with the highest incidence (four rats) occurring in rats receiving 1000 ppm AOS. Incidences of *islet cell tumors* as high as those seen in the present study have been recorded among male control CFY rats in their laboratory, and this fact, coupled with the absence of any dosage-related trends in the incidence of these tumors in the treated groups, strongly suggests that their distribution in the present study was fortuitous and was not affected by treatment with AOS. *Adrenal tumors* were found in a total of 8 male rats receiving 2500 ppm AOS. However, only 2 males were thus affected in the high dietary level group, which was the same number as those with adrenal tumors in the control group. For this reason it is considered that the high frequency of adrenal tumors in males fed 2500 ppm AOS, was not attributable to treatment. The highest incidence of *thyroid tumors* was found among females receiving 2500 ppm AOS, but again, it is very unlikely that this represents an effect of treatment with AOS, as the incidence in the high dietary level group was the same as that recorded for the control females. *Pituitary tumors* were common in both control and treated rats, with the lowest incidence among males of the high dietary level group. *Mammary gland tumors* were very common in females of all groups, and there were no significant intergroup differences. The numbers of females bearing multiple *mammary gland tumors* (2 or more tumors) expressed as a fraction of the total number of tumor-bearing females in each group, were 14/30 (47%) in controls, 17/33 (52%) in 1000 ppm, 16/38 (42%) in 2500 ppm, and 18/35 (51%) in 5000 ppm. The apparent group mean induction periods (the average time taken for the first mammary tumor in each rat to appear) for the female groups were 84 weeks in controls and 1000 ppm, 83 weeks in 2500 ppm, and 85 weeks in 5000 ppm. It is clear from these data that treatment with AOS had no effect on the frequency of multiple-tumor formation or on tumor induction time. The incidence of other tumor types was low, and the intergroup distributions gave no evidence that treatment with AOS had any effect. There were no significant intergroup differences in the overall incidence of benign tumors, malignant tumors or tumors of any type, as shown in Table 1.

There have been several reports that some kinds of lower molecular weight sultones, particularly propane sultone, may be potential carcinogens in laboratory animals (152-154). Due to the risk that AOS may contain some sultones, long-term feeding and skin-painting studies were carried out with AOS and some sultones.

The International AOS Group (155) asked the Eppley Institute for Research in Cancer to conduct a skin painting study in mice and a feeding study in rats using AOS and C_{16}1,4-sultone. The studies were coordinated by AKZO, BASF A.G., Chemische Werke Huls A.G., Colgate-Palmolive, Ethyl Corporation, Gulf Oil, Henkel & Cie, Hoechst A.G., I.C.I., Lion Corporation, Proctor & Gamble, Shell, Ulilever Limited, and Witco(France, USA). Dosage levels were as follows:

TABLE 1 Total Numbers of Rats with Tumors in 2-Year AOS Feeding Study

Sex:	Male								Female							
AOS (ppm):	0		1000		2000		5000		0		1000		2000		5000	
	D	T	D	T	D	T	D	T	D	T	D	T	D	T	D	T
Type of tumor:																
Benign	7	14	8	13	12	17	6	15	23	18	16	28	26	20	19	26
Malignant	1	1	4	1	5	0	0	0	3	3	0	5	7	2	2	4
Total no. with tumors	8	15	10	14	16	17	6	15	23	19	16	28	26	20	20	27
No. examined	19	28	23	27	24	23	20	30	25	25	19	31	26	23	20	30

D, at death; T, at termination.

Group	
I	20% AOS (Based on C_{14-18} α-olefin from Mitsubishi Chemical Industries, sulfonated and hydrolyzed by Lion Corp.)
II	25% AOS (as I above)
III	20% AOS (Based on C_{14-16} α-olefin from Ethyl, sulfonated and hydrolyzed by Colgate-Palmolive)
IV	25% AOS (as III above)
V	6.7% C_{16}1,4-sultone (Gulf/Henkel) in acetone
VI	8.3% C_{16}1,4-sultone (Gulf/Henkel) in acetone
VII	Untreated control (shaved only)
VIII	Water Control
IX	Acetone Control

Treatments were carried out three times weekly for 92 weeks with a volume of 0.02 ml of the test material being applied to approximately 1 cm^2 of the interscapular region. Necropsies were conducted at a mean survival of 30% per group. Gross and microscopic examinations of 99% of the animals revealed no obvious carcinogenic effects from skin painting C_{14-18} AOS, C_{14-16} AOS and the sultone. No unusual tumor types developed. A low incidence of skin fibrosarcomas and hemangiosarcomas developed in both treated and control groups with no increase in incidence in the treated mice. High incidences but nontreatment-related to lung adenomas, liver hemangiomas and malignant lymphomas were diagnosed in all treated and control groups. Geriatric pathology including diseases of the heart, lung, and liver were equally prevalent in the test and control groups. No toxic or carcinogenic effects of the two AOS products and sultone were observed in the skin painting study (155).

The Eppley Institute also carried out a rat-feeding study. Five hundred male and 500 female MRC rats (Wistar derived) were distributed into 11 groups, with 100 control males and females and 40 males and females in each treatment group and the extra control group. The rats were fed the materials in the diet as shown below.

Group	
I	Untreated Control
II	1.0% AOS (Based on C_{14-18} α-olefin from Mitsubishi Chemical Industries, sulfonated and hydrolyzed by Lion Corp.)
III	0.75% AOS (as II above)
IV	0.5% AOS (as II above)
V	1.0% AOS (Based on C_{14-16} α-olefin from Ethyl, sulfonated and hydrolyzed by Colgate-Palmolive)

Group (continued)	
VI	0.75% AOS (as V above)
VII	0.5% AOS (as V above)
VIII	0.33% C_{16}1,4-sultone
IX	0.25% C_{16}1,4-sultone
X	0.16% C_{16}1,4-sultone
XI	Extra Control

The experiment was terminated when the mean survival point of 50% was reached.

Gross and microscopic examinations of 99.7% of the animals revealed no carcinogenic effects from $C_{14\text{-}18}$ AOS, $C_{14\text{-}16}$ AOS or sultone treatment. A high incidence of pituitary adenomas, Leydig cell tumors and uterine polyps were diagnosed in treated and control groups, but were not treatment related. Many tumors occurred at 2.5% incidence or below. Tumors of the kidney, urinary bladder, stomach, intestines and other organs were distributed randomly at low levels (0.5% AOS, 0.16% sultone) in treated and control groups and were not treatment induced. Geriatric pathology including disease conditions of the kidney, spleen, mammary gland, testes, adrenals, and heart were equally prevalent in treated and control groups. No toxic or carcinogenic effects of the two AOS products and sultone were observed.

The carcinogenic potentials of several AOS samples including sultone were also assessed in 400 Long-Evans rats divided into four treatment groups (50 animals/sex/group) as follows (156):

1. Vehicle controls (deionized water)
2. 30.9% Active ingredient AOS (essentially hydrolyzed, composite sample of $C_{14\text{-}16}$ AOS and $C_{16\text{-}18}$ AOS)
3. 30.9% Active ingredient AOS plus sultone (partially hydrolyzed sample of AOS (30.9%) identical to (2) but containing a residual level of sultone)
4. 38.9% Active ingredient AOS (commercial $C_{14\text{-}16}$ AOS)

A 10% active ingredient (v/v) aqueous solution was applied twice weekly to the clipped dorsal surface at a dosage level of 1 ml kg^{-1} for 24 months. Mean body weight, food consumption, hematology, urinalysis, mortality, and gross post-mortem observations were comparable in all groups. Gross and histopathological examination revealed a similar incidence of neoplastic lesions in both test and control rats, predominantly of the types commonly found in aging rats, with no carcinogenic effects attributable to the percutaneous application of the AOS test materials. These studies showed no indication that AOS or higher molecular weight sultones resulting from the synthesis of AOS were carcinogenic.

No treatment related changes in skin or organ structures, other than occasional dermatitis, were noted in Swiss-Webster mice (21 animals/group) following 2 years of twice-weekly dermal applications of 5% aqueous solutions of either $C_{15\text{-}18}$ AOS (90% active ingredient), hexadecane 1,4-sultone, or a sultone concentrate (64% active ingredient) extracted from the process stream during the sulfonation of $C_{15\text{-}18}$ α-olefin (157).

Tomizawa (158) carried out lifetime studies in Wistar rats using AOS and AOS-based detergent. Aqueous solutions of the AOS and AOS base (1.0, 10, and 30% for the former and 50% for the latter) were administered. Twenty-four hours after application, each site was washed with warm water. No abnormal gross and microscopic histopathologic findings were noted.

2. Sulfate-type anionic surfactants

No indications of increased tumor incidence were noted in two feeding studies in rats given 1.5% AES in the diet or 0.1% AES in drinking water for 2 years. Skin-painting studies in mice with 5% C_{12} AES or with a 10% AES/LAS mixture for 2 years or in rabbits for 24 months with 1.25% AES did not induce any carcinogenic response. These representative studies follow.

Sodium lauryl glyceryl ether sulfonate and sodium lauryl trioxyethylene sulfate (AES) were studied for their carcinogenic potential using a 105-week skin-painting test in mice and a 2-year feeding test in rats (18). For skin tumorigenicity evaluation, groups of 30 Swiss female mice were used and test materials applied twice weekly in 5% aqueous solution to the shaven skin. Two drops (about 0.1 ml) were used for each application, so that the total quantity of test material applied to each survivor during 105 weeks was about 1 g. No papillomas or other skin tumors were observed in any of the treated animals during the course of the experiment. Mortality in the experimental groups did not differ substantially from that of the control animals of the same strain maintained under the same conditions. In feeding studies, rats received the two anionic surfactants at levels of 0.1 and 0.5% in the diet for 105 weeks. No adverse effects were observed in rats, and microscopic examination revealed no pathogenic changes resulted from ingestion of the test materials.

In another 2-year study, rats (20 animals/sex/group) received C_{12} AES (25% active ingredient) in drinking water at a concentration of 0.1% (of active ingredient). At termination, survival, growth, food consumption, body weight, clinical laboratory findings, hematology and urinalysis in control and treated animals were all comparable. The only unusual findings were a slight but consistently higher water consumption in all rats receiving the test compound in drinking water and a significant difference in the empty cecum to body weight ratio in females. Absolute organ weights in treated animals were all comparable to controls and no consistent gross or histopathology was found. Generally, pathological findings for control and treated rats after two years on test were varied and consisted predominantly of incidental findings attributable to advanced age. Various types of benign and malignant tumors were found in both groups; the frequency of tumors in the treated group was not significantly different from that in control animals (159).

Using rabbits, Suzuta and co-worker (160) conducted a chronic dermal application study using AES and AES-based detergent lasting 2 years. No deleterious effects with respect to food consumption, body weights, clinical laboratory findings, blood chemistry, or urinalysis were noted and all parameters of control and treated animals were comparable. Histopathological findings after 1 and 2 years in the treated groups were not significantly different from those in control animals.

Smyth et al. (19) carried out 2-year chronic toxicity tests using sodium 2-ethylhexyl sulfate (AS: Tergitol anionic 08) in rats and dogs. Thirty six CFE rats of each sex per group were used. Levels of 0, 0.01, 0.04, 0.16, and 0.64% in the rat diet resulted in average dosages of 0, 0.0043, 0.017, 0.069, and 0.29 g kg^{-1} day^{-1}, respectively, for males; 0, 0.0052, 0.021, 0.081, and 0.34 g kg^{-1} day^{-1} for females. Concentrations in the diet were not changed as the animals matured so dosages during the first 6 months were about 25% greater. No dose-related differences from the controls were found at any dosage level for any of the criteria examined, including incidence of neoplasms, mortality and lifespan, and incidence of histopathological lesions. In the dog study, each group consisted of three male and three female Beagle dogs. Dosage levels of 0, 0.04, 0.16, and 0.64% in the diet resulted in average dosages of 0, 0.013, 0.048, and 0.17 g kg^{-1} day^{-1}, respectively. No dose-related adverse effect in any of the criteria examined, including histopathology of major organs, was detected.

Fitzhugh and Nelson (20) conducted a 2-year chronic toxicity test on sodium lauryl sulfate (AS) in rats at dietary levels of 0, 0.25, 0.5, and 1.0%. There was no description of any evidence of a possible carcinogenic potential.

Miyaji and Miyamoto (161) carried out a lifetime skin-painting experiment using AS at levels of 5% and 20% in rats. Groups of SD-rats (10 animals/sex/group) were treated twice weekly on the shaved dorsal skin (2×2 cm). There was no obvious carcinogenic response. Following high mortalities due to pulmonary infection the authors considered a repetition of the study necessary.

Additionally, tests for any cancer-promoting action of AS were carried out (35,162). The addition of C_{12} AS to the drinking water of rats (0.25%) resulted in enhancement of nitro-*N*-nitrosoguanidine-induced gastric tumor; however, it is unclear whether this effect was due to enhanced carcinogen absorption or some other physiological mechanisms (35).

3. Carboxylate-type anionic surfactants (soaps)

Both oral and dermal exposures of laboratory animals to soap (potassium soap; soap-K) gave negative results in carcinogenicity tests (162-164).

Yamamoto (162) and Ito (164) examined the carcinogenic and cocarcinogenic effects following oral administration of soap-K with either 3,4-benzpyrene or *N*-methyl-*N'*-nitrosoguanidine (MNNG) in mice and rats. The soap-K showed neither carcinogenic nor cocarcinogenic action with 3,4-benzpyrene. In the case of the carcinogen, MNNG, the incidence of stomach adenocarcinoma in the treated groups was comparable to that in the control group, and the differences were not statistically significant. Ito (164), however, noted that soap-K per se had no carcinogenic property and that no clear conclusion could be made concerning the cocarcinogenic potential.

4. Formulated detergents based on anionic surfactants

Many anionic surfactants have been used in household and industrial detergent formulations. Of the detergents now in use, most are based on anionic surfactants exclusively or in combination with other types of surfactants such as nonionic and amphoteric surfactants. From a practical point of view, it seems important to also review studies of the carcinogenic potential of formulated detergents.

As shown in Appendix 1, many detergents and surfactant mixtures were evaluated for prolonged toxicity and carcinogenicity, and no carcinogenic potential is reported. Representative studies were reviewed as follows.

Snyder et al. (21) conducted 2-year chronic toxicity and carcinogenicity tests in rats using two types of granular household detergent. Each consisted of about 20% anionic surfactant, 50% condensed phosphate, 13% sodium sulfate, and 6% sodium silicate; the remaining 11% was made up of moisture and various minor ingredients, such as perfumes, colors, and optical brighteners. The surfactant in detergent type I was ABS, and that in detergent type II was a mixture of approximately equal parts of ABS and AS. Groups of 40 male and 40 female rats were assigned to three groups receiving 0, 0.5% detergent type I and 0.5% detergent type II in the diet. There was no indication that tumors or other signs of pathology, gross or microscopic, occurred more frequently in any of the experimental groups than in the controls. The types of tumors encountered were varied and consisted mostly of incidental findings common to old rats of this strain.

Furthermore, these two detergents were studied for carcinogenic potential by topical and subcutaneous application to Swiss mice, C57 black mice, and rabbits (22). No tumors occurred in mice or rabbits treated repeatedly with aqueous solutions of the detergent products, with or without an initiating dose of DMBA (9,10-dimethyl-1,2-benzanthracene) or in mice that received a single subcutaneous injection of detergent type II.

A two-year oral toxicity study was conducted with groups of albino rats fed a mixture (50:50) of LAS/AES at dietary levels of 0, 0.1, 0.5, or 1.0% (active ingredient). The results obtained from all groups for the following parameters were all within normal ranges for rats of this age and strain: mortality, reaction, hematology, organ weights and ratios, gross and microscopic pathology and tumor incidences (159).

Percutaneous application of aqueous solutions of a detergent containing 15.6% LAS and 18.6% tallow alkylethoxylate sulfate at concentrations of 0.1, 1.0, or 10.0%, three times weekly for 18 months, to Swiss ICR mice, did not result in any carcinogenic response. At the 10% level, acanthosis and/or hyperkeratosis of the treated skin was noted, along with the occurrence of one single benign papilloma in a group of 50 test animals (159).

As described in Section 1 of Philp (17) under the heading "Alkyl Benzene Derivatives," the noncarcinogenic potential of a surfactant mixture which was used as a vehicle in testing tumorigenicity of an optical brightener (FWA) using C57 black mice is indicated. Although the surfactant mixture (80% LAS, 18% sodium sulfate, and 2% water) was not in a very high concentration, it is still relevant to the type of contamination that might occur under practical terms.

To indicate the toxicity or carcinogenicity of many commonly used cosmetics, detergents, solvents, and fixatives, long-term toxicity studies were performed on their effects in mice's and rabbits' skin (23). Each sample was applied topically with a graduated pipette (0.02 ml, twice weekly, for life) to the backs of Swiss mice and to the inner side of the rabbit ear. Autopsies were performed on all animals, and gross lesions as well as all tumors studied histologically. A very small number of skin tumors occurred; two squamous cell papillomas, one squamous cell carcinoma, and a few tumors of dermal origin (dermatofibromas, neurofibromas, and fibrosarcomas) occurred. The incidence of spontaneous tumors did not increase. There was no report of a detergent being carcinogenic.

Occasionally compounds are reported to be carcinogenic on the basis of inadequate data. Thus it is necessary that all such tests use a proper number of randomly selected test animals, in experiments that are adequately controlled and in which the lesions obtained are established by proper pathological examination.

Judging from this viewpoint, many studies done in the proper way stress that the anionic surfactants now in use are essentially without carcinogenic activity. This fact gives us full assurance that there would be no cancer hazard from contact with these surfactants in daily life, in environment and in common usage.

B. Cocarcinogenic Action

An agent is called a cocarcinogen when it has no carcinogenic action per se but has the ability to increase tumor incidence when spplied with a carcinogen; this action is called cocarcinogenic action. In other words (1), cocarcinogenic action can be defined as an action in which both the carcinogen and the cocarcinogen are applied concomitantly and the latter augments the action of the former. When the augmenting agent is applied at the same time as the carcinogen, many factors, notably solvent or vehicle action, may play an important role.

Since it is difficult to induce adenocarcinoma in the glandular stomach of rats by the oral administration of carcinogens, gastric carcinogensis using a carcinogen combined with a surfactant has been attempted by many investigators. In order to produce undifferentiated adenocarcinoma in the rat glandular stomach, Takahashi and co-workers have made a series of experiments using such known carcinogens as 4-mitroquinoline-1-oxide (4-NQO) and *N*-methyl-*N'*-nitro-*N*-nitrosoguanidine (NG): The induction of undifferential adenocarcinoma in the glandular stomach of the experimental animal with metastasis seems to be of paramount importance because of the similarities of the induced tumors to the cancer found in human stomach (24-38).

An attempt was made to clarify the effect on gastric carcinogenesis of 4-NQO combined with ABS (26). The combination of 4-NQO with ABS produced six cases of adenocarcinoma of the glandular stomach and other malignant neoplasms of the stomach. The experimental group exhibited more prominent tumor formation than the control group, i.e., carcinogenic action. The surface-active properties of the surfactant would cause the carcinogen to penetrate the protective mucous barrier of the stomach and attack the surface epithelium of the stomach directly.

The authors reported a similar experiment. Esophageal intubation of 8% ABS with 1-mg 4-NQO twice weekly for 18 weeks resulted in a substantial incidence of tumors of the glandular stomach and squamous cell carcinomas of the forestomach in comparison to the absence of these tumors in animals receiving the carcinogen alone (24-29). The incidence of forestomach papillomas induced by 4-NQO was not influenced by LAS (27).

Coadministration of 1000 mg liter^{-1} LAS with 100 mg liter^{-1} NG in the drinking water of rats for 63 weeks did not increase the incidence of adenocarcinomas of the glandular stomach when compared to animals receiving NG alone. Other papers reported the induction of tumors in rats by oral administration of NG in combination with various types of surfactants (30-32, 34). Cocarcinogenesis was revealed in tests in which rats received 50 ppm NG in combination with 0.25% AS, 0.5% *N*-lauroyl sarcosinate, or other types of surfactants in their drinking water for 6-7 months (33,35).

Those results were considered by these authors to suggest the possibility of surfactants being capable of increasing the activity of harmful chemical substances in the living environment, although there have never been reports indicating that surfactants themselves cause gastric carcinogenesis.

Borneff (39,40) conducted two large-scale experiments, taking account of the possible interaction of surfactants with the carcinogenic hydrocarbons that might exist in our natural environment. His concern was that surfactants might increase the concentration of such carcinogenic materials dissolved in water supplies. The first experiment demonstrated that the incidence of stomach tumors and carcinomas in the forestomach increased in mice which received normal food and water containing 3% detergent Pril (a mixture of anionic and nonionic compounds) and about 0.001% of a known carcinogen, 3,4-benzpyrene, for about 400 days. Three-hundred mice were allocated into five groups.

The second experiment was carried out on 400 male mice over a period of 500 days. Carcinomas of the phrenic ampulla of the esophagus developed only in those animals that had received benzpyrene solubilized by the detergent Pril in their drinking water (total dose per animal approximately 10 mg of benzpyrene and 35-40 g of Pril detergent). Benzpyrene in the soiled food, independently of the presence of mineral oil and detergent, induced only hyperkeratoses and papilloma.

Borneff concluded, in view of these results, that the use of household detergents presents no hazard, neither from enhancement of the carcinogenic action of other materials in the diet nor from toxicity of the detergent itself. As for his concern about potential hazards from drinking water derived from detergent containing river water, he recognized that the solubilization of carcinogenic hydrocarbons could not take place at the surfactant concentrations found in sewage and river waters, since these are far below the critical micelle concentration (CMC) (39-54). At that time (1964) the so-called nonbiodegradable tetrapropylene derived ABS was used predominantly in detergent formulations; Borneff recommended detergents that could be decomposed should be used. Such biodegradable detergent materials as LAS, AS, AES, and AOS have been used since the mid-1960s.

Sakharov et al. (55) studied Sulfonol NP-1, a detergent mixture of ABS, sodium sulfate and unreacted organic materials, for carcinogenicity and cocarcinogenicity. A 20% benzene solution of Sulfonol was applied to the shaved backs of mice (27 animals per group) twice weekly for 35 weeks. Other groups received 20-methylcholanthrene (0.5% in benzene) with or without Sulfonol. Although Sulfonol itself did not induce tumors, the high incidence (70.4%) of papillomas observed in mice treated with 20-methylcholanthrene and Sulfonol together, indicated a cocarcinogenic activity.

There have been several reports on the augmentive or depressive action of sodium dodecyl sulfate (AS) and other surfactants in tumorigenesis (56-61).

C. Solubilization of Carcinogens in the Environment

Janicke (62) pointed out the possibilities of incorporating toxic and/or carcinogenic materials into living organisms, or eluting harmful materials from the soil into ground water by the physicochemical properties of detergents.

As mentioned earlier, Borneff (39,40) also conducted studies on ABS for cocarcinogenic potential, taking account of the possible interaction of surfactants with the carcinogenic hydrocarbons which might be present in our environment. These matters are relevant to the fundamental physicochemical properties of surfactants themselves and are not confined exclusively to anionic surfactants. Although this is possible in theory, experimental evidence has been presented indicating that this should not take place at the surfactant concentrations found in drinking water, and river water, etc. The solubilization of polycyclic aromatic compounds by various kinds of surfactants can occur at surfactant concentrations above their own CMC (4,5,59). The surfactant concentrations encountered in our environment are far below the CMC.

From a practical viewpoint, this phenomenon is not a causative factor in our environment, because the possible hazards from this source are negligible.

II. MUTAGENIC/GENETIC TOXICITY PROPERTY

A mutation is a change in the information content of the genetic material that is propagated through subsequent generations of cells or individuals. Mutations arise spontaneously or may be induced by a variety of physical and chemical agents and can occur in either somatic or germ cells. The primary function of genetic toxicology testing is to investigate, using test cells or organisms, the potential of chemical substances to induce mutations in man that may be transmitted via the germ cells to future generations. Scientific data generally support the hypothesis that DNA damage in somatic cells is a critical event in the initiation of cancer. Such damage can result in mutations, and tests to detect mutagenic activity may also identify chemicals that have the potential to lead to carcinogenesis. Thus, some of the tests are useful for the investigation of putative carcinogenic activity (165).

The problem of man's exposure to the increasing numbers of substances in the environment has led to increased interest in the development of methods for detecting the possible mutagenic effects of these substances. Detailed surveys of mutation test systems, from bacteriophage to mammals, and in the literature [WHO Scientific Group (67,68), Ad Hoc Committee of the Environmental Mutagen Society and the Institute for Medical Research (69), Minister of Health and Welfare, Canada (70), OECD (165), IPCS (166), U.S. Environmental Protection Agency (167).].

The main types of anionic surfactants have been shown to be devoid of mutagenic activity when investigated by these tests. Useful reviews on the mutagenic and genotoxic potentials of surfactants are available (168-170).

A. Assays for Gene Mutation

1. Reverse mutation assay

The most widely used assays for detecting chemically-induced gene mutations are those using the bacteria *Salmonella typhimurium* or *Esherichia coli*. They are relatively simple to perform, reproducible, and give reliable data on the ability of a chemical to interact with DNA and produce mutations (165).

The *S. typhimurium* histidine (his) reversion system is a microbial assay which measures $his^- \rightarrow his^+$ reversion induced by chemicals which cause base changes or frameshift mutations in the genome of this organism. A reverse mutation assay using *E. coli* detects mutation in a gene of a tryptophan-requiring bacterial strain to produce a tryptophan-independent strain.

Four protocols for reverse mutation assays have been used to test anionic surfactants: the spot test, the plate incorporation assay, the pre-incubation assay, and the host-mediated assay. Generally, a liver S-9 metabolic activation system is incorporated in the first three, in vitro assays.

The negative results and the test conditions of anionic surfactants in reverse mutation assays are summarized in Table 2. Representative studies from Table 2 are reviewed below according to the type of anionic surfactant.

(a) Alkyl benzene derivatives. Using *S. typhimurium* TA 1535, 1536, 1537, and 1583, and *E. coli* B/r WP 2 try^- and try^- hcr^- as the tester strains, Negi et al. (72) examined the mutagenic potential of linear alkyl benzene sulfonates (LAS), α-olefin sulfonates (AOS), alkyl ethoxy sulfates (AES), and alkyl sulfates (AS) by a reversion plate test using the spot application of 100, 1000, and 10000 ppm test material disk papers. No mutagenic potential was found in the surfactants tested, though the need for careful colony observation was emphasized. With AES and AOS, *S. typhimurium* TA 1536 and 1538 showed smaller colonies than did the his^+ strains. These small colonies were further examined by a replicating method and were found not to be his^+-revertant colonies. They considered that the small colonies occurred only at higher concentrations of the test material because of the resultant inhibition of bacterial growth.

Studies were also conducted to examine the sensitivity of *S. typhimurium* TA 98, 100, 1535, 1537, and 1538 strains toward LAS, AOS, and AS using the plate assay procedure described by Ames and his colleagues (71). Negi et al. (74) found all these anionic surfactants to be inactive in the presence and/or absence of rat microsomal activating systems induced by the administration

TABLE 2 Summary of Results of Mutagenicity Tests Using Surfactant Materials in Various Strains of *Salmonella typhimurium* and *Escherichia coli*

Surfactant	Material	Remarks: Max. conc., giving negative	Ref.
LAS	*S. typh.* TA 1535/1536/1537/1538	10,000 ppm	72
LAS	*E. coli* WP2try, try hcr	10,000 ppm	72
LAS	*S. typh.* TA 98/100/1535/1537/1538	100 μg/plate	74
LAS	*S. typh.* TA 98/100	200 μg/plate	173
LAS, $C_{10\text{-}13}$	*S. typh.* TA 98/100/1535/1537/1538	50 μg/plate	177
LAS	*S. typh.* TA 98/100/1535/1537/1538	2000 μg/plate	172
LAS, $C_{12\text{-}13}$			
LAS, $C_{12\text{-}13}$	*E. coli* WP2, WP2uvrA	100 μg/plate	172
LAS	*S. typh.* TA 98/100	100 μg/plate	178
LAS	*S. typh.* TA 98/100	5.0 mg/plate	179
LAS	*E. coli* WP2try, try hcr	4.0 mg/plate	179
LAS	*S. typh.* TA 98/100	500 μg/plate	171
LAS-Ca	*S. typh.* TA 98/100	1000 μg/plate	171
LAS	*S. typh.* TA 1535/1537 (host-mediated assay)	Mouse i.m. 0.1 ml/1%	72
AOS	*S. typh.* TA 1535/1536/1537/1538	10,000 ppm	72
AOS	*E. coli* WP2try, try hcr	10,000 ppm	72
AOS	*S. typh.* TA 98/100/1535/1537/1538	100 μg/plate	74
AOS	*S. typh.* TA 98/100	100 μg/ml	178
AOS (7 preparations)	*S. typh.* TA 1535/1536/1537/1538	2 mg/plate	156
AOS (4 compounds)	*S. typh.* TA 1535	2 mg/plate	156
AOS, $C_{14\text{-}16}$ (AI, 28.4%)	*S. typh.* TA 1530/1534	283 mg/kg	159
AOS	*S. typh.* TA 1535/1537 (host-mediated assay)	Mouse i.m. 0.1 ml/1%	72
AOS (4 compounds)	*S. typh.* TA 1535 (host-mediated assay)	Mouse 5000 mg/kg	156
AOS, $C_{14\text{-}16}$	*S. typh.* TA 1530[a]/1534 (host-mediated assay)	Rat 283 mg/kg	159
AS	*S. typh.* TA 1535/1536/1537/1538	10,000 ppm	72
AS	*E. coli* WP2try, try hcr	10,000 ppm	72
AS	*S. typh.* TA 98/100/1535/1537/1538	100 μg/plate	74
AS	*S. typh.* TA 98/100	100 μg/ml	178
Sodium lauryl sulphate	*S. typh.* TA 98/100	1000 μg/plate	180
Sodium lauryl sulfoacetate	*S. typh.* and saccharomyces, six strains	1000 μg/plate	181

TABLE 2 (Continued)

Surfactant	Material	Remarks: Max. conc., giving negative	Ref.
lauryl sulphate-triethanolamine	*S. typh.* TA 98/100	1000 μg/plate	171
AS	*S. typh.* TA 1535/1537 (host-mediated assay)	Mouse i.m.i. 0.1 ml/1%	72
AES	*S. typh.* TA 1535/1536/1537/1538	10,000 ppm	72
AES	*E. coli* WP2try, try hcr	10,000 ppm	72
AES	*S. typh.* TA 98/100/1535/1537/1538	100 μg/plate	74
AES	*S. typh.* TA 98/100	100 μg/ml	178
$AE_{2.5}S$, C_{12-13}	*S. typh.* TA 98/100	1000 μg/plate	173
AE_3S, C_{12-15}	*S. typh.* TA 98/100/1535/1537/1538	2000 μg/plate	172
AE_3S, C_{12-15}	*E. coli* WP2, WP2 uvrA	500 μg/plate	172
AE_3S, C_{12}	*S. typh.* TA 98/100	1000 μg/plate	180
AES-triethanolamine C_{12}	*S. typh.* TA 98/100	1000 μg/plate	180
AES	*S. typh.* TA 1535/1537 (host-mediated assay)	Mouse i.m.i. 0.1 ml/1%	72
Mama Lemon (LAS base)	*S. typh.* TA 98/100 *E. coli* WP2try, try hcr	×1	73
Wonderful K (LAS base)	*S. typh.* TA 98/100 *E. coli* WP2try, try hcr	×1	73
Family (AES base)	*S. typh.* TA 98/100 *E. coli* WP2try, try hcr	×1	73
Soap	*S. typh.* TA 98/100 *E. coli* WP2try, try hcr	×1	73

[a]Positive Result.

of polychlorobiphenyl (PCB), 3-methylcholanthrene (3-MC), and phenobarbital (PB). Negative responses were observed at surfactant concentrations of 20, 50, and 100 μg per plate. At higher concentrations growth inhibition of the tester strains was observed, making it impossible to calculate the frequency of revertants.

The mutagenic potentials of sodium LAS, calcium LAS and triethanolamine alkylsulfate were examined using *S. typhimurium* TA 98 and 100 after activation with various liver homogenates (S-9). Polychlorinated biphenyl (PCB), phenobarbital (PB) and 3-methylcholanthrene (3-MC) were used as inducers, with the S-9 fractions of livers of rats, hamsters and guinea pigs.

It was suggested that all materials were not mutagenic in this system (Sunakawa, et al., 171).

Other bacterial assays with *S. typhimurium* strain TA 100 and 1535 (base-pair mutants) and TA 98, 1537, and 1538 (frameshift mutants) exposed to C_{12-13} LAS at concentrations up to 2000 μg/plate were negative with and without added microsomal activation (172). Negative results were also observed for C_{10-14} LAS (200 μg/plate) in *S. typhimurium* TA 98 and 100 (Inoue et al., 173) and for C_{12-13} LAS in *E. coli* WP2 and WP2 uvrA (172).

In the host-mediated assay, no indications of mutagenicity were noted with LAS, as shown in Table 2. Negi et al. (72) conducted host-mediated assays on LAS, AS, AES, and AOS. *S. typhimurium* TA 1535 or 1537 was injected intraperitoneally into the host mammal ddY mouse, as an indicator organism. The animal was treated with a 1% solution (0.1 ml) of the test sample by intramuscular injection immediately following the injection of the tester strain. Two doses of the test compound were given at 1-hr intervals. About 30 min after the final injection, 1 ml of saline was injected intraperitoneally, and the fluid was withdrawn from the peritoneal cavity of the host. The mutation frequency in the recovered microorganisms was measured by counting of the viable organisms. The frequencies of the mutants of *S. typhimurium* TA 1535 were within the range $10\text{-}29/10^8$ with all test compounds. In the case of *S. typhimurium* TA 1537, the frequencies were $40\text{-}70/10^6$. The order of mutation incidence found was similar to that in the control group, indicating that those anionic surfactants were not mutagenic in the host-mediated assay.

(b) α-Olefin derivatives. As shown in Table 2, a number of in vitro and host-mediated mutagenicity tests with AOS have produced negative results. A single set of experiments showed a positive response in a host-mediated assay with rats. A concentration of 283 mg/kg of a particular batch of C_{14-16} AOS (28.4% active ingredient) was found to be mutagenic in a host-mediated assay with rats when tested with *S. typhimurium* TA 1530 (a point mutant), but not mutagenic with *S. typhimurium* TA 1534 (a frameshift mutant) (159). In vitro plate assays with AOS concentrations up to 1% were negative for both strains. Host-mediated assays with equivalent AOS products from other suppliers were also negative (159). Other batches of AOS tested, in various reverse mutation assays were also negative. The single positive response may be due to materials unrelated to the surfactant.

(c) Sulfate-type anionic surfactants. None of the sulfate-type anionic surfactants, AS and AES, yielded positive results in reverse mutation assays (Table 2). Exposure in culture to C_{12-15} AES also did not increase the incidence of mutations in the bacteria, *S. typhimurium* (strains TA 98, 100, 1535, 1537 or 1538; 2000 μg/plate) and *E. coli* (WP2 and WP2 uvrA; 500 μg/plate) with or without liver microsomal activation (172).

(d) Formulated detergents based on anionic surfactants. Oda et al. (73) performed a reversion plate test on three types of commercial light-duty liquid detergents (Mama Lemon-LAS base, Wonderful K-LAS base, and Family-AES base) and a soap powder, using *S. typhimurium* TA 98, and 100 and *E. coli* B/r WP2 try^- and try^- hcr^- strains. The results were negative for all undiluted materials and for materials diluted ten-fold and one hundred-fold.

B. Assays for Chromosomal Aberration

Various test systems have been developed for the detection of agents that cause microscopically visible damage to mammalian chromosomes.

The simplest and most sensitive assays for investigating clastogenic (i.e., chromosomal breaking) effects are those involving cultured mammalian cells. These tests identify chemicals that are capable of damaging mammalian chromosomes in an in vitro experimental system. To investigate the ability of a chemical to produce chromosome damage in the whole mammal, two well-established in vivo procedures, the micronucleus test and the mammalian cytogenetic test, are available (165).

1. In vitro cytogenetic tests

The in vitro cytogenetic test is a mutagenicity test system for the detection of chromosomal aberrations in cultured mammalian cells. No chromosomal aberrations were noted in rat liver cells exposed to $C_{12\text{-}13}$ LAS at concentrations up to 100 μg/ml (172). The frequency of chromatid and chromosome aberrations in rat liver cells exposed in culture to 100 μg/ml of $C_{12\text{-}15}$ AES for 24 hr did not differ significantly from that of control cultures (172).

2. In vivo cytogenetic tests

The most important procedure for detecting chromosome aberration is the in vivo test, permitting the direct examination of cells from treated animals. In vivo cytogenetic studies commonly use mice or rats, and of many tissues, bone marrow is preferred for the examination of effects on the metaphase chromosome in mitosis. Inoue et al. (76) and Kunita (182) performed in vivo cytogenetic tests on LAS, ABS, and two types of commercial light-duty liquid detergent (Wonderful K and Mama Lemon, both LAS based detergents). ICR/JCL male mice received a single dose of each of the test compounds, except LAS, orally. LAS was administered orally for 5 days. The dosage levels for each compound were 1/2, 1/4 and 1/8 of the LD50. Bone marrow cells were collected up to 48 hr after treatments and the cells in metaphase were examined for chromosome abnormality. The frequency of chromosome aberration from animals sacrificed after 24 hr was nearly zero in all cases. A frequency of 35.3% was obtained with the positive control (5 mg kg^{-1} of mitomycin C). The types of chromosomal exchange considered to be of genetic importance were only observed in the mitomycin C treated animals.

Masubuchi et al. (77) reported a cytogenetic study on LAS in combination with the dominant lethal assay described below. The 9-month feeding study in ICR/JCL mice, in Wistar and in SD strain rats was to study the potential effect of LAS on the hemorrhagic death from butylated hydroxytoluene (BHT) in rats and mice. In this study, an experiment with LAS alone was included, which afforded useful evidence for supporting its inability to produce chromosomal aberration. Mice and rats were fed 0.9% LAS in the diet for 9 months and 50 cells in metaphase per animal were studied for chromosome aberration. As shown in Table 3, the chromosomal breaks were observed only in one cell in a mouse, and none of the other types of chromosome aberration, such as exchange, ring chromosome, pulverization, etc. were seen.

Four anionic surfactants were tested for their potential to induce chromosome damage in vivo after chronic administration in the diet at the maximum tolerated dose, (1.13% active ingredient in each case) in Colworth/Wistar rats for 90 days (Hope, 174). The four surfactants were sodium lauryl sulfate (SLS), DOBS 055, nominally dodecyl benzene sulfonate but with a chain length of 10 to 15 carbon atoms, Dobanol 25 sulfate LUC and Dobanol 25 sulfate HCB. The two Dobanol sulfates are aliphatic alcohol sulfates with a chain length of 10-15 carbon atoms. SLS and DOBS 055 were also fed at half the above concentration. Chromosome preparations were made from the bone marrow and scored for the presence of rearrangements, chromatid and isochromatid gaps and breaks. The four surfactants were found to have no effect on the chromosomes of rat bone marrow cells.

Similarly, no significant differences in chromosomal anomalies were found in the bone marrow cells of male rats given 40, 500 or 1000 mg/kg of a surfactant mixture (55% AES:45% LAS) orally, then killed at 18, 24, or 48 hr post-dosing (159).

3. Micronucleus tests

The micronucleus test is a mammalian in vivo test that detects chemically induced damage to the chromosome or to the mitotic apparatus. Polychromatic erythrocytes in the bone marrow of rodents are used. When an erythroblast develops into an erythrocyte, the main nucleus is extruded and may leave micronuclei in the cytoplasm. Visualisation of micronuclei is facilitated in these cells because they lack a nucleus. Micronuclei form under normal conditions. The assay is based on increase in the frequency of micronucleated polychromatic erythrocytes in the bone marrow of treated animals (165).

Kishi et al. (175) studied surfactants using mouse bone marrow cells. The surfactants used were ABS, LAS, dodecyl sulfate, Tween 60, benzalkonium chloride, and four types of shampoo. Mice received intraperitoneal injections at a dosage level of 100 mg kg^{-1}. At 24 hr post-dosing, polychromatic erythrocytes with micronuclei and polychromatic erythrocytes in all erythrocytes were observed. No adverse effect on the frequency of micronucleated polychromatic erythrocytes was observed.

Koizumi et al. (176) used the micronucleus test to investigate the mutagenicity of LAS administered in early pregnancy in order to evaluate effects on the following generation. On day 3 of gestation, LAS was administered in a single oral dose of 2 mg/mouse and on day 18 of gestation, each animal was sacrificed. On day 17, each animal received a subcutaneous dose of 1, 2, or 10 mg/mouse and was sacrificed 24 hr later. There was no difference among treated groups in the incidence of polychromatic erythrocytes with micronuclei in the maternal bone marrow and in the fetal liver and blood. No mutagenicity was found.

4. Rodent dominant lethal assay

Dominant lethal effects cause embryonic or foetal death. Induction of a dominant lethal event after exposure to a test substance indicates that the substance has affected germinal tissue in the test species. Dominant lethal effects are generally accepted to be the result of chromosomal aberration (structural and numerical anomalies), but gene mutation and toxic effect cannot be excluded (165). The common dominant lethal test procedure is to treat male mice or rats with a subtoxic dose of the test substance. The treated male animals are then mated with groups of untreated females. The females are sacrificed about 14 days after mating and scored for corpora lutea, for early fetal deaths, for late fetal deaths, and for total implantations.

As described previously, Masubuchi et al. (77) conducted a dominant lethal test using LAS in the course of their 9-month feeding study on the effect of LAS on BHT induced hemorrhagic death in the mouse and rat. Two female mice were mated with each male mouse which had been receiving 0.6% LAS in the diet for 9 months. The females were sacrificed on day 13 of pregnancy, dissected, and scored for corpora lutea, implantations, dead embryos, and live fetuses. No statistically significant differences were observed between the treated and control groups.

The potential mutagenicity of a commercial detergent containing 16% LAS was also studied in the dominant lethal assay, using CDF1 mice (159). This assay was performed in mice at single dosages up to 1000 mg kg^{-1} of detergent and at 200 mg kg^{-1} for 5 consecutive days. No increase in mutagenic index was found in the detergent-treated groups, while treatment with trimethylphosphate as a positive control resulted in a significant increase in the mutagenic index.

In a series of studies with a 55% AES:45% LAS mixture, no significant differences from control values were noted. Male mice received either 100, 150 or 200 mg kg^{-1} subacutely or 500, 750, or 1000 mg kg^{-1} acutely, no significant differences between treated animal and the water dosed vehicle control groups were observed in the mutagenic parameters examined (159).

TABLE 3 Chromosome Studies of Bone Marrow Cells After Long-Term Administration (9 Months) with Dietary BHT and/or LAS in Mice (ICR) and Rats (Wistar, SD)

Animal	Chemicals and doses, %	No. of animals	No. of cells observed	No. of cells with breaks		No. of cells with gaps		No. of cells with other aberrations[a]
				Chromatid	Isochromatid	Chromatid	Isochromatid	
Mice (ICR)	Control	5	250	2	0	5	0	0
	BHT(1.5)	5	250	0	0	3	0	0
	LAS(0.9)	3	150	1	1	4	0	0
	BHT(1.5) + LAS (0.9)	5	250	1	0	4	1	0
Rats (Wistar)	Control	5	250	1	0	3	0	0
	BHT(1.5)	5	250	0	0	2	1	0
	LAS(0.9)	5	250	0	0	3	0	0
	BHT(1.5) + LAS (0.9)	5	250	0	0	1	0	0
Rats (SD)	Control	5	250	0	0	4	0	0
	BHT(1.5)	5	250	0	0	2	0	0
	LAS(0.9)	5	250	0	0	4	0	0
	BHT(1.5) + LAS (0.9)	5	250	0	0	2	0	0

[a]Exchange, ring chromosome, pulverization.

5. Cell-transformation tests

Cell-transformation tests measure the ability of a chemical to cause changes associated with the tumorigenic process in cultures of mammalian cells. Anionic surfactants (LAS, AES) have been tested in assays measuring the morphological transformation of hamster cells (Inoue et al. 173). The hamster cells used were secondary cultures derived from embryo cells of the Syrian golden hamster. No morphological transformation of the hamster embryo cells exposed in culture to $C_{10\text{-}14}$ LAS at 0.5, 1, 5, 10, 20, or 50 μg/ml were noted; cytotoxic effects occurred at the 50 μg/ml level. Additionally no morphological cell transformations were observed at concentrations up to the 50 μg/ml level of $C_{12\text{-}13}$ AES.

C. Assays for DNA Effects

1. Bacterial DNA repair tests

Bacterial DNA repair or damage is determined as differential cell killing or as growth inhibition in a series of repair proficient and deficient bacterial strains. These tests do not measure mutagenic events per se, but indicate the interaction of a chemical with genetic amterial and potential genotoxicity (167).

Two bacterial DNA repair tests have been used to study surfactants: the *E. coli polA* test and the *Bacillus subtilis rec*-assay. Anionic surfactants and surfactant mixtures have been tested in the *rec*-assay as shown in Table 4. None of the surfactants produced DNA damage in bacteria.

Negi et al. (72) studied two types of AOS, AS, LAS, and AES for mutagenic potential in a recombination assay using *Bacillus subtilis* H-17 (*Rec*$^+$) and M-45 (*Rec*$^{-1}$). They were negative at dosage levels of 100 and 1000 ppm. Oda et al. (73) also tested the three commercial light-duty liquid detergents and the soap powder mentioned above. The negative results obtained undiluted products, and ten-fold and one hundred-fold dilutions.

2. Sister chromatid exchange assay

The sister chromatid exchange (SCE) assay detects the ability of a chemical to enhance the exchange of DNA between two sister chromatids in a duplicating chromosome. The test may be performed in vitro using, for example, rodent or human cells, or in vivo using mammals, for example, rodents such as mice, rats and hamsters (167).

Only one series of SCE experiments using anionic surfactants is available. Inoue et al. (75) and Kunita (182) examined the effect of tetrapropylene-derived ABS on the frequencies of SCE by incubating lung cells of the Chinese hamster for 45 min in media containing 5 μg ml^{-1} of 5-bromodeoxyuridine and 0, 3.9, 15.6, or 62.5 μg ml^{-1} of ABS. Mitomycin C (1 g ml^{-1}) was used as a positive control. The chromosome preparations obtained from the cultured

TABLE 4 Summary of Results of DNA Repair Tests in *Bacillus subtilis* After Exposure to Surfactants Materials (Rec assay)

Surfactant	Remarks Max. conc., giving negative	Ref.
LAS	1000 ppm	72
LAS	20 μg/disk	178
LAS, $C_{10\text{-}14}$	800 μg/disk	173
LAS, $C_{12\text{-}13}$	100 μg/ml	172
LAS	4 mg/disk	179
AOS (AI, 79%)	1000 ppm	72
AOS (AI, 29.1%)	1000 ppm	72
AOS	20 μg/disk	178
AS	1000 ppm	72
AS	20 μg/disk	178
AS	10 mg/ml	180
AES	1000 ppm	72
AES	20 μg/disk	178
AES, $C_{12\text{-}13}$	2000 μg/disk	173
AES, C_{12}	10 mg/ml	180
AES-Triethanolamine	10 mg/ml	180
Sodium-*N*-cocoyl taurate	10 mg/ml	180
Mama Lemon (LAS base)	×1	73
Wonderful K (LAS base)	×1	73
Family (AES base)	×1	73
Soap	×1	73

cells were stained with Hoechst 33258 and then with fluorescent plus Giemsa staining. The frequencies of SCE were counted in 30 cells in metaphase in each concentration. The calculated frequencies were 6.9, 6.1, and 6.0 per cell at ABS concentrations of 3.9, 15.6, and 62.5 $\mu g\ ml^{-1}$, respectively. No statistically significant difference in frequency was observed between the treated groups and the negative control group in which the frequency was 6.5 per cell.

In conclusion, as mentioned in this chapter, many kinds of mutagenic and genetic toxicity tests using anionic surfactants were conducted in submammalian and mammalian systems, and all test results indicate that the principal anionic surfactants now in use have a negligible potential to cause gene mutation and chromosomal damage.

III. TERATOGENIC PROPERTY

Possible adverse influences on the development of the fetus include such effects as lethality, malformation, and retardation of growth. These are called embryotoxic effects. A teratogen or teratogenic agent is an agent that causes malformation of the fetus. The type of embryotoxic effect may vary with dosage level and the time at which the fetus is exposed. Information about the effects of chemicals on neonatal morbidity, mortality, and teratogenesis can be obtained from studies evaluating effects on fertility and general reproductive performance as well as from studies on teratogenesis direct. To evaluate chemically induced effects on fertility and reproduction, the so-called multigeneration test is recommended. This test is used to investigate effects on fertility, on gestation, and on the offspring, e.g., toxicity in the parent male and/or female, the development of the fetus, its normality including teratogenic and mutagenic effects, intrauterine mortality, lactation together with growth and development of the offspring.

To study anionic surfactants for these properties, many studies using the major anionic surfactants, ABS and LAS in particular, have been performed in rats, mice, and rabbits. The results indicate no danger of teratogenicity. In addition, single and multigeneration studies in rats revealed no effects on the general reproductive process attributable to the chronic intake of these compounds.

However, since 1969, Professor Mikami and his colleagues (78-98,183-186) (hereafter called the Mikami group) of Mie University (Department of Anatomy, Faculty of Medicine, Tsu city, Japan) have reported that commercial light-duty liquid detergents and surfactants caused significantly increased malformations in mice and rats follfowing oral, dermal, and subcutaneous administration. As a result, other laboratories, especially in Japan, have devoted much effort to checking various aspects of their results. The adoption of higher dosages during organogenesis, aiming particularly at teratogenic induction, has failed to substantiate such findings in mice, rats, or rabbits, although dosages causing maternal toxicity also induced some marginally significant reduction in intrauterine growth and increased rates of resorption.

Those studies with negative results are tabulated in Appendix 2, and the Mikami group's work is summarized in Appendix 3. Resulting from the innovatory nature of the Mikami group's reports, much public attention was attracted to the possibility that dermal applications of commercial detergents might be teratogenic. To scrutinize these reports, the Environmental Sanitation Bureau of the Japanese Ministry of Health and Welfare planned and promoted a series of group studies during 1973-1976. It must be noted that the results of the final joint project studies of the teratogenicity of LAS, in which investigators from four universities, including Dr. Mikami following precisely the same protocol, failed to support the reports made by the Mikami group. Reference details are provided below.

Useful reviews on the teratogenicity of the alkyl benzene sulfonates are available (99,169-170,187-191).

A. Teratogenicity

1. Sulfonate-type anionic surfactants

(a) Alkyl benzene derivatives.

1. Oral administration. The embryotoxicity of ABS was first investigated by Tanaka (100,101). The median fetal dose (FLD 50) in ICR/JCL mice was found to be 120 mg kg^{-1} by oral administration on day 4 of pregnancy, 60 mg kg^{-1} on day 7 and 130 mg kg^{-1} on day 7 in dd mice.

Omori et al. (102) published a more complete study of the effects of ABS in ICR/JCL mice. Dosages of 0, 24, and 240 mg kg^{-1} day^{-1} were administered by gavage to groups of 22-24 animals from days 7-13 of pregnancy. Slight body weight reduction and diarrhoea were seen in test dams during treatment, but body weight had recovered to levels comparable with controls by day 18. Gavage was associated with reduced food consumption but was not thought to have interfered with placental development. At 240 mg kg^{-1}, fetal death and resorption were increased, and there were some macerated fetuses. The incidence of abnormalities was similar in both treated and control animals, and even when maternal toxicity was induced the development of the young was not adversely affected. Pregnant rats were also exposed to levels of 0, 0.025, 0.1, 0.5, 1.0, and 2.0% ABS in the diet throughout pregnancy. The groups consisted of 14-16 rats, with the exception of the group fed the 2.0% dietary level, which contained only five animals. The malformations found were club foot in a fetus at a dosage level of 0.025% and exencephaly in three of the offspring of one dam fed 1.0% ABS. As the rest of the litter from this dam showed poor development and survival, the authors felt this finding was not significant.

Stoklosa et al. (103,104) studied the teratogenicity of keryl benzene sulfonate (KBS). KBS was administered to Wistar rats intraventricularly at doses of 0 and 40 mg kg^{-1} day^{-1} throughout pregnancy. KBS had no effect upon pregnancy or the number, sex, and weight of the litter but caused hepatotoxicity in the dams and newborn rats.

LAS teratogenicity following oral administration has also been studied by many investigators (see Appendices 2 and 4). No reports revealed teratogenicity except some from the Mikami group.

Among others, Palmer et al. (105,106) carried out comprehensive teratogenicity tests using LAS in CD-1 mice, CD rats, and New Zealand White rabbits and found that the incidence of major malformations in all species was unaffected. Signs of material toxicity were observed at the highest dose of 600 mg kg^{-1} day^{-1} in rats, and 300 mg kg^{-1} day^{-1} in mice and in rabbits, and maternal susceptibility was ranked in the order rabbits > mice > rats. Similarly, tolerance can be achieved in rats exposed to higher levels of LAS as shown by Kotani et al. (107) and Tiba et al. (108,109), who found no adverse effects in maternal, embryonic, and neonatal rats of the SD-JCL strain following the ingestion of 1.0% LAS in the diet throughout pregnancy.

Pregnant Wistar rats and New Zealand White rabbits received 0.1% LAS in drinking water (Endo et al., 148). No significant differences in maternal histopathology, numbers of corpora lutea or in the number of implantations were noted in dams killed on day 20 in rats and on day 30 in rabbits. Except for retarded ossification of the sternebrae in treated rabbits, embryonic development and litter parameters were normal in both species.

The magnesium salt of LAS (LAS-Mg) was evaluated by gavage in rabbits from days 6-13 of pregnancy, at dosages of 60, 125, and 250 mg kg^{-1} day^{-1}. Uterine contents were examined on day 29. LAS-Mg had no adverse effects upon fetal morphology, although at dosages of 125 or 250 mg kg^{-1} day^{-1} survival in utero was impaired. At dosages of 60 mg kg^{-1} day^{-1} or above, there was some impairment of maternal economy, but no effect upon the fetus (194).

Common findings reported in most of the LAS studies were suppression of the maternal body weight gain and a reduced pregnancy rate observed at higher dosage levels. Even at maternally toxic levels, no statistically significant differences in malformation rates between control and treated groups were observed. The types of malformations manifested in treated groups were the same as normally occurring malformations observed in untreated control animals. One of the most important points in the evaluation of a teratogenicity study is whether the malformation rate in the control group falls within the laboratory malformation incidence range. Information on spontaneous malformations in those test animal strains can be found in Refs. 110-116.

2. Dermal application. Absorption through the skin should be considered as in daily usage unprotected hands may often soak in detergent solution. The potential hazardous effect on the embryo from a trace amount of environmental chemicals invading the human body, had hardly been considered (117). The Mikami group introduced teratogenicity testing, in commercial light duty liquid detergents for kitchen use and LAS, using skin painting in ICR/JCL mice. They stressed that the test materials showed definite teratogenic potential, as they had observed in the oral studies.

Administration of 0.5 ml of 1% LAS solution to an area 3 × 4 cm of clipped skin daily from day 0 to day 11 of pregnancy resulted in incidences in the fetus of 100% subcutaneous congestion, 49.5% cleft palate, 100% bleeding, and 9.9% club foot or similar limb anomalies (98).

On the other hand, no teratogenic effect was observed in skin-painting studies using LAS in mice (ddY-S, ICR/JCL, CD-1), rats (CD, Wistar), or New Zealand White rabbits (118-123,145,192,195-196). For example, Masuda et al. (119) administered 0.5 ml LAS solution dermally in ICR/JCL mice at

0, 0.85, 1.7, 2.55, and 3.4% from day 1 to day 13 of pregnancy, and in ddY strain mice at 0, 0.017, 0.17, and 1.7% from day 2 to day 14 of gestation. The results were summarized as follows: (1) Neither suppression of body weight gain nor incidence of visceral defects of the dams were observed. However, a reduction in pregnancy rate was observed in the 3.4% group, where considerable skin irritation occurred at the site of application. (2) Slight growth suppression of the live fetuses in ICR/JCL mice was observed in the 0.85, 2.55, and 3.4% groups. Cleft palate (in two fetuses from one dam receiving 2.55% and in one fetus from one dam receiving 3.4%) and open eyelids (two fetuses from one dam and one fetus from one dam receiving 0.85% and 3.4%, respectively) occurred. In ddY mice, cleft palate was observed in one fetus from each of two dams that received 0.017 and 0.17% LAS. There was no statistically significant difference in fetal abnormalities between the test and control groups. (3) Retarded ossification of sternebrae in ICR/JCL mouse fetuses was observed more frequently in both 2.55 and 3.4% groups compared to the control. No skeletal abnormalities attributable to LAS application were observed. Masuda et al. (119) determined that there was no conclusive evidence of teratogenic effects attributable to LAS application, even at high concentrations.

The influence of topical LAS-Mg upon organogenesis in pregnancy was assessed using the Charles River CD rats (195). LAS-Mg solution (1.75, 3.5, and 7.0% in 3% polyethylene glycol (PEG) was administered to the female dorsal skin at a dosage level of 0.5 ml per animal, from day 7 to day 17 of gestation inclusive. On day 20, two-thirds of the females were killed and their uterine contents examimed; the remainder gave birth naturally and reared their young to weaning. Randomly selected F1 offspring were then reared, untreated, to maturity at approximately ten weeks of age, when they were mated within groups. The F1 females were killed and examined on day 20 of gestation, at which point the study was terminated.

A dose-related local erythematous reaction to treatment was observed during the dosing period, but this regressed completely on completion of treatment. With the exception of slight, nonsignificant depressions in body weight gain, there were no adverse effects on the general condition, mating performance, fertility or development attributable to LAS-Mg treatment.

Using pregnant rabbits, topical LAS-Mg was applied from day 6 to day 18 of gestation at concentrations of 0.75, 1.5, or 3.0% in 3% PEG at a constant dosage volume of 5 ml/animal. On day 29 of gestation all animals were killed to permit examination of the uterine contents. Reaction to treatment took the form of localised erythema and hyperkeratinization at the site of application, but no systemic effects were observed. Maternal body weight gain and food and water consumption showed no treatment related inter-group differences. Three rabbits (two in the low dosage group (0.75%)

and one in the intermediate dosage group (1.5%)) aborted, but this was not considered to be treatment-related. Litter response and fetal development were similar in all groups. It was concluded that the repeated topical application of LAS-Mg at concentrations of up to 3% in pregnant rabbits during organogenesis was without adverse effect upon the progress and outcome of the pregnancy (196).

As mentioned previously, a reduction in pregnancy rate was observed in animals treated dermally with higher concentrations of LAS. This effect was common to various routes of administration (122). Masuda et al. (119) also indicated in the skin-painting studies in ICR/JCL mice mentioned above that the percentages of pregnancy achieved were 69.0, 84.0, 53.3, 56.7, and 33.3%, at LAS levels of 0 (water), 0.85, 1.7, 2.55, and 3.4%, respectively. In addition, various degrees of skin reaction such as erythema, edema, and escharification were observed in the dams at the site of application, depending upon the LAS level. Such skin-irritation responses were reported by Palmer et al. (120,121), who conducted teratology studies on LAS and soap in rats, mice, and rabbits: with 3.0% LAS, severe local reaction occurred in mice and rabbits; a lesser degree of reaction may have occurred in rats but this could not be conclusively assessed because of accumulation of the material administered. Initial reaction consisted of erythema and edema with peak response being attained by day 6 in mice, by days 4-5 in rats, and by days 6-7 in rabbits. In the latter species, the peak response was a cracking and bleeding of the skin, in mice it was dead skin with the accumulated material forming a layer over the entire area of administration. When this layer was shed, after the end of the dosing period, new skin was apparent. Recovery was also evident in rats and rabbits after the peak response had been attained. In rats and mice treated with soap solution the local reactions at 30.0% were identical with or slightly more severe than those following treatment with 3.0% LAS. Clearly defined local reactions were not apparent at lower concentrations of LAS or soap, but milder reactions could have been masked by the accumulation of test material. Accompanying the local reactions at these high dosages, animals became increasingly irritable, with irritability peaking at the same time as the local reaction. In rats and mice irritability was also observed with LAS, 0.3% and soap, 3.0%. Neither local reaction nor irritability was observed in animals receiving LAS, 0.03% or soap, 0.3%.

Many factors have been shown to initiate abnormal development of the fetus in laboratory animals in addition to chemical agents when introduced during pregnancy. These include dietary deficiencies, viral infection, hyperthermia, hormonal imbalance, and various stress conditions (137,138,140). Thus, severe skin irritation resulting from the dermal application of higher concentrations of anionic surfactants may well cause secondary effects on pregnancy because of stress.

One notable point from the skin-painting studies was the problem of slight live-fetus body-weight reduction (119,123). Further studies will reveal whether the primary factor lies in direct effect on the dams, such as stress from skin irritation and disturbance of the cutaneous respiratory function, or in direct effects on the fetus from percutaneously absorbed LAS. Another notable point was the definitely dose-related retarded ossification of sternebrae in mice reported by Masuda et al. (119). Sato et al. (118) and Takahashi et al. (122) also reported an increased incidence of retarded ossification at higher dosage levels in other strains of mice; however, Imahori et al. (123) did not find any such effect even using 3.0% LAS.

In the dermal application studies reported by Palmer et al. (120,121), rats showed the highest tolerance to exposure to anionic surfactants in comparison to mice and rabbits. Only slight skin irritation was seen in rats treated with 3.0% LAS dermally. No statistically significant differences in the incidences of skeletal variants and internal or external major malformations were observed between treated and untreated groups of CD rats. No conspicuous fetal effects were found in rabbits given 3.0% LAS dermally, though rather severe skin irritation occurred at the site of application in the dams. The results of these studies indicated that embryotoxicity occurred largely where there was maternal toxicity, and the litter-oriented pattern of response suggested this was secondary to maternal toxicity.

3. *Subcutaneous injection.* No definite evidence of teratogenic potential and adverse effects on fetuses and newborn pups was obtained in ICR/JCL mice given 50 mg kg^{-1} day^{-1} of LAS by subcutaneous injection (124). The same result was reported by Takahashi et al. (122). External abnormalities and subcutaneous congestion were not evident in mouse fetuses from dams previously injected with 20 ml kg^{-1} day^{-1} of a 1.0% LAS solution, in contrast to the definite congestion reported by Mikami et al. (89) in Wistar rat fetuses following the subcutaneous injection of a 1.0% LAS solution at 10 ml kg^{-1} to the dam.

To examine the reduction in pregnancy rate, Takahashi et al. (122) administered LAS to mice by subcutaneous injection during the early stages of gestation (from day 0 to day 3) or during the period of organogenesis (from day 8 to day 11),, and in the former case confirmed that a significantly reduced pregnancy rate occurred.

(b) α-Olefin derivatives. A series of comprehensive teratogenicity studies on C_{14}-C_{18} α-olefin sulfonates (AOS) was carried out in rats, mice, and rabbits. Palmer et al. (125,126) studied AOS for teratogenic potential in CD-1 mice, CD rats, and New Zealand White rabbits by oral administration and found no evidence of teratogenic potential.

Oral administration of a magnesium salt of AOS (AOS-Mg) gave no indication of any embryotoxic or teratogenic effects in rats (197) and rabbits

(198). In rats of the Charles River CD strain, AOS-Mg was administered by gavage at dosages of 150, 300 and 600 mg kg^{-1} day^{-1} from day 7 to day 17 of gestation. Control animals received the vehicle, distilled water alone. On day 20 of gestation, two-thirds of the females from each group were killed and the fetuses recovered. The remaining dams were allowed to deliver normally and to rear their young to weaning. Randomly selected offspring of the F1 generation were then reared, untreated, to maturity, and at approximately ten weeks of age, mated within groups. The F1 females were killed and examined on day 20 of gestation, when the study was terminated. At dosages up to 600 mg kg^{-1} day^{-1}, the oral administration of AOS-Mg to pregnant female rats during organogenesis was without adverse effect upon pre- and postnatal survival and development, and upon the subsequent mating performance and fertility of the F1 offspring. It was concluded that the daily oral administration of AOS-Mg to pregnant rabbits during organogenesis, at dosage levels up to 300 mg kg^{-1} day^{-1}, had no adverse effect upon the dam or the developing fetus.

In addition, AOS-Mg was administered to pregnant rabbits, by gavage, from day 6 to day 18 of gestation inclusive, at dosage levels of 75, 150 and 300 mg kg^{-1} day^{-1}. The animals were killed on day 29 for examination of their uterine content (198). One female receiving 75 mg kg^{-1} day^{-1} exhibited total resorption and one at each dose level of 150 and 300 mg kg^{-1} day^{-1} aborted. No adverse effects attributable to treatment were recorded in litter responses or in fetal development studies. Daily oral administration of AOS-Mg to pregnant rabbits during organogenesis, at dose levels up to 300 mg kg^{-1} day^{-1}, had no adverse effect upon the dam and the developing fetuses.

Following oral administration studies, Palmer et al. (127,128) conducted skin-painting experiments using the same AOS material in the same strains of rats and mice. Concentrations of 0, 1, 5, and 20% of AOS active ingredient were administered topically during days 1-16 of pregnancy at a rate of 0.25 ml per mouse per day. Twenty four hours after each application, the skin was washed with warm water and blotted dry. Additionally, a further group of mice was similarly treated with 20% AOS except that the skin was washed and blotted dry 30 min after each application. The exposure to 20% AOS for 24 hr each day was associated with the occurrence of well-defined erythema and edema, a marked retardation of maternal weight gain, and failure to maintain pregnancy, with total litter loss occurring just before or just after implantation. However, exposure to 20% AOS limited to 30 min each day reduced the above effects by a considerable degree. Among dams with viable young there were no significant effects on litter size, litter weight, mean fetal weight, nor incidences of major malformations, minor visceral anomalies, minor skeletal anomalies, or skeletal variants. Exposure to 1 or 5% AOS for 24 hr each day was associated with a minimal retardation of maternal weight gain until day 17 of pregnancy, but there were no significant differences from control values with respect to local reactions, pregnancy rate, total litter losses, litter size, postimplantation loss, litter weight, mean fetal weight, or incidences of major malformations, minor visceral anomalies, minor skeletal anomalies, or skeletal variants.

In the rat, the same experimental procedures and conditions as in the mouse experiment were adopted, except the administration period, which lasted from day 1 to 19 of pregnancy. Parent animals showed no signs of systemic reaction and there were no deaths. Mild local reaction (slight erythema) occurred in all test groups, the duration of effect (number of days) and the number of animals affected being related to the degree of exposure, viz. a well-defined erythema or edema persisting not more than 1 or 2 days occurred in, respectively, 2/20 and 3/20 animals at the highest exposure (20% for 24 hrs). Body weight gain in all groups was marginally lower than that in controls but differences were unrelated to the degree of exposure. The pregnancy rate was essentially similar for all groups and intergroup differences in preimplantation loss were not statistically significant. No evidence of teratogenic potential was observed.

In another study, mice were treated percutaneously from day 0 to day 14 of gestation with a 0.5, 5 or 25% aqueous solution of AOS. The treatment had no toxic effect on the pregnant dams, or on litter parameters such as fetal body weight and sex ratio. AOS did not induce any teratogenic response (209).

Topical application of AOS-Mg resulted in no indications of any teratogenic or embryotoxic effect in rats (199) or rabbits (200). AOS-Mg was administered topically at concentrations of 1.75, 3.50, and 7.00% to female rats from day 7 to day 17 of pregnancy, and to pregnant rabbits from day 6 to day 18 of gestation at concentrations of 0.75, 1.50, and 3.00%.

On day 20 of gestation in the rat, two-thirds of the females from each group were killed for examination of their uterine content; the remainder were allowed to give birth naturally and to rear their young to weaning. Randomly selected offspring of the F1 generation were then reared, untreated, to maturity at approximately ten weeks of age, when they were mated within groups. The F1 females were killed and examined on day 20 of gestation, at which point the study was terminated. Rabbits were killed on day 29 of gestation, when their uterine horns were opened, and the contents examined.

It was concluded from the rat study that topical administration of AOS-Mg to F0 females during organogenesis, at concentrations up to 7.00%, was possibly associated with some small changes in fetal morphology, but these were not confirmed in the F1 offspring from females allowed to litter. In all other respects treatment with AOS-Mg was without significant effect upon the course or outcome of pregnancy, or upon the subsequent growth, development and fertility of the F1 offspring. In the rabbit, AOS-Mg applied topically, at concentrations up to 3% during organogenesis, had no adverse effects on fetal development or survival (200).

2. Sulfate-type anionic surfactants

In the case of alkyl sulfates (AS), positive indications of teratogenic potential have been reported only by the Mikami group. Mikami et al. (91) found definite maternal and fetal toxicity in ICR/JCL mice receiving AS (sodium lauryl sulfates, "Hyren Powder") both orally and dermally. Oral administration was performed by gavage at a dosage of 10 ml kg^{-1} of a 2.0% AS solution on days 6-13 of pregnancy. Dermal application was by skin painting at a dosage of 0.25-0.3 ml per animal (3 × 3 cm clipped area) of a 2% AS solution on days 7-13 of gestation and a 10 or 20% solution on days 1-16 of gestation. According to the limited data available, significant increases in the numbers of dead fetuses, live fetuses with hemorrhage on the back, foot malformations, and a significant depression of fetal body weights resulted from both oral and dermal administrations.

Sodium dodecyl sulfate (AS) was examined for potential teratogenic and embryotoxic activity (105,106). The material was administered by gavage during days 6-15 of pregnancy in CD rats and CD-1 mice, days 6-18 of pregnancy in CD rats and CD-1 mice, and days 6-18 of pregnancy in New Zealand White rabbits. Dosages employed were 0 (water), 0.2, 2, 300, and 600 mg kg^{-1} day^{-1}. Slight to moderate maternal toxicity was observed in rats receiving 300 and 600 mg kg^{-1} day^{-1}, and in rabbits and mice receiving 300 mg kg^{-1} day^{-1}. Marked maternal toxicity in the form of anorexia, death, and abortion (and/or total resorption) was evident only in rabbits and mice receiving 600 mg kg^{-1} day^{-1}. No effects on morphogenesis were observed. Effects on litter parameters were generally related to dosages causing marked maternal toxicity, the principal effects being higher fetal loss (with consequent reduction in litter size) arising from the increased incidence of total litter losses. When dams showing total litter loss were excluded from the calculations, the litter parameters were not unduly different from those of controls. At dosages that were either nontoxic or only slightly to moderately toxic to the dam, litter parameters were essentially unaffected.

Takahashi et al. (129) studied AS teratogenic potential by skin painting in ICR/JCL mice. Open eyelids were observed in all groups including the control. The incidence of cleft palate at dosage levels of 4.0 and 6.0% attracted attention even though not statistically significant. A statistically significant depression of fetal body weight and a slight retardation of fetal growth were noted at the two higher dosages. Since a definite reduction in pregnancy rate was observed at the highest dosage level of 6.0%, an additional test at a dosage level of 6.0% was carried out. Pregnancy rates of 100% and 78% were achieved in dosing from days 7-14 and days 8-15 of pregnancy, respectively.

The results of teratogenic and embryo toxicities studied using alkyl ethoxy sulfate (AES) also were reported.

In the assessment of AES-Mg on pregnancy in the rat and on development and behavior of the F1 generation to maturation, daily dosages of 0, 130, 260, and 520 mg kg^{-1} were administered orally to F0 females during days 7-17 of gestation (201).

Two thirds of the females in each group were sacrificed on day 20 of pregnancy to investigate possible embryo toxicity and teratogenicity; those remaining were allowed to litter normally and investigated for possible latent defects. At weaning, 1 male and 1 female pup per litter were selected to form the basis of the F1 generation and were maintained untreated until at least 15 weeks of age. During this period, development and behavior were assessed. Finally the reproductive performance of the F1 generation was assessed. There was no evidence of teratogenic or adverse effects on the F0 generation, or on development of the F1 generation to maturity as assessed by growth, performance in behavior tests, mating, rate of pregnancy and subsequent litter values at day 20 of pregnancy.

Percutaneous studies using AES and AES-Mg were also carried out in rats and rabbits (117,203-204) showing no increase in the incidences of any teratogenic effect.

AES-Mg was also administered orally to the pregnant New Zealand White rabbit at dosage levels of 0, 75, 150, and 300 mg kg^{-1} day^{-1} from days 6-18 of gestation. Embryonic development and litter parameters were normal and the incidences of major malformations, minor visceral and skeletal anomalies and skeletal variants were considered to be unaffected by treatment (202).

3. Carboxylate-type anionic surfactants (soaps)

Palmer et al. (120,121) compared soap (potassium salt of coco fatty acid) with LAS for potential teratogenic and embryotoxic activity in CD rats and CD-1 mice following dermal application. Dosages employed were 0, 0.3, 3.0, and 30.0%, and no teratogenic potential was evident. Systemic and local effects were manifest in mice at the two higher levels and the numbers of dams bearing live fetuses were accordingly markedly reduced (0/20 animals and 7/20 animals at concentrations of 30.0, and 3.0%, respectively). Morphological effects on the litters in treated groups were similar to those in the control group.

A commercial soap powder, Nissan powder soap, and sodium α-sulfoalkyloyl lactylate (TMS) were studied for embryotoxic and teratogenic potential (90). The dosages adopted were 10 ml kg^{-1} of the 0.2% soap solution and 0.2 and 1% of the TMS solutions. Subcutaneous injections were given to mice of an unspecified strain on days 11-13 of pregnancy (except the 1% TMS group, which was dosed on days 8-11). Curiously, no negative control such as normal saline was used in the experiment and the results obtained with treated groups were compared with those from reference groups treated with the commercial detergent Mama Lemon. No embryotoxic potential was reported for either soaps or TMS.

4. Formulated detergents based on anionic surfactants

(a) Oral administration. Studies using commercial detergents in ICR/JCL mice reported by the Mikami group can be put in order as follows:

1. Two groups of 15 mice were treated by gavage with a 1.5% solution of a commercial light-duty liquid detergent for kitchen use (LAS type) of undefined specification at volumes of 0.021 or 0.042 ml g^{-1} on days 6, 8, and 10 of pregnancy (78,82,95-97). The actual dosage levels could be as little as 2.75 and 5.5 mg kg^{-1}, or as much as 104 and 208 mg kg^{-1} according to a comment made elsewhere. Some maternal toxicity was observed and, in comparison with the nine control animals, the test groups showed an incidence in fetal resorption, a reduction in fetal weight, and a definite increase in the incidence of cleft palate and of exencephaly.

2. An undefined commercial light duty liquid detergent for kitchen use (based on fatty alcohol derivatives) was administered orally at doses of 210 and 630 mg kg^{-1} on days 6-11 of pregnancy. The result showed similar abnormalities (94).

3. Four commercial household detergents for kitchen use (My Lemon, Mama Lemon, Family, and Wonderful K soft) were administered at dosages of 0.63 ml kg^{-1} to groups of 13-18 mice on days 6-11 of pregnancy. Thirteen control mice were treated with 0.6 ml normal saline. The author's concluding remarks indicate that the doses were of the order 104 mg ABS kg^{-1}. Food consumption of the dams was reduced in the test groups, maternal toxicity was reported, and increases in fetal resorption, of abnormalities of the ribs and sternebrae, and a high incidence of cleft palate were observed. It was concluded that these detergent products were teratogenic. A supplementary report described further pathological examinations of preserved fetuses revealing definite fetal anomalies such as dorsal hemorrhage (80,82,93,97).

Sometimes it is not clear as to what, if any, statistical evaluation was made. The numbers of animals used and the numbers of dosage levels tested were below those normally regarded as necessary for adequate interpretation of this kind of experiment.

On the other hand, following oral administration of several types of formulated detergents in rats, mice, and rabbits, no teratogenicity was recorded (105,106,130-132,147). Palmer et al. (105,106) examined a commercial light duty liquid detergent for kitchen use in teratogenicity and embryotoxicity studies using CD rats, CD-1 mice, and New Zealand White rabbits. The test detergent, Mama Lemon, contained 17% LAS and 7% AES, and dosages were 0, 0.8, 8.0, 1200, and 2400 mg kg^{-1} day^{-1} in rats and rabbits and 0, 1.064, 10.64, 1600, and 3200 mg kg^{-1} day^{-1} in mice. The material was markedly toxic to rabbits at 1200 and 2400 mg kg^{-1} day^{-1} and to mice at 3200 mg kg^{-1}. Slight to moderate maternal toxicity was observed in rats receiving 2400 mg kg^{-1}, and in mice receiving 1600 mg kg^{-1}; maternal susceptibility was ranked in the order rabbits > mice > rats. Yamamoto et al. (130) also reported no teratogenicity using a commercial light duty liquid detergent, Wonderful K (LAS-type), in ICR/JCL mice at oral dosages up to 3000 mg kg^{-1} day^{-1} and observed depression of body weight and retarded ossification in viable fetuses at dosages above 1200 mg kg^{-1} day^{-1}. With these results in mind, the authors suggest that the possibility of adverse fetal effects might be small, since amounts some 1300-1600 times greater than the estimated daily human intake would be necessary before embryotoxicity was manifest in the test animals.

A further teratology study, in two generations of rats exposed to a mixture of tallow alkyl ethoxylate sulfate (55%) and LAS (45%) at dietary levels providing an intake of 0, approximately 80, 400, and 800 mg kg^{-1} day^{-1} was published (131,132). Treatment was given either continuously or on days 6-15 of each of the six pregnancies involved. No fetal toxicity nor abnormalities attributable to treatment were observed. Similarly, administration of the same mixture to New Zealand White rabbits by gavage at dosages of 0, 50, 100, and 300 mg kg^{-1} on days 2-16 of pregnancy produced no increase in fetal abnormalities. According to the authors' conclusion, these animals received doses of the surfactant mixture that were up to 6000 times the estimated human exposure (0.13 mg kg^{-1} day^{-1}) in a "worst case" situation. None of these levels had deleterious effects on any of the reproductive parameters in the rats, nor were there any effects on the development or viability of the embryo or fetus. Although the rabbits were given somewhat lower doses, they were still exposed to many times the human exposure level without deleterious effects.

(b) Dermal application. The Mikami group (79-84,98) also reported that the topical application of kitchen detergents induced high rates of malformation and anomalies. Six dosage levels (0.5 ml per animal of approximately 0.1, 1.0, 1.8, 9.1, 13.0, and 16.7% solutions) of a LAS based detergent were tested by application on days 1-13 of pregnancy to the shaved skin of groups of three to five ICR/JCL mice. Results are available for only one group of animals (at the dosage level of 9.1%), and the observations in this group (reported as percentage results derived from the offspring of three dams) included closure of the central canal of the spinal cord (70.2%), edema (39.0%), exencephaly (2.5%), cleft palate (40.7%), and anomalies of the extremities (club foot) (36.3%). Results from the control group were not given (81).

The Mikami group claimed that high incidences of subcutaneous hemorrhage (100%), spina bifida (100%), anomalies of the extremities (60%), cleft palate, and exencephaly were observed in all treated groups (79). High incidences of fetal hemorrhage (90.7%), anomalies of the extremities (44.2%), and other types of anomalies were found in Wistar rats which received 0.5 ml per animal of a 10% solution of the commercial (LAS type) detergent Mama Lemon, on days 0-16 of pregnancy (98). One commercial detergent, called fatty alcohol type B, was reported to have no teratogenic effect on mice (84).

Shortcomings in the experimental design, such as the use of insufficient animals and inadequate statistical analysis, together with the unusual descriptions of malformations and nonprecise information about the experimental conditions, lead to difficulties in interpretation.

The alarm they sounded about the potential teratogenicity and embryotoxicity of commercial household detergents in daily use attracted public attention because there had been little previous information of this kind. Tremendous efforts have been made by many investigators in examining potential teratogenicity, but however, none of them as yet have succeeded in substantiating the Mikami group's reports.

Takahashi et al. (122) studied two commercial LAS-based detergents for kitchen use (designated M and R) for teratogenic potential using mice. The topical application of up to 26.0% detergent in solution caused no teratogenic effects but a definitely reduced pregnancy rate in those animals with severe local irritation at the application site was found. This reduced pregnancy rate, especially in mice at higher dose levels, was also reported by most observers using different commercial detergents for kitchen use (LAS-type detergent (118); Mama Lemon, LAS 17%, AES 7% (133); commercial detergent A, LAS 17%, ethanol 7%, urea 15%; and commercial detergent B, polyoxyethylene alkyl ether 8%, AES 7%, ethanol 4% (134); Cherina, LAS-type (135)). Other effects, often observed at higher dose levels, were depression of the fetal body weight and retarded ossification (118,122).

(c) Subcutaneous injection. Mikami et al. (89,90,98) also reported high incidences of malformations and adverse effects on embryos in mice that received 10 ml kg^{-1} of 1.0% solution of commercial liquid detergent (Mama Lemon) for kitchen use.

Inoue and Masuda (134) studied two commercial liquid detergents for kitchen use by subcutaneous injection in mice. No statistically significant differences in the incidence of malformations were found between the control and treated groups at dosages up to 150 or 500 mg kg^{-1} day^{-1}. There was neither embryotoxicity (no reduction in pregnancy rate occurred) nor teratogenicity attributable to treatment.

B. Fertility and Reproduction

As mentioned in the introductory part of Section III, data derived from a test designed to evaluate the effects of a chemical on general reproductive performance and on offspring, usually include information about neonatal morbidity, mortality, and teratogenesis.

Together with teratogenic studies on anionic surfactants and detergents, there have also been important reproductive studies on representative anionic surfactants and formulated detergent products. These are summarized in Appendix 4.

1. Sulfonate-type anionic surfactants

Tusing et al. (136) carried out a reproduction study on ABS in rats, using dosages of 0, 0.1, and 0.5% in the diet. Bornmann et al. (137,138) studied ABS for effect on reproduction in the rat using 0.01% ABS in drinking water for two generations. Freeman et al. (139) also studied the effect of ABS on rat fertility. These multigeneration studies revealed no adverse effects attributable to treatment.

The rat multigeneration test by Bornmann et al. (138) and the rat reproduction study by Buehler et al. (140) supported the negative effect found with LAS. The former adopted dosage levels of 0 and 0.01% LAS in drinking water, and the latter, 0, 0.02, 0.1, and 0.5% in the diet. General reproductive parameters including fertility, gestation, parturition, neonatal viability, lactation, and postweaning growth were normal in all test groups and did not vary from controls. No gross abnormalities were noted.

LAS-Mg (205) or AOS-Mg (206) were studied in a two-generation study using CD rats. LAS-Mg or AOS-Mg was administered continuously, via the diet, at levels of 1250, 2500, and 5000 ppm, to male and female rats of the F0 generation for 13 weeks prior to mating and throughout the mating, gestation and lactation of two litters (F1A, F1B). Using selected animals of the second litter as the parental base, the process was repeated to produce two further litters (F2A and F2B). Animals were selected from the F2B litters to form an F2 generation, which was treated for 13 weeks and then subjected to histopathological examination.

It was concluded that the continuous dietary administration of LAS-Mg to male and female rats over two generations at concentrations of 2500 ppm and 5000 ppm, was associated with slight retardation of somatic growth but was without adverse effects upon reproductive performance and fertility. The responses of animals receiving LAS-Mg at 1250 ppm were essentially similar to those of the controls. The administration of AOS-Mg in the diet, at concentrations of up to 5000 ppm, was without adverse effect upon growth, reproductive performance and the litter responses of male and female rats exposed to the compound for two generations.

2. Sulfate-type anionic surfactants

Tusing et al. (141) conducted rat reproduction studies using both sodium lauryl glyceryl ether sulfonate and sodium lauryl trioxyethylene sulfate (AES). Dietary concentrations of 0 and 0.1% of the surfactants were selected. The reproductive studies were started after 14 weeks of treatment when 10 male and 10 female rats from each treated group and the control group were paired. The first generation offspring (F1) were maintained on the same diet as their parents. When approximately 100 days old, the F1 rats were mated to produce the F2 generation. The F2 rats were maintained on the appropriate diet for 5 weeks after weaning, at which time they were sacrificed. Ingestion of 0.1% of the sulfonate or the sulfate (AES) in the diet had no adverse effects on fertility, litter size, lactation, or the survival of offspring. The surfactants induced no changes in the blood picture or urine content in the F1 and F2 generations, and there were no gross or microscopic findings attributable to the test compounds.

Similarly, 10 rats/sex/group fed diets containing 0.1% of C_{12} AES were mated after 14 weeks on test. The F1 generation was maintained on the parental diet and mated when 100 days old. The F2 generation was fed the same diet for 5 weeks, then killed. No adverse effects on fertility, lactation, litter size or survival and growth of the offspring were seen. Hematological, biochemical and histopathological findings were comparable to controls (159). In order to measure the antifertility action of sodium dodecyl sulfate, Hensworth (211) reported that male mice fed diets containing 0.1% for 6 weeks or 1.0% for 2 weeks experienced no impairment of epididymal spermatozoa.

3. Formulated detergents based on anionic surfactants

Three multigeneration studies in rats were carried out on formulated detergent products (132,142,143). Nolen et al. (132) conducted a combined teratogenicity and two-generation study on a mixture of 55% AES (tallow alkyl ethoxylate sulfate) and 45% LAS in CD rats.

The test material was mixed into ground commercial feed at levels of 0, 0.1, 0.5, and 1.0% and fed to two generations of male and female rats continuously, or only to six females during each period of organogenesis (days 6-15) of pregnancy. The food consumed was measured throughout. Body weights of the parent animals were recorded weekly for the first 8 weeks in each generation, but afterwards were recorded only at each mating phase. After becoming sexually mature, five rats of each sex per group were sacrificed for histological examination during each generation. The remaining animals were mated on a one-to-one basis three times during each generation. The first two pregnancies (F1A,, F1B, F2A and F2B) in each generation were allowed to proceed to natural birth, while the third pregnancies in each generation (F1C and F2C) were used for teratology.

No adverse effects were noted on conception, on fetal viability or on postnatal survival. There were no statistical differences among the groups of the rat fetuses taken by Caesarian section and examined for anomalies. In 1210 rat fetuses, the overall incidence of abnormal young was 9.0%. In addition, pregnant New Zealand White rabbits were given 50, 100 or 300 mg kg^{-1} doses of the surfactant mixture by gavage on days 2-16 of gestation of one pregnancy. No adverse effects were seen in rabbits treated with the surfactant mixture. In 855 rabbit fetuses, 5.7% were abnormal, but the incidences of abnormal fetuses in the test groups were not significantly different from those in the control group. Thus, no treatment-related effects on reproduction or embryonic development were seen in either species.

Palmer (142) performed a three-generation study in CD rats using the commercial light-duty liquid detergent for kitchen use, Mama Lemon, consisting of 17% LAS and 7% AES. Dietary concentrations were 0, 0.08, 0.4, and 2.0%. There were no signs of malreaction to treatment in parent animals, and incidences of sporadic deaths and total litter losses were unrelated to dosage. The pregnancy rate and duration of gestation were unaffected and food consumption and body weight changes showed no consistent relationship to dosage over the three generations. Among litter parameters, statistically significant differences were occasionally recorded but, as these showed no consistent dosage-related trends over the three generations, they were considered unrelated to treatment. The incidence of malformations was unaffected by treatment. Additional organ weight analysis, histopathology, and skeletal staining of representative young from the F3B generation revealed no changes that could be conclusively related to treatment.

A dishwashing agent containing 15% LAS and 13% ammonium fatty alcohol polyglycol ether sulfate was given to male and female rats in drinking water at levels of 0.5 and 1.0 ml kg^{-1} from 12 to 7 days before copulation (143). Females were also treated from days 1-20 of gestation. The preparation had no effect on fertility and embryo development.

C. Japanese Governmental Research on the Embryotoxic and Teratogenic Potential of Dermally Applied Detergents

1. Research sponsored by the Ministry of Health and Welfare

(a) Background. Several reports made by the Mikami group during 1973-1975 stated that the skin application of various household detergents or their surfactants to pregnant mice and rats resulted in definite teratogenicity. Since there had been little previous work of this kind, public concern about the use of the household detergents was aroused.

As mentioned previously, data reported by the Mikami group were quite difficult to interpret and sometimes seemed conflicting. To be acceptable for publication, a teratological study report should include the following: (1) adequate specification of the test material, (2) a sufficiently precise description of test conditions (e.g., route and timing of administration and dosage levels) to enable repetition. Additionally, the following criteria must be met: (1) use of adequate numbers of animals, (2) adoption of at least three dosage levels and a suitable control group to yield a dose-response relationship, (3) proper statistical analysis, (4) descriptions of anomalies and malformations based on generally acceptable standard terminology, (5) comparison with background data. From these points of view, the Mikami reports cannot be considered adequate.

(b) Research project. In order to clarify the reports of the Mikami group, the Environmental Sanitation Bureau of the Japanese Ministry of Health and Welfare planned and promoted a series of group studies.

The group studies consisted of the following three projects: (A) a health science project entitled "Effects of Synthetic Detergents on Embryos by Percutaneous Absorption" sponsored by the Japanese Ministry of Health and Welfare (1973); (B) a special research project entitled "Safety of Synthetic Detergents" sponsored by the Science and Technology Agency (1973); and (C) a "Joint Study on the Teratogenicity of Linear Alkylbenzene Sulfonate (LAS)" (1975). Project B consisted of three parts: (1) a study of skin injury, (2) a study of the relationship between detergency and detergents residues, and (3) a study of the effects of percutaneously absorbed detergents on a living organism, including chronic, teratogenic, and metabolic studies on LAS, AES, or AOS and these formulated detergents based on LAS, AES or AOS. Professor Mikami of Mie University studied LAS and its formulated detergent in ICR/JCL mice and Wistar rats, Professor Nishimura of Kyoto University studied AES and its formulated detergent in ICR/JCL mice, and Professor Sawano of Hiroshima University studied AOS and its formulated detergent in ICR/JCL mice.

In the Joint Study on the Teratogenicity of LAS (project C), these three professors were joined by Professor Kameyama of Nagoya University. The purposes of project C were: (1) to determine whether the topical application of LAS to rats throughout pregnancy induces teratogenic and/or morbid effects on the fetus attributable to LAS or its metabolites and, if any, to evaluate the effective dose in comparison with the maternal nontoxic dose or with the practical daily human intake, and (2) to confirm if the results obtained at the four laboratories concurred, thus establishing the reliability of the group study.

(c) Results of the group study. Professor Nishimura, the coordinator of the group study, has published part of the results of projects A and B, and also a summary and discussion of the results of project C (117,144).

Project A. Nishimura et al. administered dermally a 5.0% solution of a commercial liquid detergent for kitchen use (Mama Lemon) containing 17% LAS and 7% AES to Wistar rats during the early and middle stages of pregnancy. No adverse effects on fetal growth and development, neonatal growth or on behavior, and fertility in the second generation attributable to treatment were observed. Sawano conducted a similar experiment in mice and also found no evidence of teratogenic potential. In addition, Nishimura et al. tried to immerse the caudal surface of pregnant rats in a 0.5 or 5.0% Mama Lemon solution for 2 hr every day. Although some fetal mortality was observed at both levels and some dams died at the 5.0% level, no malformations were induced.

Project B. In this group study, all participants applied the test material daily during most of the gestation period. Two dosage levels were chosen. The low dosage level was 10 times the concentration recommended for normal daily use in man, and the high dosage level was the maximum tolerated in the dam. AES and its formulated detergent (19.6% AES, 5% ethanol, 8% urea) (Kyoto University) and AOS and its formulated detergent (20% AOS, 5% ethanol, 8% urea) (Hiroshima University) were not teratogenic in mice. On the other hand, both LAS and its formulated detergent (19.9% LAS, 5% ethanol, 8% urea) were reported to manifest definite teratogenicity in rats and mice at concentrations above 1.72% (Mie University).

Project C. It should be noted that the results of project C, in which all investigators including Mikami followed precisely the same protocol and used rats of the same strain, failed to support the claims of the Mikami group. The details of the protocol, compiled by all the participants, were published by Nishimura (117,144). In short, the experimental procedures were as follows: LAS was applied daily to the clipped skin of groups of more than 20 pregnant Wistar strain rats from day 0 to day 20 of pregnancy. 0.5 ml per animal per day of aqueous solution (1, 5, and 20%) were applied and removed after 24 hr exposure by washing with wet gauze. Three control groups were included: one received no treatment, the second was skin clipped only, and the third was skin clipped and the vehicle (tap water) was applied to the exposed epidermis as in the test groups. The dams were sacrificed on day 21 of pregnancy, and the intrauterine contents were examined as detailed. A summary and interpretation of the results were made available by Nishimura (117,144). Important items reported are as follows:

1. It was found that the results obtained at the four laboratories generally agreed.
2. The dermal application of detergents to the dams did not induce external and internal (including skeletal) malformations.
3. The dermal application of detergents did not induce subcutaneous or other types of hemorrhage or skin edema in the fetus. Moreover, it was decided that some of the previously found hemorrhages may have been artifacts. After making a careful investigation into the possible reasons for the hemorrhage, Kameyama and his colleagues reported the presence of abundant

blood vessels in the fetal nape region, and suggested this could explain reports of hemorrhage in that area.

4. Standardized skeletal examination revealed conclusively that the dermal application of detergents did not affect the incidence of skeletal variants.

5. Observations on the sex ratio (males/females × 100) of live fetuses can be used to judge whether there is any sex-related tendency in fetal death. An increase in the sex ratio was observed in groups treated with 1 and 5% LAS solution only at Mie University. However, because the average number of live fetuses was not reduced in these two groups and because no difference in the sex ratio was observed between the control group and the 20% LAS group, it is unlikely that the increase at 1 and 5% was attributable to the treatment with LAS.

6. The 20% LAS group demonstrated interference with the progress of pregnancy or depression of the fetal body weight. A statistically significant retardation of ossification, observed only in the 20% LAS group, related to the depression of fetal body weight, was observed at Kyoto, Mie, and Nagaya Universities. These findings almost correlate with the results of other previous dermal application studies, and this would indicate a potential hazard to fetal development in humans should the corresponding conditions apply. However, it should be borne in mind that a striking primary contact dermatitis was induced quickly (after 3-5 days) at the application site in dams treated with 20% LAS. Also, a few days after the commencement of dermal application, a reduction in food intake and depression of maternal body weight throughout pregnancy were observed; clearly under these conditions maternal toxicity occurred. Thus, the problem does not concern fetal effects induced together with maternal toxicity, but does concern fetal effects in the absence of maternal toxicity. Furthermore, it is difficult to decide whether the possible cause of the fetal death and/or the depression of fetal body weight is due to secondary effects, such as skin irritation stress in dams, or to direct toxic effects in the fetus induced by LAS or its metabolites entering fetal tissues. The corresponding conditions would not apply in humans because the concentration of 20% LAS is some 600 times higher than the recommended concentration for practical household use, indicating a satisfactory "Safety in Use".

2. Research sponsored by the Ministry of Education and the Ministry of Health and Welfare

There has been some discussion concerning environmental pollution with detergents and effects in embryo cells in Japan. Concern about embryo cells resulted from Nomura's work (207-208) on the preimplantation embryo.

Nomura studied the lethal effects of LAS, AS, kitchen detergent, and hair shampoo on the mouse embryo. Test materials were applied to the dorsal skin of pregnant JCL/ICL mice during the preimplantation period (days 0-2). Significant numbers of embryos collected from the oviducts and uteri on day 3 showed severe deformity or remained at the morula stage. Most abnormal embryos were fragmented or remained at the 1-8 cell stages, and were either dead or dying. Even when cultivated in a detergent-free medium, they did not recover, while most growth-retarded embryos (morula) would grow with one or two days lag. Similar results were obtained with commercial kitchen detergent and hair shampoo, although embryocidal effects were not detected with natural soap and distilled water.

As a consequence of this, the Ad Hoc Research Group on the Biological Effects of Detergents was established by the Ministry of Education (189, 190), to clarify the research concerning the biological effects of detergents.

The research group concerning LAS teratology recognized that there was much data on the biological effects of detergents and most researchers concluded synthetic detergents cause no biological problem such as teratogenicity in the consumer's normal conditions of use.

In developmental toxicity, effects on the embryo are not always clear. Although there was no teratogenic potential, some developmental anomalies in the embryo and a decrease of birth rate at very low concentrations of detergent were reported.

Thus the effect on the embryo cell of detergent concentration, the egg membrane condition, the application route and distribution in the embryo, etc. were found to require further study.

Based on the results of the research project sponsored by the Ministry of Education, the Ministry of Health and Welfare planned a study to evaluate LAS disturbance of implantation in mice. Koizumi et al. (176) sponsored by the Ministry, studied implantation disturbance, distribution in the body, and mutagenicity of LAS administered in early pregnancy in rats. Approximately 14, 70, or 350 mg kg^{-1} of LAS was administered orally once on day 3 of gestation or once daily for 3 consecutive days from day 1-3 of gestation. There was no significant difference in the incidence of implantation disturbance. A single dose of 350 mg kg^{-1} of LAS was administered orally to investigate the distribution of LAS in the liver, kidney, reproductive organs (uterus and ovaries), and gastrointestinal tract 4 and 8 hr after administration on day 3 of gestation. In the liver, C_{10}-C_{11} components were more rapidly metabolized and C_{12}-C_{13} components remained unchanged. LAS was not detected in the reproductive organs. For mutagenicity, as previously described, there was no difference in the incidence of polychromatic erythrocytes with micronuclei in the maternal bone marrow, the fetal liver, and blood. No mutagenicity was found.

Simultaneously with this study, the Ministry of Health and Welfare planned to summarize the literature (some 600 publications) of seven types of surfactants and to compile all available data. The book *Detergent Toxicity and Evaluation* was published in 1983 (210). The absence of teratogenic concern in normal consumer usage of detergents was summarized in the book.

D. Voluntary Work Supplementary to the Governmental Joint Group Study

The Japanese governmental study (project C) revealed that the topical application of a 20% LAS solution for a 24 hr exposure resulted in definite interference with pregnancy and in depression of fetal body weight, in addition to severe skin irritation in the dams.

In order to examine the potential embryotoxic and teratogenic activity of LAS, the Japan Soap and Detergent Association planned a voluntary study, similar to the protocol adopted by the joint group study sponsored by the Japanese government, except that increased dosage levels and a shorter exposure period (30 min) were used. Concentrations of 0.05, 0.1, and 0.5% LAS were applied and allowed to remain on the Wistar rat skin; concentrations of 1, 5, and 20% were applied and removed after a 30 min exposure period. Also, three types of control group were included. The results reported by Rapp and Killeen (145) indicated that no adverse effects were observed on pregnancy, on the dams, or on the fetuses. The only effects attributable to LAS were local dermal changes in the dams at the two highest concentrations. These effects included initial erythema followed by dermal thickening and fissuring, marked at the 20% concentration level and very slight at the 5% level. LAS did not affect fetal parameters; there was no evidence of any teratogenic or embryotoxic response to treatment.

APPENDIX 1

Prolonged Toxicity and Carcinogenicity Studies of Anionic Surfactants and Formulated Detergents

Test material	Species, sex, number/group	Dose	Duration	Ref.
ABS	Mouse, Swiss, female, 30	5,10% dermal 2 times/week	101-102 wk	1
Petroleum ether extract of ABS	Mouse, Swiss, female, 30	1% in acetone dermal 2 times/week	92-101 wk	1
ABS	Mouse, C_{57} black, male and female, 30	5 mg in 0.5 ml of tricaprylin s.c.single	95 wk	1
ABS (Santomerse No. 3)	Rat, Carworth Farms, male and/or female, 20-40	0, 200, 1000, 2000 ppm food daily	2 yr	12
ABS	Rat, Wistar, male and female, 80	0, 0.1, 0.5% food daily	2 yr	13
ABS	Rat, Wistar, male and female, 40	0, 0.05% water daily	2 yr	13
ABS (Marlon TP 42)	Rat, male and female, 120	0, 100 ppm water daily	100 wk	10,11
ABS	Rat, male, 60	150 mg liter^{-1} water daily	1 yr	14
ABS	Rabbit, New Zealand White, 3	10% dermal, 2 times/week	115 wk	1
LAS	Mouse, SLC-ICR, male and female, 16-18	0, 0.07, 0.2, 0.6% water daily	9 mo	63
LAS	Mouse, SLC-ICR, male and female, 16-18	0, 0.6% food daily	9 mo	63
LAS, C_{12}	Mouse, C_{57} black, male and female, 40	1 mg in 0.2 ml water s.c. 2 times/week	80 wk	17
LAS (Marlon BW 2043)	Rat, male and female, 120	0, 100 ppm water daily	100 wk	10,11

LAS	Rat, Charles River, male and female, 100	0, 0.02, 0.1, 0.5% food daily	2 yr	15
LAS (AI, 34.55%)	Rat, Wistar, male, 21-23	0, 0.01, 0.05, 0.10% water	2 yr	16
LAS	Rat, Wistar, JCL, male and female, 16-18	0, 0.07, 0.2% water daily	9 mo	63
LAS	Rat, Wistar, JCL, male and female, 16-18	0, 0.6, 1.8% food daily	9 mo	63
LAS	Rat, SLC-Wistar, male and female, 100	0, 0.04, 0.16, 0.6% food daily	30 mo	149, 150
LAS + BHT	Rat, SLC-Wistar, male and female, 80	LAS: 0 ~ 0.2% BHT: 0 ~ 0.6% food daily	30 mo	149
LAS (AI, 38.74%)	Rat, Wistar, male and female, 74,124	0, 0.1% water daily	26 mo	148
LAS (AI, 19.7%)	Rat, SLC-Wistar, male and female, 50	0 (water), 0.005, 0.025, 0.125 ml dermal 3 times/week (0.25 ml/animal)	24 mo	151
AOS, C_{15-18} (AI, 90%) Hexadecane 1,4-sultone	Mouse, Swiss-Webster, 21	5% dermal 2 times/week	2 yr	157
AOS, C_{14-18}	Mouse, Swiss-Webster, male and female, 80	0 (water, acetone) non-treatment, 20, 25% dermal, 3 times/week (0.02 ml/animal)	92 wk	155
AOS, C_{14-16}	Mouse, Swiss-Webster, male and female, 80	0 (water, acetone) non-treatment, 20, 25% dermal, 3 times/week (0.02 ml/animal)	92 wk	155

APPENDIX 1 (Continued)

Test material	Species, sex, number/group	Dose	Duration	Ref.
C_{16}1,4-sultone	Mouse, Swiss-Webster, male and female, 80	6.7, 8.3% dermal, 3 times/week (0.02 ml/animal)	92 wk	155
AOS, C_{14-18} (AI, 92.90%)	Rat, CFY, male and female, 100	0, 1000, 2500, 5000 ppm food daily	2 yr	2
AOS, C_{14-18}	Rat, X-MRC male and female, 80	0, 0.5, 0.75, 1.0% food daily	24-27 mo	155
AOS, C_{14-16}	Rat, X-MRC male and female, 80	0, 0.5, 0.75, 1.0% food daily	24-27 mo	155
C_{16}1,4-sultone	Rat, X-MRC male and female, 80	0.16, 0.25, 0.33% food daily	24-27 mo	155
AOS, C_{14-19}	Rat, Wistar, male and female, 20	0 (water), 1, 10, 30% dermal 3 times/week (0.5 ml/animal)	70 wk	158
AOS, C_{14-16}/C_{16-18} (AI, 30%)	Rat, Long-Evans, male and female, 100	0 (water), 10% dermal 2 times/week	24 mo	156
AOS, C_{14-16} (AI, 38.9%)	Rat, Long-Evans, male and female, 100	0 (water), 10% dermal 2 times/week (1.0 ml/kg)	24 mo	156
Secondary alkane sulfonate	Rat	100 ppm water daily	1 yr	65
Secondary alkane sulfonate	Rat	Food	1 yr	66
Sodium lauryl glyceryl ether sulfonate	Mouse, Swiss, female, 30	0, 5% dermal 2 times/week	105 wk	18

Sodium lauryl trioxyethylene sulfate	Mouse, Swiss, female, 30	0, 5% dermal 2 times/week	105 wk	18
AES, C_{12}	Mouse, Swiss-ICR, 30	5% dermal, 0.1 ml 2 times/week	2 yr	159
Sodium lauryl trioxyethylene sulfonate	Rat, Carworth Farms, male and female, 60	0, 0.1, 0.5% food daily	105 wk	18
Sodium lauryl glyceryl ether sulfonate	Rat, Carworth Farms, male female, 60	0, 0.1, 0.5% food daily	105 wk	18
AES, C_{12} (AI, 25%)	Rat, male and female, 40	0.1% water daily	2 yr	159
AES (AI, 20.1%)	Rabbit, male, 7	0 (water), 0.05, 0.5, 1.25% dermal 3 times/week	24 mo	160
AS AS + 3,4-Benzpyrene (Bp)	Mouse, ddY/SLC male and female, 20-30	0 (benzene), AS (10%), AS (10%) + Bp (0.3%), Bp (0.3%), Benzene, Untreated dermal, 2 times/week	1 yr	162
AS, C_{12}	Rat, Osborne-Mendel, male, 10-12	0, 0.25, 0.5, 1.0% food daily	2 yr	20
Sodium 2-ethylhexyl sulfate (Tergitol anionic 08)	Rat, CFE, male and female, 36	0, 0.01, 0.04, 0.16, 0.64% food daily	2 yr	19
AS	Rat, SD male and female, 20	0 (water), 5, 20% dermal 2 times/week	Lifetime	161
AS	Dog	2 g oral	10 mo	64

APPENDIX 1 (Continued)

Test material	Species, sex, number/group	Dose	Duration	Ref.
Sodium 2-ethylhexyl sulfate (Tergitol anionic 08)	Dog, Beagle, male and female, 6	0, 0.04, 0.16, 0.64% food daily	2 yr	19
Soap-K, Soap-K + 3,4-Benzpyrene (Bp)	Mouse, ddY/SLC male and female, 20-30	0 (benzene), Soap (15%), Soap (15%) + Bp (0.3%), Bp (0.3%), Benzene, Untreated, dermal 2 times/week	1 yr	162
Soap-K, Soap-K + MNNG	Rat, Wistar, male, 20, 40	Soap (2%), Soap (2%) + MNNG (50 mg/*l*), water daily	26 wk	164
Soap-K (AI, 30%)	Rat, Fischer male and female, 21-25	0 (water), 25, 100% dermal	24 mo	163
Surfactant mixture (LAS base)	Mouse, C_{57} black, male and and female, 30	0.5% food daily	80 wk	17
Household detergent type I (ABS base)	Mouse, Swiss, female, 30	10% dermal 2 times/week	101 wk	22
Household detergent type II (ABS and AS base)	Mouse, Swiss, female, 30	5% dermal 2 times/week	115 wk	22
Detergent	Mouse, Swiss, and rabbit	dermal 2 times/week	lifetime	23
Mixture of LAS (15.6%)/Tallow alcohol Ethoxy sulfate (18.6%)	Mouse, Swiss ICR	0.1, 1.0, 10.0% dermal, 3 times/week	18 mo	159

Household detergent type II (ABS and AS base)	Mouse, C_{57}, female, 30	1% s.c. once	83 wk	22
Liquid detergent for kitchen use	Rat, male, 60	0.15, 0.3 ml liter^{-1} water daily	1 yr	14
Household detergent type I (ABS base)	Rat, Carworth Farms, male and female, 80	0, 0.5% food daily daily	2 yr	21
Household detergent type II (ABS and AS base)	Rat, Carworth Farms, male and female, 80	0, 0.5% food daily daily	2 yr	21
Mixture of LAS and AES	Rat	0, 0.1, 0.5, 1.0% food daily	2 yr	159
LAS base detergent (LAS, 19.9%)	Rat, SLC-Wistar, male and female, 50	0 (water), 0.025 ml, dermal 3 times/week (0.25 ml/animal)	24 mo	151
AOS base detergent (AOS, 22%)	Rat, male and female, 20	0 (water), 50% 3 times/week 0.5 ml	70 wk	158
AES base detergent (AES, 19.6%)	Rabbit, male, 7	0 (water), 0.25% dermal 3 times/week	24 mo	160
Household detergent type II (ABS and AS base)	Rabbit, 3	5% dermal 2 times/week	122 wk	22

APPENDIX 2

Teratogenicity Studies of Anionic Surfactants and Formulated Detergents (Negative Results)

Test material	Species (no. of animals/group)	Route	Dose	Dosing period (days of gestation)	Sacrificed day of gestation	Ref.
ABS	Mouse, ICR/JCL (3-6)	per os (intragastric)	FLD_{50}: 120 mg kg^{-1}	4	18	100
			FLD_{50}: 60 mg kg^{-1}	7	18	
ABS	Mouse, dd	per os (intragastric)	FLD_{50}: 130 mg kg^{-1}	7	18	101
ABS	Mouse, ICR/JCL (22-24)	per os (intragastric)	0 (water), 24,240 mg kg^{-1} day^{-1}	7-13	18	102
ABS	Rat, Wistar (5-16)	per os (in diet)	0, 0.025, 0.1, 0.5, 1.0, 2.0%	0-19	20	102
ABS	Rat, Wistar (5-6)	per os (in diet)	0, 0.025, 0.1, 0.5, 1.0%	0-19	Allowed to deliver	102
KBS	Rat, Wistar (20)	intraventricularly	0, 40 mg kg^{-1}	During pregnancy	22	103, 104
LAS	Mouse, CD-1 (20)	per os (intragastric)	0 (water), 0.2, 2, 300, 600 mg kg^{-1} day^{-1}	6-15	17	105, 106
LAS (AI, 99.5%)	Mouse, ICR/JCL (13-14)	per os (intragastric) 10 ml kg^{-1} day^{-1}	0 (water), 0.4, 4.0%	0-6	18	122
LAS (AI, 99.5%)	Mouse, ICR/JCL (13-14)	per os (intragastric) 10 ml kg^{-1} day^{-1}	0 (water, 0.4, 4.0%	7-13	18	122
LAS	Mouse, ICR/SLC (25-33)	per os (intragastric)	0 (water), 10, 100, 300 mg kg^{-1} day^{-1}	6-15	17	146

LAS + Hg-methyl	Mouse, ICR/JCL (10-11)	per os (drinking water)	LAS 500 ppm + Hg-methyl 15 mg/kg	6-18	18	193
LAS (AI, 99.5%)	Mouse, ddY/S (16)	Dermal (skin painting 0.5ml/4 × 4 cm)	2.22%	0-13	18	118, 149
LAS, $C_{10\text{-}14}$ (AI, 20.66%)	Mouse, ICR/JCL (25-30)	Dermal (skin painting 0.5ml/4 × 4 cm)	0 (water), 0.85, 1.7, 2.55, 3.4%	1-13	17	119
LAS, $C_{10\text{-}14}$ (AI, 20.66%)	Mouse, ddY (4-10)	Dermal (skin painting 0.5ml/4 × 4 cm)	Nontreatment, 0 (water), 0.017, 0.17, 1.7%	2-14	17	119
LAS	Mouse, CD-1 (20)	Dermal (skin painting 0.5ml/2 × 3 cm)	0 (water), 0.03, 0.3, 3.0%	2-13	17	120, 121
LAS	Mouse, CD-1 (5)	Dermal (skin painting 0.5ml/2 × 3 cm)	0 (water), 0.03, 0.3, 3.0%	2-13	Allowed to deliver	120
LAS (AI, 99.5%)	Mouse, ICR/JCL (19)	Dermal (skin painting 0.5ml/4 × 4 cm)	0 (water), 2.86, 5.72%	0-13	18	122
LAS, $C_{10\text{-}14}$ (AI, 48.6%)	Mouse, ICR/JCL (27-28)	Dermal (skin painting 0.5ml/4 × 4 cm)	0 (water), 0.03, 0.3, 3.0%	6-15	17	123
LAS, $C_{10\text{-}14}$ (AI, 20.66%)	Mouse, ICR/JCL (21-24)	Dermal (injection)	0 (water), 0.4, 2, 10, 50 mg kg^{-1} day^{-1}	7-13	18	124
LAS	Mouse, ICR/JCL (6-7)	Dermal (injection)	0 (water), 0.4, 2, 10, 50 mg kg^{-1} day^{-1}	7-13	Allowed to deliver	124

APPENDIX 2 (Continued)

Test material	Species (no. of animals/group)	Route	Dose	Dosing period (days of gestation)	Sacrificed day of gestation	Ref.
LAS (AI, 99.5%)	Mouse, ICR/JCL (13)	Dermal (injection: dorsal portion 20 ml kg^{-1} day^{-1})	0 (water), 0.35, 1.00%	0-3	18	122
LAS (AI, 99.5%)	Mouse, ICR/JCL (14-15)	Dermal (injection: dorsal portion 20 ml kg^{-1} day^{-1})	0 (water), 0.35, 1.00%	8-11	18	122
LAS (AI, 99.5%)	Mouse, ICR/JCL (15-19)	Dermal (injection: abdominal portion 20 ml kg^{-1} day^{-1})	0 (water), 0.35, 1.00%	0-3	18	122
LAS (AI, 99.5%)	Mouse, ICR/JCL (12)	Dermal (injection: abdominal portion 20 ml kg^{-1} day^{-1})	0 (water), 0.35, 1.00%	8-11	18	122
LAS	Rat, CD (20)	per os (intragastric)	0 (water), 0.2, 2, 300, 600 mg kg^{-1} day^{-1}	6-15	20	105, 106
LAS (AI, 40.68%)	Rat, SD-JCL (10)	per os (in diet)	0, 0.1, 1.0%	0-20	21	108,109
LAS (AI, 40.68%)	Rat, SD-JCL (5-6)	per os (in diet)	0, 0.1, 1.0%	0-20	Allowed to deliver	108,109
LAS	Rat, Wistar (40)	per os (drinking water)	0, 0.1%	6-15	20	148

LAS	Rat, CD (20)	Dermal (skin painting 0.5ml/4×4 cm)	0 (water), 0.03, 0.3, 3.0%	2-15	20	120, 121
LAS	Rat, CD (5)	Dermal (skin painting 0.5ml/4×4 cm)	0 (water), 0.03, 0.3, 3.0%	2-15	Allowed to deliver	120
LAS, C_{10-13}	Rat, Wistar (20)	Dermal (skin painting 0.5ml/4×6 cm)	Nontreatment, 0 (water), 1, 5, 20%	0-20	21	192
LAS, C_{11-17} (AI, 20.5%)	Rat, Wistar (20-21)	Dermal (skin painting)	Rinsing after 30 min: Nontreatment, 0 (water), 1, 5, 20% Non-Rinsing: 0.05, 0.1, 0.5%	0-20	21	141, 145
LAS-Mg	Rat, CD (20-21)	Dermal (skin painting 0.5ml/8×4 cm)	0 (PEG), 1.75, 3.5, 7.0%	7-17	20	195
LAS-Mg	Rat, CD (11-12)	Dermal (skin painting 0.5ml/8×4 cm)	0 (PEG), 1.75, 3.5, 7.0%	7-17	Allowed to deliver	195
LAS	Rabbit, New Zealand White (22)	per os (drinking water)	0, 0.1%	6-18	30	148
LAS	Rabbit, New Zealand White (13)	per os (intragastric)	0 (water), 0.2, 2, 300, 600 mg kg^{-1} day^{-1}	6-18	29	105, 106
LAS-Mg	Rabbit, New Zealand White (14)	per os (intragastric)	0 (water), 60, 125, 250 mg kg^{-1} day^{-1}	6-18	29	194
LAS	Rabbit, New Zealand White (13)	Dermal (skin painting 10 ml/12×20 cm)	0 (water), 0.03, 0.3, 3.0%	1-16	29	120, 121

APPENDIX 2 (Continued)

Test material	Species (no. of animals/group)	Route	Dose	Dosing period (days of gestation)	Sacrificed day of gestation	Ref.
LAS	Rabbit, New Zealand White (6)	Dermal (skin painting 10 ml/12×20 cm)	0 (water), 0.03, 0.3, 3.0%	1-16	Allowed to deliver	120
LAS-Mg	Rabbit, New Zealand White (14)	Dermal (skin painting) 5ml/10×10 cm)	0 (PEG), 0.75, 1.5, 3.0%	6-18	29	196
AOS, $C_{14\text{-}18}$	Mouse, CD-1 (20)	per os (intragastric)	0 (water), 0.2, 2.0, 300, 600 mg kg^{-1} day^{-1}	6-15	17	125, 126
AOS, $C_{14\text{-}19}$	Mouse (1-4)	Dermal (skin painting 0.5ml/animal)	0 (water), 0.1, 1, 5%	0-14	18	209
AOS based detergent	Mouse (1-3)	Dermal (skin painting 0.5ml/animal)	0.5, 5, 25%	0-14	18	209
AOS, $C_{14\text{-}18}$	Mouse, CELP (31)	Dermal (skin painting 0.25ml/2×4 cm)	0 (water), 1, 5, 20% (exposure: 24 hr day^{-1}), 20% (exposure: 30 min day^{-1})	1-16	17	128
AOS, $C_{14\text{-}18}$	Rat, CD (20)	Per os (intragastric)	0 (water), 0.2, 2, 300, 600 mg kg^{-1} day^{-1}	6-15	20	125, 126
AOS-Mg, $C_{14\text{-}18}$	Rat, CD (20-22)	Per os (intragastric)	0 (water), 150, 300, 600 mg kg^{-1} day^{-1}	7-17	20	197

AOS-Mg, $C_{14\text{-}18}$	Rat, CD (10-12)	Per os (intragastric)	0 (water), 150, 300, 600 mg kg^{-1} day^{-1}	7-17	Allowed to deliver	197
AOS, $C_{14\text{-}18}$	Rat, CD (20)	Dermal (skin painting 0.25ml/4×6 cm)	0 (water), 1, 5, 20% (exposure: 24 hr day^{-1}), 20% (exposure: 30 min day^{-1})	1-19	20	127
AOS-Mg, $C_{14\text{-}18}$ (AI, 94.4%)	Rat, CD (21-22)	Dermal (skin painting 0.5ml/8×4 cm)	0 (water), 1.75, 3.5, 7.0%	7-17	20	199
AOS-Mg, $C_{14\text{-}18}$ (AI, 94.4%)	Rat, CD (10-11)	Dermal (skin painting 0.5ml/8×4 cm)	0 (water), 1.75, 3.5, 7.0%	7-17	Allowed to deliver	199
AOS, $C_{14\text{-}18}$	Rabbit, New Zealand White (13)	Per os (intragastric)	0 (water), 0.2, 2, 300, 600 mg kg^{-1} day^{-1}	6-18	29	125, 126
AOS-Mg, $C_{14\text{-}18}$	Rabbit, New Zealand White (14-18)	Per os (intragastric)	0 (water), 75, 150, 300 mg kg^{-1} day^{-1}	6-18	29	198
AOS-Mg, $C_{14\text{-}18}$ (AI, 94.4%)	Rabbit, New Zealand White (14-16)	Dermal (skin painting 5ml/10×10 cm)	0 (water), 0.75, 1.5, 3.0%	6-18	29	200
AS	Mouse, CD-1 (20)	Per os (intragastric)	0 (water), 0.2, 2, 300, 600 mg kg^{-1} day^{-1}	6-15	17	105, 106
AS (sodium dodecyl sulfate)	Mouse, ICR/JCL (21-48)	Dermal (skin painting 1.5ml kg^{-1} day^{-1}, 3×3cm)	Nontreatment, 0 (water), 0.4, 4.0, 6.0%	6-13	18	129
AS (sodium dodecyl sulfate)	Mouse, ICR/JCL (9-10)	Dermal (skin painting 1.5ml kg^{-1} day^{-1}, 3×3cm)	6.0%	7-14 8-15	18	129

APPENDIX 2 (Continued)

Test material	Species (no. of animals/group)	Route	Dose	Dosing period (days of gestation)	Sacrificed day of gestation	Ref.
AS	Rat, CD (20)	Per os (intragastric)	0 (water), 0.2, 2, 300, 600 mg kg^{-1} day^{-1}	6-15	20	105, 106
AS	Rabbit, New Zealand White (13)	Per os (intragastric)	0 (water), 0.2, 2, 300, 600 mg kg^{-1} day^{-1}	6-18	29	105, 106
AES	Mouse, ICR/JCL (5-10)	Dermal (skin painting 0.25ml/2×3 cm)	0.8, 4, 8, 20%	0-18 8-15	18	117, 136
AES based detergent (AI, 19.6%)	Mouse, ICR/JCL (13-15)	Dermal (skin painting 0.25ml/2×3 cm)	4, 20, 40%	0-18 8-15	18	117, 136
AES-Mg	Rat, CFY (20)	Per os (intragastric)	0 (water), 130, 260, 520 mg kg^{-1} day^{-1}	7-17	20	201
AES-Mg	Rat, CFY (10)	Per os (intragastric)	0 (water), 130, 260, 520 mg kg^{-1} day^{-1}	7-17	Allowed to deliver	201
AES-Mg	Rat, CFY (20)	Dermal (skin painting 0.2ml/100g)	0 (water), 2.5, 5.0, 10.0%	7-17	20	203
AES-Mg	Rat, CFY (10)	Dermal (skin painting 0.2ml/100g)	0 (water), 2.5, 5.0, 10.0%	7-17	Allowed to deliver	203
AES-Mg	Rabbit, New Zealand White (13-14)	Per os (intragastric)	0 (water), 75, 150, 300 mg kg^{-1} day^{-1}	6-18	29	202

AES-Mg	Rabbit, New Zealand White (13-15)	Dermal (skin painting 1.0ml/kg)	0 (water), 0.8, 1.6, 3.3%	6-18	29	204
Soap	Mouse, CD-1 (20)	Dermal (skin painting 0.5ml/2×3 cm)	0 (water), 0.3, 3.0, 30.0%	2-13	17	120, 121
Soap	Rat, CD (20)	Dermal (skin painting 0.5ml/4×4 cm)	0 (water), 0.3, 3.0, 30.0%	2-15	20	120, 121
Commercial liquid detergent (LAS base) Mama Lemon	Mouse, CD-1 (20)	Per os (intragastric)	0 (water), 1.064, 10.64, 1600, 3200 mg kg^{-1} day^{-1}	6-15	17	105, 106
Commercial liquid detergent (LAS base) Wonderful K	Mouse, ICR/JCL (18-23)	Per os (intragastric)	Nontreatment, 0 (water), 800, 1200, 1500, 3000 mg kg^{-1} day^{-1}	6-15	18	130
Commercial liquid detergent (LAS base) Wonderful K	Mouse, ICR/JCL (9-13)	Per os (intragastric)	0 (water), 125, 250, 500, 1000, 2000, 4000 mg kg^{-1} day^{-1}	6	17	147
Commercial liquid detergent (LAS base)	Mouse, ddY/S (21)	Dermal (skin painting 0.5ml/4×4 cm)	15%	0-13	18	118
Commercial liquid detergent	Mouse, ICR/JCL (27-39)	Dermal (skin painting 0.5ml/4×4 cm)	Nontreatment, 0 (water), 15, 20%	0-12	18	133
Commercial liquid detergent (LAS base) R	Mouse, ICR/JCL (15-19)	Dermal (skin painting 0.5ml/4×4 cm)	0 (water), 13.0, 17.0, 20.0, 25.0%	0-13	18	122

APPENDIX 2 (Continued)

Test material	Species (no. of animals/group)	Route	Dose	Dosing period (days of gestation)	Sacrificed day of gestation	Ref.
Commercial liquid detergent (LAS base) M	Mouse, ICR/JCL (10)	Dermal (skin painting 0.5ml/4 × 4 cm)	0 (water), 13.0, 26.0%	0-13	18	122
Commercial liquid detergent (LAS base) A	Mouse, ICR/JCL (11-20)	Dermal (skin painting 0.5ml/4 × 4 cm)	0 (water), 0.5, 5, 15%	1-13	17	134
Commercial liquid detergent (AES base) B	Mouse, ICR/JCL (17-18)	Dermal (skin painting 0.5ml/4 × 4 cm)	0 (water), 0.5, 5, 15%	0-13	18	134
Commercial liquid detergent	Mouse, ddY/S (21)	Dermal (skin painting 0.5ml/4 × 4 cm)	13%	0-13	18	191
Commercial liquid detergent (LAS base) Cherina	Mouse, ICR/JCL (17-50)	Dermal (skin painting 0.5ml/3 × 4 cm)	0 (water), 5, 10, 20%	0-13	18	135

Commercial liquid detergent (LAS base) A	Mouse, ICR/JCL (16-17)	Dermal (injection)	0 (water), 30, 150 mg kg^{-1} day^{-1}	7-13	18	134
Commercial liquid detergent (LAS base) A	Mouse, ICR/JCL (17)	Dermal (injection)	0 (water), 30, 150 mg kg^{-1} day^{-1}	0-13	18	134
Commercial liquid detergent (AES base) B	Mouse, ICR/JCL (15-16)	Dermal (injection)	0 (water), 5, 50, 500 mg kg^{-1} day^{-1}	7-13	18	134
Commercial liquid detergent (LAS base) Mama Lemon	Rat, CD (20)	Per os (intragastric)	0 (water), 0.8, 8, 1200, 2400 mg kg^{-1} day^{-1}	6-15	20	105, 106
Mixture of LAS and AES	Rat, CD (4-10)	Per os (in diet)	0, 0.1, 0.5, 1.0%	6-15	13,21	131, 132
Commercial liquid detergent (LAS base) Mama Lemon	Rabbit, New Zealand White (13)	Per os (intragastric)	0 (water), 0.8, 8, 1200, 2400 mg kg^{-1} day^{-1}	6-18	29	105, 106
Mixture of LAS and AES	Rabbit, New Zealand (22-25)	Per os (intragastric)	nontreatment, 0 (water), 50, 100, 300 mg kg^{-1} day^{-1}	2-16	28	131, 132

APPENDIX 3

Teratogenicity Studies of Anionic Surfactants and Formulated Detergents (Mikami Group)

Test material	Species (no. of animals/group)	Route	Dose	Dosing period (days of gestation)	Sacrificed day of gestation	Ref.
LAS	Mouse, ICR/JCL (7)	Dermal (skin painting 0.5ml/3 × 4 cm)	1%	0-11	16	98
LAS, $C_{10\text{-}13}$ (AI, 19.7%)	Mouse, ICR/JCL (5-20)	Dermal (skin painting 0.25ml/3 × 3 cm)	1.75, 20, 40, 60, 100%	All	18	86
LAS, $C_{10\text{-}13}$ (AI, 19.7%)	Rat, Wistar (8)	Dermal (skin painting 0.25 ml/3 × 3 cm)	1.75, 60%	All	20	89
LAS	Rat, Wistar (20)	Dermal (skin painting 0.5ml/4 × 6 cm)	0 (water), 1, 5, 20%	All	21	87
LAS	Rat, Wistar (8)	Dermal (injection) 10 ml kg^{-1}	1%	8,10,12,14	20	89
AS (sodium lauryl sulfate)	Mouse, ICR/JCL (10,15)	per os (intragastric)	10 ml kg^{-1} of 0 (water), 2%	6-13	18	91
AS (sodium lauryl sulfate)	Mouse, ICR/JCL (15)	Dermal (skin painting 0.25ml/2 × 3 cm)	0 (water), 2%	7-13	18	91
AS (sodium lauryl sulfate)	Mouse, ICR/JCL (5-10)	Dermal (skin painting 0.25-0.3ml/3 × 3cm)	0 (water), 10, 20%	1-16	18	91

AES	Mouse	Dermal (skin painting 0.25ml/ animal)	4%	0-18	18	88
Soap (Nissan powder soap)	Mouse (16)	Dermal (injection 10 ml kg^{-1})	0.2, 1%	11-13	17	90
Sodium α-sulfo-alkyloyl lactylate	Mouse (16,10)	Dermal (injection 10 ml kg^{-1})	0.2, 1%	11-13(0.2%) 8-11(1%)	17	90
Commercial liquid detergent (ABS base) Wonderful K	Mouse, ICR/JCL (9-15)	Per os (intragastric)	0 (bio.sali.), 0.021, 0.042 ml g^{-1} of 15 ml $liter^{-1}$ bio. sali.	6,8,10	17	78, 82, 95-97
Commercial liquid detergent (fatty alcohol derivatives base)	Mouse, ICR/JCL (13-18)	Per os (intragastric)	0 (bio.sali), 210, 630 ml kg^{-1}	6-11	17	94
Commercial liquid detergent (4 types)	Mouse, ICR/JCL (13-18)	Per os (intragastric)	(bio.sali.), 0.63 ml kg^{-1}	6-11	17	80, 82, 93, 97
Commercial liquid detergent (LAS base)	Mouse, ICR/JCL	Dermal (skin painting 0.5ml/4×4 cm)	0.1, 1, 5, 10, 15, 20 ml in 100 ml water	1-13	17	79, 81
Commercial liquid detergent (LAS base)	Mouse, ICR/JCL	Dermal (skin painting)	0.1, 0.99, 4.8, 9.1, 13, 17%	0-11	16	83
LAS base detergent (AI, 19.9%)	Mouse, ICR/JCL (5-20)	Dermal (skin painting 0.25ml/ animal)	1.75, 20, 80, 100%	All	18	86

APPENDIX 3

Test material	Species (no. of animals/group)	Route	Dose	Dosing period (days of gestation)	Sacrificed day of gestation	Ref.
LAS base detergent (AI, 19.9%)	Mouse, ICR/JCL (8-20)	Dermal (skin painting 0.25ml/animal)	1.75, 60%	All	20	86
Commercial liquid detergent (LAS base) Mama Lemon	Mouse, ICR/JCL (3-4)	Dermal (injection 10 ml kg^{-1})	1, 5%	11-13(1%) 6-9 (5%)	17	89, 90, 98
Commercial liquid detergent (LAS) base)	Rat, Wistar (13)	Dermal (skin painting 0.5 ml/4×4 cm)	0, 10%	0-16	20	84
Commercial liquid detergent (LAS base) Mama Lemon	Rat, Wistar (5,7)	Dermal (skin painting 0.5ml/4×4 cm)	0 (bio. sali.), 10%	0-16	20	98

APPENDIX 4

Fertility and Reproduction Studies of Anionic Surfactants and Formulated Detergents

Material	Animal species (no. of animals[a]/group)	Dose	Generations	Ref.
ABS	Rat, Wistar (F10,M10)	0, 0.1, 0.5% food	3(F2)	13
ABS	Rat (F10,M10)	0, 0.01% water	2(F1)	11
ABS (alkylaryl)	Rat (12)	5000 ppm food	2	139
LAS	Rat, Charles River (F10,M10)	0, 0.01% water	3(F2)	11
LAS, $C_{10\text{-}16}$	Rat, Charles River (F20,M20)	0, 0.02, 0.1, 0.5% food	4(F3B)	15
LAS (AI, 38.74%)	Rat, Wistar (F20,M20)	0, 0.1% food	4	148
LAS-Mg, $C_{10\text{-}13}$	Rat, CD (F24,M12)	0, 1250, 2500, 5000 ppm food	3(F2)	205
AES, C_{12} (AI, 25%)	Rat (F10,M10)	0, 0.1% food	3(F2)	159
AOS-Mg, $C_{14\text{-}18}$	Rat, CD (F24,M12)	0, 1250, 2500, 5000 ppm food	3(F2)	206
Sodium lauryl glyceryl ether sulfonate	Rat, Carworth Farms (F10,M10)	0, 0.1% food	3(F2)	18
Sodium lauryl trioxy-ethylene sulfate	Rat, Carworth Farms (F10,M10)	0, 0.1% food	3(F2)	18
Mixture of tallow alkyl ethoxylate sulfate and LAS	Rat, CD-SD (F25,M25)	0, 0.1, 0.5, 1.0% food	3(F2C)	132
Liquid detergent for kitchen use, Mama Lemon	Rat, CD (F22,M11)	0, 0.08, 0.40, 2.00% food	4(F3B)	142
Dishwashing agent	Rat (F78,M24)	0, 0.5, 1.0 mg kg^{-1} water	1	143

[a]F = female, M = male.

REFERENCES

1. U. Saffiotti, P. Shubik, and D.L. Opdyke, *Toxicol. Appl. Pharmacol., 4*(6), 763-769(1962).
2. B. Hunter and H.G. Benson, *Toxicology, 5,* 359-370(1976).
3. R.D. Swisher, Soap and Detergent Association Scientific and Technical Report No. 4, Soap and Detergent Ass., New York, 1966.
4. H. Lange and H. Raay, *Kolloid-Z.Z. Polym., 189*(1), 55-57(1963).
5. Böhm-Gössl and R. Krüger, *Kolloid-Z.Z. Polym., 206*(1), 65-70 (1965).
6. C. Gloxhuber, *Med. Welt., 18,* 46-55(1967).
7. C. Gloxhuber, *Arch. Toxicol., 32,* 245-270(1974).
8. A.W. Neumann, *Tenside, 3*(9), 326-327(1966).
9. R.D. Swisher, *Arch. Environ. Health, 17,* 232-246(1968).
10. G. Bornmann and A. Loeser, *Z. Lebensm.-Unters Forsch, 118*(1), 51-52(1962).
11. G. Bornmann, A. Loeser, and M. Stanišić, *Fette, Seifen, Anstrichm., 65*(10), 818-824(1963).
12. O.E. Paynter and R.J. Weir, *Toxicol. Appl. Pharmacol., 2,* 641-648(1960).
13. T.W. Tusing, O.E. Paynter, and D.L. Opdyke, *Toxicol. Appl. Pharmacol., 2,* 464-473(1960).
14. Y. Tokoro, M. Hirasawa, and S. Moteki, *Japan J. Public Health, 11*(1), 29-36 (1964).
15. E.V. Buehler, E.A. Newmann, and W.R. King, *Toxicol. Appl. Pharmacol., 18,* 83-91(1971).
16. S. Tiba, *J. Food Hyg. Soc. Jap., 13*(6), 509-516(1972).
17. J.M. Philp, Fluorescent Whitening Agents, MVC Rep. 2. Statens naturvetenskapliga forskningsrad Swedish Natural Science Research Council, Stockholm, 1973, pp. 123-128.
18. T.W. Tusing, O.E. Paynter, D.L. Opdyke, and F.H. Snyder, *Toxicol. Appl. Pharmacol., 4,* 402-409(1962).
19. H.F. Smyth, C.P. Carpenter, C.S. Weil, and J.M. King, *Toxicol. Appl. Pharmacol., 17,* 53-59(1970).
20. O.G. Fitzhugh and A.A. Nelson, *J. Amer. Pharm. Ass., Sci. Ed., 37,* 29-32 (1948).
21. F.H. Snyder, D.L. Opdyke, J.F. Griffith, H.L. Rubenkoenig, T.W. Tusing, and O.E. Paynter, *Toxicol. Appl. Pharmacol., 6,* 133-140(1964).
22. D.L. Opdyke, F.H. Snyder, and H.L. Rubenkoenig, *Toxicol. Appl. Pharmacol., 6,* 141-146(1964).
23. F. Stenback and P. Shubik, *Toxicol. Appl. Pharmacol., 25,* 447(1973).
24. M. Takahashi and H. Sato, *Proc. Japanese Cancer Ass. 26th Ann. Meet.,* 1967, p. 187.
25. M. Takahashi and H. Sato, *Proc. Japanese Cancer Ass. 27th Ann. Meet.,* 1968, pp. 89-90.
26. M. Takahashi and H. Sato, *Gann Monogr., 8,* 241-261(1969).
27. M. Takahashi, H. Sugiura, and H. Sato, *Proc. Japanese Cancer Ass. 28th Ann. Meet.,* 1969, p. 80.
28. M. Takahashi, *Gann, 61,* 27-33(1970).

29. M. Takahashi, H. Sugiura, M. Tatematsu, and H. Sato, *Proc. Japanese Cancer Ass. 29th Ann. Meet.*, 1970, p. 47.
30. M. Takahashi, S. Fukushima, M. Hananouchi, T. Shirai, and H. Sato, *Proc. Japanese Cancer Ass. 29th Ann. Meet.*, 1970, p. 74.
31. M. Takahashi, S. Fukushima, and H. Sato, *Gann, 64,* 211-218(1973).
32. M. Takahashi, S. Fukushima, and M. Hananouchi, *Gann Monograph. Cancer Res., 17 Recent Top. Chem. Carcinog.*, 1975, pp. 255-267.
33. S. Fukushima, M. Takahashi, M. Hananouchi, H. Sugiura, T. Shirai, M. Tatematsu, and H. Sato, *Proc. Japanese Cancer Ass. 31st Ann. Meet.*, 1972, p. 20.
34. S. Fukushima, *Nagoya Med. J.*, *18*(3), 143-156(1973).
35. S. Fukushima, M. Tatematsu, and M. Takahashi, *Gann, 65*(4), 371-376(1974).
36. H. Sugiura, S. Fukushima, Y. Endo, and M. Takahashi, *Proc. Japanese Cancer Ass. 30th Ann. Meet.*, 1971, p. 16.
37. H. Sugiura, M. Hananouchi, S. Fukushima, T. Shirai, and M. Takahashi, *Proc. Japanese Cancer Ass. 32nd Ann. Meet.*, 1973, p. 174.
38. T. Shirai, S. Fukushima, M. Tatematsu, T. Kikutani, and M. Takahashi, *Proc. Japanese Cancer Ass. 32nd Ann. Meet.*, 1973, p. 141.
39. J. Borneff, *Arch. Hyg.*, *144*(4), 249-262(1960).
40. J. Borneff, *Arch. Hyg.*, *147*(1), 28-40(1963).
41. J. Borneff, *Zentralblatt Bakteriologie, Parasitenkunde, Infektions-krankheiten und Hygiene, Gustav Fischer Verlag, Stuttgart, I.* orig., *176*, 193-194(1959).
42. J. Borneff and R. Knerr, *Arch. Hyg.*, *143*(5), 390-404(1959).
43. J. Borneff and R. Knerr, *Arch. Hyg.*, *143*(6), 405-415(1959).
44. J. Borneff and R. Knerr, *Arch. Hyg.*, *144*(2), 81-94(1960).
45. J. Borneff and R. Fischer, *Arch. Hyg.*, *145*(1), 1-11(1961).
46. J. Borneff and R. Fischer, *Arch. Hyg.*, *145*(4), 241-255(1961).
47. J. Borneff and R. Fischer, *Arch. Hyg.*, *145*(5), 334-339(1961).
48. J. Borneff and R. Fischer, *Arch. Hyg.*, *146*(1), 1-16(1962).
49. J. Borneff and R. Fischer, *Arch. Hyg.*, *146*(3), 183-197(1962).
50. J. Borneff and R. Fischer, *Arch. Hyg.*, *146*(5), 334-345(1962).
51. J. Borneff and R. Fischer, *Arch. Hyg.*, *146*(6), 430-437(1962).
52. J. Borneff and R. Fischer, *Arch. Hyg.*, *146*(8), 572-585(1963).
53. J. Borneff, *Arch. Hyg.*, *148*(1), 1-11(1964).
54. J. Borneff and H. Kunte, *Arch. Hyg.*, *148,* 585-597(1964).
55. Yu.I. Sakharov, Z.A. Pyleva, and N.V. Sigina, *Gig. Tr. Prof. Zabol.*, *17*(9), 56-58(1973).
56. L. Merenmies, *Acta Pathologica Microbiol. Scandinavica, 46* Suppl. 130, 1-107 (1959).
57. M.E. Hodes, A.K. Warren, and C.G. Palmer, *Nature 188*(4745), 157-158(1960).
58. K. Dammert, *Acta Pathologica Microbiol. Scandinavica Suppl.*, *124*(1), 1-139 (1957).
59. K. Satälä, *Nature, 174,* 873-875(1954).
60. H. Hanson, *Z. Krebsforsch.*, *71,* 51-58(1968).
61. D. Guerritore, *Z. Krebsforsch.*, *63,* 142-148(1959).
62. W. Janicke, *Gesundheitswesen und Desinfektion, 54*(6), 81-84(1962).

63. M. Yoneyama, K. Mabuchi, M. Ikawa, H. Kobayashi, and H. Ichikawa, *Ann. Rep. Tokyo Metr. Res. Lab. P.H., 27-2,* 105-112(1976).
64. S.J. Fogelson and D.E. Shoch, *Arch. Int. Med., 73,* 212-216(1944).
65. J.M. Quack and A.K. Reng, *Fette. Seifen Anstrichm., 78*(5), 200-206(1976).
66. G. Schneider, *Soap and Chemical Specialties,* Nov. 1970, pp. 56-64.
67. WHO Scientific Group, *Wld. Hlth. Org. Tech. Rep. Ser. No. 482* (1971).
68. WHO Scientific Group, *Wld. Hlth. Org. Tech. Rep. Ser. No. 546* (1974).
69. Ad Hoc Committee of the Environmental Mutagen Society and the Institute for Medical Research, *Toxicol. Appl. Pharmacol., 22,* 269-275(1972).
70. Minister of Health and Welfare Canada, *Testing of Chemicals for Carcinogenicity, Mutagenicity, and Teratogenicity,* Health and Welfare Canada, 1975.
71. B.N. Ames, J. McCann, and E. Yamasaki, *Mutation Res., 31,* 347-364(1975).
72. T. Negi, K. Sugiyama, K. Inai, and T. Hayashi, *Ann. Res. Rep.* (*S-49*), Lion Fat & Oil Co., p. 1763(1974).
73. Y. Oda, A. Kuwano, K. Inoue, Y. Hamano, H. Yamamoto, B. Mitsuda, and N. Kunita, *Res. Rep. Osaka Pref. Inst. Public Health,* Ed. of Food Sanitation, 8, 11-16(1977).
74. T. Negi, K. Sugiyama, N. Omoto, and S. Mukaiyama, *Ann. Res. Rep.* (*S-52*), Lion Fat & Oil Co. (1977).
75. K. Inoue, T. Shibata, H. Yamamoto, Y. Hamano, Y. Oda, A. Kuwano, B. Mitsuda, and N. Kunita, *Res. Rep. Osaka Pref. Inst. of Public Health,* Ed. of Food Sanitation, (8), 25-28(1977).
76. K. Inoue, T. Shibata, Y. Hamano, Y. Oda, A. Kuwano, H. Yamamoto, B. Mitsuda, and N. Kunita, *Res. Rep. Osaka Pref. Inst. of Public Health,* Ed. of Food Sanitation, (8), 17-24(1977).
77. M. Masubushi, A. Takahashi, O. Takahashi, and K. Hiraga, *Ann. Rep. Tokyo Metr. Res. Lab. P.H., 27-2,* 100-104(1976).
78. Y. Mikami, H. Nagai, Y. Sakai, S. Fukushima, and T. Nishino, *Cong. Anom.* (*Jap.*), *9*(4), 230(1969): Abstract of paper presented at 9th Ann. Meet. Congenital Anomalies Res. Assoc. Japan, 1969/6, Kyoto.
79. Y. Mikami, Y. Sakai, and I. Miyamoto, *Cong. Anom.* (*Jap.*), *13*(3), 174-175 (1973): Abstract of paper presented at 13th Ann. Meet. Congenital Anomalies Res. Assoc. Japan, 1973/7, Hiroshima.
80. Y. Mikami and Y. Sakai, *Acta Anat. Nippon., 48*(1), 77-78(1973): Abstract of paper presented at 78th Meet. of the Societas Anatomica Nipponica, 1973/4, Chiba.
81. Y. Mikami, Y. Sakai, and I. Miyamoto, *Teratology 8*(1), 98(1973): Abstract of paper presented at 13th Ann. Meet. Congenital Anomalies Res. Assoc. Japan, 1973/7, Hiroshima.
82. Y. Mikami, Y. Sakai, I. Miyamoto, S. Kitamura, H. Hujishima, Y. Wakai, S. Akiyoshi, H. Sekiguchi, I. Agata, M. Hatori, S. Imaizumi, N. Tani, and N. Nishimura, *Report of Department of Anatomy, Mie University School of Medicine, 21,* 1-20(1973).
83. Y. Mikami and Y. Sakai, *Acta Anat. Nippon., 49*(2), 144(1974): Abstract of paper presented at the 33rd Central District Meet. of the Societas Anatomica Nipponica, 1973/9, Kanazawa.
84. Y. Mikami and Y. Sakai, *Acta Anat. Nippon., 49*(1), 33(1974): Abstract of paper

presented at 79th Meet. of the Societas Anatomica Nipponica, 1974/4, Kanazawa.
85. Y. Mikami, *Japan Med. J., 2731,* 33-34(1976).
86. Y. Mikai, Y. Sakai, S. Iwata, S. Kitamura, I. Miyamoto, I. Agata, S. Akiyoshi, S. Imaizumi, N. Nishimura, and H. Nishimura, *Report of Department of Anatomy, Mie University School of Medicine, 24,* 1-38(1976).
87. Y. Mikami, Y. Sakai, S. Iwata, I. Agata, S. Akiyoshi, H. Nishimura, N. Nishimura, and S. Imaizumi, *Report of Department of Anatomy, Mie University School of Medicine, 24,* 39-50(1976).
88. Y. Mikami, S. Iwata, Y. Sakai, S. Imaizumi, H. Nishimura, S. Kitamura, and I. Miyamoto, *Report of Department of Anatomy, Mie University School of Medicine, 24,* 51-62(1977).
89. Y. Mikami, S. Iwata, Y. Sakai, S. Imaizumi, N. Nishimura, S. Kitamura, and I. Miyamoto, *Report of Department of Anatomy, Mie University School of Medicine, 24,* 63-68(1977).
90. Y. Mikami, S. Iwata, Y. Sakai, S. Imaizumi, H. Nishimura, S. Kitamura, and I. Miyamoto, *Report of Department of Anatomy, Mie University School of Medicine, 24,* 69-80(1977).
91. Y. Mikami, S. Iwata, Y. Sakai, S. Imaizumi, and H. Nishimura, *Report of Department of Anatomy, Mie University School of Medicine, 24,* 81-91(1977).
92. Y. Mikami, *Report of Department of Anatomy, Mie University School of Medicine, 24,* 93-95(1977).
93. H. Iseki, *Report of Department of Anatomy, Mie University School of Medicine, 20,* 119-142(1972).
94. N. Ishimori, *Report of Department of Anatomy, Mie Prefecture University School of Medicine, 19,* 29-44(1971).
95. A. Nagai, *Report of Department of Anatomy, Mie Prefecture University School of Medicine, 18,* 61-74(1970).
96. Y. Sakai, H. Nishimura, S. Imaizumi, N. Nishimura, S. Akiyoshi, and I. Agata, *Report of Department of Anatomy, Mie University School of Medicine, 22,* 61-67(1975).
97. N. Tani, *Report of Department of Anatomy, Mie University School of Medicine, 22,* 1-14(1975).
98. Y. Wakai, *Report of Department of Anatomy, Mie University School of Medicine, 22,* 41-60(1975).
99. F.A. Charlesworth, *Information Bulletin BIBRA, 14*(7), 330-333(1975); *Food Cosmet. Toxicol., 14,* 152-154(1976).
100. R. Tanaka, *Japan J. Hygiene, 20*(3), 84-85(1965): Abstract of paper presented at 35th Meet. Japanese Society for Hygiene, 1965/5, Hiroshima.
101. R. Tanaka, *Japan J. Public Health, 13*(7), 230(1966): Abstract of paper presented at 22nd Meet. Japanese Society of Public Health, 1965/10, Osaka.
102. Y. Omori, T. Kuwamura, S. Tanaka, K. Kawashima, and S. Nakaura, *J. Food Hyg. Soc. Jap., 9*(6), 473-480(1968).
103. E. Stoklosa, S. Kosmider, and J. Kosiewicz, *Medycyna Pracy, 23*(4), 373-381 (1972).
104. E. Stoklosa, S. Kosmider, A. Brzegowy, and J. Kosiewicz, *Patol. Po., 24*(3), 391-401(1973).

105. A.K. Palmer, M.R. Lovell, A.M. Neuff, M.A. Readshaw, D.D. Cozens, and C.N. Roberts, *Jap. J. Public Health, 19*(10), 349(1972): Abstract of paper presented at 31st Meet. Japanese Society of Public Health, October 1972, Sapporo.
106. A.K. Palmer, M.A. Readshaw, and A.M. Neuff, *Toxicology, 3,* 91-106(1975).
107. S. Kotani, S. Tiba, A. Imahori, S. Shiobara, T. Kitagawa, H. Tiba, Y. Yanagida, and K. Yamamoto, *Jap. J. Public Health, 20*(10), No. 165 (1973): Abstract of paper presented at 32nd Meet. Japanese Society of Public Health, 1973/10, Hiroshima.
108. S. Tiba, S. Shiobara, T. Kitagawa, and S. Kotani, *Jap. J. Public Health, 21* (10): 337(1974): Abstract of paper presented at 33rd Meet. Japanese Society of Public Health, 1974/10, Fukushima.
109. S. Tiba, S. Shiobara, A. Imahori, and T. Kitagawa, *J. Food Hyg. Soc. Jap., 17*(1), 66-71(1976).
110. K. Takano, H. Yamamura, M. Yada, and F. Ariyuki, *Acta Anat. Nippon., 40* (1), 32(1965).
111. M. Yasuda, H. Yamamura, and T. Katsuya, *Cong. Anom.* (*Jap.*), *4*, 97-98 (1965).
112. K. Kaziwara, *Pharmacometrics* (*Jap.*), *2*(1), 1-12(1968).
113. K. Esaki and Y. Takioka, *Cong. Anom.* (*Jap.*), *10*, 216-217(1970).
114. A.N. Worden and C.N. Roberts, *Abstracts of Japan-British Medical Symposium Sponsored by British Embassy,* Tokyo, Japan Med. Ass., Feb. 28-March 6, pp. 128-146, 1972.
115. H. Nishimura and T. Tanimura, Examination on side-effects of drugs on fetuses, embryonic pharmacology seminar course, Nippon Keieikagaku Kyokai, p. 6 (1973).
116. T. Katsuya, Abstract of paper presented at the Joint Committee of 20th Meet. Japan Society of Human Genetics and 15th Meet. Congenital Anomalies Res. Assoc. Japan, p. 81, Tokyo, 1975/11.
117. H. Nishimura, *Cong. Anom.* (*Jap.*), *16*(3), 175-185(1976).
118. K. Sato, H. Ando, K. Yazawa, and K. Hiraga, *Ann. Rep. Tokyo Metr. Res. Lab. P.H., 24,* 441-448(1972).
119. F. Masuda, K. Okamoto, and K. Inoue, *J. Food Hyg. Soc. Jap., 14*(6), 580-582(1973); *15*(5), 349-355(1974).
120. A.K. Palmer and M.A. Readshaw, *J. Food Hyg. Soc. Jap., 14*(6), 638(1973): Abstract of paper presented at 26th Meet. Food Hygienic Society Jap., November 1973, Nagoya.
121. A.K. Palmer, M.A. Readshaw, and A.M. Neuff, *Toxicology, 4,* 171-181(1975).
122. A. Takahashi, K. Sato, H. Ando, Y. Kubo, and K. Hiraga, *Ann. Rep. Tokyo Metr. Res. Lab. P.H., 26-2,* 67-78(1975).
123. A. Imahori, T. Kitagawa, and S. Shiobara, *Japan J. Public Health, 23*(2), 68-72(1976).
124. F. Masuda and K. Inoue, *Cong. Anom.* (*Japan*), *14*(4), 309-318(1974).
125. A.K. Palmer, M.R. Lovell, A.M. Neuff, M.A. Readshaw, D.D. Cozens, and C.N. Roberts, *Japan J. Public Health, 19*(10), 350(1972): Abstract of paper presented at 31st Meet. Jap. Society Public Health, 1972/10, Sapporo.

126. A.K. Palmer, M.A. Readshaw, and A.M. Neuff, *Toxicology, 3,* 107-113 (1975).
127. A.K. Palmer and D. Bailes, Huntingdon Research Centre Report, LFO/31/76860, Dec. 14, 1976: A Report to Lion Corporation, Japan, Unpublished Data.
128. A.K. Palmer and D. Bailes, Huntingdon Research Centre Report, LFO/32/77265, April 21, 1977: A Report to Lion Corporation, Japan, Unpublished Data.
129. A. Takahashi, H. Ando, Y. Kubo, and K. Hiraga, *Ann. Rep. Tokyo Metr. Res. Lab. P.H., 27-2,* 113-118(1976).
130. H. Yamamoto, K. Inoue, Y. Hamano, Y. Oda, A. Kuwano, B. Mitsuda, and N. Kunita, *Res. Rep. Osaka Pref. Inst. of Public Health,* Vol. 7, 141-145(1976).
131. G.A. Nolen, Cong. Anom. (Jap.), *14*(3), 194(1974): Abstract of paper presented at 14th Meet. Congenital Anomalies Res. Assoc. Japan, 1974/7, Sendai.
132. G.A. Nolen, L.W. Klusman, L.F. Patrick, and R.G. Geil, *Toxicology, 4,* 231-243(1975).
133. M. Iimori, S. Inoue, and K. Yano, *Yukagaku, 22*(12), 807-813(1973).
134. K. Inoue and F. Masuda, *J. Food Hyg. Soc. Japan, 17*(2), 158-169(1976).
135. M. Nakahara, M. Ishibashi, T. Asao, T. Tsukamoto, K. Shinagawa, Y. Kinoshita, H. Tamamoto, B. Mitsuda, and N. Kunita, *Res. Rep. Osaka Pref. Inst. of Public Health,* 7, 131-135(1976).
136. T.W. Tusing, O.E. Paynter, and D.L. Opdyke, *Toxicol. Appl. Pharmacol., 2*(4), 464-473(1960).
137. J.M. Manson, *Casarett and Doull's Toxicology, The Basic Science of Poisons,* Macmillan Publishing Company, New York (1986).
138. H. Nishimura and T. Tanimura, *Clinical Aspects of the Teratogenicity of Drugs,* American Elsevier Publishing Company Inc., New York (1976).
139. S. Freeman, M.W. Burrill, Li Tsan-Wen, and A.C. Ivy, *Gastroenterology, 4,* 332-343(1945).
140. J.L. Schardein, *Drugs as Teratogens,* CRC Press Inc., Ohio (1976).
141. I.W. Daly, R.E. Schroeder, and J.C. Killeen, *Fd. Cosmet. Toxicol., 18,* 55-58 (1980).
142. A.K. Palmer, D.D. Cozens, P. Batham, and C.P. Cherry, LFO 10/731029, March 25, 1974: A Report to Lion Corporation, Japan, Unpublished Data.
143. F. Leuschner, R. Poche, and C. Gloxhuber, *Fette Seifen Anstrichm., 71*(7), 570-580(1969).
144. H. Nishimura, Taiji Kagaku, *Ondori-sha Co.,* Tokyo, pp. 95-122(1976).
145. W.R. Rapp and J.C. Killeen, Report of Bio/dynamics Inc. No. 75-1275, March 29, 1976; A Report to Japanese Soap and Detergent Association, Japan, Unpublished data.
146. S. Shiobara and A. Imahori, *J. Food Hyg. Soc. Japan, 17*(4), 295-301(1976).
147. Y. Hamano, H. Yamamoto, K. Inoue, Y. Oda, A. Kuwano, B. Mitsuda, and N. Kunita, *Res. Rep. Osaka Pref. Inst. Public Health,* Vol. 7, 147-152(1976).
148. T. Endo, Y. Furuido, K. Namie, N. Yamamoto, H. Hasunuma, and K. Ueda, *Annual Report of the Tokyo Metropolitan Res. Inst. for Environ. Protection,* 1980, pp. 236-246.

149. K. Hiraga, Ministry of Education, ad hoc Research Group of Environmental Science S-510, "Biological Effects on Synthetic Detergents," pp. 81-94(1981).
150. K. Hiraga, and T. Fujii, *Gann, 68,* 368-370(1977).
151. S. Taniguchi, A. Yamada, S. Morita, S. Ogaki, and T. Noda, *Result of STudies on Synthetic Detergents,* Science and Technology Agency, Research Coordination Bureau (Ed.), 1978, pp. 18-54.
152. H. Druckrey, H. Kruse, and P. Preussman, *Naturwissenschaften, 59*(9), 448 (1968): *Z. Krebsforsch, 75,* 69-84(1979).
153. B.L. Van Duuren, S. Melchionne, et al., *J. National Cancer Inst., 6*(1), 143-149(1971).
154. T.J. Slaga, G.T. Bowden, et al., *Cancer Research, 33,* 769-776(1973).
155. The Eppley Institute for Research in Cancer, University of Nebraska Medical Center, Final Report "Evaluation of the Carcinogenicity of Two Alpha Olefin Sulphonates and One Sultone," Sponsored by the International AOS Group, August 15, 1980 (unpublished data).
156. Colgate Palmolive Co. Unpublished data cited from: Arthur D. Little Inc., Human Safety and Environmental Aspects of Major Surfactants: A report to the Soap and Detergent Association, May 31, 1977; Arthur D. Little Inc., Human Safety and Environmental Aspects of Major Surfactants (Supplement), Feb. 20, 1981: A Report to the Soap and Detergent Association.
157. Chevron Chemical Co. Unpublished data cited from: Arthur D. Little Inc., Human Safety and Environmental Aspects of Major Surfactants. A report to the Soap and Detergent Association, May 31, 1977; Arthur D. Little Inc., Human Safety and Environmental Aspects of Major Surfactants (Supplement), Feb. 20, 1981: A Report to the Soap and Detergent Association.
158. S. Tomizawa, *Result of Studies on Synthetic Detergents,* Science and Technology Agency, Research Coordination Bureau (Ed.), 1978, pp. 55-88.
159. Proctor & Gamble. Unpublished data cited from: Arthur D. Little Inc., Human Safety and Environmental Aspects of Major Surfactants. A report to the Soap and Detergent Association, May 31, 1977; Arthur D. Little Inc., Human Safety and Environmental Aspects of Major Surfactants (supplement), Feb. 20, 1981: A Report to the Soap and Detergent Association.
160. T. Suzuta, T. Muramoto, Y. Sakatani, and H. Makita, *Result of Studies on Synthetic Detergents,* Science and Technology Agency, Research Coordination Bureau (Ed.), 1978, pp. 89-111.
161. T. Miyaji, and M. Miyamoto, *Chronic Toxicity of Sodium Alkyl Sulfate (AS), Studies on Surface Active Agents for Synthetic Detergents,* Ministry of Health and Welfare, Environmental Health Bureau, Food Chemistry Division (Ed.), 1981, pp. 43-47.
162. H. Yamamoto, *J. Nara Med. Ass., 28,* 307-327(1977).
163. R. Fukunishi, *Result of Potassium Fatty Acids—Chronic Toxicity, Studies on Surface Active Agents for Synthetic Detergents,* Ministry of Health and Welfare, Environmental Health Bureau, Food Chemistry Division (Ed.), 1981, pp. 101-111.
164. N. Ito, *Result of Sucrose Fatty Acid Ester, Co-carcinogenicity, Studies on Surface Active Agents for Synthetic Detergents,* Ministry of Health and Welfare

Environmental Health Bureau, Food Chemistry Division (Ed.), 1981, pp. 96-99.
165. OECD, OECD Guideline for Testing of Chemicals, Genetic Toxicology, No. 471-478, 1980-1982.
166. The WHO Environmental Health Criteria No. 51, Guide to Short-Term Tests for Detecting Mutagenic and Carcinogenic Chemicals.
167. U.S. Environmental Protection Agency, Code of Federal Regulations 40, Part 798 Health effects testing guidelines.
168. J. Yam, K.A. Booman, W. Broddle, L. Geiger, J.E. Heinze, Y.J. Lin, K. McCarthy, S. Reiss, V. Sawin, R.I. Sedlak, R.S. Slensinski, and G.A. Wright, *Fd. Chem. Toxic., 22*(9), 761-769(1984).
169. Arthur D. Little, Inc. Human Safety and Environmental Aspects of Major Surfactants: A Report to the Soap and Detergent Association (May 31, 1977).
170. Arthur D. Little, Inc. Human Safety and Environmental Aspects of Major Surfactants (Supplement): A Report to the Soap and Detergent Association (Feb. 20, 1981).
171. T. Sunakawa, K. Inoue, and K. Okamoto, *Eisei Kagaku, 27*(4), 204-211(1981).
172. Shell Toxicology Laboratory. Unpublished data cited from: Arthur D. Little, Inc., Human Safety and Environmental Aspects of Major Surfactants (Supplement), Feb. 20, 1981: A Report to the Soap and Detergent Association.
173. K. Inoue, T. Sunakawa, and S. Takayama, *Fd. Cosmet. Toxicol., 18,* 289-296 (1980).
174. J. Hope, *Mutation Research, 56,* 47-50(1977).
175. M. Kishi, S. Sato, Y. Horiguchi, and K. Ito, *Bull. Kanagawa P. H. Lab.,* (14), 57-58(1984).
176. N. Koizumi, R. Ninomiya, Y. Inoue, T. Tsukamoto, M. Fujii, and Y. Yamamoto, *Arch. Environ. Contam. Toxicol., 14,* 73-81(1985).
177. R.D. Swisher, W.E. Gledhill, R.A. Kimmerle, T.A. Tauli, *Carboxylated Intermediates in the Biodegradation of Linear Alkylbenzene Sulfonate* (*LAS*). VII. International Congress on Surface Active Substances. Proceedings, Moscow, 1976, *4*, 218-230.
178. Y. Oda, H. Yamamoto, and S. Kunida, *Proc. 9th Res. Conf. Environ. Mutagen Soc. Japan,* Osaka, 1980, pp. 37.
179. Tokyo Metropolitan Government, Public Health Bureau, "Studies of Toxicity on Sodium Alkylbenzene Sulfonate (Abstract)", (1980/March).
180. K. Morita, M. Ishigaki, and T. Abe, *Proc. 10th Soc. Cosmet. Chem.,* 1981, pp. 45-52.
181. Cosmetic Ingredient Review, *J. Amer. Coll. Toxicol., 6*(3), 261-277(1987).
182. S. Kunita, Effects of detergents on the living body, *Research on Human Effects,* Ministry of Education (Ed.), Reports of Studies on 'Environmental Science,' B73 R20-4," 1981, pp. 59-68.
183. Y. Mikami, *Japan Medical Journal,* (2795), 29-30(1977).
184. Y. Mikami, *Igaku Hyoron,* (66), 16-18(1980).
185. Y. Mikami, Ministry of Education, ad hoc Research Group of Environmental Science S-500, 1981, pp. 29-50.
186. Y. Mikami, H. Kawabe, and I. Kobayashi, *Goseisenzaikenkyukaishi, 4*(1), 107-122(1981).

187. J.G. Wilson and F.C. Fraser, *Handbook of Teratology,* Plenum Press, 1977, pp. 371-372.
188. H. Nishimura, Ministry of Education, ad hoc Research Group of Environmental Science S-510, *Biological Effects on Synthetic Detergent,* 1981, pp. 15-22.
189. K. Yoshida, *Biological Effects on Synthetic Detergent,* Ministry of Education, ad hoc Research Group of Environmental Science S-510, 1981, pp. 1-14.
190. K. Yoshida, Effects of detergents on the living body, *Research on Human Effects,* Ministry of Education (Ed.), Reports of Studies on 'Environmental Science,' B73 R20-4,'' 1981, pp. 25-26.
191. K. Hiraga, Ministry of Education, ad hoc Research Group of Environmental Science S-500, 1981, pp. 81-94.
192. Y. Kameyama, J. Sawano, H. Nishimura, and Y. Mikami, *Results of Studies on Synthetic Detergents,* Science and Technology Agency, Research Coordination Bureau (Ed.), 1981, pp. 123-168.
193. A. Kuwano, Y. Hamano, K. Inoue, Y. Oda, H. Yamamoto, and S. Kunida, *Res. Rep. Osaka Pref. Inst. of Public Health, 8,* 7-10(1977).
194. J.M. Tesh, and F.W. Ross, Life Science Research Report, 78/LIF29/479, December 21, 1978, A Report to Lion Corporation, Japan (unpublished data).
195. J.M. Tesh, S.M. Wilson, and S.A. Tesh, Life Science Research Report, 79/LIF27/010, January 9, 1979: A Report to Lion Corporation, Japan (unpublished data).
196. J.M. Tesh, R.W. Ross, and A. Moss, Life Science Research Report, 78/LIF33/309, August 31, 1978: A Report to Lion Corporation, Japan (unpublished data).
197. J.M. Tesh, P.A. McAnulty, and S.A. Tesh, Life Science Research Report, 79/LIF036/328, June 20, 1979: A Report to Lion Corporation, Japan (unpublished data).
198. J.M. Tesh, F.W. Ross, E.K. Budgen, and S.A. Tesh, Life Science Research Report, 79/LIF040/381, August 8, 1979: A Report to Lion Corporation, Japan (unpublished data).
199. J.M. Tesh, F.W. Ross, A.L. Pritchard, and S.A. Tesh, Life Science Research Report, 80/LIF038/056, March 3, 1980: A Report to Lion Corporation, Japan (unpublished data).
200. J.M. Tesh, F.W. Ross, E.K. Budgen, and S.A. Tesh, Life Science Research Report, 79/LIF042/258, May 14, 1979: A Report to Lion Corporation, Japan (unpublished data).
201. A.K. Palmer and M.E. Killick, Huntingdon Research Centre Report, LFO 58/78580, January 26, 1979: A Report to Lion Corporation, Japan (unpublished data).
202. A.K. Palmer and M.E. Killick, Huntingdon Research Centre Report, LFO 62/771038, January 31, 1978: A Report to Lion Corporation, Japan (unpublished data).
203. A.K. Palmer, P.A. Allen, and R. Clark, Huntingdon Research Centre Report, LFO 60/79146, March 26, 1979: A Report to Lion Corporation, Japan (unpublished data).

204. A.K. Palmer and P.A. Allen, Huntingdon Research Centre Report, LFO 64/7819, February 16, 1978: A Report to Lion Corporation, Japan (unpublished data).
205. J.M. Tesh and P.A. McAnulty, Life Science Research Report, 80/LIF032/493, November 26, 1980: A Report to Lion Corporation, Japan (unpublished data).
206. J.M. Tesh, P.A. McAnulty, and J.C. Whitney, Life Science Research Report, 80/LIF044/508, December 4, 1980: A Report to Lion Corporation, Japan (unpublished data).
207. T. Nomura, S. Kimura, S. Hata, T. Kanzaki, and H. Tanaka, *Life Sci., 26,* 49-54(1980).
208. T. Nomura, S. Hata, K. Shibata, and T. Kusafuka, *Mutation Research, 190,* 25-29(1987).
209. J. Sawano, "Result of Studies on Synthetic Detergents," Science and Technology Agency , Research Coordination Bureau (Ed.), 1978, pp. 112-113.
210. *The Detergent Toxicity and Evaluation,* Ministry of Health and Welfare, Environmental Health Bureau, Food Chemistry Division (Ed.), 1983, Japan.
211. B.N. Hemsworth, *Environ. Biol. Med., 9,* 243-244(1981).

8

Daily Human Intake

W. STERZEL Henkel KGaA, Düsseldorf, Germany

I. ROUTES OF INTAKE

The widespread use of anionic surfactants both by industry and in consumer products such as detergent or cleaning formulations, cosmetics, shampoos, and toothpastes means that these materials reach the environment and that every individual is exposed to these substances. The magnitude of material released into the environment, mostly in waste systems, is indicated by the U.S. International Trade Commission's (1) estimate of 1500 million kilograms of anionic surfactants produced in 1985 for the United States or 990 million kilograms for the European community (2).

In the 1950s-1960s, a problem became evident. The synthetic surfactants began to be noticeable in wastewaters, treated sewage, and the receiving waters due to the same property that had led to their success—they retained their foaming properties in natural waters. The major surfactant in general use at that time, namely alkylbenzenesulfonates (ABS) derived from tetrapropylene (tp), was quite resistant to microbial attack. This led to marked concentrations of tpABS in sewage up to 10-20 ppm (3), sewage plant effluent up to about 5-10 ppm, and in rivers in the range of 0.2 ppm, with peaks up to 1 ppm (4-6).

Concern for both the toxic properties and aesthetic value of the waterways prompted the discontinuation of hard-to-degrade tpABS in favor of the "softer" (easier to degrade) surfactants such as linear alkylbenzenesulfonates (LAS). The comparison of the 1964 and the 1987 values (4) demonstrates a significant drop in anionic surfactant levels with increased biodegradeability of the surfactants in use (down to 0.05 ppm).

Both the acute and chronic toxicity of LAS and other anionic surfactants have been reviewed by Swisher (8), Gloxhuber (9), and in this volume by Potokar (10). Most anionics have very low chronic toxicity, as demonstrated by Bornmann (11), Kay (12), Fitzhugh (13), and Tusing (14) in 2-year feeding experiments. Up to daily doses of 200 mg per kilogram bodyweight, no adverse effects were observable.

Thus, if human consumption is significantly lower than levels necessary to cause toxic signs in animal feeding studies, the safety of such compounds can be assured.

This chapter considers only the anionic surfactants. The major synthetic compounds are linear alkylbenzenesulfonates (LAS), alkyl sulfates (AS), alcohol ethoxy sulfates (AES), α-olefin sulfonates (AOS), and secondary alkane sulfonates (SAS). Since the tetrapropylene-derived alkylbenzenesulfonates are no longer in use, they are not included. The natural "soaps" and those made by saponification of fatty acids are also anionic surfactants of some importance.

Human intake of surfactants can be divided into several routes. One route is dermal absorption from cosmetics, shampoos, and bar soaps, and cleaning, dishwashing, and laundry detergents. Detergent production workers and possibly consumers would be subject to inhalation of surfactants. Oral ingestion of surfactants includes the acute dosages, such as a child's or an adult's accidental ingestion, and the not-so-obvious areas, such as low levels of surfactants in or on drinking water and foods.

This chapter attempts to cover relevant data up to 1989. Although there is considerable literature on biochemical, toxicological, and environmental effects of anionic surfactants, there is limited material on the presence, concentration, or release of these materials into the human environment. An excellent review on that subject is the monograph by Swisher (8). Also noteworthy are some excellent reviews by Swisher (3), Gloxhuber (9), Bartnik and Künstler (15) and Gerike et al. (4). Therefore, the data presented here will be an update of these previous reviews with some overlapping.

II. ORAL INTAKE OF ANIONIC SURFACTANTS BY HUMANS

Oral ingestion appears to be the only viable measure of daily human intake. As inhalation of anionic surfactants is limited mainly to a small number of subjects who might accidentally come in contact with these materials (consumers, production workers), inhalation is not considered to be relevant with respect to daily human intake. The second route of exposure occurs through the skin with the use of cosmetics and toiletries. The skin comes into contact with surfactants also during dishwashing or when washing hands. Furthermore, an indirect absorption of surfactants is also conceivable through textiles that may have adsorbed tensides. Published data suggest that only small amounts of anionic surfactants are resorbed through the intact skin of laboratory animals (16-18). As human skin is generally less permeable to chemicals than animal skin (20-23), the amounts of anionic surfactants absorbed cutaneously are probably even smaller. This was proved by Blank et al. (19) in in-vitro studies with excised human skin. Therefore this route was not taken into consideration.

Therefore, it appears reasonable to use the oral intake as the best estimate of daily human intake of anionic surfactants. Oral ingestion of anionic surfactants can be covered in three areas. Surfactants in food include residues on vegetables from washing, food content such as fish and meats, and the residue picked up by food from dish and cookware. Personal products such as toothpaste, mouthwash, and oral cosmetics contain surfactants that can be ingested. The final source is the drinking water, most of which comes from ground and surface waters that contain some recycled waste waters.

The area with the least amount of information is that of personal products. Toothpaste should be the major contributor. Average daily dentifrice usage ranges between 1.7 and 2.8 g (24,25). Adults ingest 1-2% of this material or 0.04 g day^{-1}. For a child one has to consider that 30-40% of dentifrice can be swallowed (25-27), which amounts to an average of 0.6 g day^{-1}. As suggested by Swisher (8), most toothpaste contains from 0.5 to 1.5% surfactants. The toothpaste sold in the United States and Europe usually contains an anionic surfactant. Therefore the amounts ingested can be estimated to be between 0.01 and 0.9 mg kg^{-1} day^{-1}.

A number of anionic surfactants have been approved in the United States for use with fruit and vegetable by the U.S. Food and Drug Administration (28). Similar regulations have been enacted in other countries. Although this method for cleaning has been used for several years, few data are available on the surfactant residues remaining on vegetables after rinsing. Swisher (8) estimated a daily intake of 1 mg surfactants from food-washing residues. Schmity (29), using radio-tagged surfactants, suggested an intake of 0.3 mg day^{-1}. Hiraga (30) and Taira et al. (31) estimated LAS intake from fruit and vegetable residues to be 9-14 mg day^{-1}. However, a 1974 review by the Japanese Soap and Detergent Industry Association (32) summarized the level of intake from washing foods in surfactants to be about 1.0 mg human^{-1} day^{-1} or 0.016 mg kg^{-1} day^{-1}.

Processed foods such as dairy and baked goods contain surfactants. Most of those used in the baking industry are nonionic and therefore not applicable to this study (33). Lanciotti et al. (34) reported up to a mean of 10 ppm anionic surfactants in milk powder which may result for infants in an estimated intake of no greater than 0.2 mg kg^{-1} day^{-1}. Although surfactants are used in food processing, mostly in cleaning systems, no data were obtainable. This is the result of extended rinsing procedures following the usage of cleaning products (35). Wakabayashi et al. (36) demonstrated that the major part of anionic surfactants accumulated in carp fish is located in the liver and gall bladder, e.g., parts of fish that are discarded during food processing. Comotto et al. (37) reported that the muscle tissue of fish in water containing 0.5 ppm LAS accumulated 20 μg g^{-1}. As the concentrations of residual anionic surfactants in rivers are ten times lower (0.05 ppm), one can assume 2 μg g^{-1} of muscle tissue. This can give rise to an intake of 5 μg kg^{-1} body-weight day^{-1} (150 g fish day).

A final source of anionic surfactants in food would come from its contact with residues remaining on service or cookware after washing. The major portion of anionic surfactants are removed from these dishes with rinsing (38). The remaining detergent is not homogenously absorbed at the dishes. Schmitz (38) demonstrated that approximately 50% of the detergent is located at the rim of the plates examined and therefore is not ingestable under realistic use conditions. According to Wedell (39) the maximal ingestable amount of anionic surfactant per day from residues on tableware totals to 0.38 mg or 0.006 mg kg^{-1} day^{-1}, respectively. This value is similar to that noted by Swisher (8), Schlüßler (40), and Borneff (41).

The final oral route of anionic surfactants is in the drinking water. The concentrations normally met in public water supplies as a result of re-use of wastewater are very low. It was for aesthetic reasons, for example, that the revised drinking water standards of the U.S. Public Health Service (1962) set an upper limit of 0.5 ppm on ABS; foaming is not noticeable below that concentration. This same upper limit of 0.5 ppm appears in the drinking water quality standards of several other countries and the World Health Organization (WHO) standards of the same period set a limit of 0.2 ppm. European countries have similar regulations for example in the German Drinking Water Regulations with an upper limit of 0.2 ppm. In 1973, the European Economic Community did a directive (42) which prohibits use of detergents that degrade less than 90%. This biodegradability resulted in a significant drop of the amount of anionic surfactants in the waters of those countries (43-45). A U.S. Geological Survey (46) of drinking water in New York (November 1970 to April 1972) found mean values in water of less than 0.05 ppm, far below these national standards. In 1990, no anionic surfactants were detectable (detection limit 0.01 ppm) (47) in German municipal drinking waters. Using the German standard of 0.2 mg $liter^{-1}$ and assuming a daily intake of two liters of water, this would add only 0.007 mg kg^{-1} day^{-1} to the total intake of anionic surfactants.

III. CONCLUSION

The major and perhaps only route of intake of the anionic surfactants by humans is the oral route. Absorption through the skin and inhalation are minimal, often acute, and restricted to a relatively small population. There are several areas in which anionic surfactants have been measured. These include personal products such as toothpaste. Food may be the major source of surfactant through residues from fruit and vegetable washing, residues from cooking or processing, and bioaccumulation in meat tissue such as fish. Drinking water is the final area for ingestion of surfactants, most of which comes from the household detergents via the waste sewage systems. Each of these separate areas may be totaled: 0.01-0.9 mg Kg^{-1} day^{-1} for personal products, 0.016 mg Kg^{-1} day^{-1} in fruit-vegetables, 0.005 mg Kg^{-1} day^{-1} in meat products, 0.006 mg Kg^{-1} day^{-1} from residues on tableware, and 0.007 mg Kg^{-1} day^{-1} in drinking water. The total daily intake for humans thus can be estimated to be between 0.044 and 0.944 mg Kg^{-1} bodyweight of anionic surfactants per day, where the high level only fits for small children (ingestion of toothpaste). Even this high level is small in comparison to the established no effect level for anionic surfactants of 200 mg Kg^{-1} day^{-1}.

The data presented show beyond any reasonable question that the margin of safety is very great, and there is no indication that hazard exists.

REFERENCES

1. U.S. International Trade Commission, Synthetic Organic Chemicals, U.S. Production and Sales of Surface Active Agents, January 1975.
2. Comité Européen des Agents de Surface et leurs Intermédiaires Organiques [CESIO], Cesio Statistics 1990.
3. R.D. Swisher, *Surfactant Biodegradation,* Marcel Dekker, New York, 1970.
4. P. Gerike, K. Winkler, W. Schneider, and W. Jakob, *Tenside Surfactants Detergents, 26,* 136-140(1989).
5. T. Waldmeyer, *Water Poll. Contr., 67,* 66-79(1968).
6. H.F. Ludwig and A.S. Sekaran, *Wat. Res., 2,* 257-262(1988).

7. C. Hammerton, *J. Proc. Inst. Sew. Purif., 1,* 48-53(1961).
8. R.D. Swisher, *Arch. Environ. Health, 17,* 232-246(1968).
9. Ch. Gloxhuber, *Arch. Toxicol., 3,* 245-270(1970).
10. M. Potokar, Chap. 3 of this volume.
11. G. Bornmann and A. Loeser, *Fette, Seifen, Anstrichmittel, 65,* 814-824(1963).
12. J.H. Kay, F.E. Kohn, and J.C. Calandra, *Toxic. Appl. Pharmacol., 7,* 812-818 (1965).
13. O.G. Fitzhugh and A.A. Nelson, *J. Amer. Pharm. Assoc., 37,* 29-32(1948).
14. T.W. Tusing, *Toxic. Appl. Pharmacol., 4,* 402-409(1962).
15. F. Bartnik and K. Künstler, in *Surfactants in Consumer Products,* Springer Verlag, Berlin, Heidelberg, 1987, pp. 475-503.
16. D. Howes, *J. Soc. Cosmet. Chem., 26,* 47-63(1975).
17. C. Prottey and T. Ferguson, *J. Soc. Cosmet. Chem., 26,* 29-46(1975).
18. J.G. Black and D. Howes, *J. Soc. Cosmet. Chem., 30,* 157-165(1979).
19. J.H. Blank and E. Gould, *J. Invest. Dermatol., 37,* 311-315(1961).
20. D.P. Rice, *Toxicol. Appl. Pharmacol., 39,* 377-381(1979).
21. R.C. Wester and H.J. Maibach, *Occup. Ind. Dermatol.* Chicago: Year Book. Med., 201(1982).
22. S. Tomiyama, *J. Amer. Oil Chem. Soc., 52,* 135-138(1975).
23. K. Minegishi, M. Osawa, and T. Yamaha, *Ch. Pharm. Bull., 25,* 821-825(1977).
24. Comité de Liaison des associations Européennes de l'Industrie de la Perfumerie, des Produits Cosmetiques et de Toilette (Colipa), Use Levels of Cosmetics—1981.
25. W.E. Barnhart, L.K. Hiller, G.J. Leonhard, and S.E. Michaels, *J. Dent. Res., 53,* 1317-1322(1974).
26. J. Ekstrand and M. Ehrnebo, *Caries Res., 14,* 96-102(1980).
27. H. Naccache, P.L. Simard, L. Trahan, M. Demers, C. Lapointe, and J.-M. Brodeur, *Caries Res., 24,* 359-263(1990).
28. U.S. Food Drug Administration, *Fed. Reg., 29,* 3472(1964).
29. J. Schmity, *Tenside, 10,* 11-13(1973).
30. K. Hiraga, in *On the Safety of Detergents,* Nippon Hair Society, Japan, 1974, pp. 55-67.
31. H. Taira, T. Eguchi, T. Mori, and M. Aoyama, *Food Sanit. Res., 25,* 155-161 (1975).
32. Literature Survey on the Safety of Synthetic Detergents, Nippon Soap and Detergent Industry Ass., 1975, pp. 6-7.
33. B.W. Landfried, Cereal Foods World, *22*(8), 338-340(1974).
34. E. Lanciotti, G. Grasso, and L. Borselli, *Min. Ped., 28*(30), 2413-2418(1976).
35. D.G. Dunsmore, *Residue Reviews, 86,* 1-63(1983).
36. M. Wakabayashi, M. Kikuchi, H. Kojima, and T. Yoshida, *Chemosphere, 11,* 917-924(1978).
37. R.M. Comotto, R.A. Kimerle, and R.D. Swisher, 2nd Ann. Aquatic Toxicology Symposium by ASTM, 1977.
38. J. Schmitz, *Tenside Detergents, 10,* 11-13(1973).

39. H. Wedell, in *Chem. Phys. Appl. Surface Active Subst.,* Proc. 4th Int. Congr., Vol. 3, 237-244. C. Paquot (Ed.), Gordon Breach, London, 1964.
40. H.-J. Schlüßler, *Chem. Ing. Tech., 52,* 246-247(1980).
41. J. Borneff, *Arch. Hyg. Bekt., 141,* 578-595(1957).
42. Official Journal of the European Communities No. L247/51, Directive 73/404/ EEC, 1973.
43. L. Huber, *Z. Wasser-Abwasser-Forsch., 2*(5), 183-191(1969).
44. R. Wickbold, *Tenside, 11,* 137-144(1974).
45. T. Waldmeyer, *Water Poll. Contr., 67,* 66-79(1968).
46. U.S. Geological Survey, New York MBAS Study, November 1970-April 1972 (1975).
47. Personal communication from the Public Water Supply Station Düsseldorf.

9

Pharmacological Properties

WERNER KÄSTNER Henkel KGaA, Düsseldorf, Germany

I. INTRODUCTION

In view of the pharmacological effects observed in both cationic and nonionic surfactants, it is not surprising that the pharmacological properties of anionics have been studied in some detail. As all of these compounds show a high degree of surface activity, it is believed that they influence the absorption of substances from skin and mucous membranes. Bile acids, which may be regarded as anionics, justify the assumption that anionics exhibit properties in the gastrointestinal tract.

II. EFFECTS ON THE SKIN

With dermatological products, it is usually desirable that the active principle reach a certain layer of the epidermis or dermis, but does not enter the systemic circulation. The aim, then, is penetration into the skin. Transdermal drug delivery systems require that a maximum amount of the topically administered drug reach the systemic circulation; their objective is, therefore, an optimal permeation through the skin. The obstacle to percutaneous absorption is the cutaneous barrier. The barrier function is almost entirely in the stratum corneum, a thin membranous layer that is mechanically strong and capable of resisting chemical attack. Once a molecule has transversed the layers of dead cells, the stratum corneum, there is no way to prevent its further diffusion into and across the viable cell layers into the blood and lymph capillaries. The magnitude of this systemic uptake of drug depends on the physicochemical characteristics of the drug, the formulation, and the skin conditions. Penetration (going into) and permeation (going through) can be combined into the collective term absorption (1).

Transepidermal delivery of drugs for systemic action has gained considerable interest because the transdermal route of drug application has some significant advantages in comparison to oral intake. These advantages are avoidance of the variable absorption and first-pass metabolism with oral therapy, continuous drug input, permitting use of drugs with short elimination half-lives, ability to terminate drug effect rapidly and improved patient compliance. However, because of the skin's impermeability only relatively potent agents can be used by this route. If the skin permeation could be enhanced, more drugs could be applicated dermally. Absorption promoters are defined as substances that have no pharmacological properties of their own in the amount/concentration used, but which can improve the penetration of drugs into the skin or their permeation through the skin (or mucosa) by reducing barrier resistance. Other terms as accelerants, penetration enhancers, absorption-enhancing agents, penetration carriers are used synonymously with absorption promoters.

Surfactants were supposed to be such penetration enhancers, in connection with skin and mucosa absorption of drugs, because of their surface active properties. Promoting experiments on skin were conducted with regard to allergies and with the aim to enhance percutaneous absorption of topical administered pharmaceutical substances. Skin penetration enhancers reversibly decrease the barrier resistance of the stratum corneum and allow drugs to penetrate more readily to the viable tissues and the systemic circulation.

Percutaneous absorption generally is studied by two types of model systems: those employing excised skin and those using living animals or humans in situ.

Chowhan and Pritchard (2) evaluated the relative effect of cationic, nonionic, amphoteric, and some anionic (0.05% dioctyl sodium sulfosuccinate, 4 or 2% sodium lauryl sulfate, 4 or 2% sodium laurate in an aqueous gel preparation or in an oil-in-water cream formulation) surfactants on percutaneous naproxen absorption using excised skin from rats, rabbits, and humans. The effect of nonionic, cationic, and amphoteric surfactants on the relative flux of naproxen was small. The anionic surfactants, sodium laurate (SL), and sodium lauryl sulfate (SLS), increased the in vitro flux of naproxen appreciably. Differences in the magnitude of the effect of surfactants in promoting percutaneous absorption of naproxen through skin of different species were observed. Excised human skin, which was least permeable in control experiments, showed a larger increase in the in vitro relative flux due to the presence of 2% SL and SLS, both in creams. Excised rabbit skin, which was most permeable in the control experiments, showed a moderate increase in the relative flux of naproxen when creams containing these surfactants were tested. The excised rat skin showed a relative small increase in naproxen flux from creams containing the same surfactants.

Using rat in vivo models, Chowhan et al. (3) repeated these studies regarding the effect of surfactant SLS and SL, 2% in cream each, on percutaneous absorption of naproxen. Comparisons between the in vivo normalized relative absorption rates and the in vitro relative flux indicate good correlation between the two model systems. These results suggest that the in vitro skin penetration model provides a good index of the in vivo percutaneous naproxen absorption in the rat. Surfactants only showed a small increase in percutaneous absorption. The 2% SL caused a higher plasma concentration of naproxen in the rat than the 2% SLS. With a bioassay model using rats for determining antipyretic effects of naproxen the effect of the 2% SLS in a naproxen cream in enhancing percutaneous absorption of naproxen could not be demonstrated. The creams with and without SLS showed similar onset, intensity, and duration of antipyretic activity. This result is probably because of the lack of sensitivity of the bioassay model to show small enhancement in percutaneous absorption in the rat from creams containing surfactants.

The effect of certain surface active agents on the diffusion characteristics through a cellophane membrane of acetylsalicylic acid, salicylamide, and phenacetin were studied by El-Sourady et al. (4). SLS increased salicylamide diffusion amount below and at its critical micellar concentration (0.1 and 0.3%), but above the critical micellar concentration (0.6%) decreased the amount of the drug diffused, and had no significant effect on the acetylsalicylic acid diffusion. By increasing the concentration of SLS (0.3 and 0.6%), the amount of phenacetin diffused was subsequently reduced. Only 0.1% SLS slightly increased the amount of phenacetin. Other authors (5) who examined the enhancement of naloxone penetration through human skin in vitro using various vehicles and penetration enhancers also found that SLS, SL, and sodium oleate (10% in propylene glycol) only slightly increased flux. The penetration enhancing effects of SLS were also examined using vehicles other than propylene glycol. 10% SLS in Isopropanol doubled naloxone flux. Naloxone skin penetration rate was also evaluated as a function of enhancer concentration. The increase of penetration rate reached a maximum (for example 20% of lauric acid) and at higher concentrations a decreasing effect appeared. The authors came to the conclusion that the skin penetration enhancing effects are specific and dependent on the drug, vehicle, enhancer concentration and probably other factors.

Another publication by Ashton (6) has the same topic. Some information about percutaneous absorption response to surfactants in humans are given.

Transdermal permeation of vasopressin were determined using rat epidermis by Banerjee and Ritschel (7). The anionic surfactant SLS was used to study the enhancement effect on skin penetration of surfactants. Five concentrations of SLS, 0.92, 1.92, 4.62, 7.98, and 19.2 mM containing vasopressin (105.44 μg/ml) were studied. SLS showed no significant effect on vasopressin skin permeation.

The above-described examples of studies regarding the search of possible absorption promoters the anionic surfactants containing a lauryl alkyl moiety have received much attention. The concentration of surfactants in the absorption vehicle may be of critical importance to percutaneous absorption. Concentrations below the critical micelle concentration (CMC) may reduce drug permeability through the skin, whereas concentrations slightly above the CMC may enhance it. Possible explanations offered for this limiting concentration value include drug-surfactant complexation below the CMC, hydration of epidermis at higher surfactant concentrations, and actual damage to the epidermis (8).

Barry (9) presents a general theory for accelerant activity based on possible alterations at the molecular level of the components of the horny layer. Anionic surfactants swell the horny layer, uncoiling and extending protein helices and thus opening up the protein-controlled polar route. But SLS (1%) does not entirely destroy the protein or lipid structure. The grossly swollen protein region probably imbibes a considerable amount of water and this allows better drug permeation.

Summarizing the topic of enhancing effects of anionic surfactants regarding cutaneous drug absorption it can be stated: The penetration rate of drugs depends on the chemical structure (8) and the concentrations of surfactants (2,4,5,8) and also on the vehicle or preparation (5) used for the dermal application. Considering these and other factors, which may play a role, it is understandable that some investigators found an increase in flux (2-5), others saw no effects (4,7), and also a reduction of absorption was reported (4). But it can be assumed that anionic surfactants enhance the drug permeation through the skin, generally.

III. EFFECTS ON THE EYE

Not only the enhancement of dermal penetration is of interest but also the question: Which substances are able to promote the absorption from the surface of mucous membranes? A special problem consists on the eyes, because in many cases of eye diseases a topical application of drugs is necessary. Some substances are known to enhance the absorption rate through the cornea. Surfactants belong to these substances, but the problem is that these chemicals have a high irritation potential and only low concentrations are tolerated from the eye's mucous membranes.

Various absorption studies on the eye under simultaneous application of different concentrations of surfactants have been performed. In these investigations the pharmacuetical substance studied was applied with or without surfactant. After a certain period of contact the amount of the drug which had been absorbed was determined, usually by estimating the concentration of the drug in the aqueous humor. Obviously, anionics should not possess irritating properties at the concentrations used.

In connection with the use of surfactants in ophthalmic pharmaceutical preparations for adults or for infants and juveniles, where they function as bactericidal agents or facilitate the uptake of the drugs by increasing the permeability of the cornea, Clayton et al. (10) studied, whether surfactants penetrate into the eye and whether the properties of the juvenile and adult eye differ significantly. Observations on two labeled anionic detergents (sodium lauryl sulfate and an ethoxylated (C_{18})alkyl sulfate), one 10-μl drop of the 10% test solution applied to the upper quadrant of the right eye of adult and young rabbits, were reported. Both detergents readily entered the eyes of adult and juvenile rabbits. Within 0.5 hr of a single application, they were found in all ocular tissues, and in the aqueous, vitreous and plasma. In all cases the cornea showed the highest levels, followed by choroid, retina, and iris. The younger animals had higher levels. The juvenile cornea seems to be more permeable than the adult one to surfactants.

Green et al. (11) studied the same questions as described above (10), surfactant penetration into young and adult rabbit eyes, and compared the ocular penetration of the two anionic surfactants sodium lauryl(C_{12}) sulfate (SLS), ethoxylated alkyl(C_{18}) sulfate, and the cationic surfactant benzalkonium chloride with those of other ophthalmic drugs (prednisolone phosphate, pilocarpine, epinephrine, dexamethasone, etc.). They found significant differences between young and adult rabbits, both in terms of the mass of surfactant entering the various tissue and fluid compartments of the eye. The differences between C_{12} and C_{18} chain anionic surfactants are pronounced, with the C_{18} compound being retained for longer times. Young animals amass greater quantities of surfactants per unit tissue weight. Both, anionic and cationic detergents show a very slow turnover once within the ocular tissues, while most ophthalmic drugs are lost from ocular fluids and tissues within a relatively short time. Both anionic and cationic surfactants are avidly retained by the cornea, in particular the epithelium, which acts as a reservoir for further diffusion of material to other sites.

Multiple drop administration of SLS led to the accumulation of substantial amounts of the detergent with most tissues (cornea, epithelium, lens, and all systemic tissues) reaching an apparent equilibrium after 13 drops (4 days) administration, but some tissues (e.g., iris) showing continued accumulation even after 22 drops (7 days) (12). The data for the ocular and systemic distribution of SLS in neonatal and juvenile rabbits after multiple drop administration show clearly that the tissues of neonates accumulate more detergent in the treated eye than do tissues of juveniles. Despite the fact that the difference in age is only a matter of between 2 and 3 weeks between neonates and juveniles, ocular tissues from neonatal animals accumulate from 1.5- to 7-fold more SLS than juvenile tissues (13).

Eyes that are treated with drugs often show injuries. Therefore, the question had to be clarified as to whether these anionics interfere with the epithelialization of the cornea. There was also the opinion that penetration can be achieved only by damaging the epithelium of the cornea. Experiments have shown that no penetration-promoting effect could be seen in eyes which had corneas with destroyed epithelium (14,15).

Absorption-promoting studies were done by Ginsburg and Robson (14). They studied the effect of sodium dodecyl sulfate (SDS) on the penetration of sulphazetamide. These tests were done on isolated corneas as well as on intact rabbit eyes. After a certain period of contact (15 min with the isolated cornea and 6 min with the intact eye), the concentration of the drug in the cornea or in the fluid of the anterior chamber of the eye was estimated. It was found that 1% Aerosol OT raises the penetration about 260% while the corresponding increases with 0.1 and 1% SDS were 125 and 330%, respectively. Aerosol OT has no penetration-promoting effect on the isolated cornea. The amounts applied of the surfactant were, however, not without irritating properties. A mucous effluence was observed on the surface of the cornea when concentrations were applied which elevated the penetration by 200% or more. After six applications of only 0.1% SDS solution (applied in intervals of 1 hr), damages of the epithelium of the eye in the form of edema

TABLE 1 Effect of Various Wetting Agents on the Penetration of Sulfathiazole

Agent	Sulfathiazole conc., mg/100 ml
Aerosol OT 1%	11.2
Aerosol OS 1%	14.2
Sodium lauryl sulfate 1%	8.1
Tergitol 7 1%	5.6

and a coloring of the cornea could be observed. These investigations showed that SDS raises the penetration. This effect depends on a damage to the cornea, possibly by a dissolution of intercellular cement. The authors suggest applying low concentrations.

The penetration-promoting effects of Aerosol OT, Aerosol OS, Tergitol 4, Tergitol 7, Tergitol 08, and sodium lauryl sulfate were investigated by Bellows and Gutman (16). These anionics were applied together with sulfonamides into the eyes of rabbits with the result that the concentration of the sulfonamide in the anterior chamber of the eye was higher than in the absence of surfactant. The effect was dependent on the concentration of the surfactant. It is supposed that the surfactants lower the surface tension between the cornea epithelium and its neighborhood. The investigations were done with young adult rabbits. Aerosol OS showed a clear effect with a concentration of 0.2%. A limitation of the penetration-promoting effect was reached with 10%. The effect on the penetration of sulfathiazole of a 1% solution of the anionics is shown in Table 1. With higher concentrations of the anionic, higher concentrations of the sulfonamides could be observed in the fluid of the chamber of the eye (the surfactant used as Aerosol OS). Table 2 demonstrates the dependence of the concentration of sulfathiazole from the

TABLE 2 Effect of Different Concentrations of Tergitol 7 on the Penetration of Sulfathiazole

Conc. Tergitol 7, %	Sulfathiazole, mg/100 ml
0.5	1.4
1.0	5.6
5.0	7.4
10.0	10.6
Undiluted	17.0

concentration of Tergitol 7. Both anionics showed a remarkable influence on the penetration of sulfathiazole.

A similar penetration experiment was done with sodium sulfazetamide on rabbit eyes (with or without cornea epithelium) by Ginsburg and Robson (15). The surfactant used was Duponol ME dry. In these tests right and left eyes were compared. Sulfonamide (10% solution) was applied to one eye, while the same solution of the sulfonamide together with 0.1% of the anionic was applied to the other. Similar tests were performed on eyes of which the corneal epithelium had been removed and also on eyes that had been excised. The results of these investigations (also with a 10% solution of the sulfonamide and 0.1% of the anionic) are summarized in Table 3.

The drug concentration is seen to be elevated in all types of tissues by the addition of the anionic. There were no significant differences between the results of in vivo and in vitro experiments with the exception of the conjunctiva, which showed a higher drug concentration in the case of the isolated eye. The latter observation is probably due to absorption and subsequent removal by the blood vessels of the conjunctiva of a considerable fraction of the drug in the case of the living eye. The results obtained with eyes with denuded corneas show that the anionics do not influence the permeability of the cornea if the corneal epithelium is removed. However, with regard to the conjunctiva, the influence of the anionic on drug penetration remains unchanged. According to this result, sulfonamides penetrate the denuded cornea more easily, and it may be concluded that the anionics support penetration across the barrier which is formed by the epithelium of the cornea.

The ability of certain anionics to promote drug penetration in the rabbit eye was also studied by Feldman et al. (17). With 1% solutions (buffered) of Triton X-30, "ultrawet 30-DS," and other surfactants, the authors found

TABLE 3 Concentrations of Sulfacetamide in Ocular Tissue (mg/100 g) Following Application of 10% Sodium Sulfacetamide with or without Detergent

	In vivo normal cornea		In vivo denuded cornea		In vitro normal cornea	
Tissue	S + D	S	S + D	S	S + D	S
Conjunctiva	391.0	127.0	531.0	353.0	1330.0	400.0
Cornea	349.0	191.0	1250.0	1190.0	370.0	145.0
Aqueous humor	29.1	6.4	336.0	422.0	16.3	4.8
Iris	20.7	14.0	184.0	238.0	42.9	18.5
Sclera	170.0	29.3	132.0	180.0	202.0	59.0

S + D = 10% sodium sulfacetamide/0.1% Duponol ME dry. S = 10% sodium sulfacetamide alone.

that the penetration of different types of drugs can be augmented. Best results were observed at a pH value between 8 and 9. Occasionally, the application of the surfactant together with the drug led to eye irritation.

Swan (18) found that the minimum concentrations of wetting agents which produce conjunctival irritation are higher than the concentrations necessary for ophthalmological drugs, provided that application of the drug is not too frequent. Of a group of surfactants, Grant (19) could identify those in experiments on rabbit eyes that increase drug penetration at low concentrations. According to Grant, irritating effects must be tolerated in certain cases if the penetration of a necessary drug can be increased to effective tissue levels.

Finally, investigations of Leopold (20) should be mentioned. Leopold studied various types of surfactants which had been considered for use in ophthalmological drugs, with regard to irritancy upon single and repeated applications on the intact eye and eyes with denuded cornea, respectively, to determine the concentration levels of lowest irritancy. It was found that the low concentration of 0.1% is not tolerated sufficiently well if the applications are repeated and that the regeneration of the eye epithelium is retarded. It was concluded that concentrations above 0.1% should not be used in ophthalmological drugs because it must be supposed that the epithelium of the cornea may be damaged and show infected ulcera. The studies relating to epithelium regeneration and to the effects of anionic surfactants were done on rabbits on both eyes of which the corneal epithelium had been removed on a surface of 6 mm diameter. One eye was then treated with a 0.1 or 0.5% solution of the surfactant (4 × 4 drops). The other eye was treated with physiological saline solution. The daily evaluation of the regeneration was done by aid of fluorescein. The median regeneration periods (in days) are given in Table 4. It is seen that all anionics tested give rise to a retardation of the reepithelialization at concentrations of both 0.5 and 0.1%.

TABLE 4 Average Regeneration Periods of the Epithelium in Days After Four Applications of a Surfactant Solution within 24 hr

Substance	Concentration 0.1%	Concentration 0.5%
Tergitol 4	5.33	6.83
Tergitol 7	4.50	5.50
Tergitol 08	4.50	4.67
Duponol ME dry	3.50	5.50
Aerosol OS	5.17	7.33
Aerosol OT	5.00	8.00
Sodium chloride (isotonic solution)	3.47	3.28

IV. EFFECTS ON THE NASAL MUCOUS MEMBRANES

Some drugs are applied as sprays or drops directly on the nasal mucous membranes. The absorption rate shall be determined with the aim of enhancing absorption in cases where systemic effects are intended, or of delaying the absorption and forming a depot in cases where topical effects are intended.

It is well known that insulin, a pancreatic peptide hormone, is barely absorbed after administration by nonparenteral routes because of its high molecular weight and its degradation by proteolytic enzymes. A high nonparenteral dose of insulin is necessary to produce the same pharmacological effect as a dose given parenterally.

Hirai et al. (21) showed that the nasal administration of insulin to dogs and humans resulted in a significant increase of the blood immunoreactive insulin level, with remarkable hypoglycemia. This nasal absorption of insulin could be enhanced by the addition of a promoter such as sodium glycocholate to the insulin solution. Later, Hirai et al. (22) investigated in more detail the nasal absorption of insulin and screened for absorption-promoting agents for the nasal absorption of insulin among various kinds of surfactants. They used the model of a dose-dependent hypoglycemia after nasal administration of an insulin preparation to rats. The absorption rate of insulin through nasal mucosa was measured by the decrements of the plasma glucose level. The plasma glucose level was reduced significantly by the addition of 1% sodium lauryl sulfate (SLS), potassium laurate or bile acid salts as sodium taurocholate, and sodium cholate at pH 7.4, and a dose of 10 U insulin/kg. The absolute bioavailability of insulin estimated by the hypoglycemic effect was 5% after nasal administration of the pH 3.1 insulin solution without promoter. However, the bioavailability was increased to about 30% by addition of the above-mentioned anionic surfactants at a concentration of 1%.

The mechanisms of the promoting effect of surfactants on the nasal absorption of insulin was investigated in rats by the same authors (23). The promoting effect of SLS was paralleled by its ability to lyze the rabbit erythrocyte and to release protein from nasal mucosa, whereas the promoting effect of bile acid salts was attributed not only to their direct effect on the nasal mucosa, but also to the inhibitory effect on proteolytic enzymes. Observations by scanning electron microscopy revealed that the mucosal alterations caused by a promoter were reversible and the membrane was relatively rapidly restored. A good correlation was observed between the effects on the biomembrane and the absorption-promoting effect of the surfactants, especially the nonionic surfactants, SLS and saponin. This suggests that the effect of the surfactant on the permeability of the nasal mucosa to insulin may be due to the perturbation or disorder of the structural integrity of the nasal mucosa caused by the surfactant.

In a further publication, Hirai (24) summarized the results in connection with nasal absorption of insulin and enhancing effects of surfactants. The purpose of the studies was to find a way to improve the bioavailability of drugs, especially of drugs that are without, or only of small, effects by oral application. Hirai (24) demonstrated that the nasal mucosa can be a new application route of drugs for systemic use. The bioavailability of nasally administered insulin with 1% sodium glycocholate in rats, dogs, and humans was estimated at about 30% by comparing the hypoglycemic effect with that after intravenous (i.v.) administration. The results of the studies suggest that the surfactants appear to increase the permeability of the nasal mucosa, and thereby promote the nasal absorption of a high molecular weight polypeptide such as insulin.

From these findings, it becomes evident that the nasal cavity is a useful application site for both lipophilic and hydrophilic drugs including polypeptides, and that nasal administration is superior to the oral administration, especially for drugs poorly absorbable through the gastrointestinal tract.

The nasal absorption of the antihypertensive pharmaceutical substance hydralazine and its influencing by the surfactant sodium glycocholate was studied by Kaneo (25) in rats. Although the drug is quickly and almost completely absorbed, the bioavailability of hydralazine is relatively low after oral administration due to the extensive first-pass metabolism. In order to improve the bioavailability of hydralazine, the nasal administration was considered to be a nonparenteral route. The nasal absorption of hydralazine was markedly enhanced by sodium glycocholate (concentration: 1 mM hydralazine + 30 mM = 1,5% surfactant) in the in situ nasal recirculation experiments. The in vivo experiment showed that the drug was well absorbed. However, the absorption rate observed at pH 3.0 was relatively slow ($t_{½}$ = 8.4 min). Drugs administered intranasally are removed from the nasal area by the normal pattern of mucus flow. Therefore, the rapid absorption is preferable in order to attain good bioavailability in the actual state. The absorption rate of hydralazine could be significantly enhanced by adding a surfactant ($t_{½}$ = 2.9 min) or adjusting the pH to 6.5 ($t_{½}$ = 4.0 min).

V. EFFECTS ON THE GASTROINTESTINAL TRACT

A. Acceleration of Absorption

As described in the earlier passage about the enhancement of nasal absorption of drugs, most investigations in connection with studying the effects of surfactants on the intestinal absorption rate had been carried out with the aim to find substances that enhance the absorption rate of poorly absorbable drugs. If the intestine permeability could be improved by special adjuvants, such drugs could be applied orally. Because surfactants are supposed to have an effect on the increasing of absorption, tests on the gastrointestinal tract were performed in order to find out whether anionics promote absorption or not.

The tests to study the promoting effects of different anionic surfactants on the absorption rate of various drugs were carried out on nearly all segments of the gastrointestinal (GI) tract. The oral frenulum of dogs (26), the keratinized oral mucosa of hamster cheek pouch (27), the buccal mucosa in rats (28), the gingival mucosa of humans (29), the loop of rat stomach (30) was used as well as the hamster jejunal segments (31), the rat jejunum, ileum and colon (32-35), and the human intestinum (36-39).

Regarding the used anionic surfactants that were tested for effects of GI absorption, sodium dodecyl or sodium lauryl sulfate and sodium dioctyl sulfosuccinate (DSS) were the most frequent ones.

The results of the studies have shown that it cannot be claimed that the anionic surfactants enhance the absorption rate of drugs in every case. The promoting or inhibition effects depend on several properties as on their own chemical structures and on the structures of the drugs (28) that will be absorbed, the concentration of the surfactants—lower concentration can have an increase, higher a decrease of absorption (36)—the pH-values of the medium (27,28), and the presence of other substances (40). Not only the absorption rate of drugs or organic compounds are influenced by anionic surfactants but also other changes of gastrointestinal mucosa were observed in connection with absorption studies, for example affecting the intestinal water (31), glucose and cation (Na^+, K^+) (32,41) transport, and protein and phospholipid release from the mucosal surface and liberation of cholesterol (33).

In most studies a promoting effect on absorption rate was observed, in some studies retarding effects were established, and there are also publications that report about both increasing and decreasing effects in dependence of concentrations, pH, ions, and other substances.

Appel et al. (42) found that lauryl sulfate in a concentration of only 0.01% increases the diffusion of substances through membranes and cell walls. Oral therapy with strophantine has been desired for a long time by clinicians. According to Krause's investigations (43), the addition of sodium lauryl sulfate (SLS) to strophantine-containing tablets should result in an absorption of 10% of the drug present in the intestine. A similar result was obtained by Eybl (44) after rectal application of strophantine. Briseid and Bergersen (45) again studied the effect of lauryl sulfate on the absorption of various drugs. As a model, the isolated loop of intestine was used. It was found that SLS in concentrations exceeding the critical micelle concentration elevates the absorption of ouabain, phenolphthalein, and pralidoxim. It is supposed that the anionics act directly on the membrane of the intestine.

The absorption of certain dyes (phenol red, methyl violet, Ponceau 6R, and Lichtgrün SF gelblich) from the large intestine in the presence of tetrapropylene benzene sulfonate or dodecyl benzene sulfonate has been studied by Bormann and Stanisic (46), who concluded that anionics can accelerate as well as retard the absorption by complex formation.

Similar observations were reported by Lish and Weikel (47), who found that SLS and DSS—unlike a nonionic surfactant—increase the amount of absorption of phenol red from the colon of rats. In contrast, the absorption of basic methyl violet is not influenced by these anionics. In the case of phenol red the accelerating effect can be counteracted by application of atropin.

Because bile salts, which are physiological anionic surfactants, play a significant role in absorption processes in the intestinum, some investigators examined whether they enhance the absorption of certain drugs.

The effect of oral administration of 300 and 600 mg doses of a bile salt (sodium deoxycholate) on absorption of a poorly absorbable drug, phenolsulfonphthalein, in humans is reported by Gouda et al. (36). The 300-mg dose caused an increase in the initial phenolsulfonphthalein absorption rate, suggesting a direct action of the bile salt on membrane permeability. The 600-mg dose caused a decrease in drug bioavailability as measured by the total amount excreted in 24 hr. This decrease was attributed to micellar entrapment of the drug molecule.

Gouda et al. (37) also studied the influence of synthetic surfactants on the absorption of a poorly absorbable drug, phenolsulfonphthalein. Twelve healthy subjects were administered at weekly intervals 100-ml water, or 250 mg DSS, or 240 mg dioctyl calcium sulfosuccinate (DCS) as capsules with 100-ml water each, and 1 hr later 20-mg phenolsulfonphthalein dissolved in 50-ml water. Urine were collected 1-24 hr after the application of surfactant. The intake of both anionic surfactants, DSS and DCS, 1 hr before dye administration, increased the initial rate of dye absorption, but no significant change in the extent of drug absorption was observed. It appears that both anionic surfactants cause a temporary change in membrane permeability, resulting in an increase in the rate of drug absorption. These findings must be considered in connection with the administration of ionic fecal-softener laxatives as DSS with nonabsorbable peristaltic stimulants. It may be possible that the absorption promoting effect of surfactants change a seemingly innocent nonabsorbable laxative to a toxic agent.

The effects of various nonionic and ionic surfactants and two bile salts (sodium cholate and sodium taurocholate) on the gastrointestinal absorption of the β-lactam antibiotic cefazolin were investigated using the in situ rat gastrointestinale perfusion technique (48). SLS (5-30 mM) and sodium di-2-ethylhexyl sulfosuccinate (10 mM) promoted the cefazolin absorption. Bile salts revealed no significant promotional effect.

Among cationic and nonionic surfactants, the effect of SLS on the permeability of oral frenulum removed from anaesthetized dogs was determined in vitro (26). Permeability to 12 organic compounds as propanol, acetamide, urea, glycerol, mannitol, sucrose, inulin, dextran was measured in the presence and absence of surfactants. SLS at concentrations from 0.025-1.0% caused dose-related increases in permeability to each of the solutions tested.

Shah et al. (38) demonstrated that anionic surfactants do not enhance the GI absorption rate of drugs in any case. A pilot bioavailability study was carried out to evaluate the drug interaction and influence of DSS on the absorption/bioavailability of tetracycline. Three tetracycline products—a fast-dissolving capsule, a slow-dissolving capsule, and a suspension were used in the study. DSS was administered (a) 30 minutes before tetracycline administration, and (b) on days 3, 2, and 1 in the evening before tetracycline administration to three volunteers each. Frequent urine samples were collected up to 48 hr and were analyzed by a microbiological method. Although not statistically significant in this small study, the results suggest that there is a reduction in tetracycline bioavailability due to DSS. The indiscriminate use of surface-active agents to increase the dissolution rate of solid oral dosage forms in the belief that the resulting increased dissolution improves product bioavailability must be questioned due to these test results.

The effect of surfactants on the diffusion rate of aspirin, salicylamide, and phenacetin through the rat's ileum was studied by El-Sourady et al. (34). The diffusion rate of phenacetin was increased clearly, but only slightly of aspirin, and was decreased of salicylamide by a 0.3% SLS solution.

The influence of surfactants (a cationic surfactant, SLS, and DSS) on sulphamethoxazole (SMZ) bioavailability was studied on healthy human volunteers by Roa and Rambhau (39). After overnight fasting, the subjects were administered a suspension, containing 800-mg SMZ and 25-mg surfactant in 20-ml water, followed by 250-ml water. From 0 to 12 hr after intake of the suspension blood levels of SMZ were measured. SLS and DSS considerably reduced the bioavailability of SMZ. Anionic surfactants retarded the bioavailability to a greater extent compared to cationic surfactant. This reduction in the absorption of SMZ due to anionic surfactants is explained with the micellar effects by rendering drug molecules unavailable for absorption.

The effects of various adjuvants on absorption of buccally administered insulin in rats were examined by Aungst and Rogers (28). The goal of the study was to determine which adjuvants are the most potent absorption promoters for buccal insulin. Insulin absorption was estimated from the cumulative response of plasma glucose concentrations and by comparison to an i.m. dose/response curve. In the absence of an absorption promoter, buccal insulin was less than 4% as effective as i.m. insulin. The effects of some of the great number of tested substances should be demonstrated. The effects of the steroidal detergent sodium glycocholate were dependent on the concentration, and concentrations exceeding 1% were required to reach a significant increase of buccal insulin absorption. Sodium deoxycholate had a similar effect as sodium glycocholate in a concentration of 5% each. SLS (5%) significantly increased buccal insulin efficacy. Sodium laurate (5%) was not effective at pH 7.4, but this could have been due to its very limited solubility at that pH. A clear 5% sodium laurate solution was made at pH 8.9, and this vehicle increased buccal insulin delivery with a similar efficacy as SLS. The effects of anionic surfactants on buccal absorption were very similar to those on nasal insulin absorption described by Hirai et al. (22). SLS and potassium laurate enhanced nasal insulin absorption.

In an attempt to develop an effective bioadhesive device for timolol, a β-adrenoceptor blocking agent, Deasy and Neill (29) described the preparation of the drug base and the in vitro and in vivo evaluation of the final composite device (compact). The compacts, containing a core with the drug, were located on the gingival mucosa below the first molar of five human volunteers. After periodical removal, each compact was assayed for residual drug content. SLS was added to the core formulation of certain compacts as a penetration enhancer. An average of 34% of the drug loading was absorbed over 3 hr. Addition of SLS 0.1% to the core increased the mean amount absorbed over 3 hr to 61% of the drug loading.

Finally, it should be mentioned that very low concentrations of DSS significantly increased the uptake of Na-pentobarbital by the goldfisch (49). The effects of this anionic surfactant on the fish reached a maximum level within minutes and were reversible. Maximum effect occured at a concentration of 0.003%. Increasing the concentration up to 0.005% did not result in increased absorption of the drug by the fish. No interaction between pentobarbital and the surfactant was observed. The results are consistent with a mechanism that involves a direct effect of DSS on the biologic membrane of the fish resulting in alteration of membrane permeability.

In connection with the investigations of adjuvant effects on enhancement of absorption rate of drugs other effects of anionic surfactants were reported as the influence on glucose, water, and cation transport through the intestinal mucosa. Optiz (50) observed that the absorption of magnesium sulfate can be raised by alkylaryl sulfonate. A similar result was published by Kozlik (51), who reported that the absorption of magnesium sulfate is accelerated by dodecyl sulfate. The toxicity of magnesium sulfate in mice was used as a parameter of this effect. Narcotic symptoms were observed earlier in animals which had received a surfactant in addition to the magnesium sulfate.

The intestinal absorption of glucose by addition of SLS was studied by Kozlik (52), who observed an acceleration. This finding has been confirmed by Mosinger (53) in a similar experiment with SLS concentrations of 2.5-5%

(we suppose that such high concentrations led to irritations of the mucous membranes).

The surface active properties of anionic surfactants influence in different kind the cells of the instetinum. The relationship of the surface properties of a group of anionic surfactants to their effects on intestinal water transport was studied by Gullikson et al. (31). Dose-response inhibition of water transport in everted hamster jejunal segments was obtained with two long-chain surfactants (sodium dodecyl sulfate (SDS) and DSS 0.5 mM each), and dihydroxy bile salts (0.5 mM deoxycholate, 1.0 mM chenodeoxycholate, and 4.0 mM taurodeoxycholate), whereas no activity was seen with trihydroxy and tri-keto bile salts. The relative effects on water transport were paralleled by their abilities to lyse the erythrocytes. These two biological effects were related to the surface properties of the agents, as determined by critical micelle concentration and surface tension reduction.

Sund (35) studied the effect of SDS upon net sodium and water transport in vivo. Into tied jejunal loops in anaesthetized rats, solutions of sodium chloride or sulfate were instilled, with or without addition of SDS. The sodium salt concentration was varied, so that there was either a sodium concentration gradient from the plasma to the lumen or a gradient from the lumen to the plasma. In the presence of the surfactant, net secretion took place when there was a significant gradient from the plasma to the lumen, and net absorption when the gradient was reversed. With only a small gradient from the lumen to the plasma, there was complete inhibition of net sodium transport. In principle, in the presence of SDS, sodium was transported along its concentration gradient. In control experiments, sodium absorption essentially independent of sodium concentration occurred. The results indicate that with SDS, normal carrier-mediated net sodium absorption is successively substituted by passive diffusion, as the surfactant dose-dependently causes increased epithelial permeability to sodium. Less than 1.7 mM (0.05%) dodecylsulfate seemed to cause this effect.

Other surfactants, for example DSS, affected jejunal transport of sodium and water in the same way as SDS: From hypotonic sodium chloride solutions, they inhibited net water absorption or induced secretion in a dose dependent manner; from hypotonic sodium salt solutions net sodium absorption was enhanced (54). These results also indicate that, with increasing doses of surfactant, diffusion becomes increasingly important as the determinant of net sodium transport, suggesting that the surfactants cause increased passive permeability of the intestinal epithelium.

The effects of anionic surfactants on glucose and cation transport in rat jejunum, ileum, and colon in vivo were studied by Sund and Matheson (32). Osmotically balanced solutions of glucose (0.5-300 mM) and sodium chloride, containing surfactants, were instilled into the small or large intestine of anaesthetized rats. Net absorption or secretion of glucose, sodium, and potassium was studied. The surfactants tested were SDS (3.4-17 mM), DSS (1.8-11 mM) and deoxycholate (2.5 mM). The 17 mM SDS seemed to abolish completely physiological glucose transport in the jejunum and ileum. At a lower concentration, and with the other surfactants, normal glucose transport was affected to an intermediate extent.

Similar effects were observed by Sund and Olsen (41). Control cation and glucose transport were changed by DSS and SDS in the way described above (32). The authors came to the conclusion that the surfactants mainly affect paracellular sodium transport, and do not support the theory that the effect of DSS on the intestinal transport is related to an activation of the mucosal adenylate cyclase/cAMP system, at least not in short term experiments. The in vivo reversibility of the jejunal glucose and cation transport alteration caused by intraluminal surfactants in the rat was studied by Sund and Jacobsen (55). They pre-exposed tied jejunal loops in anaesthetized rats under standardized conditions for 30 min with Tyrode solutions containing surfactants (5.6 mM/liter DSS or 8.5-17 mM/liter SDS). 5, 20, or 150 min after wash out of bulk surfactant, the loops were reinstilled with Tyrode containing glucose at 5-15 mM/liter. Net glucose, sodium, and potassium transport were studied for 15 min by changes in intraluminal amounts, and compared with results obtained in control rats. The experiments showed that, in most cases, the alteration in glucose transport capacity is rapidly reversed, much of the restoration taking place shortly after the bulk surfactant had been removed, and was normal or fairly normal after 150 min. The changes in net cation transport tended to be less easily reversible than the alteration in glucose transport.

In connection with the discussion, whether the reduction of glucose absorption after treatment with surfactants could be a result of cellular damage of the intestinal villi, the authors concluded from these results that surfactants may interfere with biomembrane transport in a way that is not necessarily linked to histological and structural damage.

Nakagawa et al. (56) examined the digestion of sucrose and the absorption of glucose from rabbit small intestine pretreated with several detergents and found that some nonionic surfactants inhibited, but SLS did not inhibit, both the absorption of glucose and the activity of sucrase. Treatment with 5% SLS had no marked effect on glucose absorption after the administration of sucrose, but after 50 (not after 60-90) min, minor decreases of glucose levels were observed, which might be a result of damage caused by the surfactant.

In some cases, investigators examined whether the promoting effect of surfactants can be influenced by other surfactants, for example by the bile salts or by presence or absence of certain cations.

Kurosaki et al. (27) investigated the effects of surfactants on the absorption of salicyclic acid from keratinized oral mucosa by hamster cheek pouch method in vivo at pH-values 3.0, 4.0, and 7.0. Four surfactants, SLS, one cationic and one nonionic surfactant, and sodium taurocholate, were examined as adjuvants. The interaction between salicyclic acid and each surfactant was determined by the molecular sieve method using Sephadex G-25.

In presence of 20 mM SLS, the absorption of salicyclic acid was not affected in the lower pH region, but was significantly increased at pH 7.0 where almost all the salicyclic acid molecules exist in the ionized form. The cationic and the nonionic had an decreased effect in the lower pH region. 20 mM STC showed no effect on salicyclic acid absorption in any conditions examined. After the pretreatment of the cheek pouch with 20 mM solution of SLS the absorption of salicyclic acid increased significantly in all pH conditions examined. Although the effects of SLS and the cationic surfactant were nearly identical to each other at pHs 3.0 and 4.0, the enhancement caused by SLS was significantly greater than that by the cationic at pH 7.0. The pretreatment with 1.0 mM solution of ionic surfactants did not bring about a significant increase in permeability in comparison with that of a simple buffer solution (expressed as control). However, when the pretreatment was carried out with the solution of concentration higher than 5.0 mM surfactant, the absorption increased significantly. These findings indicate that SLS can lower the barrier function of the keratinized oral mucosa to polar substances and that this effect is dependent on the surfactant concentration.

Kim et al. (30) examined the mechanism of taurine-induced enhancement of drug absorption by comparison between the actions of taurine and SLS, using the loop of rat stomach in situ. The gastric absorption of aspirin and sulfanilic acid was enhanced by a concentration of 20 mM SLS and salicylamide was enhanced even by a concentration of 0.05 mM SLS. No effect had a concentration of 1 mM and a decreasing effect had a concentration of 20 mM on salicylamide absorption. The SLS-induced enhancement seems to be Na^+-dependent. SLS failed to enhance the absorption of aspirin and salicylamide when NaCl was completely replaced by mannitol or in the presence of ouabain even in the NaCl medium. A concentration of 1 mM taurine had similar enhancing effects on aspirin and salicylamide absorption, but had no effect on sulfanilic acid. The actions of SLS and taurine on drug absorption are considered to have some similar points.

The effects of three typical synthetic surfactants, a nonionic and a cationic surfactant and SLS in combination with the physiologic one, sodium taurocholate (STC), on the intestinal absorption of four model drugs were examined in rats by using the in situ recirculation technique (57). SLS and STC were used in concentrations of 20 mM alone or in combination. The intestinal absorption of sulfaguanidine, a poorly absorbable drug, which is in unionized form at the physiological pH region of the small intestine, was enhanced by SLS and STC alone and by combination of both. But the combination had no additional effect on the enhancing effect of SLS. Quinine is cationic at pH 6.5, lipophilic and is moderately absorbed from rat small intestine. Imipramine is cationic and highly lipophilic at pH 6.5 and well absorbed from rat small intestine. SLS alone and in combination with STC had an inhibitory effect on quinine and imipramine absorption. Sulfadimethoxine is partly anionic at pH 6.5, it is relatively lipophilic and well absorbable from rat small intestine. SLS alone and in combination with STC decreased the absorption rate significantly. Summarizing, it can be stated that SLS alone and in combination with STC had an increasing effect only on the poorly absorbable drug, but had no or a decreasing effect on moderately or well absorbed drugs. The other tested surfactants had similar effects. The authors explained these results by the micellar interactions between drugs and surfactants.

Briseid et al. (40) reported that the increases in the absorption of oubain, phenolsulfonphthalein, and pralidoxime caused by 17 mM SLS from jejunal loops of anaesthetized rats were significantly reduced if sodium and chloride or chloride and bicarbonate were replaced by other ions in the loop fluid. Separate substitution of sodium, chloride, or bicarbonate did not significantly alter the SLS-caused absorption.

In connection with the investigations that were conducted with the aim to find out how the surfactants increase the absorption of poorly absorbable substances, some other effects as protein release, change of enzyme activities, and biochemical and histological alterations on intestinal mucosa were reported.

Studies of the action of deoxycholate on hamster small intestine in vivo showed that water secretion was accompanied by increase in permeability of the mucosa to inulin, dextran, and albumin (31). These secretory and permeability changes were accompanied by both biochemical and histological alterations: exfoliation (DNA release), membrane effects (sucrase release), and shortened villi. Electron microscopy revealed extensive alteration of the brush border membrane with a decrease in binding of lanthanum and the development of permeability to tracer in villus tip cells. In contrast, taurocholate, which did not alter water transport, did not effect intestinal permeability or the brush border membrane. The surface properties of anionic surfactants apparently cause changes in absorptive cell membranes that result in intestinal secretion.

Whitmore et al. (33) also found that among other surfactants SLS in concentrations of 0.05, 0.5, 5, 10, and 50 mM increased the intestinal absorption rates of both L-valine and salicylate. These authors compared the actions of surfactants on the absorptive capability of rat jejunal tissue in vivo with their effects on the amounts of protein and phospholipid release from the mucosal surface under the same conditions. Release of a comparatively small amount of protein was accompanied by large increases in the absorption rates; much larger quantities of phospholipid were released before any increase in absorption was seen. The liberation of cholesterol by high concentrations of anionic surfactants suggested that some disruption of the mucosal membrane occurred under those conditions. The relative potency of the surfactants in stimulating both absorption of the drugs and the release of polypeptides and lipids followed the order: anionic > nonionic > cationic.

Feldman et al. (58) reported that sodium taurodeoxycholate, a physiological surface-active agent, accelerates the release of total phosphorus, lipid phosphorus, and protein from the everted rat small intestine. The results indicate that the interaction of the surfactant with the biological membrane accelerates the loss of structural integrity of the preparation and increases membrane permeability. Feldman and Reinhard (59) determined the effect of the chain length of a homologous series of anionic surfactants on the biological membrane. The effect of a series of sodium alkyl sulfates from C_6 to C_{14} on the loss of protein from the everted rat small intestine was measured. At a surfactant concentration of 10 mM, a peak effect on protein release was noted with sodium decyl sulfate (C_{10}). Measurement of protein release as a function of SLS (C_{12}) concentration resulted in the observation that the effect appears to be due to the micellar phase of the surfactant solution. At a concentration of C_{12} above the CMC (0.6 mM), the loss of protein from the intestinal preparation increased as the concentration of surfactant was increased. But there is a maximum amount of protein that can be released from the everted rat small intestinal sacs by surface active agents. At equivalent micellar concentrations of C_8-C_{14} (12.5 times the CMC), there was no difference in the amount of protein released in the presence of the individual alkyl sulfates.

The increases in the absorption of phenolsulfonphthalein and pralidoxime caused by SLS were potentiated by theophyllin (25 mM) and reduced by imidazole (25 mM). The addition of dibutyryl cyclic AMP (2.5 mM) to the loop fluid increased the absorption of the test substances (40). This effect was reduced by imidazole, but under the experimental conditions it was not potentiated by theophylline.

Determination of cyclic AMP (cAMP) in the rat intestinal mucosa showed that the level of this substance was significantly higher in the presence than in the absence of SLS (40). It is concluded that the data obtained support the idea of an increased level of cAMP as the main basis for the effect of SLS on the absorption. This assumption is supported by further investigations of Briseid et al. (60). They found that the level of cAMP in the jejunal mucosa from tied loops of anaesthetized rats was significantly increased

(27-50%) when SLS was added to the loop fluid (2-27 mM). Imidazole (25 mM) did not significantly alter the resting level of cAMP, but reduced the increase caused by SLS (17 mM). Theophylline (25 mM) significantly increased the intestinal level of cAMP, and potentiated the increase caused by SLS. Ouabain (2.5 mM) did not alter the level of cAMP in the presence or in the absence of SLS. It is concluded that the SLS-effect on absorption can only partly be ascribed to its effect on the intestinal level of cAMP.

The difficulties of using surfactants as promoters for enhancement of the absorption rate of poorly absorbable drugs consist in their irritation of mucous membranes in higher concentrations. The use of surfactants for acceleration of drug absorption is only possible in such cases, in which the promoting concentration is below the irritant one. Since there seems to be a connection between both concentrations, the practical use of surfactants as adjuvants of oral drug application is limited.

B. Laxative Properties

Dioctyl sodium sulfosuccinate (DSS) is used as a contact-laxative and promoted as a stool softener. DSS softens the stool by net water accumulation in the intestine. It is suggested that this is accomplished by inhibition of Na-K-ATPase, stimulation of adenyl cyclase and a decrease in mucosal prostaglandin E_2 (61).

While Eickholt and White (62) and Hyland (63) supposed that the laxative effect of DSS is due to its mild irritating and secretion-promoting effects on the mucosa of intestine, wrote Estler (64) that the influence on the water and electrolyte transport in the intestinal mucosa is the reason for the laxative effect. DSS inhibits the sodium and water transport from the intestine into the blood vessels and is able to invert the physiological transport direction. This process is called the hydragogen effect.

In contrast, Chapman et al. (65) have found that the maximum recommended dose of DSS (300 mg/day) had no effect on stool weight, stool frequency, stool water, or mean transit time. The dose of 300 mg/day did not increase ileal or colonic output of solids or water, and did not alter the water and electrolyte balance of small or large intestine in healthy human subjects. The test results were compared with the results of a placebo group in a randomized, single blind crossover study.

In connection with the laxative effect some authors (66,67) reported that the application of DSS showed side effects (liver damage).

Jenssen et al. (68) studied the release of gastrointestinal regulatory peptides after a soap enema. Nine healthy volunteers received a 1-liter soap enema (6 ml spiritus saponis kalinii dissolved in water to 1 liter), which is frequently used in routine preparation of the colon before roentgenographic or endoscopic procedures. All persons defecated within 9 min. Compared with the basal values, the systolic blood pressure rose significantly at 9, 12, and 15 min, and the diastolic blood pressure at 9 min after start of the enema infusion. The plasma levels of vasoactive intestinal polypeptide rose significantly at 10, 15, 20, and 25 min and the plasma cholecystokinin at 10 and 15 min after start of administration of the soap enema. Somatostatin peaked significantly only at 10 min after start of the enema infusion, and plasma motilin and plasma pancreatic polypeptide, a sensitive indicator of vagal tone, were not affected at all by the soap enema. The authors suggested that the peptide releases are of colonic origin and that vasoactive intestinal polypeptide and cholecystokinin may play mediatory roles in the enema-induced defecation.

C. Treatment of Ulcers

Some publications discuss gastric and intestinal ulcer treatment in view of the observed inactivation of pepsin by sodium alkyl sulfate (69), the inhibition of lysozyme by sodium hexadecyl sulfate (70), and the retardation of histamine-induced stomach secretion by SDS (71,72). Alkyl sulfate is said to possess therapeutic properties in the treatment of stomach ulcers (73). Surfactants have been reported to clear wounds (74), and Shey et al. (75) report the formation of mucous as an effect of alkyl sulfate in dogs, which could be of therapeutic value in the treatment of ulcers. On the other hand, sodium alkyl sulfate has been reported to inhibit stomach motility (76).

In contrast to these favorable effects, Davenport (77) describes that—depending on the concentration—decyl sulfate is able to destroy the mucous membranes of the stomach. This effect is connected with the destruction of cells and layers of lipoprotein.

An overview of the above-cited publications about the effects of surfactants on ulcers shows that most studies were carried out in the 1940s and in the early 1950s. Only a few publications followed that time. But in the last two decades, no adequate publications are to be found in the literature. From this fact it can be concluded that more effective anti-ulcer drugs are on the market and that surfactants used for treatment of ulcers do not play any role today.

VI. OTHER PHARMACEUTICAL PROPERTIES

A. Antimicrobial Effects

In contrast to quarternary ammonium compounds, which are in some cases excellent antimicrobials, anionics show relatively weak effects. They are moderately active against gram-positive organisms, but have little or no activity against gram-negatives, presumably because the outer membranes of gram-negative bacteria do not allow penetration by the surfactant to the cytoplasmic membrane.

Other than the investigations conducted with the aim of examining the antibacterial, antifungal, or antiviral effects of special surfactants, studies were carried out to examine the synergistic effects of anionic surfactants on known antimicrobial pharmaceutical substances. In some publications special cations are described that are capable of enhancing the bactericidal activity of surfactants.

Voss (78) reported that the effect of alkyl (C_{12}) benzene sulfonate on gram-negative germs is negligible, while some effects are seen in the case of gram-positive germs.

The inhibition of *S. aureus* by alkyl benzene sulfonate has also been reported by Oba and Tomiyama (79). According to Damm (80), the alkali salts of palmitic acid kill bacterium coli at a concentration of 0.72%. It is further reported that sodium dodecyl sulfate (SDS), sodium dodecyl amino-propionic acid, sodium ricinoleate, bile acids, and alkylaryl sulfonates are effective against gram-positive germs. Alkylaryl sulfonates are even effective against staphylocci; sodium lauryl sulfate (SLS) and deoxycholate against viruses of influenza A, smallpox, and psittacosis. The inactivating and hemolytic concentrations are in the same range.

A 0.5% solution of sodium polyanethanolsulfonate (SPS), a polyanionic detergent, possesses antibacterial activity against some isolates of *Neisseria meningitidis, Neisseria gonorrhoeae,* and *Peptostreptococcus anaerobius* (81). But hemoglobin can inactive SPS at an optimum SPS/hemoglobin ratio of 1:6.

Schaller et al. (82) reported that sodium lauryl ether sulfate can inhibit *Microsporum gypseum, Aspergillus niger,* and *Epidermophyton floccosum.*

Ward and Ashley (83) reported that the effects of detergents on stability of enteric viruses was highly dependent on pH. Sodium dodecylsulfate (SDS) was shown to be a potent virucidal agent for reovirus, but the effects of 0.1% SDS were minimal near neutrality and much more pronounced at low than at high pH values. SDS was extremely virucidal at pH values below 5. Qualitative similar effects were found with the stabilities of *coliphages T4, f2* and *Q.beta.* In earlier studies by Mandel (84) was also found that poliovirus is destabilized by SDS at pH values below about 5.

In a further study of Ward and Ashley (85) about the inactivation of rotavirus by SDS, it was shown that treatment of a diluted lysate of rotavirus with 0.1% SDS for 60 min at 21°C caused a reduction in infectivity of more than three orders of magnitude. A similar result was found when the purified 1.35-g/ml rotavirus particles obtained after CsCl gradient centrifugation were treated under the same conditions.

The effect of alkyl (C_{12}) benzene sulfonate against *Staphylococcus aureus* is intensified by bivalent inorganic cations (86).

In connection with the synergetic bactericidal activity and surfactants Voss (78) already reported that the bactericidal effectiveness of several anionic surfactants against *Staphylococcus aureus* is increased in the presence of low concentrations of divalent cations, especially alkine earth and metals of group IIB of the period table. However, these combinations did not result in increased activity against *Escherichia coli.*

Berg and Zimmerer (87) described the broad spectrum of antibacterial activity and proposed mechanism of action of rare earth (lanthanides) metal salts of anionic surfactants. Seven anionic surfactants, among others sodium decyl and dodecyl sulfate and dioctyl sodium sulfosuccinate (DSS) were tested. As test organisms *E. coli* and *Staph. aureus* were used. All tested surfactants were inactive against the gram-negative *E. coli* in a concentration of 0.2%, and moderately active against the gram-positive *Staph. aureus*. However, when the surfactants were combined with $LaCl_3$, the antibacterial activity against both microorganisms were greatly increased. The combinations exhibited efficacy comparable to that of quaternary ammonium desinfectants.

Hartmann et al. (88) investigated the antimicrobial long-term efficacy of the combined application of an anionic detergent (Dermofug, containing 20% ammoniumdodecylsulfate and 12,75% dodecylsulfonbenzolsulfonacid 2,2′,2″-nitrilotriethanol-salt), either with 50% isopropanol or FABRY's tincture (3% salicyclic acid, 1% liquified phenol containing 50% isopropanol) on the resident flora of the human forehead skin after repeated applications using the forehead skin test. The test results show that both combinations tested are able to penetrate the infrainfundibulum of the sebaceous glands. The combined application of 10% Dermofug® solution (D) and FABRY's tincture (F) showed better antibacterial efficacy than the combination of D and 50% isopropanol (I) against the coagulase-negative staphylococci in the stratum corneum/acroinfundibulum as well as in the infrainfundibulum of the sebaceous glands. Against *Propionibacterium spp.*, D and F showed better efficacy than D and I in the stratum corneum/acroinfundibulum, and equal efficacy in the infrainfundibulum of the sebaceous glands. In a further investigation the same authors (89) studied the antibacterial effect on human skin resident flora of 10% Dermofug solution, combined either with a 3% salicyclic acid tincture or 1% phenol acid tincture using again the forehead skin test. The combination with salicyclic acid tincture showed better antibacterial efficacy against the coagulase-negative staphylococci and *Propionibacterium spp.* in the stratum corneum/acroinfundibulum as well as in the infrainfundibulum of the sebaceous glands. The combination with phenol acid tincture had no antibacterial efficacy against *Propionibacterium spp.* in the infrainfundibulum of the sebaceous glands. In contrast, the combined application of Dermofug solution and 3% salicyclic acid tincture led to a long-term reduction of the counts of *Propionibacterium spp.* in the infrainfundibulum of the sebaceous glands.

The propionibacteriae and the staphylococci are supposed to be the main reason for the acne disease. The anionic surfactants are considered to have an anti-acne effect (90) because of their bactericidal effects against these bacteria as shown in the publication of Hartman et al. (88,89).

Iwata and Yamaguchi (91) tested the antifungal activity of some anionic and nonionic surfactants against *Candida albicans* and the synergistic fungicidal action of these surfactants in combination with the imidazole antimycotic clotrimazole. Of the six anionic and five nonionic agents, i.e., sodium lauryl polyoxyethylene sulfate (Sannol NES) and sodium deoxycholate (DOC), and all of the nonionic agents had almost no effect on the growth of *Candida albicans* at concentrations ranging from 0.001-0.4%. DSS, sodium laurylbenzene sulfonate (LBS), SLS, and sodium tetradecanoyl sulfonate (Lipolan 1400) were not inhibitory at concentrations of 0.02% or less. However, they showed varying extents of fungistatic and/or fungicidal activities at higher concentrations.

The combination studies were performed to examine whether or not a given surfactant would potentiate (or counteract) the anti-Candida action of chlotrimazole. With the only exception of Sannol NES, all anionic surfactants were more or less capable of potentiating the anti-Candida action of chlotrimazole, when combined with 0.13 or 0.5 μg/ml of the antimycotic. Substantial synergy was attained by 0.01% LBS, 0.01% SLS, 0.025% Lipolan 1400, 0.05% DOC, and 0.025% DSS. In general, combinations of chlotrimazole with LBS or SLS exhibited fungistatic synergy, whereas those with Lipolan 1400, DOC, or DSS were effective in fungicidal synergy. A similar effect was observed with a half-lower concentration of each surfactant in combination with 0.5 μg/ml chlotrimazole. Of all the combinations tested, chlotrimazole plus DSS showed the most fungicidal activity.

In contrast, a polyoxyethylene surfactant of anionic type, Sannol NES, counteracted chlotrimazole. This antagonistic phenomenon was particularly prominent when relatively high concentrations of the surfactant (0.1%) and of chlotrimazole (10 μg/ml) were combined. A similar degree of antagonism was also observed with four nonionic polyoxyethylene surfactants.

Another synergistic effect of surfactants was found by Amtmann et al. (92). Xanthate compounds have been shown to exhibit antiviral activity against various DNA and RNA viruses under acid pH conditions. By simultaneous administration of certain detergents in a concentration of 40 μg/ml it is now possible to utilize the unique broad range antiviral spectrum of Xanthates under physiological pH conditions (pH 7.4). When used in conjugation with tricyclodecan-9-yl-xanthate, sodium deoxycholate, SDS, and certain fatty acids, which have no antiviral activity of their own, inhibit the replication of various DNA and RNA viruses, such as herpes simplex, vesicular stomatitis and Coxsackie B 4, in vitro at pH 7.4.

As Barkvoll et al. (93) showed, it can not be assumed that in all cases of combination between antimicrobial substances and anionic surfactants synergistic effects are observed. About the interaction between chlorhexidine digluconate (CH) and SLS in vivo is reported (93). CH is a cationic and it forms salts of low solubility with anionics such as phosphate, sulfate and carboxyl. Toothpastes contain anionic surfactants, one of the most widely used being SLS. Barkvoll et al. (93) examined the possible interaction between CH and the anion SLS in vivo. The interference of SLS on the antiplaque potential of CH (in concentrations of 0.2% each) was investigated. The results showed that even a 30-min interval between SLS and CH-rinsing gave a significantly reduced antiplaque effect of CH, whereas the neutralizing effect of SLS disappeared after 2 hr. The authors concluded that SLS is not compatible with CH, even when these compounds are introduced seperately in the oral cavity. The time between toothbrushing and a CH rinse should at least be 30 min, if a reduction in the antimicrobial effect is to be avoided.

B. Miscellaneous

In an experimental investigation, Scheinberg et al. (94) studied the effect of sodium lauryl sulfate (SLS) on wound healing in guinea pigs. They observed an accelerated epithelialization.

In contrast, Green et al. (95) reported that 1.3% SLS reduced the rate of corneal epithelial regrowth after mechanical and heptanol-generated lesions in rabbit corneas and increased epithelial sucrose permeability. SLS caused an expansion of the lesion after some degree of healing and caused damage to the epithelial cells with disruption and expansion of the extracellular space. It may be critisized that the tested concentration of 1.3% is far above the compatible concentration on the eyes and that this concentration causes also irritant reactions.

In subtoxic doses sodium dodecyl sulfate (SDS) can raise the permeability to dextrose in the frog heart (96). In investigations with alkyl benzene sulfonate and *n*-olefin sulfonate on isolated frog muscle, Ogura and Tamura (97) found that the acetylcholine concentration can be suppressed. These authors also found a negative ionotropic effect on the heart. The incubation of mast cells of the peritoneal cavity of the rat with alkyl sulfate results in liberation of histamine according to Prottey and Ferguson (98).

In some publications influences of surfactants on biological substances are described. Determination of cyclic AMP (cAMP) in the rat intestinal mucosa showed that the level of cAMP was significantly higher in the presence than in the absence of SLS (40,60).

SDS inhibited the activities of carboxypeptidases A, B, and P. The inhibitory effects of SDS on these enzyme activities depended not on its concentration but on its absolute amount (99).

About the effect of surfactants on the level of blood serum lipids in rats treated with detergents is reported by Szymaniec et al. (100). Sodium 2-ethylhexylsulfosuccinate in a dose of 10 mg/kg was applied intraperitoneally once per week for 12 weeks. The content of total cholesterol, β-lipoproteins and total lipids in serum were estimated and compared with the levels of control rats, which were injected with 0.9% saline solution. Total cholesterol was increased, β-lipoproteins were decreased, total lipid content was not changed significantly. A similar effect is reported by Elges et al. (101). The addition of anionic surfactants at 1 mg/kg body weight to rabbits fed a diet containing cholesterol and at 4 mg/kg to rabbits fed a diet containing suflower oil resulted in an increase of esterified cholesterol, fatty acids, total lipids, and triglycerides compared to controls fed the same diet without surfactants. The study showed the even small amounts of surfactants can effect lipid metabolism.

Not only lipid metabolism but also carbohydrate metabolism can be affected by surfactants as it was described by Antal (102). Sodium dodecyl benzene sulfonate was fed, mixed in a diet, rats for 4 weeks in a dose of 0.25 g/kg daily. This application caused an increase of liver weights and of glucose blood levels.

Finally, investigations of Havermann and Rüger (103) should be mentioned; these authors studied the influence of alkyl sulfate and alkyl benzene sulfonate addition on the utilization of food using pigs as test animals. The test animals showed greater weight increases than the controls.

REFERENCES

1. W.A. Ritschel, *Angew. Chem. Intl.,* 699-710(1960).
2. Z.T. Chowhan and R. Pritchard, *J. Pharm. Sci., 67,* 1272-1274(1978).
3. Z.T. Chowhan, R. Pritchard, W.H. Rooks, II, and A. Tomolonis, *J. Pharm. Sci., 67,* 1645-1647(1978).
4. H.A. El-Sourady, F.S. Habib, and S.I. Mohamed, *Pharmazie, 40,* 557-558 (1985).
5. B.J. Aungst, N.J. Rogers, and E. Shefter, *Int. J. Pharm., 33,* 225-234(1986).
6. P. Ashton, *Diss. Abstr. Int. B, 49,* 1101(1988).
7. P.S. Banerjee and W.A. Ritschel, *Int. J. Pharm., 49,* 189-197(1989).
8. W.A. Ritschel and O.L. Sprockel, *Drugs of Today, 24,* 613-628(1988).
9. B.W. Barry, *Int. J. Cosmet. Sci., 10,* 281-293(1988).
10. R.M. Clayton, K. Green, M. Wilson, A. Zehir, J. Jack, and L. Searle, *Fd. Chem. Toxic., 23,* 239-246(1985).
11. K. Green, J.M. Chapman, and L. Cheeks, *J. Toxicol. Cut. Ocul. Toxicol., 6,* 89-107(1987).

12. K. Green, L. Cheeks, and J.M. Chapman, in *Ophthalmic Drug Delivery,* M.F. Saettone, M. Bucci and P. Speiser (Eds.), Fidia Research Series 11, Liviana Press, Springer Verlag, 1987, 171-177.
13. K. Green, J. Chapman, L. Cheeks, and R.M. Clayton, *Concepts Toxicol., 4,* 126-132(1987).
14. M. Ginsburg and J.M. Robson, *Brit. J. Ophthalmol., 33,* 574-579(1949).
15. M. Ginsburg and J.M. Robson, *Brit. J. Ophthalmol., 29,* 185-193(1945).
16. J.G. Bellows and M. Gutmann, *Arch. Ophthalmol., 30,* 352-357(1943).
17. J.B. Feldmann, P. de Long, and C.P. Brown, *Arch. Ophthalmol., 40,* 668-679 (1948).
18. K.C. Swan, *Amer. J. Ophthalmol., 27,* 1118-1122(1944).
19. W.M. Grant, *Toxicology of the Eye,* CC Thomas, Springfield, Ill., 1962.
20. J.H. Leopold, *Arch. Ophthalmol., 34,* 99-102(1945).
21. S. Hirai, T. Ikenaga, and T. Matsuzawa, *Diabetes, 27,* 296-299(1978).
22. S. Hirai, T. Yashiki, and H. Mima, *Int. J. Pharm., 9,* 165-172(1981).
23. S. Hirai, T. Yashiki, and H. Mima, *Int. J. Pharm., 9,* 173-184(1981).
24. S. Hirai, J. Takeda, *Res. Lab., 41,* 246-286(1982).
25. Y. Kaneo, *Acta. Pharm. Sci., 20,* 379-388(1983).
26. I.A. Siegel and H.P. Gordon, *Toxicology Letters, 26,* 153-157(1985).
27. Y. Kurosaki, S. Hisaichi, C. Hamada, T. Nakayama, and T. Kimura, *Int. J. Pharm., 47,* 13-19(1988).
28. B.J. Aungst and N.J. Rogers, *Int. J. Pharm., 53,* 227-235(1989).
29. P.B. Deasy and C.T. O'Neill, *Pharm. Acta Helv., 64,* 231-235(1989).
30. K.S. Kim, M. Kurokawa, T. Kimura, and H. Sezaki, *J. Pharm. Dyn., 5,* 509-514(1982).
31. G.W. Gullikson, W.S. Cline, V. Lorenzsonn, L. Benz, W.A. Olsen, and P. Bass, *Gastroenterol., 73,* 501-511(1977).
32. R.B. Sund and I. Matheson, *Acta Pharmacol. et Toxicol., 42,* 253-258(1978).
33. D.A. Whitmore, L.G. Brookes, and K.P. Wheeler, *J. Pharm. Pharmacol., 31,* 277-283(1979).
34. H.A. El-Sourady, F.S. Habib, and S.I. Mohamed, *Drug. Dev. Ind. Pharm., 12,* 889-898(1986).
35. R.B. Sund, *Acta Pharmacol. et Toxicol., 37,* 382-296(1975).
36. M. Wafik Gouda, N. Khalafalah, and S.A. Khalil, *J. Pharm. Sci., 66,* 727-728 (1977).
37. M. Wafik Gouda, S. El Gamal, A.R. Ebian, and S.A. Khahil, *Can. J. Pharm. Sci., 13,* 16-18(1978).
38. V.P. Shah, J.P. Hunt, W.R. Fairweather, V.K. Prasad, and G. Knapp, *Biopharm. and Drug Disposition, 7,* 27-33(1986).
39. Y.M. Rao and D. Rambhau, *Indian J. Pharm. Sci., 49,* 8-12(1987).
40. G. Briseid, K. Briseid, and K. Kirkevold, *Naunyn-Schmiedeberg's Arch. Pharmacol., 292,* 137-144(1976).
41. R.B. Sund and G. Olsen, *Acta Pharmacol. et Toxicol., 49,* 65-71(1981).
42. W. Appel, H. Schievelbein, and E. Werle, *Arzneimittel-Forsch., 7,* 742-745 (1957).
43. D. Krause, *Arzneimittel-Forsch., 5,* 428-432(1955).

44. V. Eybl, *Pharmazie, 12,* 30-33(1957).
45. G. Briseid, K. Briseid, and B. Bergersen, *Naunyn-Schmiedeberg's Arch. Pharmacol., 282,* 45-57(1974).
46. G. Bormann and M. Stanisic, *Naunyn-Schmiedeberg's Arch. Exp. Path. Pharmak., 243,* 420-428(1962).
47. P.M. Lish and J.H. Weikel, Jr., *Toxicol. Appl. Pharmacol., 1,* 501-504(1959).
48. E. Miyamoto, A. Tsuji, and T. Yamana, *J. Pharm. Sci., 72,* 651-654(1983).
49. M. Wafik Gouda, *Can. J. Pharm. Sci., 9,* 37-40(1974).
50. K. Opitz, G. Bormann, and A. Loeser, *Arzneimittel-Forsch., 11,* 1066-1067 (1961).
51. V. Kozlik and V. Eybl, *Pharmazie, 10,* 477-480(1955).
52. V. Kozlik and B. Mosinger, *Pharmazie, 11,* 22-24(1956).
53. B. Mosinger and V. Kozlik, *Pharmazie, 11,* 539-542(1956).
54. R.B. Sund, *Acta Pharmacol. et Toxicol., 37,* 297-308(1975).
55. R.B. Sund and D.N. Jacobsen, *Acta Pharmacol. et Toxicol., 43,* 339-345(1978).
56. S. Nakagawa, K. Takanashi, H. Mitsui, M. Kobayashi, M. Yonezawa, T. Aizawa, T. Ohsumi, and G. Mukai, *Nihon Univ. J. Med., 21,* 107-116(1979).
57. T. Kimura, J. Nakamura, and H. Sezaki, *J. Pharmacobio-Syn., 8,* 661-668 (1985).
58. S. Feldmann, M. Reinhard, and C. Willson, *J. Pharm. Sci., 62,* 1961-1964(1973).
59. S. Feldmann and M. Reinhard, *J. Pharma. Sci., 65,* 1460-1462(1976).
60. G. Briseid, I. Oye, and K. Briseid, *Naunyn-Schmiedeberg's Arch. Pharmacol., 298,* 263-266(1977).
61. W.G. Thompson, *Current Therapeutics, 21,* 109-118(1980).
62. Th.H. Eickholt and W.F. White, *Drugstandards, 28,* 154-162(1960).
63. C.M. Hyland, *The Practitioner, 200,* 698-699(1968).
64. C.-J. Estler, *Münch. med. Wschr., 125,* 95-96(1983).
65. R.W. Chapman, J. Sillery, D.D. Fontana, C. Matthys, and D.R. Saunders, *Gastroenterology, 89,* 489-493(1985).
66. H. Godfrey, *JAMA, 215,* 643(1971).
67. Editorial, Arznei-telegramm 6/77, *zit. nach Medical Letter, 19,* 45(1977).
68. T.G. Jenssen, P.G. Burhol, and R. Jorde, *Scand. J. Gastroenterology, 20,* 762-766(1985).
69. O.E. Lobstein, B.J. Hull, and S.J. Fogelson, *Gastroenterology, 20,* 474-484 (1952).
70. J.F. Prudden, *Gastroenterology, 15,* 426-437(1950).
71. H. Shay, S.A. Komarov, and H. Siplet, *Science, 105,* 2718(1947).
72. S.J. Fogelson and O.E. Lobstein, *Proc. Soc. Exp. Biol. Med., 75,* 334(1950).
73. J.B. Kirsner and R.A. Wolff, *Gastroenterology, 2,* 93-101(1944).
74. S.A. Portis, C.L. Block, and H. Necheles, *Gastroenterology, 3,* 106(1944).
75. H. Shay, S.A. Komarov, H. Siplet, and S.S. Fels, *Science, 103,* 50-52(1946).
76. H. Necheles and J. Sporn, *Amer. J. Gastroenterol., 46,* 481(1966).
77. H.W. Davenport, *Gastroenterology, 54,* 175-181(1968).
78. J.G. Voss, *J. Bacteriol., 86,* 207-211(1963).
79. K. Oba and S. Tomiyama, *Yukagaku, 14,* 364(1965); cited from *Fette, Seifen, Anstrichm., 12,* 1096(1966).

80. H. Damm, 3rd Int. Kong. Grenzflächenaktive Stoffe, Band IV, Sektion D/1, Nr. 29, 256-259(1960).
81. S.C. Edberg and M.K. Edberg, *J. Clin. Microbiol., 18,* 1047-1050(1983).
82. F.K. Schaller, G. Shubach, and H. Rieth, *Mykosen, 11,* 419-422(1968).
83. R.L. Ward and C.S. Ashley, *Appl. Environ. Microbiol., 38,* 314-322(1979).
84. B. Mandel, *Virology, 17,* 288-294(1962).
85. R.L. Ward and C.S. Ashley, *Appl. Environ. Microbiol., 39,* 1154-1158(1980).
86. I. Nikodemusz, F. Fodor, and L. Sujbert, *Forum Städte-Hygiene, 28,* 176-178(1977).
87. R.W. Berg and R.E. Zimmerer, J. Ind. Microbiol., 1, 377-381(1987).
88. A.A. Hartmann, P. Elsner, W. Lutz, M. Pucher, and H. Hackel, *Zbl. Bakt. Hyg. B, 186,* 526-535(1988).
89. A.A. Hartmann, P. Elsner, K. Kremer, and H. Hackel, *Zbl. Bakt. Hyg. B, 186,* 536-544(1988).
90. M. Gloor and A.A. Hartmann, *Hautarzt, 38,* 702-703(1987).
91. K. Iwata and H. Yamaguchi, *Antimicrob. Agents Chemother., 12,* 206-212 (1977).
92. E. Amtmann, K. Müller-Decker, A. Hoss, G. Schalasta, C. Doppler, and G. Sauer, *Biochem. Pharmacol., 36,* 1545-1549(1987).
93. P. Barkvoll, G. Rolla, and A.K. Svendsen, *J. Clin. Periodontol., 16,* 593-595 (1989).
94. S. Scheinberg, S.P. Bralow, and N. Necheles, *Surgery, 24,* 972(1948).
95. K. Green, R.E. Johnson, J.M. Chapman, E. Nelson, and L. Cheeks, *J. Toxicol. Cut. Ocul. Toxicol., 8,* 253-269(1989).
96. E. Klotzbücher, *Dtsch. Z. Verdauungs-Stoffwechselkrankh., 9,* 222(1949).
97. Y. Ogura and J. Tamura, *Ann. Rep. Inst. Food Microbiol.* (*Chiba Univ.*), *20,* 95-99(1967).
98. C. Prottey and T.F.M. Ferguson, *Food Cosmet. Toxicol., 14,* 425-430(1976).
99. M. Kamo and A. Tsugita, *J. Biochem., 102,* 243-246(1987).
100. J. Szymaniec, H.I. Trzeciak, H. Machalska, and B. Turczynski, *Med. Pr., 28,* 189-192(1977).
101. E. Elges, H. Straus, and M. Popovici, *Clujul Med., 51,* 52-57(1978).
102. M. Antal, *Zeitschrift für Ernährungswirtschaft, 10,* 35-39(1970).
103. H. Havermann and H. Rüter, *Arch. Tierernähr. Suppl, 5,* 60-76(1954).

Index